Handbook of Dendritic Cells
Volume 2

Edited by
Manfred B. Lutz, Nikolaus Romani,
and Alexander Steinkasserer

D1348253

Related Titles

R. A. Meyers (Ed.)

Encyclopedia of Molecular Cell Biology and Molecular Medicine, 2nd Edition

2005

ISBN 3-527-30542-4

http://meyers-emcbmm.de

H. Kropshofer, A. Vogt (Eds.)

Antigen Processing Cells – From Mechanisms to Drug Development

2005

ISBN 3-527-31108-4

A. Hamann, B. Engelhardt (Eds.)

Leukocyte Trafficking – Molecular Mechanisms, Therapeutic Targets, and Methods

2005

ISBN 3-527-31228-5

A. Meager (Ed.)

The Interferons – Characterization and Application

2005

ISBN 3-527-31180-7

K. M. Pollard (Ed.)

Autoantibodies and Autoimmunity – From Mechanisms to Treatments

2005

ISBN 3-527-31141-6

S. H. E. Kaufmann (Ed.)

Novel Vaccination Strategies

2004

ISBN 3-527-30523-8

D. Wedlich (Ed.)

Cell Migration in Development and Disease

2005

ISBN 3-527-30587-4

Handbook of Dendritic Cells
Volume 2

Biology, Diseases, and Therapies

Edited by
Manfred B. Lutz, Nikolaus Romani,
and Alexander Steinkasserer

With an Introduction by
Ralph M. Steinman

WILEY-VCH Verlag GmbH & Co. KGaA

The Editors

PD Dr. Manfred B. Lutz
Department of Dermatology
University Hospital of Erlangen
Hartmannstr. 14
91052 Erlangen
Germany

Prof. Dr. Nikolaus Romani
Department of Dermatology
Innsbruck Medical University
Anichstrasse 35
6020 Innsbruck
Germany

Prof. Dr. Alexander Steinkasserer
Department of Dermatology
University Hospital of Erlangen
Hartmannstr. 14
91052 Erlangen
Germany

Cover
Colored scanning electron microscope
pictures of dendritic cells. (Courtesy of
Prof. Dr. Kristian Pfaller, Innsbruck
University, Austria.)

■ All books published by Wiley-VCH are carefully
produced. Nevertheless, authors, editors,
and publisher do not warrant the information
contained in these books, inclding this book, to
be free of errors. Readers are advised to keep in
mind that statements, data, illustrations,
procedural details or other items may
inadvertently be inaccurate.

Library of Congress Card No.
applied for

British Library Cataloguing-in-Publication Data
A catalogue record for this book is available from
the British Library.

**Bibliographic information published by
Die Deutsche Bibliothek**
Die Deutsche Bibliothek lists this publication
in the Deutsche Nationalbibliografie; detailed
bibliographic data is available in the Internet at
http://dnd.ddb.de.

© 2006 WILEY-VCH Verlag GmbH & Co. KGaA,
Weinheim

Composition   Fotosatz Detzner, Speyer
Printing   Strauss GmbH, Mörlenbach
Bookbinding   Litges & Dopf Buchbinderei
GmbH, Heppenheim

Printed in the Federal Republic of Germany
Printed on acid-free paper

ISBN-13: 978-3-527-31109-5
ISBN-10: 3-527-31109-2

Preface

Time flies.... Life, including research, is becoming ever faster. Papers are published online long before they appear in print. So, why such an "old-fashioned" book? Ralph Steinman and Jacques Banchereau gave a good answer to this concern in the "predecessor book" to this volume: the first edition of the Dendritic Cell book edited by Michael Lotze and Angus Thomson in 1999 [1]. Books may serve as a comprehensive historical record of the state-of-the-art at a given time. Even books published as long ago as 1991, such as Gerold Schuler's volume on Langerhans Cells [2], still provide a valuable source of knowledge, not only for novices to the field. For these reasons we felt that our attempt to assemble yet another such book was justified and would provide a useful service to the scientific community.

Five years have elapsed since the second edition of above-mentioned dendritic cell book. Dendritic cell research has proceeded tremendously during this period, and has increasingly been linked with clinical research. Clearly, the time is ripe to have another such reference volume on our shelves and desks to browse through, search, find and sometimes perhaps even remember. We have encouraged the contributors to also look back in time and to try and put recent data into a wider perspective.

Most of the relevant issues in dendritic cell biology have been covered in this book. Of course, nothing is perfect, and some interesting and important areas have not been discussed. To obtain a critical synopsis of the scope of this volume we recommend the reader to enjoy Ralph Steinman's introductory chapter as an "apéritif".

We thank all contributors to this book for their great efforts and for their time. All of us have multiple commitments, and "besides this", many of us have children who'd often like to do something else than just watch their mothers or fathers writing papers..... Therefore, we appreciate these efforts even more, and we express our thanks to all those who encouraged us to undertake this endeavour. We also thank our editorial partner, Dr. Andreas Sendtko from Wiley-VCH and Brigitte Wölfel from Erlangen, who managed the transformation of a pile of individual manuscripts into a nice book smoothly and efficiently.

Handbook of Dendritic Cells. Biology, Diseases, and Therapies.
Edited by M. B. Lutz, N. Romani, and A. Steinkasserer.
Copyright © 2006 WILEY-VCH Verlag GmbH & Co. KGaA, Weinheim
ISBN: 3-527-31109-2

Finally, we are indebted to our teachers and mentors, above all Ralph Steinman and Gerold Schuler, who made it possible that we are now in the position to edit such a book.

Enjoy the book, and enjoy dendritic cell research!

Manfred B. Lutz
Niki Romani
and Alexander Steinkasserer

Erlangen and Innsbruck, January 2006

References

1 Lotze, M.T. and Thomson, A.W. (Eds.), *Dendritic Cells: Biology and Clinical Applications.* Academic Press, San Diego, London, **1999** (1st edn) and **2001** (2nd edn).
2 Schuler, G. (Ed.), *Epidermal Langerhans Cells.* CRC Press, Inc., Boca Raton, Florida, **1991**.

Contents

Volume 1

Handbook of Dendritic Cells. Biology, Diseases, and Therapies.
Edited by M. B. Lutz, N. Romani, and A. Steinkasserer.
Copyright © 2006 WILEY-VCH Verlag GmbH & Co. KGaA, Weinheim
ISBN: 3-527-31109-2

Volume 3

List of Contributors

Marion Abt
Institute for Virology and
Immunobiology
University of Würzburg
Versbacherstrasse 7
97078 Würzburg
Germany

Luciano Adorini
BioXell
Via Olgettina 58
20132 Milano
Italy

Toni Aebischer
Department of Molecular Biology
Max-Planck-Institute for
Infection Biology
Schumannstr. 21/22
10117 Berlin
Germany

Julio Aliberti
Department of Immunology
Duke University Medical School
DVMV Box 3010
Durham, NC 27710
USA

Paola Allavena
Department of Immunology and
Cell Biology
Mario Negri Institute
Via Eritrea 62
20157 Milano
Italy

Jacques Banchereau
Baylor Institute for Immunology
Research
3434 Live Oak
Dallas, TX 75204
USA

Georg Bartsch
Department of Urology
Medical University of Innsbruck
Anichstrasse 35
6020 Innsbruck
Austria

Ottavio Beretta
Department of Biotechnology and
Bioscience
University of Milano-Bicocca
Piazza della Scienza 2
20126 Milano
Italy

Handbook of Dendritic Cells. Biology, Diseases, and Therapies.
Edited by M. B. Lutz, N. Romani, and A. Steinkasserer.
Copyright © 2006 WILEY-VCH Verlag GmbH & Co. KGaA, Weinheim
ISBN: 3-527-31109-2

Nina Bhardwaj
New York University
School of Medicine
550 1st Ave
New York, NY 10016
USA

Gennady Bocharov
Mathematics Department
University College Chester
Parkgate Road
Chester CH1 4BJ
UK

Laura C. Bonifaz
Research Unit on
Autoimmune Diseases
The Rockefeller University
1230 York Ave
New York, NY 10021
USA

André Boonstra
Division of Immunoregulation
The National Institute for Medical
Research (NIMR)
The Ridgeway
Mill Hill
London NW7 1AA
UK

Barbara Bottazzi
Department of Immunology and
Cell Biology
Mario Negri Institute
Via Eritrea 62
20157 Milano
Italy

Wei Cao
M.D. Anderson Cancer Center
University of Texas
1515 Holcombe Blvd.
Houston, TX 77030
USA

Giusy Capuano
Department of Biotechnology and
Bioscience
University of Milano-Bicocca
Piazza della Scienza 2
20126 Milano
Italy

Esther C. de Jong
Department of Cell Biology and
Histology
University of Amsterdam
Meibergdreef 15
1105 AZ Amsterdam
The Netherlands

Annalisa Del Prete
Mario Negri Institute
Via Eritrea 62
20157 Milano
Italy

Madhav V. Dhodapkar
Laboratory of Tumor Immunology
and Immunotherapy
The Rockefeller University
1230 York Avenue
New York, NY 10021
USA

Andrea Doni
Department of Immunology and Cell
Biology
Mario Negri Institute
Via Eritrea 62
20157 Milano
Italy

Jan Dörrie
Department of Dermatology
University Hospital of Erlangen
Hartmannstrasse 14
91052 Erlangen
Germany

Gaelle Elain
Hopital Necker-Enfants Malades
University of Paris René Descartes
149, rue de Sevres
75015 Paris Cedex 15
France

Fabio Facchetti
Department of Pathology
University of Brescia
Vle Europa 11
25123 Brescia
Italy

Paul J. Fairchild
Sir William Dunn School of
Pathology
University of Oxford
South Parks Road
Oxford OX1 3RE
UK

Claudia Falkensammer
Department of Urology
Medical University of Innsbruck
Anichstrasse 35
6020 Innsbruck
Austria

Joseph Fay
Baylor Institute for
Immunology Research (BIIR)
3434 Live Oak
Dallas, TX 75204
USA

Olivera J. Finn
Department of Immunology
University of Pittsburgh
Biomedical Science Taver
Pittsburgh, PA 15260
USA

Véronique Flamand
Institute for Medical Immunology
Université Libre de Bruxelles
8 rue Adrienne Bolland
6041 Gosselies
Belgium

Darin Fogg
Hopital Necker-Enfants Malades
University of Paris René Descartes
149, rue de Sevres
75015 Paris Cedex 15
France

Maria Foti
Department of Biotechnology and
Bioscience
University of Milano-Bicocca
Piazza della Scienza 2
20126 Milano
Italy

I. Frank
Center for Biomedical Research
Population Council
1230 York Avenue
New York, NY 10021
USA

K. Gamerdinger
Max-Planck Institut for
Immunobiology
Stuebeweg 51
79108 Freiburg
Germany

Hubert Gander
Department of Urology
Medical University of Innsbruck
Anichstrasse 35
6020 Innsbruck
Austria

Cecilia Garlanda
Department of Immunology and Cell
Biology
Mario Negri Institute
Via Eritrea 62
20157 Milano
Italy

Teunis B.H. Geijtenbeek
Department of Molecular Cell
Biology and Immunology
Vrije Universiteit Medical Center
Amsterdam
v.d. Boechorststraat 7
1081 BT Amsterdam
The Netherlands

Frederic Geissmann
Hopital Necker-Enfants Malades
University of Paris René Descartes
149, rue de Sevres
75015 Paris Cedex 15
France

Michel Gilliet
M.D. Anderson Cancer Center
University of Texas
1515 Holcombe Blvd.
Houston, TX 77030
USA

Dale I. Godfrey
Department of Microbiology and
Immunology
University of Melbourne
Parkville, Victoria 3010
Australia

Michel Goldman
Institute for Medical Immunology
Université Libre de Bruxelles
8, rue Adrienne Bolland
6041 Gosselies
Belgium

Romina Goldszmid
Laboratory of Parasitic Diseases
National Institute of Allergy and
Infectious Diseases
50 South Drive
Bethesda, MD 20892-8003
USA

Siamon Gordon
Sir William Dunn School of
Pathology
University of Oxford
South Parks Road
Oxford OX1 3RE
UK

Francesca Granucci
Department of Biotechnology and
Bioscience
University of Milano-Bicocca
Piazza della Scienza 2
20126 Milano
Italy

Hamida Hammad
Department of Pulmonary Medicine
Erasmus Medical Center
Dr. Molewaterplein 50
3015 GE Rotterdam
The Netherlands

Leonhard Heinz
Institute of Immunology
Medical University of Vienna
Lazarettgasse 19
1090 Wien
Austria

Christine Heufler
Department of Dermatology and
Venereology
Medical University of Innsbruck
Anichstrasse 35
6020 Innsbruck
Austria

Hubertus Hochrein
Institut für Medizinische
Mikrobiologie, Immunologie und
Hygiene
TU München
Trogerstrasse 9
81675 München
Germany

Lorenz Höltl
Department of Urology
Medical University of Innsbruck
Anichstrasse 35
6020 Innsbruck
Austria

Tomoki Ito
M.D. Anderson Cancer Center
1515 Holcombe Blvd.
University of Texas
Houston, TX 77030
USA

Dragana Jankovic
Laboratory of Parasitic Diseases
National Institute of Allergy and
Infectious Diseases
50 South Drive
Bethesda, MD 20892-8003
USA

Almut Jörgl
Institute of Immunology
Medical University of Vienna
Lazarettgasse 19
1090 Wien
Austria

Holger Kanzler
M.D. Anderson Cancer Center
University of Texas
1515 Holcombe Blvd.
Houston, TX 77030
USA

Martien L. Kapsenberg
Department of Cell Biology and
Histology/Dermatology
University of Amsterdam
Meibergdreef 15
1105 AZ Amsterdam
The Netherlands

Franz Koch
Department of Dermatology and
Venereology
Medical University of Innsbruck
Anichstrasse 35
6020 Innsbruck
Austria

Sophie Koutouzov
Baylor Institute for Immunology
Research (BIIR)
3434 Live Oak
Dallas, TX 75204
USA

Bart N. Lambrecht
Department of Pulmonary Medicine
Erasmus Medical Center
Dr. Molewaterplein 50
3015 GE Rotterdam
The Netherlands

Antonio Lanzavecchia
Institute for Research in Biomedicine
Via Vincenzo Vela 6
6500 Bellinzona
Switzerland

Adriana T. Larregina
Departments of Dermatology and
Immunology
University of Pittsburgh
Medical Center
Pittsburgh, PA 15213
USA

Yong-Jun Liu
M.D. Anderson Cancer Center
University of Texas
1515 Holcombe Blvd.
Houston, TX 77030
USA

Burkhard Ludewig
Research Department
Kantonal Hospital St. Gallen
Building 09
9007 St. Gallen
Switzerland

Manfred B. Lutz
Department of Dermatology
University of Erlangen
Hartmannstrasse 14
91052 Erlangen
Germany

Andrew S. MacDonald
Institute of Immunology and
Infection Research
University of Edinburgh
West Mains Road
Edinburgh EH9 3JT
UK

Gordon MacPherson
Sir William Dunn School of
Pathology
University of Oxford
South Parks Road
Oxford OX1 3RE
UK

Alberto Mantovani
Department of Immunology and Cell
Biology
Mario Negri Institute
Via Eritrea 62
20157 Milano
Italy

Markus G. Manz
Institute for Research in Biomedicine
(IRB)
Via Vincenzo Vela 6
6500 Bellinzona
Switzerland

Luisa Martinez-Pomares
School of Molecular Medical Sciences
Faculty of Medicine and
Health Sciences
University of Nottingham
Queens Medical Centre
Nottingham NG7 2UH
UK

Alexis Mathian
Baylor Institute for Immunology
Research (BIIR)
3434 Live Oak
Dallas, TX 75204
USA

Emma J. McKenzie
Sir William Dunn School of
Pathology
University of Oxford
South Parks Road
Oxford OX1 3RE
UK

Simon Milling
Sir William Dunn School of
Pathology
University of Oxford
South Parks Road
Oxford OX1 3RE
UK

Francesca Mingozzi
Department of Biotechnology and
Bioscience
University of Milano-Bicocca
Piazza della Scienza 2
20126 Milano
Italy

Heidrun Moll
Institute for Molecular Biology of
Infectious Diseases
University of Würzburg
Röntgenring 11
97070 Würzburg
Germany

Adrian E. Morelli
Department of Surgery
Thomas e. Starzl Transplantation
Institute
200 Lothrop St.
Pittsburgh, PA 15213-2582
USA

G. Morrow
Center for Biomedical Research
Population Council
1230 York Avenue
New York, NY 10021
USA

Nora Mueller
Institute for Virology and
Immunobiology
University of Würzburg
Versbacherstrasse 7
97078 Würzburg
Germany

Subhankar Mukhopadhyay
Sir William Dunn School of
Pathology
University of Oxford
South Parks Road
Oxford OX1 3RE
UK

Christian Münz
Laboratory of Viral Immunobiology
The Rockefeller University
1230 York Avenue
New York, NY 10021-6399
USA

Francis M. Ndungu
The National Institute for
Medical Research (NIMR)
The Ridgeway
Mill Hill
London NW7 1AA
UK

Dirk M. Nettelbeck
Department of Dermatology
University Hospital of Erlangen
Hartmannstrasse 14
91052 Erlangen
Germany

Kathleen F. Nolan
Sir William Dunn School of
Pathology
University of Oxford
South Parks Road
Oxford OX1 3RE
UK

Anne O'Garra
Division of Immunoregulation
The National Institute for Medical
Research (NIMR)
The Ridgeway
Mill Hill
London NW7 1AA
UK

Karel Otero
Mario Negri Institute
Via Eritrea 62
20157 Milano
Italy

A. Karolina Palucka
Baylor Institute for Immunology
Research (BIIR)
3434 Live Oak
Dallas, TX 75204
USA

Virginia Pascual
Baylor Institute for Immunology
Research (BIIR)
3434 Live Oak
Dallas, TX 75204
USA

Alison Paterson
Sir William Dunn School of
Pathology
University of Oxford
South Parks Road
Oxford OX1 3RE
UK

Norman Pavelka
Department of Biotechnology and
Bioscience
University of Milano-Bicocca
Piazza della Scienza 2
20126 Milano
Italy

Edward J. Pearce
Department of Pathobiology
University of Pennsylvania
Philadelphia, PA 19104-67008
USA

Mattia Pelizzola
Department of Biotechnology and
Bioscience
University of Milano-Bicocca
Piazza della Scienza 2
20126 Milano
Italy

Giuseppe Penna
BioXell
Via Olgettina 58
20132 Milano
Italy

S. Peretti
Center for Biomedical Research
Population Council
1230 York Avenue
New York, NY 10021
USA

Barbara Platzer
Institute of Immunology
Medical University of Vienna
Lazarettgasse 19
1090 Wien
Austria

Stefan Pöhlmann
Institute for Clinical and
Molecular Virology
University Erlangen-Nürnberg
Schlossgarten 4
91054 Erlangen
Germany

Melissa Pope
Center for Biomedical Research
Population Council
1230 York Avenue
New York, NY 10021
USA

Alexander T. Prechtel
Department of Dermatology
University Hospital Erlangen
Hartmannstrasse 14
91052 Erlangen
Germany

Paolo Puccetti
Department of Experimental
Medicine and Biochemical Sciences
University of Perugia
Via del Giochetto
06122 Perugia
Italy

Thomas Putz
Department of Urology
Medical University of Innsbruck
Anichstrasse 35
6020 Innsbruck
Austria

Andrea Rahm
Department of Urology
Medical University of Innsbruck
Anichstrasse 35
6020 Innsbruck
Austria

Giorgio Raimondi
Thomas E. Starzl Transplantation
Institute
University of Pittsburgh
South 3459 Fifth Avenue
Pittsburgh, PA 15213
USA

Reinhold Ramoner
Department of Urology
Medical University of Innsbruck
Anichstrasse 35
6020 Innsbruck
Austria

Gudrun Ratzinger
Department of Dermatology and
Venereology
Medical University of Innsbruck
Anichstrasse 35
6020 Innsbruck
Austria

Peter Reisner
Institute of Immunology
Medical University of Vienna
Lazarettgasse 19
1090 Wien
Austria

Maria Rescigno
Department of Experimental
Oncology
European Institute of Oncology
Via Ripamonti 435
20141 Milano
Italy

Elena Riboldi
Section of General Pathology and
Immunology
University of Brescia
Vle Europa 11
25123 Brescia
Italy

Paola Ricciardi-Castagnoli
Department of Biotechnology and
Bioscience
University of Milano-Bicocca
Piazza della Scienza 2
20126 Milano
Italy

Nikolaus Romani
Department of Dermatology and
Venereoogy
Medical University of Innsbruck
Anichstrasse 35
6020 Innsbruck
Austria

Luigina Romani
Department of Experimental
Medicine and Biochemical Sciences
University of Perugia
Via del Giochetto
06122 Perugia
Italy

Sem Saeland
INSERM U 503
IFR 128
21, avenue Tony Garnier
69365 Lyon
France

Federica Sallusto
Institute for Research in Biomedicine
Via Vincenzo Vela 6
6500 Bellinzona
Switzerland

Niels Schaft
Department of Dermatology
University Hospital of Erlangen
Hartmannstrasse 14
91052 Erlangen
Germany

Ulrich E. Schaible
Department of Immunology
Max-Planck-Institute for
Infection Biology
Schumannstr. 21–22
10117 Berlin
Germany

Sibylle Schneider-Schaulies
Institute for Virology and
Immunobiology
University of Würzburg
Versbacherstrasse 7
97078 Würzburg
Germany

Brigitte Sénéchal
Hopital Necker-Enfants Malades
University of Paris René Descartes
149, rue de Sevres
75015 Paris Cedex 15
France

Alan Sher
Laboratory of Parasitic Diseases
National Institute of Allergy and
Infectious Diseases
50 South Drive
Bethesda, MD 20892-8003
USA

Ken Shortman
Immunoogy Division
The Walter and Eliza Hall Institute
Parkville, Victoria 3050
Australia

Mojca Škoberne
New York University School of
Medicine
550 1st Ave
New York, NY 10016
USA

Hermelijn H. Smits
Department of Parasitology
Leiden University Medical Center
Albinusdreef 2
2333 ZA Leiden
The Netherlands

Mark J. Smyth
Trescowthick Laboratories
Peter MacCallum Cancer Centre
A'Beckett Street
8006 Victoria
Australia

Silvano Sozzani
Section of General Pathology and
Immunology
University of Brescia
Vle Europa 11
25123 Brescia
Italy

Alexander Steinkasserer
Department of Dermatology
University Hospital Erlangen
Hartmannstrasse 14
91052 Erlangen
Germany

Ralph M. Steinman
The Rockefeller University
Laboratory of Physiology and
Immunology
1230 York Avenue
New York, NY 10021-6399
USA

Joan Stein-Streilein
Schepens Eye Research Institute
20 Staniford Street
Boston, MA 02114
USA

Patrizia Stoitzner
Department of dermatology and
Venereology
Medical University of Innsbruck
Anichstrasse 35
6020 Innsbruck
Austria

Herbert Strobl
Institute of Immunology
Medical University of Vienna
Lazarettgasse 19
1090 Wien
Austria

Sabine Taschner
Institute of Immunology
Medical University of Vienna
Lazarettgasse 19
1090 Wien
Austria

Magali Terme
Department of Clinical Biology
ERM0208 INSERM
39, rue Camille Desmoulins
94805 Villejuif Cedex
France

H.-J. Thierse
Max-Planck Institut for
Immunobiology
Stuebeweg 51
79108 Freiburg
Germany

Angus W. Thomson
W1544 Biomedical Science Tower
University of Pittsburth
200 Lothrop Street
Pittsburgh, PA 15213
USA

Martin Thurnher
Department of Urology
Medical University of Innsbruck
Anichstrasse 35
6020 Innsbruck
Austria

David F. Tough
Senior Group Leader & Deputy
Scientific Head
The Edward Jenner Institute for
Vaccine Research
Compton Newbury
RG20 7NN
UK

S. Trapp
Center for Biomedical Research
Population Council
1230 York Avenue
New York, NY 10021
USA

Giorgio Trinchieri
Laboratory of Parasitic Diseases
National Institute of Allergy and
Infectious Diseases
50 South Drive
Bethesda, MD 20892
USA

Christoph H. Tripp
Department of Dermatology and
Venereology
Medical University of Innsbruck
Anichstrasse 35
6020 Innsbruck
Austria

François Trottein
Centre d'Immunologie et de Biologie
Parasitaire
Institut National de la Santé et de la
Recherche Médicale
Unité 547
1, rue du Professeur Calmette
59019 Lille Cedex
France

Emma Turnbull
The Edward Jenner Institute for
Vaccine Research
Compton
Newbury
Berkshire RG20 7NN
UK

S.G. Turville
Center for Biomedical Research
Population Council
1230 York Avenue
New York, NY 10021
USA

Hideki Ueno
Baylor Institute for Immunology
Research (BIIR)
3434 Live Oak
Dallas, TX 75204
USA

Britta C. Urban
Nuffield Department of
Clinical Medicine
University of Oxford
Old Road
Oxford OX3 7LJ
UK

Matteo Urbano
Department of Biotechnology and
Bioscience
University of Milano-Bicocca
Piazza della Scienza 2
20126 Milano
Italy

L. Vachot
Center for Biomedical Research
Population Council
1230 York Avenue
New York, NY 10021
USA

Serani L.H. van Dommelen
Trescowthick Laboratories
Peter MacCallum Cancer Centre
St. Andrews Place
East Melbourne, Victoria
Australia

Yvette van Kooyk
Department of Molecular Cell
Biology and Immunology
Vrije Universiteit Medical Center
Amsterdam
v.d. Boechorststraat 7
1081 BT Amsterdam
The Netherlands

Annunciata Vecchi
Mario Negri Institute
Via Eritrea 62
20157 Milano
Italy

Jóse A. Villadangos
Immunology Division
The Walter and Eliza Hall Institute
Parkville Victoria 3050
Australia

Caterina Vizzardelli
Department of Biotechnology and
Bioscience
University of Milano-Bicocca
Piazza della Scienza 2
20126 Milano
Italy

Hermann Wagner
Institut für Medizinische
Mikrobiologie, Immunologie und
Hygiene
TU München
Trogerstrasse 9
81675 München
Germany

Herman Waldmann
Sir William Dunn School of
Pathology
University of Oxford
South Parks Road
Oxford OX1 3RE
UK

Yui-His Wang
M.D. Anderson Cancer Center
University of Texas
1515 Holcombe Blvd.
Houston, TX 77030
USA

Yi-Hong Wang
M.D. Anderson Cancer Center
University of Texas
1515 Holcombe Blvd.
Houston, TX 77030
USA

Marca H.M. Wauben
Department of Immunohematology
and Blood Transfusion
Leiden University Medical Center
Albinusdreef 2
2300 RC Leiden
The Netherlands

Hans-Ulrich Weltzien
Max-Planck Institut for
Immunobiology
Stuebeweg 51
79108 Freiburg
Germany

Mary Jo Wick
Department of Clinical Immunology
Göteborg University
Guldhedsgatan 10
413 46 Göteborg
Sweden

Eddy A. Wierenga
Department of Cell Biology and
Histology
University of Amsterdam
Meibergdreef 15
1105 AZ Amsterdam
The Netherlands

Florian Winau
Department of Immunology
Max-Planck-Institute for
Infection Biology
Schumannstr. 21-22
10117 Berlin
Germany

Felix Yarovinsky
Laboratory of Parasitic Diseases
National Institute of Allergy and
Infectious Diseases
50 South Drive
Bethesda, MD 20892-8003
USA

Stephen F. Yates
Sir William Dunn School of
Pathology
University of Oxford
South Parks Road
Oxford OX1 3RE
UK

James W. Young
Memorial Sloan-Kettering Cancer
Center
Weill Medical College of
Cornell University
New York, NY 10021-6094
USA

Ulf Yrlid
Department of Clinical Immunology
University of Gothenberg
Box 400
40530 Göteborg
Sweden

Ivan Zanoni
Department of Biotechnology and
Bioscience
University of Milano-Bicocca
Piazza della Scienza 2
20126 Milano
Italy

Martin Zenke
Institute for Biomedical Engineering
(Cell Biology)
Aachen University Medical School
Pauwelstrasse 30
52074 Aachen
Germany

Laurence Zitvogel
Immunology Unit
ERM0208 INSERM
39 rue Camille Desmoulins
94805 Villejuif Cedex
France

Color Plates

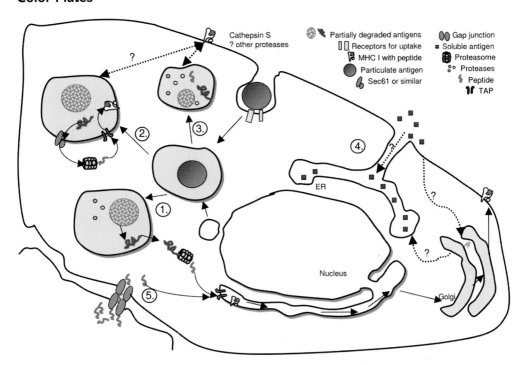

Fig. 22.1 Mechanisms of crosspresentation. Particulate antigens enter the antigen presenting cells most efficiently by receptor mediated endocytosis. Endosomes containing the ingested antigen then acquire selected endoplasmatic reticulum (ER)-derived elements and, after initial processing within the endosome, the antigens exit the late endosome and enter the cytoplasm. They then enter the classical pathway and are processed in a similar manner as endogenous antigens (1). Alternatively, the antigens may remain in the endosomes that acquire selected characteristics of ER and are fully processed in close viscinity of the phago-endosome. They re-enter the phago-endosomes and are there loaded onto MHC class I molecules (2). Whereas in the former cases the antigen processing involves transporter of antigen processing (TAP), this is not always required. Albeit less efficiently, the vacuolar processing pathway (3) may rely on vacuolar proteases, especially cathepsin S, to generate peptides that are loaded onto MHC class I molecules within the vacuole. A special pathway may be in place for soluble antigens (4) which by yet unidentified mechanisms directly access the ER. Gap junctions may permit the transfer of already processed peptides from donor to acceptor cells (5).

Handbook of Dendritic Cells. Biology, Diseases, and Therapies.
Edited by M. B. Lutz, N. Romani, and A. Steinkasserer.
Copyright © 2006 WILEY-VCH Verlag GmbH & Co. KGaA, Weinheim
ISBN: 3-527-31109-2

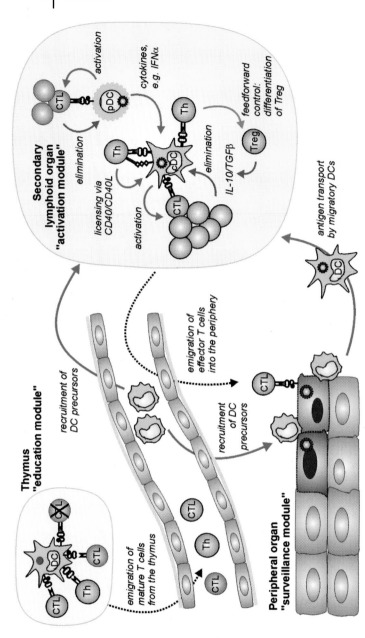

Fig. 23.1 Systems biologist's view of DC–CTL interactions. *Modularity* in the complex system of (direct and indirect) DC–CTL interactions is represented by different organizational modules: thymus ("educational module"), secondary lymphoid organs ("activation module"), and peripheral nonlymphoid organs ("surveillance module"). Different DC subsets, e.g. conventional DCs (cDC) versus plasmacytoid DCs (pDC) provide heterogenous redundancy and ensure homogenous outcome, that is, activation of CTL and secretion of stimulating cytokines. Positive *feedback* mechanisms (e.g. chemokine-driven recruitment of DC precursors into peripheral organs or secondary lymphoid organs, or CD40-mediated licensing of DCs) are indicated by green arrows. Negative feedback control loops such as elimination of DCs by effector CTL are depicted in red. Generation of regulatory Th cells (Treg) can be seen as feedforward control (blue arrow).

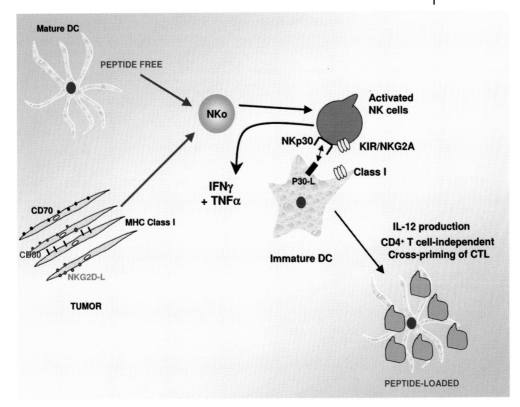

Fig. 24.1 A DC/NK/DC cross-talk leading to efficient cross-presentation of tumor antigens. Tumor cells over-expressing NKG2D ligands or CD70 or CD80 might directly trigger NK-cell activation. Alternatively, exogenous LPS-activated mature DC can turn on NK cells *in vivo*. DC-activated NK cells produce IFNγ which in turn, promotes endogenous DC activation and production of IL-12 and elicitation of CTL cross priming in the absence of CD4+ T-cell help [43]. In the human setting, DC-activated NK cells or IL-2 activated NK cells will induce DC maturation through engagement of NKp30, in a TNF-α dependent-manner.

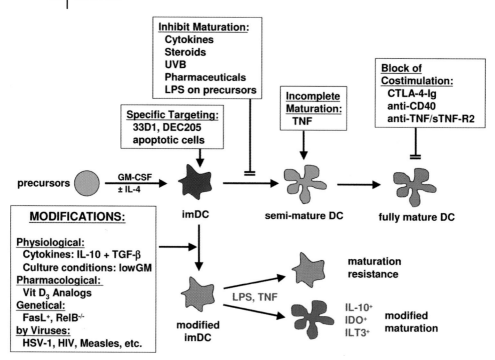

Fig. 26.1 Methods of *in vitro* generation or *in vivo* targeting of tolerogenic DC. Many methods of generating tolerogenic DC are focused around immature DC. They can be targeted specifically *in vivo* by delivery of apoptotic material or targeting to specific surface receptors. Spontaneous maturation of *in vitro* generated DC can be achieved by addition of maturation inhibitors. Inhibition of *in vivo* maturation is attempted by inhibiting co-stimulation of already matured DC. Incompletely matured DC (semi-mature DC) can still be tolerogenic. Due to the problem of maintaining stable immature DC, DC modifications are desired that result in maturation-resistant immature DC or modified/alternatively matured DC expressing inhibitory surface receptors (IDO, ILT-3) or cytokines (IL-10).

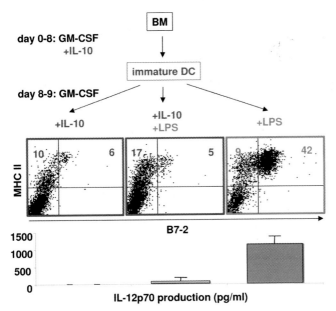

Fig. 26.3 Importance of maturation resistance by immature DC for their tolerogenicity. BM-DC were cultured for 8 days with GM-CSF (200 U ml⁻¹) plus IL-10 (10 ng ml⁻¹) to inhibit spontaneous maturation. Then the cells were washed and replated with GM-CSF and IL-10, or IL-10 plus LPS or LPS for another 24 h. FACS analysis was performed with the cells for surface MHC II (M5/114-PE) and B7-2 (FITC) and the IL-12p70 production measured by ELISA. The data show that in the absence of the inhibitory IL-10 signal, the immature DC rapidly mature on LPS.

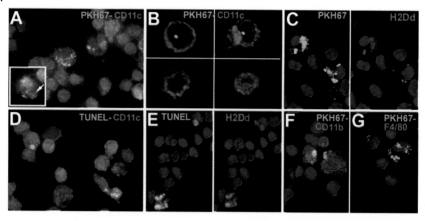

Fig. 29.3 Entrapment of apoptotic cells by splenic DC *in vivo*. Internalization of apoptotic cells by splenic DC was analyzed in cytospins of immunobead-sorted DC 1 h after injection of PKH67-labeled (green) apoptotic (BALB/c) splenocytes in (B10) mie. (A) CD11c⁺ DC with apoptotic cell fragments (green) and with DAPI⁺ intracytoplasmic inclusions, -likely DNA from ingested apoptotic cells (in blue indicated by arrow in inset). (B) Serial sections analyzed by confocal microscopy confirmed the intracellular localization of PKH67-labeled fragments in splenic CD11c⁺ DC. (C) The donor origin (BALB/c) of the intracytoplasmic inclusions in (B10) DC was confirmed by H2D^d expression (in red) in PKH67-labeled (green) fragments. (D & E) FITC-TUNEL staining in combination with Cy3-anti-CD11c or Cy3-anti-H2D^d confirmed the presence of donor (BALB/c)-derived apoptotic cells within (B10) DC. (F & G) One h after i.v. injection of apoptotic cells, DC that internalized apoptotic cells expressed CD11b^{hi} and F4/80$^{lo/-}$. Nuclei were counterstained with DAPI (1000x). Reproduced with permission from *Blood* **2003**. 101: 611–620 (A.E. Morelli et al.).

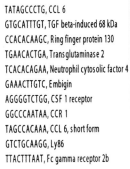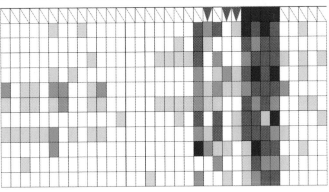

Key to Gene Expression Chart:

Abundantly Expressed Gene (>1%) ⬛
Abundantly Expressed Gene (>0.3%)
Moderately Expressed Gene (>0.1%)
Significant Expression (>=7 tags)
Positive Expression (>=3 tags)
Positive (not significant <3 tags)
Negative (undetected or no tags) ⬜

Fig. 30.6 SAGE tags associated with modulated DCs. Using SAGEClus software, tags associated with the pharmacologically-modulated DC populations have been selected based on the relatedness of their expression profiles to the idealized test pattern indicated in the top row of the clustergram. A blue square in the test pattern indicates moderate tag representation in that library, as indicated in the expression key, while an inverted grey triangle indicates no tag representation and a diagonal line no preference. Details of the comparator libraries indicated can be found in [86].

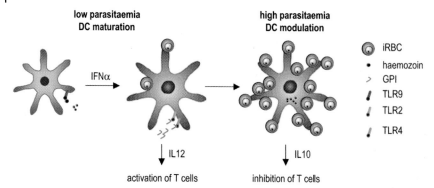

low parasitaemia
DC maturation

IFNα

high parasitaemia
DC modulation

- iRBC
- haemozoin
- GPI
- TLR9
- TLR2
- TLR4

↓ IL12

activation of T cells

↓ IL10

inhibition of T cells

Fig. 31.1 Simplified diagram on the effect of parasitemia on DC function. Early on during infection, engagement of TLR9 by hemozoin results induces plasmacytoid DC to secrete IFNα and engagement of TLR2 and TLR4 by GPI induces myeloid DC to secrete IL12.

With increasing parasitemia, more and more myeloid DC in the spleen might be modulated either directly through interaction with iRBC or through ingestion of increasing amounts of hemozoin and secrete IL10.

Lamp1

L. major

Merge

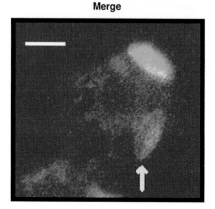

Fig. 32.2 Endosomal compartments in DC harbor *Leishmania* parasites. DC infected for one hour with *L. major* that had been pre-stained with 5-chloromethylfluorescein-diacetat (CMFDA, green) were subjected to intracellular staining for Lamp1, a marker of late endosomes/lysosomes, using a phyco-erythrin-labeled antibody. The localization of Lamp1 (red, top left) and *L. major* parasites (green, top right) was analyzed by confocal microscopy. A parasite residing in a parasitophorous vacuole that also contains Lamp1 is indicated by arrows (merge, bottom left). Bar: 5 μm.

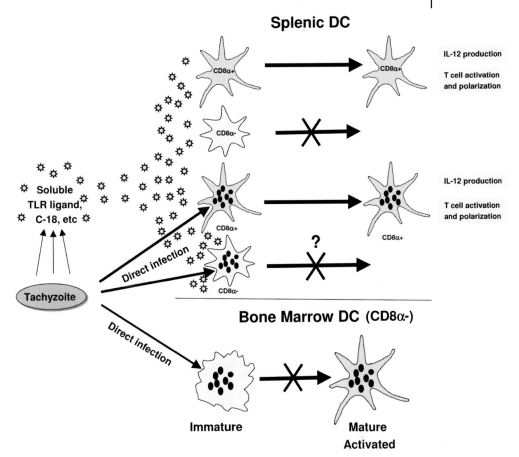

Fig. 33.1 Encounter of DC subsets with *T. gondii* leads to distinct outcomes. CD8α⁺ but not CD8α⁻ DC in spleen and other tissues respond to soluble products released by tachyzoites and present in STAg. Whether intracellular tachyzoites (as opposed to soluble parasite products) can trigger responses in CD8α⁻ DC, particularly after extended incubation, remains to be determined. While direct infection does not appear to influence the activation status of splenic DC, bone-marrow-derived DC not only fail to respond to soluble tachyzoite Ag but when infected with the parasite display a block in maturation and response to exogenous stimulation with LPS or CD40L [14].

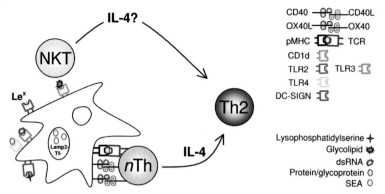

Fig. 34.2 Th2 induction by DC in response to SEA. It is likely that multiple components of SEA bind to a range of pattern recognition receptors on DC, including C-type lectins [19] and TLRs [22, 25, 27]. Although several TLRs have been implicated in this process, MyD88-deficient mice are still capable of making Th2 responses to SEA [66], suggesting either a degree of redundancy in the system or, more likely, that MyD88-independent TLR-initiated signaling is important. Once internalized SEA is distributed to Lamp2⁻, Transferrin⁻ vesicular compartments within the DC [18].

CD40 expression is critical for DC induction of Th2 response to SEA [46], suggesting downstream involvement of CD40-mediated activation events in this process, with OX-40L being one potential candidate that could fulfill this function. IL-4-deficient DC show no deficiency in Th2 induction to SEA [41], although IL-4 from a source other than the initiating DC is paramount to sustain the developing CD4 T-cell response. NKT cells may provide one source of such IL-4, as CD1d-deficient DC display impaired Th2-induction abilities to SEA [35].

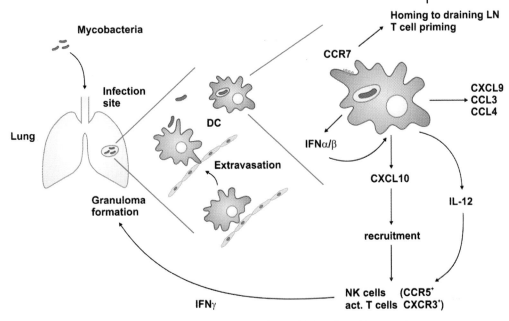

Fig. 36.1 Role of DC in tuberculosis. Inhalation of mycobacteria by aerosol leads to the infection of macrophages and DC in the lung. Infection activates DC to express CCR7 and to subsequently migrate to draining lymph nodes for T-cell priming. Concurrent release of type I IFN-α/β starts an autocrine activation loop initiating chemokine secretion (CXCL10, 9, CCL3, 4) to recruit activated T cells and NK cells to the site of infection. Onset of a protective T-cell response with IFNγ and TNFα as essential cytokines leads to macrophage/DC activation and granuloma formation to restrict spread of mycobacteria.

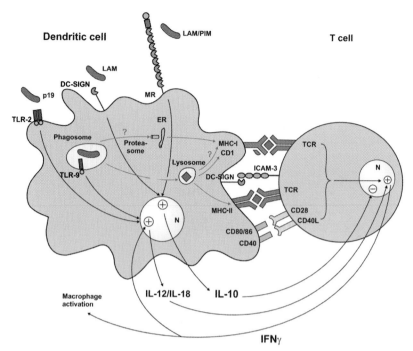

Fig. 36.2 The DC as immune-modulating APC. Mycobacteria and their pathogen associate molecular patterns (PAMP) such as lipoarabinomannan (LAM), the 19-kDa lipoprotein (p19) and low-methylated bacterial DNA (CpG) bind various pattern recognition receptors (PRR) on DC. Engagement of DC-SIGN and the mannose receptor (MR) induce release of the anti-inflammatory/immuno-suppressive cytokine IL-10. In contrast, ligands for Toll-like receptors (p19 – TLR-2, CpG – TLR-9) activate DC to secrete the pro-inflammatory cytokines IL-12/IL-18. Thereby an IFNγ-dominated T-cell response (T helper type I) is initiated leading to macrophage activation and mycobactericidal effector mechanisms. DC are potent antigen-presenting cells employing MHC-I, -II, CD1 and co-stimulatory molecules (CD80, CD86, CD40) to prime mycobacterium-specific T cells. Mycobacterial antigens are delivered to the lysosomes for processing and binding to MHC-II and CD1 molecules. Loading of lipids onto CD1 molecules involves the lipid transfer proteins saposins. The pathway leading to MHC-I presentation within infected cells – whether through processing and loading within late endosomes/lysosomes, processing by proteasomes and loading in the ER, or by both – is yet unclear.

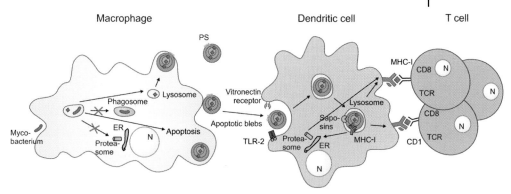

Fig. 36.3 The detour pathway of MHC-I and CD1 presentation. In macrophages, myco-bacteria are segregated within phagosomes from the classical MHC class I pathway. Moreover, macrophages do not express group I CD1 molecules (CD1a, b, c). Finally, mycobacteria-infected cells loose their ability to present antigens to T cells. These hinderances to induce proper T-cell immunity are overcome by the detour pathway in tuber-culosis. Infection-induced apoptosis leads to the release of phosphatidyl-serine (PS)-positive apoptotic blebs from infected cells. Thereby, mycobacterial antigens are carried to non-infected DC for presentation. Apoptotic blebs are engulfed by the vitronectin receptor (VR) and probably the PS-receptor and reach the endosomal system of the DC. DC matura-tion is initiated upon engagement of TLR-2 by mycobacterial PAMP (such as p19). DC subsequently prime T cells through MHC-I and CD1, but also MHC-II molecules. Processing of mycobacterial antigens is predominantly dependent on the lysosomal pathway. Saposins are involved in this process.

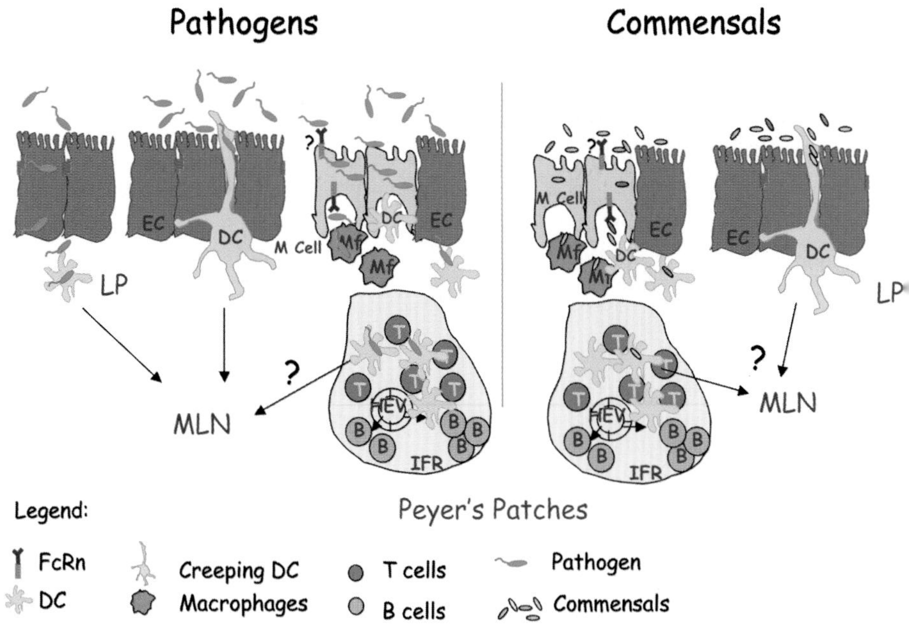

Pathogens

Commensals

Legend:

Peyer's Patches

Fig. 37.1 Mechanisms of bacterial uptake. The mechanisms of bacterial entrance depend on their pathogenicity. Most of the pathogens have developed strategies to penetrate ECs or to facilitate M-cell invasion, alternatively they are captured by creeping DCs (left). Commensal bacteria can enter mucosal surfaces either through M cells or DCs (right). M cells can release their 'cargo' to underlying phagocytic cells, including DCs, that can migrate to the interfollicular region (IFR) of Peyer's Patches for T and B-cell interactions, whereas DCs that take up bacteria directly across mucosal surfaces are likely to migrate to MLN. Alternatively, PP-DCs could migrate to MLN. An alternative mechanism for antigen entry across a mucosal surface that also targets DCs and could be used for bacterial internalization is mediated by neonatal Fc receptors (FcRn) expressed by adult human (but not mouse) intestinal epithelial cells. FcRn transport directs and delivers the antigens in the form of immune complexes directly to underlying DCs. (HEV: high endothelial venules).

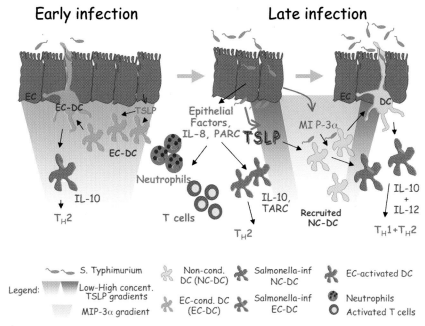

Fig. 37.2 Early *Salmonella typhimurium* infection: resident DCs are conditioned by EC-released TSLP (EC-DC). EC-DC release IL-10 after bacterial exposure and drive default T_H2 responses to *S. typhimurium*.

Late infection: since *S. typhimurium* is an invasive bacterium, it induces ECs to release pro-inflammatory chemokines like IL-8 (CXCL-8) and PARC (CCL-18), which attract neutrophils, granulocytes and activated T cells that generate an inflamed site. The binding of *Salmonella* to the basolateral membrane of ECs induces the upregulation of TSLP. TSLP at this concentration drives T_H1 rather than T_H2 promoting DCs in response to bacteria. Unidentified EC-derived factors can also activate 'bystander' DCs that have not been in contact directly with the bacteria. DCs activated in this way release IL-10 and TARC (CCL-17) but not IL-12, thus driving and recruiting T_H2 T cells. *Salmonella* also induces the release of MIP-3α (CCL-20) that recruits CCR6-expressing immature DCs.

Most likely, recruited DCs are not subjected to EC-conditioning, rather they could find increased TSLP concentrations in the infected site. Newly recruited DCs (NC-DC) can either creep between ECs to take up bacteria or they can phagocytose bacteria that have breached across the epithelial barrier and release both IL-10 and IL-12, thus promoting T_H1 and T_H2 responses. This allows the establishment of protective anti-*Salmonella* responses.

VII
Th1 and Th2 Decision

20
The Plasticity of Dendritic Cells Populations in Promoting Th-cell Responses

André Boonstra, Giorgio Trinchieri and Anne O'Garra

An effective immune system requires a high degree of flexibility in order to combat different types of pathogens. In peripheral tissues, dendritic cells (DC) sample their environment for antigens and transport them to the lymph nodes. The close interaction between DC and naïve helper T (Th) cells in the lymph nodes results in the activation of Th cells via the interaction of antigen complexed with MHC II on the antigen-presenting cell and the specific T-cell receptor on Th cells, leading to the development of appropriate effector Th-cell responses [1–3]. As will be discussed in more detail below, DC do not merely provide an "on/off" signal, but exhibit a remarkable plasticity in directing the development of Th1 or Th2 responses. This flexibility is the net result of the existence of different DC populations, their modulation by environmental constraints and conditions, their response to pathogens, their past experiences, and the differential state of maturation. Distinct DC populations may be poised to respond to different pathogens due to differential Toll-like receptor (TLR) expression. Thus, DC, because of their heterogeneity and plasticity, are able to instruct the immune system to tailor its responses to deal with a wide range of pathogens efficiently.

20.1
Effector Th-cell Populations

Activation of naïve T cells by antigen presenting cells can result in the development of Th1 or Th2 cells, which can be distinguished on the basis of the cytokines that they produce. Th1 cells produce interferon (IFN)-γ and lymphotoxin and play a central role in cell-mediated immunity important for the eradication of intracellu-

Handbook of Dendritic Cells. Biology, Diseases, and Therapies.
Edited by M. B. Lutz, N. Romani, and A. Steinkasserer.
Copyright © 2006 WILEY-VCH Verlag GmbH & Co. KGaA, Weinheim
ISBN: 3-527-31109-2

lar pathogens, such as bacteria and viruses [4, 5]. Th2 cells, producing interleukin (IL)-4, IL-5, and IL-13, are important in immunity to helminths, and contribute to eosinophilic inflammation and allergic reactions [6, 7].

The cytokines produced in the micro-environment are key factors in determining the type of Th response. The dominant cytokines are IL-12 and IL-4, which induce Th1 and Th2 responses, respectively [8, 9]. Mice deficient in IL-12, or in the IL-12-induced transcription factor STAT-4 had markedly reduced Th1 responses [10, 11], whereas mice deficient in IL-4 or in IL-4-induced STAT-6 had reduced Th2 responses [12–16]. Th1-produced IFN-γ and Th2-produced IL-4 can promote the growth or differentiation of their own respective T-cell subsets by upregulating the transcription factors T-bet and GATA-3, respectively [17–22], but they can also antagonize the development of the opposing subset [23]. For example, IL-4 directly triggers the differentiation into Th2 cells and downregulates their expression of the IL-12-receptor β2 chain [24], whereas IFN-γ upregulates the expression of the IL-12-receptor β2 chain on Th1 cells in mice [24]. Species-specific differences are observed with respect to the effects of type I IFN. IFN-α and IFN-β upregulate the expression of the IL-12-receptor β2 chain, thereby enhancing IL-12-mediated Th1-cell development in human T cells [25]. Moreover, type I IFN has been reported to induce human T cells – but not mouse T cells – to differentiate into Th1 cells even in the absence of IL-12 [26]. The basis of the species-specific activity of type I IFN is unclear, because type I IFN transiently induces STAT-4 phosphorylation both in human and mouse T cells, although the effect may only be observed at higher type I IFN concentrations in the latter [27]. Indeed, LCMV infection *in vivo* results in STAT-4 activation by IFN-α/β in murine T cells and consequently IFN-γ production. Interestingly, IFN-α/β-induced activation of STAT-1 was shown to inhibit STAT-4 phosphorylation and IFN-γ production [27]. Therefore, the relative levels of activated STAT-1 versus STAT-4 by IFN-α/β may determine the IFN-γ production during viral infection.

20.2
Factors Inducing the Development of Th1 or Th2 Cells

DC are known to induce the activation and proliferation of T cells. Moreover, the finding that DC produced IL-12 upon challenge indicated that DC were also capable of directing the type of Th-cell response [28–30], with IL-12 being the key cytokine in driving Th1 responses. The recognition that DC-derived cytokines were key in driving Th-cell responses was complemented by studies showing that the interaction of DC and Th was subject to modulation by multiple factors [31]. In this, the strength of stimulation as determined by the antigen dose and co-stimulation, genetic background, tissue-derived factors, and specific cytokines produced after encounter with pathogen-derived products, were shown to play an important role in determining whether a Th1 or Th2 response develops.

20.2.1
The Strength of DC–Th-cell Interaction

The strength of the interaction mediated through the T-cell receptor and the MHC/peptide complex and the dose of antigen were found to affect Th-cell development [31–34]. *In vitro* stimulation of OVA-peptide specific TCR transgenic CD4$^+$ T cells with low antigen doses presented by bone-marrow derived DC or splenic CD11c$^+$ DC induces an IL-4 response, whereas high antigen doses favors increased IFN-γ production [32–35]. Separation of splenic CD11c$^+$ DC into CD11c$^+$CD8α$^+$ and CD11c$^+$CD8α$^-$ DC showed that both DC populations had a similar effect of the antigen dose on Th-cell polarization [35]. The Th1-cell development at high antigen doses was only partially blocked by anti-IL-12p40 antibodies, suggesting the additional involvement of Th1-inducing factors other than IL-12 and IL-23. Similar to conventional DC, bone marrow-derived plasmacytoid DC, although less efficient in inducing T-cell proliferation at low antigen doses, showed the same flexible ability to drive the development of Th1 cells at high antigen dose, and preferentially Th2 cells at low antigen doses [35]. These findings clearly point out that intrinsically DC populations have the ability to direct both Th1 and Th2 development under specific circumstances. The enhanced Th1-cell development at high antigen doses can be explained, at least in part, by enhanced activation of T cells resulting in strong upregulation of CD40-ligand on Th cells, which delivers a positive feedback signal to DC to augment its IL-12p70 production [34]. When the CD40/CD40-ligand interaction was prevented using anti-CD40-ligand antibodies or DC from CD40 knock-out mice, Th1 polarization was inhibited at high antigen doses. Similar results were obtained using IL-12p35 knock-out mice [34]. Th2-cell development at low antigen doses was dependent on IL-4, since neutralization of IL-4 inhibited Th2-cell development [33]. However, DC obtained from IL-4 knock-out and wild-type mice both induced Th2 responses to a similar degree, which indicates that T cells, but not DC, are the source of IL-4 (Boonstra and O'Garra, unpublished). Additionally, it was reported that a strong TCR signal resulted in sustained Erk activation and Th1 differentiation, whereas a weak signal induces transient Erk activation and an increase in IL-4 production [36].

In agreement with the mouse studies, activated human monocyte-derived DC pulsed with high or low doses of toxic shock syndrome toxin-1 (TSST-1) stimulated autologous T cells at high TSST doses to develop Th1 responses, and at low doses to develop Th2 responses [37]. In MLR responses, although the absolute dose of antigen cannot be established, it was shown that human monocyte-derived DC promote Th2-cell differentiation at low DC:T cell ratio (1:300) in a primary MLR, and Th1-cell differentiation at a high DC:T ratio (1:4) [38]. These findings were reproduced using mouse splenic CD11c$^+$ DC in an antigen-specific system, with a low DC:T ratio resulting in a mixed Th0/Th2 phenotype, and a high ratio in the development of Th1 cells [39] (Boonstra and O'Garra, unpublished).

20.2.2
Co-stimulators

Enhanced contact of the T cell with the APC or a prolonged duration of the interaction can also be achieved by co-stimulatory molecules. Indeed, in addition to CD40/CD40ligand interaction, a role for CD28/B-7 [40], OX40/OX40-ligand [41], and LFA-1/ICAM [42, 43] has been shown in Th polarization. The results of this strengthened interaction between the APC and the T cells may lower the threshold (i.e. effective antigen dose) needed for CD40-ligand upregulation and Th1-cell development [44].

20.2.3
Genetic Background

The genetic background has been shown to influence the ability of DC to develop Th1 or Th2 responses. This has been shown elegantly by comparing the responses to *Leishmania major* infection in B10.D2 mice, which develop a protective Th1 response with self-healing lesions, and in BALB/c mice, which develop a Th2 response resulting in the development of progressive lesions [45]. Although CD11c$^+$CD8α^- DC from both mouse strains primed CD4$^+$ T cells *in vivo*, CD11c$^+$CD8α^- DC obtained from BALB/c mice stimulated T cells to produce less IFN-γ and more IL-5 as compared to CD11c$^+$CD8α^- DC from B10.D2 mice. No differences in cytokine expression, including IL-12, were found when comparing DC from both strains, except for higher IL-1β expression by DC isolated from B10.D2 mice, which was shown to reduce Th2-cell responses upon injection *in vivo* [45].

20.3
Opposing Concepts: Pre-programmed versus Flexible DC Direct Th-cell Development

The ability of DC to direct Th1 or Th2-cell responses is dependent on cytokines, the strength of the stimulus, as determined by the antigen dose and co-stimulators, as well the genetic background, resulting in a high degree of flexibility of DC, which enables DC to respond to a wide variety of stimuli appropriately. It came therefore as a surprise that in 1999 almost simultaneously, in both mouse and human studies, it was suggested that specific DC subpopulations were pre-programmed to induce the development of Th1 cells, and others to induce Th2 responses, and that they were fixed in doing so [46–48].

20.3.1
Mouse Dendritic Cell Populations in Directing Th-cell Development

In vivo it was shown that splenic CD11c$^+$CD8α^+ DC, cultured overnight with GM–CSF and pulsed with KLH, induced the development of Th1 cells *in vivo*, whereas antigen-pulsed CD11c$^+$CD8α^- DC induced Th2 cells when injected into mice [47]. Th1 development induced by the transfer of CD11c⁺CD8α⁺ DC was IL-12 mediated, since CD11c⁺CD8α⁺ DC isolated from IL-12p40 knock-out mice did not promote Th1-cell development. Furthermore, IL-12p70 production was higher by CD11c⁺CD8α⁺ DC as compared to CD11c⁺CD8α⁻ DC upon stimulation [47]. These findings led to the hypothesis that specific DC populations produced IL-12 and promoted Th1 responses, whereas other DC populations did not produce IL-12 and as a result favored the development of Th2 responses (Fig. 20.1). Similar findings were reported by Pulendran et al., who showed using Flt3-ligand-treated mice, that OVA-peptide pulsed CD11c⁺CD8α⁻ DC induced higher levels of IL-4 production by T cells, while similar levels of IFN-γ were induced by T cells upon transfer of CD11c⁺CD8α⁻ DC and CD11c⁺CD8α⁺ DC into mice [48].

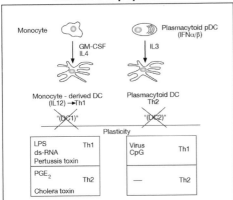

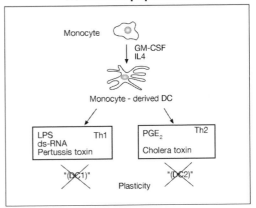

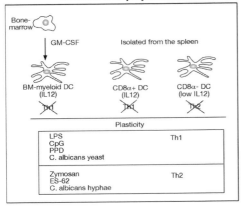

Therefore, DC are flexible in driving Th1 or Th2 development,

BUT

Many different types of DC
Different maturation state
Environmental constraints & conditions
Differential TLR expression
Antigen dose
Co-stimulators
Genetic background

Fig. 20.1 Human and mouse DC populations have a high degree of flexibility in driving Th-cell regulation, but there are many variables: different types of dendritic cell; in different maturation states; environmental constraints and conditions; different TLR expression; antigen dose; co-stimulators and genetic background.

The proposed concept that specific DC populations were fixed in promoting Th-cell responses was soon after challenged by a number of studies (Fig. 20.1). *In vivo* transfer of peptide pulsed CD8α⁺ and CD8α⁻ DC, which were not cultured, but freshly isolated from the spleen, showed that both splenic DC subsets have the potential to prime Th1 and Th2 cells when injected intravenously [49]. Also, stimulation with different pathogenic products demonstrated the flexibility of DC populations in promoting Th1 or Th2 responses. *In vitro*, both sorted CD8α⁺ and CD8α⁻ DC stimulated with purified protein derivative (PPD) from *Mycobacterium tuberculosis* showed augmented development of IFN-γ producing cells, while both subsets responded to zymosan by enhanced priming of IL-4 producing cells [39]. Also mouse bone marrow derived myeloid DC stimulated with LPS or CpG DNA induced Th1 responses [35], whereas stimulation with a filarial nematode glycoprotein ES-62 favored the development of Th2 cells *in vitro* [50, 51]. Different forms of the fungus *Candida albicans* can instruct DC to induce either Th1 or Th2 responses. At the yeast stage, the fungus induced IL-12 production by a mouse DC cell line and promoted Th1 responses, while at the hyphae stage the fungus surprisingly stimulated these DC to produce IL-4 and induce Th2 responses [52]. However, using a wide range of stimuli, many groups have been unable to detect the production of IL-4 protein or expression of IL-4 mRNA by primary mouse DC [34, 51, 53].

Limited data is available on the ability of mouse plasmacytoid DC to direct Th-cell development. Freshly isolated plasmacytoid DC from the spleen express low levels of MHC II and co-stimulators and induce weak T-cell proliferation, and consequently could not trigger Th-cell development [35, 54]. However, stimulation of plasmacytoid DC with CpG DNA induced the expression of MHC II and costimulators, resulting in enhanced CD4⁺ T-cell proliferation. Furthermore, strong Th1-cell development was observed upon activation of CD4⁺ T cells with CpG-stimulated plasmacytoid DC, which was partly IL-12-dependent [35]. In contrast, Krug et al. were unable to induce Th1-cell development even upon stimulation of mouse plasmacytoid DC with CpG, and showed a role for plasmacytoid DC in memory responses [55]. However, as will be discussed below, the same group showed that human plasmacytoid DC induced strong Th1 responses when stimulated with influenza virus [56]. Clearly, these reports demonstrate the high degree of functional plasticity of mouse DC populations upon challenge with a variety of pathogens, or due to variations in their micro-environment.

20.3.2
Human Dendritic Cell Populations in Directing Th-cell Development

Also initial studies using human DC populations suggested that human DC were pre-programmed to induce the development of a particular Th phenotype. It was reported that human CD40-ligand-activated DC, differentiated from monocytes by culturing with GM–CSF and IL-4, were found to induce Th1-cell development *in vitro*, which was at least in part IL-12 dependent, whereas CD40-ligand-activated DC derived from blood CD11c⁻CD4⁺CD3⁻ plasmacytoid cells induced Th2 cells [46]. CD40-ligand activated monocyte-derived DC produced IL-12, whereas the ability to produce IL-12 of activated plasmacytoid DC, which secrete high levels of IFN-α, remains controversial [46]. Also these findings, suggesting that specific human DC are pre-programmed to induce either Th1 or Th2 cells, have been challenged, as will be discussed below. Consequently, the proposed nomenclature for monocyte-derived DC as DC1, and plasmacytoid DC as DC2, reflecting their suggested ability to be fixed in driving Th1 or Th2 responses, respectively, is confusing and is not preferred.

Cella et al. challenged the finding that human plasmacytoid DC were fixed in driving exclusively Th2 responses by showing that, when stimulated with either CD40-ligand or influenza virus, IL-3 cultured plasmacytoid DC induced the development of strong Th1 responses [56]. Simultaneously, Kadowaki et al., showed that IL-3 cultured plasmacytoid DC induce human allogeneic T cells to induce a Th2 cytokine profile, whereas herpes simplex virus activated plasmacytoid DC induced the development of Th1 cells, producing IFN-γ and IL-10 [57]. The ability of virus-activated plasmacytoid DC to induce IFN-γ production in T cells was dependent on IFN-α/β, but independent of IL-12, whereas both IL-12 and IFN-α/β were involved for CD40-ligand-activated plasmacytoid DC [56, 57]. The factors responsible for inducing Th2-cell development are unknown. However, addition of IL-12 to IL-3-cultured plasmacytoid DC co-cultures with T cells showed an enhanced percentage of IFN-γ producing cells, but no decreased frequency of IL-4 producing cells, suggesting that the absence of IL-12 does not lead to a Th2 default phenotype [57]. A role for OX40-ligand has been suggested in promoting Th2-cell development by plasmacytoid DC, since plasmacytoid DC cultured with IL-3 expressed high levels of OX40-ligand, and neutralization of OX40-ligand significantly inhibited the ability of IL-3 activated plasmacytoid DC to induce the development of Th2 cells [58].

Similar to plasmacytoid DC, human monocyte-derived DC also induce the differentiation into both Th1 and Th2 cells depending on the stimulus or presence of pro/anti-inflammatory molecules. Pathogen-derived products, such as lipopolysaccharide, double stranded RNA, and fixed *Staphylococcus aureus* Cowan's strain (SAC), as well as T-cell-derived factors, such as CD40-ligand and IFN-γ can all stimulate monocyte-derived DC to induce Th1 development, via IL-12 dependent mechanisms [30, 59, 60]. On the other hand, activation of monocyte-derived DC in the presence of *Schistosoma* egg antigen, cholera toxin, or anti-inflammatory molecules such as IL-10, TGF-β, prostaglandin-E$_2$ and steroids promotes Th2-cell development [61–64]. An early trigger of the allergic immune cascade might be represented by human thymic stromal lymphopoietin (TSLP), a novel IL-7 like cytokine that is produced by human epithelial, stromal, and mast cells, and in particular, is highly expressed by keratinocytes of atopic dermatitis patients, but not in other types of skin inflammation. TSLP activates human blood CD11c$^+$ DC, but not monocyte-derived DC, to produce Th2-attracting chemokines but no IL-12, and to induce naïve CD4$^+$ and CD8$^+$ T-cell differentiation into effector cells with a typical pro-allergic phenotype [65].

Thus, although DC are flexible, they may be pre-programmed to drive either Th1 or Th2-cell development in response to certain pathogen-derived products.

20.4
Differential TLR Expression by Distinct Dendritic Cell Populations

The Th1-inducing capacity of DC is largely determined by their production of IL-12p70 and/or IFN-α upon stimulation by pathogen-derived products [28, 37, 46, 47, 56, 60, 66–68]. A number of studies have suggested that distinct DC populations have an intrinsic inability to produce IL-12p70 and/or IFN-α, which consequently results in a differential ability to drive Th1 or Th2 cells [46–48]. However, this lack of production of IL-12 and/or IFN-α by specific DC subpopulations could be the result of the lack of expression of the corresponding Toll-like receptor (TLR) or other pathogen recognition receptor, or due to negative regulation of IL-12 or IFN-α expression.

In human and mice, distinct DC subsets have been shown to respond to different microbial products partly due to the differential expression of different TLR. Human monocyte-derived DC express TLR2, TLR3, TLR4, TLR8 and low levels of TLR7, and respond to their respective ligands [67, 69–71]. Similarly to human monocyte-derived DC, mouse bone-marrow derived myeloid DC express TLR2 and TLR4, and respond to their respective ligands Pam3Cys and LPS, whereas mouse splenic CD11c$^+$CD8α^+, CD11c$^+$CD8α^- DC and plasmacytoid DC express little to no TLR2 and TLR4 and do not respond to their respective ligands. Both human and mouse plasmacytoid DC express TLR7 and TLR9, and respond to their ligands R-848 and CpG to produce IFN-α and IL-12 [35, 67, 72, 73]. However, it should be noted that mouse TLR9 expression is not restricted to plasmacytoid DC, but is expressed more broadly, albeit at lower levels, on other DC subsets [35]. The cytokine production induced by TLR ligation is partly determined by intrinsic properties of the specific DC population. As a consequence of TLR9 ligation, mouse plasmacytoid DC produce IL-12p70 as well as IFN-α, whereas bone marrow-derived myeloid DC produce only IL-12p70 [35]. Also, ligation of TLR7 using imiquimod or R-848 stimulation of human plasmacytoid and blood myeloid CD11c$^+$ DC induced the production of IFN-α and IL-12p70 respectively, both of which resulted in Th1-cell development [67]. This suggests that intrinsic differences between DC populations exist resulting in distinct cytokine profiles. Differential TLR expression may explain why some pathogen-derived products induce differential cytokine production in specific DC subtypes, and may thus have the potential under certain circumstances to direct Th1 or Th2 development.

20.4.1
Modulation of TLR Expression

At present, limited data is available on the regulation of TLR expression, and therefore it can not be excluded that under specific stimulation conditions, or during inflammation or pathology, TLR expression can be modulated, thereby changing the responsiveness to TLR ligation of specific DC subtypes. Some studies point out that this is indeed the case. Maturation of monocytes into monocyte-derived DC has been shown to result in enhanced TLR3 and TLR4 expression [74]. Stimulation with LPS enhanced the expression of TLR9 by human PBMC [75] and mouse DC [76]. Also, cytokines can affect TLR expression. IL-10 was shown to prevent LPS-induced TLR4 mRNA upregulation in monoytes [74], and GM-CSF enhanced the expression of TLR2 and TLR9 by human neutrophils [77].

20.5
Modulation of IL-12p70 or IFN-α Production

Stimulation of DC with TLR ligands alone generally induces low levels of IL-12p70, and often induces only IL-12p40 production. Enhanced expression of the limiting IL-12p35 subunit can be achieved by additional stimulation with various cytokines or with T-cell associated signals. A synergistic signal for IL-12p70 production is delivered by direct cell–cell contact with activated T cells via the interaction of CD40L on T cells with CD40 on DC [29, 30, 78, 79]. Various studies have shown that CD11c$^+$CD8α$^+$DC are superior over CD11c$^+$CD8α$^-$ DC to produce IL-12p70 upon stimulation with pathogen-derived products alone [47, 80, 81]. However, under optimal stimulation conditions both splenic DC populations have the ability to produce significant IL-12p70 production. Following stimulation with CpG or soluble toxoplasma antigen (STAg) alone low amounts of IL-12p70 are produced by splenic CD11c$^+$CD8α$^+$DC only, whereas both CD11c$^+$CD8α$^+$ and CD11c$^+$CD8α$^-$ DC produce IL-12p70 upon simultaneous microbial stimulation and CD40-ligation [79]. CD40-ligation is not only essential for IL-12p70, but also for IL-10 production [80] (Boonstra and O'Garra, unpublished), and augments the production of other cytokines, such as IFN-α, IL-12p40 and TNF (Boonstra and O'Garra, unpublished).

Further regulation of IL-12 production can be exerted by cytokines or other regulatory molecules present in the micro-milieu of the DC. Cytokines such as IFN-γ, GM-CSF and surprisingly also the Th2-associated cytokines IL-4 and IL-13, are potent enhancers of IL-12p70 production [68, 82–84]. These cytokines can be produced by activated T cells, but also NK cell derived IFN-γ plays an important role in promoting Th1-cell development [85–87]. On the other hand, suppression of IL-12 production has been described by IL-10, IFN-α, TGF-β, prostagladin-E$_2$, and steroids [88–92].

IL-10 has been shown to suppress IL-12p70 production by DC and macrophages, resulting in the inhibition of Th1-cell development [93, 94]. Both CD8α$^+$ and CD8α$^-$ DC obtained from IL-10 knock-out mice produce IL-12p70 *in vitro* upon stimulation with fixed *Staphylococcus aureus* and IFN-γ [53]. *In vivo*, infection of IL-10 knock-out mice with *Brucella abortus* resulted in higher IL-12p40 production by both splenic CD8α$^+$ and CD8α$^-$ DC as compared to wild type mice [95], clearly showing the ability of IL-10 to inhibit IL-12 production by DC. Recently, it was demonstrated that IL-4 enhances LPS or CpG induced IL-12p70 production by decreasing the production of IL-10 by DC. In IL-10-deficient mice the ability of IL-4 to upregulate IL-12p70 was not observed [96].

Besides IL-10, also IFN-α can inhibit the production of IL-12 [91, 92]. Both human and mouse plasmacytoid DC produce high levels of IFN-α in response to viruses, and upon stimulation with agonists of TLR9 (CpG DNA) and TLR7 (R848). The early production of IFN-β and IFN-α4 can promote the production of other IFN-α subtypes by positive feedback mechanisms. Dalod et al., showed that following infection with murine cytomegalovirus (MCMV), plasmacytoid DC (CD11c$^+$GR1$^+$CD11b$^-$CD8α$^+$) were the major source of IFN-α/β and IL-12 [97]. Conventional splenic CD11c$^+$CD11b$^-$CD8α$^+$ produced low levels of IL-12. However, blocking the production of IFN-α/β *in vivo* by depletion of plasmacytoid DC did not significantly affect IL-12 production in serum, which was shown to be due to increased IL-12 production by the conventional splenic CD11c$^+$CD11b$^-$CD8α$^+$ DC. Similarly, abrogation of IFN-α/β signaling using MCMV infected IFN-α/βR knock out mice, resulted in increased IL-12 production by splenic CD11c$^+$CD11b$^-$CD8α$^+$ DC [97]. These and other findings indicate that the major cell type responsible for IL-12 production can vary depending on the pathogen (plasmacytoid DC for MCMV, CD11c$^+$CD8α$^+$ DC for STAg), and is highly regulated by cytokines in the micro-environment. Interestingly, the same study showed that high viral dose resulted in higher levels of IL-12, but lower levels of IFN-α as compared to a lower viral dose [97].

As outlined above, the levels of IL-12 and/or IFN-α produced by a specific DC population are determined by multiple factors, such as the nature of the pathogen, the additional requirement for T cell–derived signals, and the presence of cytokines that promote or suppress IL-12 production. Moreover, also the stage of maturation of DC populations is an important factor. IL-12p70 production by human monocyte-derived DC stimulated with LPS or CD40-ligand can be strongly enhanced by IFN-γ. However, maturation of monocyte-derived DC resulted in the gradual loss of responsiveness to IFN-γ due to loss of the IFN-γR [60]. Similarly, as discussed above, DC populations at different stages of their maturation may vary in responsiveness to pathogens due to regulation of TLR expression.

It has been recently shown that ligands for different TLR strongly synergize in inducing IL-12p70 production by either human or mouse DC [98]. In particular, ligands for TLR utilizing the adaptor molecules MyD88 and TRIF can when combined induce up to two orders of magnitude more IL-12p70 than when used alone [98]. The synergism is evident particularly at the level of accumulation of transcripts of IL-12p35 that requires an autocrine effect of secreted type I IFN signaling through STAT-1. Indeed, type I IFN, which at high doses inhibits IL-12p40 production, may increase, similarly to but less potently than IFN-γ, IL-12p70 production by mouse [98] and human DC [99]. Thus, pathogens able to simultaneously trigger different TLR may be able to produce an early burst of biologically active IL-12p70 in a cell-autonomous way with the help of autocrine type I IFN. Maintenance and amplification of IL-12 production then requires the participation of IFN-γ or IL-4 produced by innate effector cells (NK, NKT, mast cells), or by antigen-specific T cells as well as co-stimulatory molecules such as CD40-ligand expressed by T cells activated in an antigen-specific or non-specific way.

Additionally, activation of DC by LPS resulted in a transient IL-12 production, followed by a refractory period to further stimulation [37]. More specifically, monocyte-derived DC stimulated with LPS for 8 h showed augmented IL-12p70 production when restimulated via CD40-ligation, whereas microbial stimulation 24 h prior to CD40-ligation did not result in detectable IL-12p70 production. As a consequence, soon after activation, DC induce Th1-cell development upon encounter of T cells, whereas the interaction with T cells at later time points preferentially promotes Th0 or Th2 responses. The inability to produced IL-12 at later time points after activation has been referred to as "exhaustion" [37] or "paralysis" [100] of IL-12 production.

The tight regulation of IL-12p70 and IFN-α/β production by DC, which is dictated by the type of pathogen, the involvement of T-cell feedback, and cytokines offers the DC a high degree of flexibility to promote Th1 or Th2 polarization.

20.6
Factors Responsible for Driving Th2-cell Development

The factors responsible for Th2-cell polarization by DC are still largely unclear. Although, IL-4 is a key factor in driving Th2 responses, there is still debate whether DC can produce IL-4. On the basis of early findings that mouse splenic CD8α^- DC do not produce IL-12 and promote Th2 responses, and reports that "Th2-stimuli" induce little IL-12 production, the hypothesis was put forward that Th2 cells develop by default in the absence of IL-12. However, recent studies showed that in the absence of IL-12, Th1-inducing pathogens did not induce Th2 responses by default [101, 102]. Using IL-12 knock-out mice on both BALB/c and C57Bl/6 genetic backgrounds, mice infected with *Mycobacterium avium* or primed with STAg from *Toxoplasma gondii* exhibited a marked reduction in *in vitro* IFN-γ production, but no significant IL-4 production in the absence of IL-12 [102]. The development of Th2 responses by stimulation of mouse splenic DC with *Schistosoma* egg antigen (SEA) correlated with unaltered or downregulation of the expression of co-stimulatory molecules and chemokines upon stimulation, and a temporary delay in T-cell cycling [103]. In humans, TSLP, produced by epithelial and stromal cells, is very potent in endowing CD11c$^+$ blood DC and tonsil DC with the ability to produce Th2-attracting chemokines and to induce Th2 responses [65].

20.7
Modulation by Tissue Factors

In the mouse, CD11c$^+$CD8α^+ and CD11c$^+$CD8α^- DC reside in distinct locations in the spleen in steady-state conditions: CD11c$^+$CD8α^+ DC reside in the T-cell areas, whereas CD11c$^+$CD8α^- DC are located mainly in the marginal zone. Therefore, these DC populations are exposed to a different micro-milieu with different cells and cytokines. CD11c$^+$CD8α^+ DC are surrounded by lymphocytes, and are therefore likely under inflammatory conditions to be exposed to T-cell-derived cytokines, such as IFN-γ. CD11c$^+$CD8α$^-$ DC are surrounded by marginal zone macrophages, which may, under specific conditions produce more IL-10, prostaglandin-E$_2$ and pro-inflammatory cytokines such as TNF. Therefore, under steady state conditions, the findings that splenic CD11c$^+$CD8α$^+$ and CD11c$^+$CD8α$^-$ DC have a differential ability to produce cytokines [47, 80, 81] may be the results of the effect of exposure to a different cytokine repertoire *in vivo*. These conditions may change during immune responses when CD11c$^+$CD8α$^-$ DC migrate from the marginal zone to the T-cell area of the spleen [104].

A role for the tissue environment in modulating Th1 and Th2 responses has been most extensively studied in mice. In the Peyer's patches, CD11c$^+$CD8α$^-$ DC produce IL-10 and little IL-12p70 upon stimulation with trimeric CD40-ligand or SAC and IFN-γ. Besides IFN-γ, these cells induced T cells to produce IL-4 and IL-10, giving it a mixed Th1/Th2 phenotype [105]. As a comparison, splenic CD8α$^+$ and CD8α$^-$ DC induced T cells to produce similar levels of IFN-γ as compared to Peyer's patches DC but low IL-4 and IL-10.

Also comparison of DC isolated from the liver and the spleen, showed functional differences with respect to IL-12p70 production, which were ascribed to the distinct anatomical location. In the spleen, CD40-ligand activated splenic CD8α$^+$ DC produce high levels of IL-12p70 as compared to CD8α$^-$ DC. However, CD40-ligand activated liver CD8α$^+$ and CD8α$^-$ DC both induce high IL-12p70 levels, which were comparable to the levels produced by splenic CD8α$^+$ DC [106]. These findings suggest that there is no intrinsic difference between CD8α$^+$ and CD8α$^-$ DC, but that the observed differences are due to distinct anatomical location of the DC with different cytokine environments. The function of respiratory tract DC isolated from rat has been shown to be suppressed by pulmonary alveolar macrophages. *In vivo* depletion of alveolar macrophages enhanced the MHC II expression and T-cell stimulatory capacity of DC [107]. Additionally, respiratory tract DC preferentially induce Th2 responses when pulsed *in vitro* and transferred *in vivo* [108]. In this context, it has recently been shown that depletion of DC, from the airways of CD11c-diphteria toxin receptor transgenic mice during ovalbumin aerosol challenge, abolished the characteristics of asthma, including eosinophilic inflammation, bronchial hyper-reactivity, and the production of Th2 cytokines (IL-4, IL-5, IL-13). Reconstitution of CD11c-depleted mice with CD11c$^+$ DC restored the asthmatic symptoms [109]. Culture of rat respiratory tract DC *in vitro* in the presence of GM–CSF resulted in the production of Th1 cytokines [108].

20.8
Concluding Remarks

Exposure of DC to various stimuli during inflammation dramatically modulates its environment, and consequently the characteristics of the DC itself.

The ability of the antigen dose to affect Th-cell development could provide an explanation why different DC subsets have been suggested to intrinsically direct Th1 or Th2-cell development [46–48]. It is plausible that due to possible differences in their stage of maturation, they express different levels of MHC II and/or co-stimulatory molecules, and therefore may have different effective antigen doses presented by the APC to the T cell. In addition, DC subpopulations at different stages of their maturation may process antigens with different efficacies and consequently present different amounts of MHC II/peptide on their surface, which affects the effective antigen dose and consequently to altered Th-cell development.

The net result of the complex interplay of the state of maturation of the DC, previous encounters, tissue factors, the effects of pathogen-derived products, the strength of signal as well as genetic factors makes the DC extremely efficient as a sensor of diverse pathogenic challenges, and highly flexible in directing the appropriate immune responses. On the other hand, the ability to differentially express TLR, and thus respond to different TLR ligands, may suggest that certain DC may be destined to respond to distinct pathogen-derived products, and thus different pathogens, to direct Th1-cell development. These issues remain to be determined, and are fundamental for vaccination strategies and therapeutic interventions.

References

1 Banchereau, J., Steinman, R.M., Dendritic cells and the control of immunity. *Nature* **1998**. 392: 245–252.

2 Steinman, R. M., The dendritic cell system and its role in immunogenicity. *Annu. Rev. Immunol.* **1991**. 9: 271–296.

3 Lanzavecchia, A. Sallusto, F., Regulation of T cell immunity by dendritic cells. *Cell* **2001**. 106: 263–266.

4 Mosmann, T. R., Cherwinski, H., Bond, M. W., Giedlin, M. A. Coffman, R. L., Two types of murine helper T cell clone. I. Definition according to profiles of lymphokine activities and secreted proteins. *J. Immunol.* **1986**. 136: 2348–2357.

5 Sher, A. Coffman, R. L., Regulation of immunity to parasites by T cells and T cell-derived cytokines. *Annu. Rev. Immunol.* **1992**. 10: 385–409.

6 O'Garra, A., Cytokines induce the development of functionally heterogeneous T helper cell subsets. *Immunity* **1998**. 8: 275–283.

7 Romagnani, S., Lymphokine production by human T cells in disease states. *Annu. Rev. Immunol.* **1994**. 12: 227–257.

8 Murphy, K. M. Reiner, S. L., The lineage decisions of helper T cells. *Nat. Rev. Immunol.* **2002**. 2: 933–944.

9 O'Garra, A. Robinson, D., Development and function of T helper 1 cells. *Adv. Immunol.* **2004**. 83: 133–162.

10 Magram, J., Connaughton, S. E., Warrier, R. R., Carvajal, D. M., Wu, C., Ferrante, J., Stewart, C., Sarmiento, U., Faherty, D. A. Gately, M. K., IL-12-deficient mice are defective in IFN-γ production and type 1 cytokine responses. *Immunity* **1996**. 4: 471–481.

11 Gately, M. K., Renzetti, L. M., Magram, J., Stern, A. S., Adorini, L., Gubler, U. Presky, D. H., The interleukin-12/interleukin-12-receptor system: role in normal and pathologic immune responses. *Annu. Rev. Immunol.* **1998**. 16: 495–521.

12 Kuhn, R., Rajewsky, K., Muller, W., Generation and analysis of interleukin-4 deficient mice. *Science* **1991**. 254: 707–710.

13 Kopf, M., Le Gros, G., Bachmann, M., Lamers, M. C., Bluethmann, H. Kohler, G., Disruption of the murine IL-4 gene blocks Th2 cytokine responses. *Nature* **1993**. 362: 245–248.

14 Kaplan, M., Schindler, U., Smiley, S. T. Grusby, M. J., Stat6 is required for mediating responses to IL-4 and for the development of Th2 cells. *Immunity* **1996**. 4: 313–319.

15 Shimoda, K., van Deursen, J., Sangster, M.Y., Sarawar, S.R., Carson, R.T., Tripp, R.A., Chu, C., Quelle, F. W., Nosaka, T., Vignali, D.A.A., Doherty, P.C., Grosveld, G., Paul, W.E., Ihle, J.N., Lack of IL-4-induced Th2 response and IgE class switching in mice with disrupted Stat6 gene. *Nature* 1996. 380: 630–633.

16 Takeda, K., Tanaka, T., Shi, W., Matsumoto, M., Minami, M., Kashiwamura, S., Nakanishi, K., Yoshida, N., Kishimoto, T. Akira, S., Essential role of Stat6 in IL-4 signalling. *Nature* 1996. 380: 627–630.

17 Zheng, W. Flavell, R. A., The transcription factor GATA-3 is necessary and sufficient for Th2 cytokine gene expression in CD4 T cells. *Cell* 1997. 89: 587–596.

18 Zhang, D. H., Cohn, L., Ray, P., Bottomly, K. Ray, A., Transcription factor GATA-3 is differentially expressed in murine Th1 and Th2 cells and controls Th2-specific expression of the interleukin-5 gene. *J. Biol. Chem.* 997. 272: 21597–21603.

19 Szabo, S. J., Kim, S.T., Costa, G.L., Zhang, X., Fathamn, C.G., Glimcher, L.H., A novel transcription factor, T-bet, directs Th1 lineage commitment. *Cell* 2000. 100: 655–669.

20 Afkarian, M., Sedy, J. R., Yang, J., Jacobson, N. G., Cereb, N., Yang, S. Y., Murphy, T. L. Murphy, K. M., T-bet is a STAT1-induced regulator of IL-12R expression in naive CD4+ T cells. *Nat. Immunol.* 2002. 3: 549–557.

21 Grogan, J. L., Mohrs, M., Harmon, B., Lacy, D. A., Sedat, J. W. Locksley, R. M., Early transcription and silencing of cytokine genes underlie polarization of T helper cell subsets. *Immunity* 2001. 14: 205–215.

22 Zhu, J., Guo, L., Watson, C. J., Hu-Li, J. Paul, W. E., Stat6 is necessary and sufficient for IL-4's role in Th2 differentiation and cell expansion. *J. Immunol.* 2001. 166: 7276–7281.

23 Abbas, A. K., Murphy, K.M., Sher, A, Functional Diversity of helper T lymphocytes. *Nature* 1996. 383: 787–793.

24 Szabo, S., Dighe, A., S. , Gubler, U. Murphy, K. M., Regulation of the Interleukin (IL)-12β2 Subunit Expression in Developing T Helper 1 (Th1) and Th2 Cells. *J. Exp. Med.* 1997. 185: 817–824.

25 Rogge, L., Barberis-Maino, L., Biffi, M., Passini, N., Presky, D. H., Gubler, U. Sinigaglia, F., Selective expression of an interleukin-12 receptor component by human T helper 1 cells. *J. Exp. Med.* 1997. 185: 825–831.

26 Rogge, L., D'Ambrosio, D., Biffi, M., Penna, G., Minetti, L.J., Presky, D.H., Adorini, L., Sinigaglia, F., The role of Stat4 in species-specific regulation of Th development by Type I IFNs. *J. Immunol.* 1998. 161: 6567–6574.

27 Nguyen, K. B., Watford, W. T., Salomon, R., Hofmann, S. R., Pien, G. C., Morinobu, A., Gadina, M., O'Shea, J. J. Biron, C. A., Critical role for STAT4 activation by type 1 interferons in the interferon-gamma response to viral infection. *Science* 2002. 297: 2063–2066.

28 Macatonia, S. E., Hosken, N. A., Litton, M., Vieira, P., Hsieh, C.-S., Culpepper, J., Wysocka, M., Trinchieri, G., Murphy, K. M. O'Garra, A., Dendritic cells produce IL-12 and direct the development of Th1 cells from naive CD4+ T cells. *J. Immunol.* 1995. 154: 5071–5079.

29 Heufler, C., Koch, F., Stanzl, U., Topar, G., Wysocka, M, M., Trinchieri, G., Enk, A., Steinman, R.M., Romani, N., Schuler, G., Interleukine-12 is produced by dendritic cells and mediates T helper 1 development as well as interferon-γ production by T helper 1 cells. *Eur. J. Immunol.* 1996. 26: 659–668.

30 Cella, M., Scheidegger, D., Palmer-Lehmann, K., Lane, P., Lanzavecchia, A., Alber, G., Ligation of CD40 on dendritic cells triggers production of high levels of interleukin-12 and enhances T cell stimulatory capacity: T-T help via APC activation. *J. Exp. Med.* 1996. 184: 747–752.

31 Constant, S. L. Bottomly, K., Induction of Th1 and Th2 CD4+ T cell responses: the alternative approaches. *Annu. Rev. Immunol.* 1997. 15: 297–322.

32 Constant, S., Pfeiffer, C., Woodard, A., Pasqualini, T. Bottomly, K., Extent of T cell receptor ligation can determine the functional differentiation of naive CD4+ T cells. *J. Exp. Med.* 1995. 182: 1591–1596.

33 Hosken, N. A., Shibuya, K., Heath, A. W., Murphy, K. M. O'Garra, A., The effect of antigen dose on CD4+ T cell phenotype development in an αβ-TCR-transgenic mouse model. *J. Exp. Med.* **1995**. 182: 1579–1584.

34 Ruedl, C., Bachmann, M. F. Kopf, M., The antigen dose determines T helper subset development by regulation of CD40 ligand. *Eur. J. Immunol.* **2000**. 30: 2056–2064.

35 Boonstra, A., Asselin-Paturel, C., Gilliet, M., Crain, C., Trinchieri, G., Liu, Y. J. O'Garra, A., Flexibility of mouse classical and plasmacytoid-derived dendritic cells in directing T helper type 1 and 2 cell development: dependency on antigen dose and differential Toll-like receptor ligation. *J. Exp. Med.* **2003**. 197: 101–109.

36 Jorritsma, P. J., Brogdon, J. L. Bottomly, K., Role of TCR-induced extracellular signal-regulated kinase activation in the regulation of early IL-4 expression in naive CD4+ T cells. *J. Immunol.* **2003**. 170: 2427–2434.

37 Langenkamp, A., Messi, M., Lanzavecchia, A. Sallusto, F., Kinetics of dendritic cell activation: impact on priming of TH1, TH2 and nonpolarized T cells. *Nat. Immunol.* **2000**. 1: 311–316.

38 Tanaka, H., Demeure, C. E., Rubio, M., Delespesse, G. Sarfati, M., Human monocyte-derived dendritic cells induce naive T cell differentiation into T helper cell type 2 (Th2) or Th1/Th2 effectors. Role of stimulator/responder ratio. *J. Exp. Med.* **2000**. 192: 405–412.

39 Manickasingham, S. P., Edwards, A. D., Schulz, O. Reis e Sousa, C., The ability of murine dendritic cell subsets to direct T helper cell differentiation is dependent on microbial signals. *Eur. J. Immunol.* **2003**. 33: 101–107.

40 Rulifson, I. C., Sperling, A.I., Fields, P.E., Fitch, F.W., Bluestone, J.A., CD28 costimulation promotes the production of Th2 cytokines. *J. Immunol.* **1997**. 158: 658–665.

41 Flynn, S., Toellner, K. M., Raykundalia, C., Goodall, M. Lane, P., CD4 T cell cytokine differentiation: the B cell activation molecule, OX40 ligand, instructs CD4 T cells to express interleukin 4 and up-regulates expression of the chemokine

receptor, Blr-1. *J. Exp. Med.* **1998**. 188: 297–304.

42 Salomon, B., Bluestone, J.A., LFA-1 interaction with ICAM-1 and ICAM-2 regulates Th2 cytokine production. *J. Immunol.* **1998**. 161: 5138–5142.

43 Luksch, C. R., Winqvist, O., Ozaki, M.E., Karlsson, L., Jackson, M.R., Peterson, P.A., Webb, S.R., Intercellular adhesion molecule-1 inhibits interleukin 4 production by naive T cells. *Proc. Natl. Acad. Sci. USA* **1999**. 96: 3023–3028.

44 Rogers, P. R. Croft, M., CD28, Ox-40, LFA-1, CD4 modulation of Th1/Th2 differentiation is directly dependent on the dose of antigen. *J. Immunol.* **2000**. 164: 2955–2963.

45 Filippi, C., Hugues, S., Cazareth, J., Julia, V., Glaichenhaus, N. Ugolini, S., CD4+ T cell polarization in mice is modulated by strain-specific major histocompatibility complex-independent differences within dendritic cells. *J. Exp. Med.* **2003**. 198: 201–209.

46 Rissoan, M.-C., Soumelis, V., Kadowaki, N., Grouard, G., Briere, F., de Waal Malefyt, R., Liu, Y-J., Reciprocal control of T helper cell and dendritic cell differentiation. *Science* **1999**. 283: 1183–1186.

47 Maldonado-Lopez, R., De Smedt, T., Michel, P., Godfroid, J., Pajak, B., Heirman, C., Thielemans, K., Leo, O., Urbain, J., Moser, M., CD8alpha+ and CD8alpha- subclasses of dendritic cells direct the development of distinct T helper cells in vivo. *J. Exp. Med.* **1999**. 189: 587–592.

48 Pulendran, B., Smith, J. L., Caspary, G., Brasel, K., Pettit, D., Maraskovsky, E. Maliszewski, C. R., Distinct dendritic cell subsets differentially regulate the class of immune response in vivo. *Proc. Natl. Acad. Sci. USA* **1999**. 96: 1036–1041.

49 Schlecht, G., Leclerc, C. Dadaglio, G., Induction of CTL and nonpolarized Th cell responses by CD8alpha(+) and CD8alpha(–) dendritic cells. *J. Immunol.* **2001**. 167: 4215–4221.

50 Whelan, M., Harnett, M. M., Houston, K. M., Patel, V., Harnett, W. Rigley, K. P., A filarial nematode-secreted product signals dendritic cells to acquire a phenotype that drives development of

Th2 cells. *J. Immunol.* **2000**. 164: 6453–6460.

51 MacDonald, A. S., Straw, A. D., Bauman, B. Pearce, E. J., CD8– dendritic cell activation status plays an integral role in influencing Th2 response development. *J. Immunol.* **2001**. 167: 1982–1988.

52 d'Ostiani, C. F., Del Sero, G., Bacci, A., Montagnoli, C., Spreca, A., Mencacci, A., Ricciardi-Castagnoli, P. Romani, L., Dendritic-cells discriminate between yeasts and hyphae of the fungus Candida albicans. Implications for initiation of T helper cell immunity in vitro and in vivo. *J. Exp. Med.* **2000**. 191: 1661–1674.

53 Maldonado-Lopez, R., Maliszewski, C., Urbain, J. Moser, M., Cytokines regulate the capacity of CD8alpha(+) and CD8alpha(–) dendritic cells to prime Th1/Th2 cells in vivo. *J. Immunol.* **2001**. 167: 4345–4350.

54 Asselin-Paturel, C., Boonstra, A., Dalod, M., Durand, I., Yessaad, N., Dezutter-Dambuyant, C., Vicari, A., O'Garra, A., Biron, C., Briere, F. Trinchieri, G., Mouse type I IFN-producing cells are immature APCs with plasmacytoid morphology. *Nat. Immunol.* **2001**. 2: 1144–1150.

55 Krug, A., Veeraswamy, R., Pekosz, A., Kanagawa, O., Unanue, E. R., Colonna, M. Cella, M., Interferon-producing cells fail to induce proliferation of naive T cells but can promote expansion and T helper 1 differentiation of antigen-experienced unpolarized T cells. *J. Exp. Med.* **2003**. 197: 899–906.

56 Cella, M., Facchetti, F., Lanzavecchia, A. Colonna, M., Plasmacytoid dendritic cells activated by influenza virus and CD40L drive a potent TH1 polarization. *Nat. Immunol.* **2000**. 1: 305–310.

57 Kadowaki, N., Antonenko, S., Lau, J. Y. Liu, Y. J., Natural interferon alpha/beta-producing cells link innate and adaptive immunity. *J. Exp. Med.* **2000**. 192: 219–226.

58 Ito, T., Amakawa, R., Inaba, M., Hori, T., Ota, M., Nakamura, K., Takebayashi, M., Miyaji, M., Yoshimura, T., Inaba, K. Fukuhara, S., Plasmacytoid dendritic cells regulate Th cell responses through OX40 ligand and type I IFNs. *J. Immunol.* **2004**. 172: 4253–4259.

59 Cella, M., Jarrossay, D., Facchetti, F., Alebardi, O., Nakajima, H., Lanzavecchia, A. Colonna, M., Plasmacytoid monocytes migrate to inflamed lymph nodes and produce large amounts of type I interferon. *Nat. Med.* **1999**. 5: 919–923.

60 Kalinski, P., Hilkens, C. M., Wierenga, E. A. Kapsenberg, M. L., T-cell priming by type-1 and type-2 polarized dendritic cells: the concept of a third signal. *Immunol. Today* **1999**. 20: 561–567.

61 King, C., Davies, J., Mueller, R., Lee, M. S., Krahl, T., Yeung, B., O'Connor, E. Sarvetnick, N., TGF-beta1 alters APC preference, polarizing islet antigen responses toward a Th2 phenotype. *Immunity* **1998**. 8: 601–613.

62 Vieira, P. L., Kalinski, P., Wierenga, E. A., Kapsenberg, M. L. de Jong, E. C., Glucocorticoids inhibit bioactive IL-12p70 production by in vitro-generated human dendritic cells without affecting their T cell stimulatory potential. *J. Immunol.* **1998**. 161: 5245–5251.

63 Gagliardi, M. C., Sallusto, F., Marinaro, M., Langenkamp, A., Lanzavecchia, A. De Magistris, M. T., Cholera toxin induces maturation of human dendritic cells and licences them for Th2 priming. *Eur. J. Immunol.* **2000**. 30: 2394–2403.

64 van der Kleij, D., Latz, E., Brouwers, J. F., Kruize, Y. C., Schmitz, M., Kurt-Jones, E. A., Espevik, T., de Jong, E. C., Kapsenberg, M. L., Golenbock, D. T., Tielens, A. G. Yazdanbakhsh, M., A novel host-parasite lipid cross-talk. Schistosomal lyso-phosphatidylserine activates toll-like receptor 2 and affects immune polarization. *J. Biol. Chem.* **2002**. 277: 48122–48129.

65 Soumelis, V., Reche, P. A., Kanzler, H., Yuan, W., Edward, G., Homey, B., Gilliet, M., Ho, S., Antonenko, S., Lauerma, A., Smith, K., Gorman, D., Zurawski, S., Abrams, J., Menon, S., McClanahan, T., de Waal-Malefyt Rd, R., Bazan, F., Kastelein, R. A. Liu, Y. J., Human epithelial cells trigger dendritic cell mediated allergic inflammation by producing TSLP. *Nat. Immunol.* **2002**. 3: 673–680.

66 de Jong, E. C., Vieira, P. L., Kalinski, P., Schuitemaker, J. H., Tanaka, Y.,

Wierenga, E. A., Yazdanbakhsh, M. Kapsenberg, M. L., Microbial compounds selectively induce Th1 cell-promoting or Th2 cell-promoting dendritic cells in vitro with diverse Th cell-polarizing signals. *J. Immunol.* **2002**. 168: 1704–1709.

67 Ito, T., Amakawa, R., Kaisho, T., Hemmi, H., Tajima, K., Uehira, K., Ozaki, Y., Tomizawa, H., Akira, S. Fukuhara, S., Interferon-alpha and interleukin-12 are induced differentially by Toll-like receptor 7 ligands in human blood dendritic cell subsets. *J. Exp. Med.* **2002**. 195: 1507–1512.

68 Vieira, P. L., de Jong, E. C., Wierenga, E. A., Kapsenberg, M. L. Kalinski, P., Development of Th1-inducing capacity in myeloid dendritic cells requires environmental instruction. *J. Immunol.* **2000**. 164: 4507–4512.

69 Kadowaki, N., Ho, S., Antonenko, S., Malefyt, R. W., Kastelein, R. A., Bazan, F. Liu, Y. J., Subsets of human dendritic cell precursors express different toll-like receptors and respond to different microbial antigens. *J. Exp. Med.* **2001**. 194: 863–869.

70 Hornung, V., Rothenfusser, S., Britsch, S., Krug, A., Jahrsdorfer, B., Giese, T., Endres, S. Hartmann, G., Quantitative expression of toll-like receptor 1–10 mRNA in cellular subsets of human peripheral blood mononuclear cells and sensitivity to CpG oligodeoxynucleotides. *J. Immunol.* **2002**. 168: 4531–4537.

71 Jarrossay, D., Napolitani, G., Colonna, M., Sallusto, F. Lanzavecchia, A., Specialization and complementarity in microbial molecule recognition by human myeloid and plasmacytoid dendritic cells. *Eur. J. Immunol.* **2001**. 31: 3388–3393.

72 Gilliet, M., Boonstra, A., Paturel, C., Antonenko, S., Xu, X. L., Trinchieri, G., O'Garra, A. Liu, Y. J., The Development of Murine Plasmacytoid Dendritic Cell Precursors Is Differentially Regulated by FLT3-ligand and Granulocyte/Macrophage Colony-Stimulating Factor. *J. Exp. Med.* **2002**. 195: 953–958.

73 Edwards, A. D., Diebold, S. S., Slack, E. M., Tomizawa, H., Hemmi, H., Kaisho, T., Akira, S. Reis e Sousa, C., Toll-like receptor expression in murine DC subsets: lack of TLR7 expression by CD8 alpha+ DC correlates with unresponsiveness to imidazoquinolines. *Eur. J. Immunol.* **2003**. 33: 827–833.

74 Muzio, M., Bosisio, D., Polentarutti, N., D'Amico, G., Stoppacciaro, A., Mancinelli, R., van't Veer, C., Penton-Rol, G., Ruco, L. P., Allavena, P. Mantovani, A., Differential expression and regulation of toll-like receptors (TLR) in human leukocytes: selective expression of TLR3 in dendritic cells. *J. Immunol.* **2000**. 164: 5998–6004.

75 Eaton-Bassiri, A., Dillon, S. B., Cunningham, M., Rycyzyn, M. A., Mills, J., Sarisky, R. T. Mbow, M. L., Toll-like receptor 9 can be expressed at the cell surface of distinct populations of tonsils and human peripheral blood mononuclear cells. *Infect. Immunol.* **2004**. 72: 7202–7211.

76 An, H., Yu, Y., Zhang, M., Xu, H., Qi, R., Yan, X., Liu, S., Wang, W., Guo, Z., Guo, J., Qin, Z. Cao, X., Involvement of ERK, p38 and NF-kappaB signal transduction in regulation of TLR2, TLR4 and TLR9 gene expression induced by lipopolysaccharide in mouse dendritic cells. *Immunology* **2002**. 106: 38–45.

77 Hayashi, F., Means, T. K. Luster, A. D., Toll-like receptors stimulate human neutrophil function. *Blood* **2003**. 102: 2660–2669.

78 Koch, F., Stanzl, U., Jennewein, P., Janke, K., Heufler, C., Kampgen, E., Romani, N., Schuler, G., High level IL-12 production by murine dendritic cells: upregulation via MHC class II and CD40 molecules and downregulation by IL-4 and IL-10. *J. Exp. Med.* **1996**. 184: 741–746.

79 Schulz, O., Edwards, A. D., Schito, M., Aliberti, J., Manickasingham, S., Sher, A. Reis e Sousa, C., CD40 triggering of heterodimeric IL-12 p70 production by dendritic cells in vivo requires a microbial priming signal. *Immunity* **2000**. 13: 453–462.

80 Edwards, A. D., Manickasingham, S. P., Sporri, R., Diebold, S. S., Schulz, O., Sher, A., Kaisho, T., Akira, S. Reis e Sousa, C., Microbial recognition via Toll-like receptor-dependent and -indepen-

dent pathways determines the cytokine response of murine dendritic cell subsets to CD40 triggering. *J. Immunol.* **2002.** 169: 3652–3660.

81 Hochrein, H., Shortman, K., Vremec, D., Scott, B., Hertzog, P. O'Keeffe, M., Differential production of IL-12, IFN-alpha, IFN-gamma by mouse dendritic cell subsets. *J. Immunol.* **2001.** 166: 5448–5455.

82 D'Andrea, A., Ma, X., Aste-Amezaga, M., Paganin, C. Trinchieri, G., Stimulatory and inhibitory effects of interleukin (IL)-4 and IL-13 on the production of cytokines by human peripheral blood mononuclear cells: priming for IL-12 and tumor necrosis factor alpha production. *J. Exp. Med.* **1995.** 181: 537–546.

83 Marshall, J. D., Robertson, S. E., Trinchieri, G. Chehimi, J., Priming with IL-4 and IL-13 during HIV-1 infection restores in vitro IL-12 production by mononuclear cells of HIV-infected patients. *J. Immunol.* **1997.** 159: 5705–5714.

84 Hochrein, H., O'Keeffe, M., Luft, T., Vandenabeele, S., Grumont, R. J., Maraskovsky, E. Shortman, K., Interleukin (IL)-4 is a major regulatory cytokine governing bioactive IL-12 production by mouse and human dendritic cells. *J. Exp. Med.* **2000.** 192: 823–833.

85 Scharton, T. M. Scott, P., Natural killer cells are a source of interferon gamma that drives differentiation of CD4+ T cell subsets and induces early resistance to Leishmania major in mice. *J. Exp. Med.* **1993.** 178: 567–577.

86 Gerosa, F., Baldani-Guerra, B., Nisii, C., Marchesini, V., Carra, G. Trinchieri, G., Reciprocal activating interaction between natural killer cells and dendritic cells. *J. Exp. Med.* **2002.** 195: 327–333.

87 Martin-Fontecha, A., Thomsen, L. L., Brett, S., Gerard, C., Lipp, M., Lanzavecchia, A. Sallusto, F., Induced recruitment of NK cells to lymph nodes provides IFN-gamma for T(H)1 priming. *Nat. Immunol.* **2004.** 5: 1260–1265.

88 van der Pouw Kraan, T. C., Boeije, L. C., Smeenk, R. J., Wijdenes, J. Aarden, L. A., Prostaglandin-E2 is a potent inhibitor of

human interleukin 12 production. *J. Exp. Med.* **1995.** 181: 775–779.

89 Blotta, M. H., DeKruyff, R. H. Umetsu, D. T., Corticosteroids inhibit IL-12 production in human monocytes and enhance their capacity to induce IL-4 synthesis in CD4+ lymphocytes. *J. Immunol.* **1997.** 158: 5589–5595.

90 Kalinski, P., Hilkens, C. M., Snijders, A., Snijdewint, F. G. Kapsenberg, M. L., IL-12-deficient dendritic cells, generated in the presence of prostaglandin E2, promote type 2 cytokine production in maturing human naive T helper cells. *J. Immunol.* **1997.** 159: 28–35.

91 Cousens, L. P., Orange, J. S., Su, H. C. Biron, C. A., Interferon-alpha/beta inhibition of interleukin 12 and interferon-gamma production in vitro and endogenously during viral infection. *Proc. Natl. Acad. Sci. USA* **1997.** 94: 634–639.

92 McRae, B. L., Semnani, R. T., Hayes, M. P. van Seventer, G. A., Type I IFNs inhibit human dendritic cell IL-12 production and Th1 cell development. *J. Immunol.* **1998.** 160: 4298–4304.

93 D'Andrea, A. D., Aste-Amezaga, M., Valainte, N. M., Ma, X., Kubin, M. Trinchieri, G., Interleukin-10 inhibits lymphocyte IFN-γ production by suppressing natural killer cell stimulatory factor/interlukin-12 synthesis in accessory cells. *J.Exp.Med.* **1993.** 178: 1041–1048.

94 Murphy, E. E., Terres, G., Macatonia, S. E., Hsieh, C.-S., Mattson, J., Lanier, L., Wysocka, M., Trinchieri, G., Murphy, K. O'Garra, A., B7 and interleukin-12 cooperate for proliferation and IFN-γ production by mouse Th1 clones that are unresponsive to B7 costimulation. *J. Exp. Med.* **1994.** 180: 223–231.

95 Huang, L. Y., Reis e Sousa, C., Itoh, Y., Inman, J. Scott, D. E., IL-12 induction by a TH1–inducing adjuvant in vivo: dendritic cell subsets and regulation by IL-10. *J. Immunol.* **2001.** 167: 1423–1430.

96 Yao, Y., Li, W., Kaplan, M. H. Chang, C. H., Interleukin (IL)-4 inhibits IL-10 to promote IL-12 production by dendritic cells. *J. Exp. Med.* **2005.** 201: 1899–1903.

97 Dalod, M., Salazar-Mather, T. P., Malmgaard, L., Lewis, C., Asselin-Paturel, C., Briere, F., Trinchieri, G. Biron, C. A., Interferon alpha/beta and interleukin 12 responses to viral infections: pathways regulating dendritic cell cytokine expression in vivo. *J. Exp. Med.* **2002**. 195: 517–528.

98 Gautier, G., Humbert, M., Deauvieau, F., Sciuller, M., Hiscott, J., Bates, E. E., Trinchieri, G., Caux, C. Garrone, P., A type I interferon autocrine-paracrine loop is involved in Toll-like receptor-induced interleukin-12p70 secretion by dendritic cells. *J. Exp. Med.* **2005**. 201: 1435–1446.

99 Heystek, H. C., den Drijver, B., Kapsenberg, M. L., van Lier, R. A. de Jong, E. C., Type I IFNs differentially modulate IL-12p70 production by human dendritic cells depending on the maturation status of the cells and counteract IFN-gamma-mediated signaling. *Clin. Immunol.* **2003**. 107: 170–177.

100 Reis e Sousa, C., Yap, G., Schulz, O., Rogers, N., Schito, M., Aliberti, J., Hieny, S. Sher, A., Paralysis of dendritic cell IL-12 production by microbial products prevents infection-induced immunopathology. *Immunity* **1999**. 11: 637–647.

101 MacDonald, A. S. Pearce, E. J., Cutting edge: polarized Th cell response induction by transferred antigen-pulsed dendritic cells is dependent on IL-4 or IL-12 production by recipient cells. *J. Immunol.* **2002**. 168: 3127–3130.

102 Jankovic, D., Kullberg, M. C., Hieny, S., Caspar, P., Collazo, C. M. Sher, A., In the absence of IL-12, CD4(+) T cell responses to intracellular pathogens fail to default to a Th2 pattern and are host protective in an IL-10(−/−) setting. *Immunity* **2002**. 16: 429–439.

103 Jankovic, D., Kullberg, M. C., Caspar, P. Sher, A., Parasite-induced Th2 polarization is associated with down-regulated dendritic cell responsiveness to Th1 stimuli and a transient delay in T lymphocyte cycling. *J. Immunol.* **2004**. 173: 2419–2427.

104 Reis e Sousa, C., Hieny, S., Scharton-Kersten, T., Jankovic, D., Charest, H., Germain, R.N., Sher, A., In vivo microbial stimulation induces rapid CD40 ligand-independent production of interleukin 12 by dendritic cells and their redistribution to T cell areas. *J. Exp. Med.* **1997**. 186: 1819–1829.

105 Iwasaki, A. Kelsall, B. L., Unique functions of CD11b+, CD8 alpha+, double-negative Peyer's patch dendritic cells. *J. Immunol.* **2001**. 166: 4884–4890.

106 O'Connell, P. J., Son, Y. I., Giermasz, A., Wang, Z., Logar, A. J., Thomson, A. W. Kalinski, P., Type-1 polarized nature of mouse liver CD8alpha− and CD8alpha+ dendritic cells: tissue-dependent differences offset CD8alpha-related dendritic cell heterogeneity. *Eur. J. Immunol.* **2003**. 33: 2007–2013.

107 Holt, P. J., Oliver, J., Bilyk, N., McMenamin, P. G., Kraal, G. Thepen, T., Downregulation of the antigen presenting cell function(s) of pulmonary dendritic cells *in vivo* by resident alveolar macrophages. *J. Exp. Med.* **1993**. 177: 397–407.

108 Stumbles, P. A., Thomas, J. A., Pimm, C. L., Lee, P. T., Venaille, T. J., Proksch, S. Holt, P. G., Resting respiratory tract dendritic cells preferentially stimulate T helper cell type 2 (Th2) responses and require obligatory cytokine signals for induction of Th1 immunity. *J. Exp. Med.* **1998**. 188: 2019–2031.

109 van Rijt, L. S., Jung, S., Kleinjan, A., Vos, N., Willart, M., Duez, C., Hoogsteden, H. C. Lambrecht, B. N., In vivo depletion of lung CD11c+ dendritic cells during allergen challenge abrogates the characteristic features of asthma. *J. Exp. Med.* **2005**. 201: 981–991.

21
Microbial Instruction of Dendritic Cells

Esther C. de Jong, Hermelijn H. Smits, Eddy A. Wierenga and Martien L. Kapsenberg

21.1
Introduction

Efficient protection against the different types of microbe that may invade our body requires tailor-made responses by natural and antigen-specific immune cells. Natural defense is mediated by stromal cells of body-lining tissues, such as epithelial cells, fibroblast and endothelial cells, as well as more specialized migratory immune cells such as NK cells, granulocytes, macrophages and dendritic cells (DC). Importantly, these epithelial DC cells are the unique natural immune cells that inititiate adaptive, antigen-specific responses mediated by T cells and, eventually, T cell-dependent B cells [1]. DC initiate antigen-specific immunity by stimulating naïve T cells with microbial antigen (signal 1). In addition, DC decide between immunity and tolerance by expressing variable patterns of T-cell co-stimulatory molecules (signal 2). Finally, DC orchestrate protective immune responses through the expression variable sets of T-cell-polarizing molecules (signal 3) that promote the development of certain classes of effector T helper cells (Th1, Th2 or regulatory T cells) [2]. In this chapter we will review the current knowledge on how DC determine the class of the antigen-specific immune response through their ability to recognize different pathogen types.

21.2
Effector Th1 and Th2 Cells and Regulatory T Cells

Polarized effector Th1 and Th2 cells develop from a common pool of naïve precursor T cells upon stimulation by DC. Th1 cells produce high levels of IFN-γ and TNF-β, cytokines that are instrumental in the induction of protective cell-mediated immune responses against intracellular pathogens such as viruses, certain types of (myco)bacteria and protozoa. Effector Th2 cells produce high levels of IL4, IL-5 and IL-13 and are involved in protection against helminths [3, 4]. Recently, a third subset of T cells has been described, the regulatory T cells. These T cells play a major

Handbook of Dendritic Cells. Biology, Diseases, and Therapies.
Edited by M. B. Lutz, N. Romani, and A. Steinkasserer.
Copyright © 2006 WILEY-VCH Verlag GmbH & Co. KGaA, Weinheim
ISBN: 3-527-31109-2

role in the maintenance of tolerance against self and harmless environmental proteins [5]. In addition, most pathogens induce variable amounts of regulatory T cells that hinder the development of effector cells and protective immunity, thereby contributing to the success of the pathogen. Paradoxically, these cells are also beneficial to the host. They prevent excessive damage as a result of full blown effector T cell-driven inflammation and ensure the development of solid memory as a result of the longer persistence of the pathogen [6, 7]. Two subsets of Treg can be distinguished: naturally occurring CD4+CD25+ regulatory T cells [8], which are primarily involved in central and peripheral tolerance to auto-antigens [9], and adaptive regulatory T cells [5], such as Tr1 cells and Th3 cells [10, 11]. Adaptive regulatory T cells originate, like Th1 and Th2 cells, from uncommitted peripheral naïve or central memory Th cells upon activation by antigen in a certain immunological context [12] and normally respond to innocuous foreign antigens and are primarily associated with mucosal tolerance to ubiquitous antigens and nonpathogenic microflora [13, 14]. The outcome of an immune response to a certain pathogen is thus determined by the balanced development of effector and/or regulatory T cells, which is orchestrated by DC.

21.3
Dendritic Cells and Pattern Recognition Receptors

Two major types of DC are classic CD11c+ myeloid DC and CD123+ (IL-3Rα) plasmacytoid DC (PDC) [15, 16]. Immature myeloid DC continuously patrol the body-lining epithelia for incoming pathogens. Upon pathogen recognition, they undergo a process of maturation, including their migration toward the draining lymphoid tissues and their acquisition of potent T-cell stimulatory capacity [17, 18]. Like all natural immune cells, DC detect pathogens with germ-line encoded "pattern recognition receptors" (PPRs), that recognize "pathogen-associated molecular patterns" (PAMPs), which are evolutionary conserved microbial molecules essential to pathogen function [19, 20]. An important group of PRR are the family of Toll-like receptors (TLR). To date, eleven different human TLR (TLR1–11) have been described [21].The classical human myeloid CD11c+ DC express TLR1-6 and TLR8 [22, 23], which allow them to recognize a broad panel of bacterial and viral compounds, as discussed in detail below.

PDC are found in immune tissues and in variable numbers in peripheral tissues, in particular during inflammation. They produce high levels of type I interferons (IFN) upon viral infection and are believed to play an important role in antiviral defense as natural immune cells [24]. The TLR expression profiles of PDC strongly differ from that of myeloid DC [22, 23]. In contrast to the wide distribution of TLR on myeloid DC, PDC selectively express intracellular TLR7 and TLR9, recognizing single stranded (ss)RNA [25, 26] and CpG-containing DNA motifs [27], respectively. At present, the role of PDC in the initiation of specific immunity is not entirely clear, but they are very likely to play a role in the activation of effector T cells at the site of infection.

TLR ligation of myeloid DC leads to their activation and immediately results in the production of pro-inflammatory cytokines, which contribute to local innate immune responses In addition, TLR signaling initiates the program of DC maturation, leading to the transformation of immature DC into potent effector DC that are uniquely capable of initiating specific immune responses by driving the development of effector Th cells from naïve Th cells [17, 28]. Amongst others, these mature DC express selective sets of T cell-polarizing molecules, either soluble or membrane-bound, that determine the balance between Th1, Th2 or regulatory T-cell development [2]. The expression profile of these T cell-polarizing molecules by DC is dependent on and imprinted by the binding of pathogen to selective PRRs (e.g. TLR) of DC in their immature state, resulting in selective programming of these DC during their maturation. Basically, DC exposed to intracellular pathogens express T-cell-polarizing molecules promoting Th1 responses, whereas certain helminths prime DC for the expression of molecules that drive the development of Th2 cells (Fig. 21.1) [29]. Furthermore, certain types of pathogens prime for DC with a strong capability to induce the development of regulatory T cells [30].

Another large family of PRRs is formed by the C-type lectins that recognize specific carbohydrate structures present in pathogens, such as mannosylated lipoarabinomannan (ManLAM) on mycobacteria [31, 32]. In contrast to ligation of TLRs, the binding of a C-type lectins results in the internalization of bound pathogens followed processing for antigen presentation [33]. Although *in vitro*-generated human immature monocyte-derived DC express many C-type lectins, only a few have been identified on myeloid blood DC [15]. In contrast, plamacytoid DC express

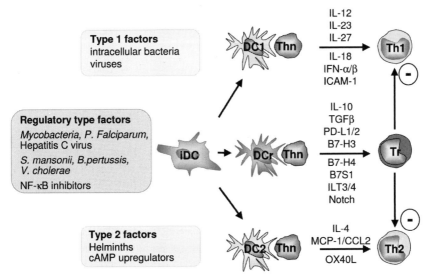

Fig. 21.1 Dendritic cell polarization. Schematic representation of factors that influence the T-cell polarizing capacity of DC and the expression of molecules (either cytokines or membrane-bound molecules) that drive the development of either Th1, Th2 or adaptive regulatory T cells. (iDC immature DC).

Blood DC Antigen (BDCA)-2, Dectin-1 and DEC205. Most C-type lectins on DC are type II transmembrane proteins containing at least one carbohydrate recognition domain which bears, depending on the amino acid sequence, specificity for mannose, galactose or fructose structures [34].

Little is known about the role of C-type lectins in the priming of DC for the induction Th1 or Th2 cells. Recognition of pathogens by C-type lectins will favor immune suppression whereas concomitant TLR triggering overrules the tolerizing function of C-type lectins, resulting in immune activation [35]. The role of C-type lectins in pathogen recognition will be discussed in more detail in Chapter 7.

Apart from these two major groups of PRRs, a number of other molecules expressed on antigen presenting cells, have been identified to bind pathogen-derived compounds, although information about the consequences for the Th-cell development is scarce. The type 3 complement receptor (CR3) is capable of binding *Mycobacterium tuberculosis*, yeast-derived zymosan and filamentous hemagglutinin (FHA) from *Bordetella pertussis* [36, 37]. Also different types of scavenger receptors are known to recognize pathogens such as *Neisseria meningitidis* via SR-A [38] and *Plasmodium falciparum* [39]. Also the Nod proteins (Nod1 and Nod2) have recently been shown to be intracellular PRRs that bacterial peptidoglycan (PGN) although Nod1 and Nod2 detect distinct motifs within this structure [40]. Interestingly, a mutation in *Nod2* resulting in the loss of the capacity to bind PGN-derived muramyl dipeptide, is associated with Crohn's disease [41, 42].

21.4
DC-derived Factors that Promote Th1, Th2 or Regulatory T-cell Responses

21.4.1
Th1 Cell-promoting Factors

Of all known Th1-promoting factors, IL-12 has been studied most extensively. The bioactive IL-12p70 molecule is a heterodimer composed of a 35-kDa light chain (p35) and a covalently bound 40-kDa heavy chain (p40). The p40 subunit is produced in large excess, but cannot induce signal transduction upon receptor ligation. Homodimers of IL-12p40 may even prevent IL-12 p70 signaling [43]. The IL-12 receptor (IL-12R) is primarily expressed on NK cells and activated T cells and consists of two chains, the IL-12Rβ1 and the signaling IL-12Rβ2 chain. IL-12p40 binds the IL-12Rβ1 chain, whereas the p35 subunit ligates the IL-12Rβ2 chain [44]. Recently, two related heterodimeric cytokines have been identified, IL-23 and IL-27, which together with IL-12 are referred to as the IL-12 family. Like IL-12, these two cytokines are mainly produced by hematopoietic phagocytic cells (monocytes and macrophages) and DC and their production is enhanced by IFN-γ [4]. IL-23 consists of the IL-12p40 subunit and a unique p19 subunit. The IL-23 receptor consists of the IL-12Rβ1 chain that binds p40, whereas p19 binds to the IL-23R, a novel gp130-like chain [45]. IL-27 is a heterodimer of p28 and the Epstein–Barr virus-induced gene 3 (EBI3), an α-receptor-like soluble chain homologous to IL-12p40 [46, 47]. In contrast to EBI3, which, like p40, is produced in large excess, the subunit p28 cannot be secreted without EBI3 [47]. EBI3 binds to a receptor chain designated T-cell cytokine receptor (TCCR) or WSX-1 and p28 ligates the gp130 chain [47, 48].

Although all three members of the IL-12 family induce the production of IFN-γ by activated T cells and NK cells, their effects on memory versus naïve Th cells and on the magnitude of the response may be different. Naïve Th cells rapidly produce IFN-γ in response to IL-12 or IL-27 and develop into Th1 cells [47, 43]. However, the effects of IL-27 are dependent on IL-12 or IL-18 [49]. Unlike IL-12, IL-23 does not have a potent effect on naïve Th cells but is particularly efficient in promoting the production of IFN-γ in memory Th cells [50]. Expression ratios of these IFN-γ-inducing cytokines depend upon distinct ligation of PRRs by pathogenic compounds as will be discussed below.

Another group of cytokines that may play a role in the development of Th1 cells is the family of type I IFNs which consists of the homologous cytokines IFN-α, β, τ and ω. These cytokines have important antiviral activities, but have also been shown to possess a wide range of immunoregulatory functions [51]. IFN-α is a subfamily of various homologous IFN species that differ in their antiviral effects and their effects on T-cell proliferation and NK cell activity [51]. Although it is widely accepted that type I IFNs drive Th1 development, the evidence is not substantial. Several studies have demonstrated that they increase IFN-γ mRNA and protein production in naïve and previously activated Th cells, but this increase is by far not as high as the increase induced by IL-12 [9]. Part of the effect of IFN-α on IFN-γ production may be explained indirectly by the transient upregulation of the IL-12Rβ2, resulting in an enhanced sensitivity for IL-12 or by inhibition of IL-4 production [52].

Not only the IL-12 family cytokines but also other factors may play a role in the induction of Th1 cells. IL-18 was first described as IFN-γ-inducing factor and was found to be circulating in mice during endotoxemia [53]. It is constitutively expressed as pro-IL-18 by monocytes, macrophages and DC, both in mice and humans. The production of biologically active IL-18 is induced by factors (such as lipopolysaccharide (LPS) or gram-positive bacteria) that stimulate the expression of IL-1β-converting enzyme that cleaves proIL-18 into mature IL-18 [54]. In contrast to IL-12, IL-18 by itself induces only low levels of IFN-γ in human CD4$^+$ T cells. However, in the additional presence of IL-12, IL-18 strongly enhances IFN-γ production [55] and IL-18 itself does not induce the development of Th1 cells from naïve precursors [56].

Finally, not only soluble factors but also membrane-bound molecules may promote Th1 responses. One such molecule is ICAM-1 that ligates LFA-1 on the T cells. ICAM-1/LFA-1 interaction supports adhesion between DC and T cells. Both in mice and human ICAM-1/LFA-1 interaction promotes the induction of Th1 responses, which is particularly evident in the absence of T cell-polarizing cytokines, such as IL-4 or IL-12 [57–59].

21.4.2
Th2 Cell-promoting Factors

In contrast to the abundant knowledge about Th1-inducing factors, little is known about the active induction of Th2 cells by DC. It is widely accepted that IL-4 is a potent inducer of Th2 cells from naïve precursors and that the main source of IL-4 is the Th-cell population itself, indicating that IL-4 acts as an autocrine factor amplifying IL-4 production by developing Th2 cells. Some reports suggest that murine DC may have the capacity to produce low levels of IL-4 in response to certain yeast hyphae or virus species but no reports on human DC-derived IL-4 are available [60, 61]. An important prerequisite for the induction of Th2 responses is that the capacity of DC to produce IL-12-family members is down-regulated, to prevent induction of IFN-γ production. Indeed, IL-12 is capable of restoring the IL-12-responsiveness in established effector Th2 cells *in vitro*, resulting in high levels of IFN-γ and downregulated IL-4 production [62]. Although it was suggested earlier that Th2 responses are the mere result of the absence of IL-12 (or IL-12 family members) [63], we [64] and others [65, 66] have clearly shown that the development of Th2 cells driven by helminth-primed DC requires active Th2 cell-polarizing factors expressed by these DC.

A molecule expressed both by murine and human DC that is clearly involved in the induction of Th2 cells from naïve precursors is membrane-bound OX40L, the ligand of OX40 (CD134) expressed on T cells. OX40L-OX40 interaction not only co-stimulates proliferation of CD4+ T cells, but was also proposed to contribute to Th2 development. Co-stimulation of murine or human naïve Th cells with an OX40L-transfected cell line enhances the expression of IL-4 and promotes the development of Th2 cells *in vitro* [67, 68].

21.4.3
Regulatory T-cell-promoting Factors

In all of the studies on the induction of regulatory T cells by DC, IL-10 has been implicated as the (co-)responsible factor. IL-10 is a pleiotropic cytokine that functions at different levels of the immune response. For example, it blocks proliferation and cytokine production by T cells, by inhibiting the phosphorylation of CD28, thereby abrogating downstream signaling [69]. *In vitro* studies showed that naïve CD4+ T cells that are cultured in IL-10 become regulatory T cells with the capacity to suppress the activation and proliferation of bystander T cells [70]. IL-10 may also induce tolerance indirectly, as it inhibits the full maturation of DC and downregulates their MHC class II and IL-12 expression [71, 72].

Apart from IL-10, several other factors expressed by DC may be involved in the induction of regulatory T cells, including TGF-β, novel inhibitory members of the B7 family, members of the Notch signaling pathway (e.g. Serrate1) [73, 74] or ILT3 and ILT4 [75, 76]. The expanding family of B7 molecules are all expressed by DC and other APC. They are ligands for co-stimulatory molecules on T cells and their expression is probably tightly regulated as they play critical roles in the control and fine-tuning of the immune response [77]. The best-characterized B7 family members are the ligands of CD28 and CTLA-4, CD80 and CD86, also known as B7.1 and B7.2. Both are upregulated on DC upon activation [20].

More recently discovered family members are the two counter-structures of programmed death1 (PD-1), i.e. the broadly expressed PD-L1 (also called B7-H1) [78] and the DC-specific PD-L2 (also called B7-DC) [79]. Yet others are B7-H3 and B7-H4. B7-H3 (also referred to as B7-RP2) binds a still unknown receptor on T cells that is distinct from CD28, CTLA-4, ICOS and PD-1 [80]. Experiments with mice deficient for either of these B7 family molecules or experiments using blocking reagents suggested negative regulatory functions in the immune response [81–85]. Although it is tempting to speculate on a role of these molecules in the induction of regulatory T cells, it can not be excluded that loss of tolerance observed in these mouse models is merely due to T-cell anergy or apoptosis.

21.5
TLR-mediated Activation of DC by Microbes and their Compounds

Accumulating data underline the concept that ligation of different PRR during microbial exposure primes for functional maturation of effector DC with distinct Th-polarizing capacities. The original belief was that TLR triggering by definition always results in the development of Th1 cells. However, as discussed below, it is now clear that differences in the expression of T-cell polarizing factors may be a reflection of TLR ligation.

21.5.1
TLR2

TLR2 can form heterodimers with TLR1 or with TLR6 and thus distinguish the subtle differences between triacyllipopeptides of gram-negative bacteria (recognized by TLR1/2), and diacyllipopeptides of mycoplasma (recognized by TLR2/6) [86, 87]. Triggering of these different TLR2-containing heterodimers translates into different cytokine patterns. Ligation of TLR1/2 by lipoproteins from *Mycobacterium* species results in low IL-12 but enhanced IL-23 production [88]. An additional nonredundant role IL-27 is suggested by a study of *M. bovis* bacillus Calmette–Guerin infection in IL-27R-deficient (WSX$^{-/-}$) mice, showing impaired early IFN-γ production and poorly differentiated granulomas, despite the availability of IL-12 and IL-23 [89].

TLR2/6 ligation, on the other hand, leads to other cytokine profiles. Apart from mycoplasmal diacyllipopeptides, this heterodimer recognizes zymosan from yeast cell walls and PGN from gram-positive bacteria [90, 91]. However, none of these compounds induces IL-12 production, but they lead to the production of IL-10, instead [62, 88, 92, 93]. Data on the consequence of TLR2/6 ligation for T-cell-polarization varies considerably between different studies. Indeed, DC activated with mycoplasmal lipopeptide 2 were reported to induce unpolarized T-cell responses whereas zymosan primed for Th2-promoting DC [92, 93] and PGN for Th1-promoting DC with an increased expression of IL-23-p19 [62]. How these findings relate to the above mentioned TLR2/6-mediated IL-10 production remains unclear. Clearly, ligation of TLR2 by schistosomal-derived lyso-phosphatidylserine (lyso-PS) induces IL-10 production, as well. In this case, TLR2 ligation leads to the development of DC that promote the development of regulatory T cells from naïve precursors in a IL-10-dependent fashion [94].

21.5.2
TLR3

TLR3 recognizes dsRNA derived from viruses as well as the eggs of the helminth *Schistosoma mansonii* [95, 96] and promotes cross-priming to virus-infected cells [97]. Virus-derived dsRNA primes *in vitro* for a human DC phenotype that strongly supports Th1 cell development without enhanced IL-12 production [64]. *In vitro* studies with human monocyte-derived DC primed with dsRNA suggest the alternative involvement of IL-27 (A.J. van Beelen, unpublished results), type I IFN [98] and/or ICAM-1 [59] in the induction of Th1 cells. IL-12-deficient mice are able to clear viral infections, also suggesting a role for other Th1-promoting factors. It should be noted, however, that dsRNA can bypass TLR3 by binding to the intracellular receptor protein kinase R (PKR), which induces high levels of IFN-α in murine CD11+ DC [99].

21.5.3
TLR4

A well-studied PAMP/TLR-interaction is the binding of LPS from Gram negative bacteria to TLR4 [100]. Ligation of TLR4 by LPS induces low levels IL-12, IL-23 and IL-27 in monocyte-derived DC. Consequently, LPS-primed DC do induce Th1 cell responses, but clearly not as potently as dsRNA-primed DC [64, 98, 101]. Indeed, the Th1-driving capacity of LPS-primed DC is readily overruled by the presence of exogenous Th-cell-polarizing factors such as IFN-γ or PGE2 that potently enhance or decrease the IL-12 production [102]. LPS derived from distinct Gram negative bacteria may induce different levels of Th1- or Th2-promoting factors. For example, LPS derived from *Escherichia coli* binds TLR4, induces IL-12 production and primes for Th1 responses whereas LPS from *Prophyomonas gingivalis*, which may not ligate TLR4, fails to induce IL-12 and primes for Th2 cell responses [103]. These findings may possibly be due to differences in TLR4 requirements. Very recently, it was demonstrated that the phosphorylcholine-containing glycoprotein from the filarial nematode *Acanthocheilonem vitae*, ES-62, binds to TLR4 and signals via Myd88 [104]. This molecule has anti-inflammatory capacities and has been described to induce both Th2 as well as regulatory T cells [105].

21.5.4
TLR5

Only one ligand has been described for TLR5, i.e. the bacterial flagellin, a major component of bacterial flagella [106]. Whether TLR5-mediated activation of DC promotes a certain type of Th-cell response, remains to be established. One study using human monocyte-derived DC *in vitro* showed that flagellin and LPS induced similar levels of IL-12 and promoted Th1 responses to the same extend [107], whereas others showed that flagellin-activated human DC did not produce IL-12p70 [108]. This latter observation is corroborated by a murine *in vivo* study with flagellin from *Salmonella typhimurium*, demonstrating that flagellin induces the production of IL-12p40, but not the biological active IL-12p70, and induces Th2 responses [109]. However, the response to flagellin may be dependent on the context of presentation, as the soluble flagellar protein FliC alone induces a Th2 response in mice *in vivo*, whereas the response to FliC on intact *Salmonella* results in a Th1 response [110].

21.5.5
TLR7/8

Although TLR7 and TLR8 largely recognize the same structures, the expression pattern of these closely related intracellular TLR is quite distinct. In humans, TLR7 is strongly expressed by PDC, but only weakly by CD11c+ blood DC, and not at all by monocyte-derived DC. TLR8, in contrast, is expressed both by monocyte-derived DC and CD11c+ DC, but not by PDC. [22, 23, 111]. TLR7 and TLR8 recognize anti-viral imidazolquinoline peptides and synthetic structures like R848 and loxorubine [107, 112, 113]. More recently it was discovered that viral ssRNA is the physiological ligand of TLR7 and TLR8. Murine PDC produce IFN-α and IL-12 in response to ssRNA, whereas macrophages and DC produce only IL-12 in these conditions [25, 26]. Although it is not yet known whether DC activated by ssRNA will indeed induce a Th1 cell response, ligation of TLR7 and TLR8 by imidazolquinolines, both in mice and man, modulates antigen-specific Th2 responses into Th1 cell responses [114, 115] which could be inhibited by the addition of a neutralizing anti-IL-12 antibody [114].

21.5.6
TLR9

TLR9 in humans is only expressed on PDC, whereas in mice it is expressed both by CD11c+ DC and by PDC [101, 116]. TLR9 recognizes unmethylated CpG-containing bacterial DNA [27]. Activation of PDC with such CpG motifs results in the strong expression of IFN-α and promotion of Th1 cells [22, 23]. The production of IL-12 by human PDC *in vivo* is controversial [111, 117], but PDC precultured in IL-3 and subsequently stimulated with TLR9-binding CpG motifs in the additional presence of CD40L are efficient producers of IL-12p70 [118]. IL-3, like the ligation of BDCA-2, inhibits the production of type I IFN and allows for the production of IL-12p70 [119]. In contrast to human PDC, murine PDC readily produce IL-12p70 in response to various viruses and TLR ligands [120].

21.5.7
TLR10/11

TLR10 is expressed in the lung and by B cells. Its ligand(s) and function are as yet unknown, although a genetic variation in TLR10 is associated with the risk of developing asthma [121]. TLR11 is widely distributed in mice, including expression by macrophages, liver cells, kidney cells and bladder epithelial cells. It recognizes a structure expressed by uropathogenic bacteria and mice lacking TLR11 are highly susceptible to infection with these bacteria [122]. As for TLR10, a role for TLR11 in T-cell polarization has not been reported. A human equivalent of TLR11 has not yet been described.

21.6
Th1 Cell-promoting DC

Many pathogenic compounds prime DC for the induction of Th1 cells, as has been described above, although the expression of the various Th1-promoting factors may differ. IL-12 is probably the best-studied Th1-promoting factor. Its production is readily induced by stimulation of immature DC by microbial compounds. However, only certain intracellular pathogens prime immature DC to develop into high-IL-12-producing mature DC. These high levels of IL-12 are produced upon CD40 ligation by the rapidly induced CD40L on activated naïve T cells in the draining lymph nodes. This indicates that the intrinsic capacity of mature effector DC to produce high levels of IL-12 (or other polarizing factors) is imprinted by previous exposure to pathogens during their immature phase in the infected tissues. Only the later CD40-CD40L interaction allows for full expression of the polarizing signals by mature DC and consequently for the selective development of Th1 or Th2 responses.

Examples of pathogens that indeed prime for high levels of IL-12 production upon subsequent CD40 ligation, are *Toxoplasma gondii, Leishmania major and Mycobacteria* species [123, 124]. The importance of IL-12 for the protection against these intracellular pathogens is corroborated by observations in patients with mutations in *IL-12p40* or *IL-12Rβ1* and, therefore, do not produce or respond to IL-12 and IL-23. Many of these patients develop chronic infections with *Mycobacteria* species and *Salmonella*. They can, however, efficiently clear viral and extracellular infections [125, 126]. Similarly, IL-12p35-deficient mice still develop polarized Th1 responses to certain mycobacterial infections, provided the p40 subunits are intact, suggesting a pronounced role of IL-23 [127, 128].

21.7
Th2 Cell-promoting DC

Several parasitic helminths as well as the hyphae from of the yeast *Candida albicans* have been reported to prime DC for the promotion of Th2 cells [60, 64–66]. Recently data are immerging about the PRRs involved in the recognition of helminths or helminth-derived structures. As has been pointed out previously several helminth-derived products may activate TLR2, TLR3 or TLR 4 [96, 104, 129] and drive the development of Th2 cells. But also sugar-containing structures, possibly binding to C-type lectins, are involved in the induction of a Th2-promoting phenotype in DC [130]. The role of C-type lectins in the induction of a Th2-promoting DC phenotype has been demonstrated in a recent study Bergman and coworkers [131] showing that a Lewis antigen positive variant of *Helicobacter pylori* escapes protective immunity by the binding to DC-specific ICAM-3 grabbing nonintegrin (DC-SIGN), which blocks the induction of protective Th1 cells, whereas the Lewis antigen negative variant induces a potent protective Th1 cell response.

Little is known about the DC-expressed factors involved in the priming for Th2 cells. Only the expression of OX40L is clearly instrumental in the development of Th2 cells [64]. Resting murine and human DC do not express OX40L, but this is rapidly induced upon CD40 ligation. However, human effector monocyte-derived DC express differential levels of OX40L following CD40 ligation, depending on their priming conditions. Th1-inducing DC, primed with IFN-γ or poly I:C, do not express OX40L, whereas Th2-promoting DC primed with PGE_2 or schistosomal egg antigen (SEA) did express detectable levels of OX40L, which were further up-regulated by subsequent CD40 ligation [64]. In addition, a critical role for OX40L-OX40 interaction in a functional Th2 response has been shown *in vivo* in a murine model of experimental leishmaniasis [132].

Interestingly, Th2-promoting DC with enhanced expression of OX40L show a low expression of ICAM-1 expression, whereas Th1-promoting DC show high levels of ICAM-1 and no OX40L [59, 64]. This may indicate possible opposing regulation of expression of these two co-stimulatory molecules.

21.8
Regulatory T-cell-promoting DC

It is now clear that DC not only play a role in the induction of effector Th1 or Th2 cells but can also promote the development of regulatory T cells that have the capacity to downregulate proliferation and cytokine production of effector Th1 or Th2 cells [133]. Paradoxically, pathogens that require Th1 or Th2 effector T cells to be successfully eliminated may actually induce immune suppression via the induction or expansion of regulatory T cells. This immune tolerance clearly benefits both the pathogen, by increasing its survival rate, and the host limiting the detrimental effects of chronic Th1 or Th2 responses [6, 134]. Pathogens may sustain immune tolerance via different mechanisms. Some pathogens, such as *P. falciparum*, hepatitis C virus and *Mycobacteria* species, prevent the maturation of DC into cells with a potent T-cell stimulatory capacity and, instead, lead to the development of tolerogenic DC [31, 135, 136]. *P. falciparum* infects erythrocytes which bind the scavenger receptor CD36 on DC which in turn inhibits LPS-induced maturation and, as a result, the T-cell stimulatory capacity of DC [135, 137]. Activation of murine DC with *Plasmodium*-infected eythrocytes induces the secretion of yet unidentified soluble inhibitory factors. In case of hepatitis C, virus-specific structures such as the core protein and nonstructural protein 3 inhibit the maturation, cytokine production and T-cell stimulatory capacity of DC, possibly resulting in the generation of the IL-10-producing CD4$^+$ T cells that have been shown to persist in patients infected with hepatitis C virus [138–140]. Yet another suppressive mechanism is the ligation of the C-type lectin DC-SIGN by ManLAM derived from *Mycobacteria* species. This interaction has been demonstrated to inhibit the LPS-induced maturation and IL-12 production by DC, but enhances their IL-10 production [141]. Whole mycobacteria do inhibit IL-12 production via DC-SIGN but, in contrast to ManLAM, do not inhibit DC maturation, suggesting that other compounds of mycobacteria overrule the inhibition of maturation [142].

Other pathogens allow the full maturation of DC but still induce the development of regulatory T cells. One of the first reported examples of this category of pathogens is *Schistosoma mansonii*. *S. mansonii*-derived lyso-PS ligates TLR2 and drives the development of regulatory DC that promote the generation of regulatory T cells. Lyso-PS does not inhibit LPS-induced maturation, but inhibits IL-12 production while enhancing the production of IL-10. This DC-derived IL-10 which contributes together with a yet unknown membrane-bound factor contributes to the development of regulatory T cells [94]. In addition, injection of the *Schistosoma egg*-derived glycoconjugate LNFPIII in mice induced the presence of an APC population that produced high levels of IL-10 and TGFβ and efficiently suppressed the proliferation of T cells *in vitro* [143]. Indeed, in patients suffering from chronic helminth infection antigen-specific regulatory T cells could be isolated and where characterized by the secretion of elevated levels of IL-10 and/or TGFβ and are associated with immunosuppression [144, 145].

Also *B. pertussis* induces the development of regulatory DC without inhibiting their maturation. The responsible virulence factor, filamentous haemaglutinin (FHA) inhibits IL-12 production but induces IL-10 production in murine DC and drives regulatory DC that promote the development of regulatory T cells capable of blocking secondary unrelated infections, e.g. with influenza virus [134]. Similar modulation of DC function has been described for another compound of *B. pertussis*, adenylate cyclase toxin, and for cholera toxin from *Vibrio cholerae* [146, 147].

21.9
Indirect Priming of DC

During primary infections, pathogens will not only affect DC directly but also indirectly via infected tissue cells. Various tissue cell types express TLR, recognize pathogens and produce inflammatory mediators that can contribute to the shaping of the immune response. These factors include pro- and anti-inflammatory cytokines, chemokines, eicosanoids, heat-shock proteins and cell surface-bound molecules. We have shown that poly I:C-activated human primary keratinocytes express several TLRs including TLR3 and prime monocyte-derived DC for promoting Th1 responses which is dependent on TNF-α, IL-18 and IFN-α [148]. In the mouse, however, two independent studies showed that murine DC induce the development memory T-cell responses only if they are directly activated via their TLR, but not if they are activated indirectly by inflammatory signals from bystander DC [149, 150]. So far, there is no clear consensus on whether or not indirect activation of DC contributes to the formation of polarized Th1 or Th2 responses.

21.10
Concluding Remarks

The increasing knowledge of the biology of DC indicates both their complexity and their flexibility. It is clear that DC have a crucial role in properly adapting the class of immune response to the type of invading pathogen through the differential expression of T-cell polarizing signals upon ligation of particular PRRs. However, details on the underlying mechanisms are still largely unknown. It is becoming increasingly evident that pronounced Th1 responses are initiated upon ligation of the intracellular TLRs (i.e. TLR3, 7, 8 and 9) resulting in high expression of IL-12 family members. Also data have recently emerged on the mechanism underlying DC-driven Th2 cell development, for example in response to helminth infections. Currently, the generation of regulatory DC and the mechanisms by which they promote the development of regulatory T cells are major research topics in many laboratories, hopefully providing important new information in the near future.

The importance of the knowledge on how different T-cell responses evolve under the influence of differentially primed DC is that it may be helpful in the design of new therapies for a variety of immune disorders.

References

1 Banchereau, J., Steinman, R.M., Dendritic cells and the control of immunity. *Nature* **1998**. 392: 245–252.

2 Kalinski, P., Hilkens, C.M.U., Wierenga, E.A., Kapsenberg, M.L., T-cell priming by type-1 and type-2 polarized dendritic cells: the concept of a third signal. *Immunology Today* **1999**. 20: 561–567.

3 Mosmann, T.R., Coffman, R.L., TH1 and TH2 cells: different patterns of lympho-kine secretion lead to different functional properties. *Annu. Rev. Immunol.* **1989**. 7: 145–173.

4 Trinchieri, G., Pflanz, S., Kastelein, R.A., The IL-12 family of heterodimeric cytokines: new players in the regulation of T cell responses. *Immunity* **2003**. 19: 641–644.

5 Bluestone, J.A., Abbas, A.K., Natural versus adaptive regulatory T cells. *Nat. Rev. Immunol.* **2003**. 3: 253–257.

6 Belkaid, Y., Piccirillo, C.A., Mendez, S., Shevach, E.M., Sacks, D.L., CD4+CD25+ regulatory T cells control Leishmania major persistence and immunity. *Nature* **2002**. 420: 502–507.

7 Montagnoli, C., Bacci, A., Bozza, S., Gaziano, R., Mosci, P., Sharpe, A.H., Romani, L., B7/CD28-dependent CD4+CD25+ regulatory T cells are essential components of the memory-protective immunity to Candida albicans. *J. Immunol.* **2002**. 169: 6298–6308.

8 Levings, M.K., Sangregorio, R., Sartirana, C., Moschin, A.L., Battaglia, M., Orban, P.C., Roncarolo, M.G., Human CD25+CD4+ T suppressor cell clones produce transforming growth factor beta, but not interleukin 10, and are distinct from type 1 T regulatory cells. *J. Exp. Med.* **2002**. 196: 1335–1346.

9 Bach, J.F., Francois, B.J., Regulatory T cells under scrutiny. *Nat. Rev. Immunol.* **2003**. 3: 189–198.

10 Levings, M.K., Bacchetta, R., Schulz, U., Roncarolo, M.G., The role of IL-10 and TGF-beta in the differentiation and effector function of T regulatory cells. *Int. Arch. Allergy Immunol.* **2002**. 129: 263–276.

11 Fukaura, H., Kent, S.C., Pietrusewicz, M.J., Khoury, S.J., Weiner, H.L., Hafler, D.A., Induction of circulating myelin basic protein and proteolipid protein-specific transforming growth factor-beta1-secreting Th3 T cells by oral adinistration of myelin in multiple sclerosis patients. *J. Clin. Invest* **1996**. 98: 70–77.

12 Groux, H., O'Garra, A., Bigler, M., Rouleau, M., Antonenko, S., de Vries, J.E., Roncarolo, M.G., A CD4+ T-cell subset inhibits antigen-specific T cell responses and prevents colitis. *Nature* **1997**. 389: 737–742.

13 Cavani, A., Nasorri, F., Prezzi, C., Sebastiani, S., Albanesi, C., Girolomoni, G., Human CD4+ T lymphocytes with remarkable regulatory functions on dendritic cells and nickel-specific Th1 immune responses. *J. Invest. Dermatol.* **2000**. 14: 295–302.

14 Khoo, U.Y., Proctor, I.E., Macpherson, A.J., CD4+ T cell down-regulation in human intestinal mucosa: evidence for intestinal tolerance to luminal bacterial antigens. *J. Immunol.* **1997**. 158: 3626–3634.

15 MacDonald, K.P., Munster, D.J., Clark, G.J., Dzionek, A., Schmitz, J., Hart, D.N., Characterization of human blood dendritic cell subsets. *Blood* **2002**. 100: 4512–4520.

16 Rissoan, M.C., Soumelis, V., Kadowaki, N., Grouard, G., Briere, F., de Waal, M., Liu, Y.J., Reciprocal control of T helper cell and dendritic cell differentiation [see comments]. *Science* **1999**. 283: 1183–1186.

17 Banchereau, J., Briere, F., Caux, C., Davoust, J., Lebecque, S., Liu, Y.J., Pulendran, B., Palucka, K., Immuno-biology of dendritic cells. [Review] [289 refs]. *Annu. Rev. Immunol.* **2000**. 18: 767–811.

18 Sallusto, F., Lanzavecchia, A., Under-standing dendritic cell and T-lymphocyte traffic through the analysis of chemokine receptor expression. *Immunol. Rev.* **2000**. 177: 134–140.

19 Janeway, C.A., Jr., Approaching the asymptote? Evolution and revolution in immunology. *Cold Spring Harb. Symp. Quant. Biol.* **1989**. 54 Pt 1: 1–13.

20 Janeway, C.A., Jr., Medzhitov, R., Innate immune recognition. *Annu. Rev. Immunol.* **2002**. 20: 197–216.

21 Akira, S., Hemmi, H., Recognition of pathogen-associated molecular patterns by TLR family. *Immunol. Lett.* **2003**. 85: 85–95.

22 Jarrossay, D., Napolitani, G., Colonna, M., Sallusto, F., Lanzavecchia, A., Specialization and complementarity in microbial molecule recognition by human myeloid and plasmacytoid dendritic cells. *Eur. J. Immunol.* **2001**. 31: 3388–3393.

23 Kadowaki, N., Ho, S., Antonenko, S., Malefyt, R.W., Kastelein, R.A., Bazan, F., Liu, Y.J., Subsets of human dendritic cell precursors express different toll-like receptors and respond to different microbial antigens. *J. Exp. Med.* **2001**. 194: 863–869.

24 Kadowaki, N., Antoneko S., Lau J.Y-N., Liu Y.J., Natural interferon a/b-producing cells link annate and addaptive immunity. *J. Exp. Med.* **2000**. 192: 219–225.

25 Heil, F., Hemmi, H., Hochrein, H., Ampenberger, F., Kirschning, C., Akira, S., Lipford, G., Wagner, H., Bauer, S., Species-specific recognition of single-stranded RNA via toll-like receptor 7 and 8. *Science* **2004**. 303: 1526–1529.

26 Diebold, S.S., Kaisho, T., Hemmi, H., Akira, S., Reis E Sousa, Innate antiviral responses by means of TLR7-mediated recognition of single-stranded RNA. *Science* **2004**. 303: 1529–1531.

27 Hemmi, H., Takeuchi, O., Kawai, T., Kaisho, T., Sato, S., Sanjo, H., Matsumoto, M., Hoshino, K., Wagner, H., Tadeka, K., Akira, S., A Toll-like receptor recognizes bacterial DNA. *Nature* **2001**. 408: 740–745.

28 Akira, S., Takeda, K., Kaisho, T., Toll-like receptors: critical proteins linking innate and acquired immunity. *Nat. Immunol.* **2001**. 2: 675–680.

29 Kapsenberg, M.L., Dendritic-cell control of pathogen-driven T-cell polarization. *Nat. Rev. Immunol.* **2003**. 3: 984–993.

30 Smits, H.H., de Jong, E.C., Wierenga, E.A., Kapsenberg, M.L., Different faces of regulatory DCs in homeostasis and immunity. *Trends Immunol.* **2005**. 26: 123–129.

31 Geijtenbeek, T.B., van Vliet, S.J., Koppel, E.A., Sanchez-Hernandez, M., Vandenbroucke-Grauls, C.M., Appelmelk, B., van Kooyk, Y., Mycobacteria target DC-SIGN to suppress dendritic cell function. *J. Exp. Med.* **2003**. 197: 7–17.

32 Tailleux, L., Schwartz, O., Herrmann, J.L., Pivert, E., Jackson, M., Amara, A., Legres, L., Dreher, D., Nicod, L.P., Gluckman, J.C., et al., DC-SIGN is the major Mycobacterium tuberculosis receptor on human dendritic cells. *J. Exp. Med.* **2003**. 197: 121–127.

33 Figdor, C.G., van Kooyk, Y., Adema, G.J., C-type lectin receptors on dendritic cells and Langerhans cells. *Nature Rev. Immunol.* **2002**. 2: 77–84.

34 Weis, W.I., Taylor, M.E., Drickamer, K., The C-type lectin superfamily in the immune system. *Immunol. Rev.* **1998**. 163: 19–34.

35 Ichikawa, H.T., Williams, L.P., Segal, B.M., Activation of APCs through CD40 or Toll-like receptor 9 overcomes tolerance and precipitates autoimmune disease. *J. Immunol.* **2002**. 169: 2781–2787.

36 Ross, G.D., Regulation of the adhesion versus cytotoxic functions of the Mac-1/CR3/alphaMbeta2-integrin glycoprotein. *Crit Rev. Immunol.* **2000**. 20: 197–222.

37 Ishibashi, Y., Claus, S., Relman, D.A., Bordetella pertussis filamentous hemagglutinin interacts with a leukocyte signal transduction complex and stimulates bacterial adherence to monocyte CR3 (CD11b/CD18). *J. Exp. Med.* **1994**. 180: 1225–1233.

38 Peiser, L., De Winther, M.P., Makepeace, K., Hollinshead, M., Coull, P., Plested, J., Kodama, T., Moxon, E.R., Gordon, S., The class A macrophage scavenger receptor is a major pattern recognition receptor for Neisseria meningitidis which is independent of lipopolysaccharide and not required for secretory responses. *Infect. Immun.* **2002**. 70: 5346–5354.

39 Krieger, M., Stern, D.M., Series introduction: multiligand receptors and human disease. *J. Clin. Invest* **2001**. 108: 645–647.

40 Chamaillard, M., Girardin, S.E., Viala, J., Philpott, D.J., Nods, Nalps and Naip: intracellular regulators of bacterial-induced inflammation. *Cell Microbiol.* **2003**. 5: 581–592.

41 Girardin, S.E., Boneca, I.G., Viala, J., Chamaillard, M., Labigne, A., Thomas, G., Philpott, D.J., Sansonetti, P.J., Nod2 is a general sensor of peptidoglycan through muramyl dipeptide (MDP) detection. *J. Biol. Chem.* **2003**. 278: 8869–8872.

42 Inohara, N., Ogura, Y., Fontalba, A., Gutierrez, O., Pons, F., Crespo, J., Fukase, K., Inamura, S., Kusumoto, S., Hashimoto, M. et al., Host recognition of bacterial muramyl dipeptide mediated through NOD2. Implications for Crohn's disease. *J. Biol. Chem.* **2003**. 278: 5509–5512.

43 Trinchieri, G., Interleukin-12 and the regulation of innate resistance and adaptive immunity. *Nat. Rev. Immunol.* **2003**. 3: 133–146.

44 Gately, M.K., Renzetti, L.M., Magram, J., Stern, A.S., Adorini, L., Gubler, U.,

Presky, D.H., The interleukin-12/ interleukin-12-receptor system: role in normal and pathologic immune responses. [Review] [169 refs]. *Annu. Rev. Immunol.* **1998**. 16: 495–521.

45 Parham, C., Chirica, M., Timans, J., Vaisberg, E., Travis, M., Cheung, J., Pflanz, S., Zhang, R., Singh, K.P., Vega, F., et al., A receptor for the hetero-dimeric cytokine IL-23 is composed of IL-12Rbeta1 and a novel cytokine receptor subunit, IL-23R. *J. Immunol.* **2002**. 168: 5699–5708.

46 Devergne, O., Hummel, M., Koeppen, H., Le Beau, M.M., Nathanson, E.C., Kieff, E., Birkenbach, M., A novel interleukin-12 p40-related protein induced by latent Epstein-Barr virus infection in B lymphocytes. *J. Virol.* **1996**. 70: 1143–1153.

47 Pflanz, S., Timans, J.C., Cheung, J., Rosales, R., Kanzler, H., Gilbert, J., Hibbert, L., Churakova, T., Travis, M., Vaisberg, E. et al., IL-27, a heterodimeric cytokine composed of EBI3 and p28 protein, induces proliferation of naive CD4(+) T cells. *Immunity* **2002**. 16: 779–790.

48 Pflanz, S., Hibbert, L., Mattson, J., Rosales, R., Vaisberg, E., Bazan, J.F., Phillips, J.H., McClanahan, T.K., de Waal, M.R., Kastelein, R.A., WSX-1 and glycoprotein 130 constitute a signal-transducing receptor for IL-27. *J. Immunol.* **2004**. 172: 2225–2231.

49 Takeshita, F., Leifer, C.A., Gursel, I., Ishii, K.J., Takeshita, S., Gursel, M., Klinman, D.M., Cutting edge: Role of Toll-like receptor 9 in CpG DNA-induced activation of human cells. *J. Immunol.* **2001**. 167: 3555–3558.

50 Oppmann, B., Lesley, R., Blom, B., Timans, J.C., Xu, Y.M., Hunte, B., Vega, F., Yu, N., Wang, J., Singh, K. et al., Novel p19 protein engages IL-12p40 to form a cytokine: IL-23, with biological activities similar as well as distinct from IL-12. *Immunity* **2000**. 13: 715–725.

51 De Maeyer, E., Maeyer-Guignard, J., Type I interferons. *Int. Rev. Immunol.* **1998**. 17: 53–73.

52 Rogge, L., D'Ambrosio, D., Biffi, M., Penna, G., Minetti, L.J., Presky, D.H., Adorini, L., Sinigaglia, F., The role of

Stat4 in species-specific regulation of Th cell development by type I IFNs. *J. Immunol.* **1998**. 161: 6567–6574.

53 Dinarello, C.A., Novick, D., Puren, A.J., Fantuzzi, G., Shapiro, L., Muhl, H., Yoon, D.Y., Reznikov, L.L., Kim, S.H., Rubinstein, M., Overview of interleukin-18: more than an interferon-gamma inducing factor. *J. Leukoc. Biol.* **1998**. 63: 658–664.

54 Dunne, A., Ejdeback, M., Ludidi, P.L., O'Neill, L.A., Gay, N.J., Structural complementarity of Toll/interleukin-1 receptor domains in Toll-like receptors and the adaptors Mal and MyD88. *J. Biol. Chem.* **2003**. 278: 41443–41451.

55 Barbulescu, K., Becker, C., Schlaak, J.F., Schmitt, E., Meyer, z.B.K., Neurath, M.F., IL-12 and IL-18 differentially regulate the transcriptional activity of the human IFN-gamma promoter in primary CD4+ T lymphocytes. *J. Immunol.* **1998**. 160: 3642–3647.

56 Takeda, K., Tsutsui, H., Yoshimoto, T., Adachi, O., Yoshida, N., Kishimoto, T., Okamura, H., Nakanishi, K., Akira, S., Defective NK cell activity and Th1 response in IL-18-deficient mice. *Immunity* **1998**. 8: 383–390.

57 Luksch, C.R., Winqvist, O., Ozaki, M.E., Karlsson, L., Jackson, M.R., Peterson, P.A., Webb, S.R., Intercellular adhesion molecule-1 inhibits interleukin 4 production by naive T cells. *Proc. Natl. Acad. Sci. U. S. A.* **1999**. 96: 3023–3028.

58 Salomon, B., Bluestone, J.A., LFA-1 interaction with ICAM-1 and ICAM-2 regulates Th2 cytokine production. *J. Immunol.* **1998**. 161: 5138–5142.

59 Smits, H.H., de Jong, E.C., Schuitemaker, J.H., Geijtenbeek, T.B., van Kooyk, Y., Kapsenberg, M.L., Wierenga, E.A., Intracellular Adhesion Molecule-1-LFA-1 ligation favors human Th1 development. *J. Immunol.* **2002**. 168: 1710–1716.

60 d'Ostiani, C.F., Del Sero, G., Bacci, A., Montagnoli, C., Spreca, A., Mencacci, A., Ricciardi-Castagnoli, P., Romani, L., Dendritic cells discriminate between yeasts and hyphae of the fungus Candida albicans. Implications for initiation of T helper cell immunity in vitro and in vivo. *J. Exp. Med.* **2000**. 191: 1661–1674.

61 Kelleher, P., Maroof, A., Knight, S.C., Retrovirally induced switch from production of IL-12 to IL-4 in dendritic cells. *Eur. J. Immunol.* **1999**. 29: 2309–2318.

62 Smits, H.H., van Beelen, A.J., Hessle, C., Westland, R., de Jong, E., Soeteman, E., Wold, A., Wierenga, E.A., Kapsenberg, M.L., Commensal Gram-negative bacteria prime human dendritic cells for enhanced IL-23 and IL-27 expression and enhanced Th1 development. *Eur. J. Immunol.* **2004**. 34: 1371–1380.

63 Jankovic, D., Kullberg, M.C., Noben-Trauth, N., Caspar, P., Paul, W.E., Sher, A., Single cell analysis reveals that IL-4 receptor/Stat6 signaling is not required for the in vivo or in vitro development of CD4+ lymphocytes with a Th2 cytokine profile. *J. Immunol.* **2000**. 164: 3047–3055.

64 de Jong, E.C., Vieira, P.L., Kalinski, P., Schuitemaker, J.H., Tanaka, Y., Wierenga, E.A., Yazdanbakhsh, M., Kapsenberg, M.L., Microbial compounds selectively induce Th1 cell-promoting or Th2 cell-promoting dendritic cells with diverse Th cell-polarizing signals. *J. Immunol.* **2002**. 168: 1704–1709.

65 MacDonald, A.S., Straw, A.D., Bauman, B., Pearce, E.J., CD8– dendritic cell activation status plays an integral role in influencing Th2 response development. *J. Immunol.* **2001**. 167: 1982–1988.

66 Balic, A., Harcus, Y., Holland, M.J., Maizels, R.M., Selective maturation of dendritic cells by Nippostrongylus brasiliensis-secreted proteins drives Th2 immune responses. *Eur. J. Immunol.* **2004**. 34: 3047–3059.

67 Flynn, S., Toellner, K.M., Raykundalia, C., Goodall, M., Lane, P., CD4 T cell cytokine differentiation: the B cell activation molecule, OX40 ligand, instructs CD4 T cells to express interleukin 4 and upregulates expression of the chemokine receptor, Blr-1. *J. Exp. Med.* **1998**. 188: 297–304.

68 Ohshima, Y., Yang, L.P., Uchiyama, T., Tanaka, Y., Baum, P., Sergerie, M., Hermann, and Delespesse, G., OX40 costimulation enhances interleukin-4 (IL-4) expression at priming and promotes the differentiation of naive human CD4(+) T cells into high IL-4-producing effectors. *Blood* **1998**. 92: 3338–3345.

69 Joss, A., Akdis, M., Faith, A., Blaser, K., Akdis, C.A., IL-10 directly acts on T cells by specifically altering the CD28 co-stimulation pathway. *Eur. J. Immunol.* **2000**. 30: 1683–1690.

70 Levings, M.K., Sangregorio, R., Galbiati, F., Squadrone, S., de Waal, M.R., Roncarolo, M.G., IFN-alpha and IL-10 induce the differentiation of human type 1 T regulatory cells. *J. Immunol.* **2001**. 166: 5530–5539.

71 de Waal, M.R., Haanen, J., Spits, H., Roncarolo, M.G., te, V.A., Figdor, C., Johnson, K., Kastelein, R., Yssel, H., de Vries, J.E., Interleukin 10 (IL-10) and viral IL-10 strongly reduce antigen-specific human T cell proliferation by diminishing the antigen-presenting capacity of monocytes via downregula-tion of class II major histocompatibility complex expression. *J. Exp. Med.* **1991**. 174: 915–924.

72 Steinbrink, K., Wolfl, M., Jonuleit, H., Knop, J., Enk, A.H., Induction of toler-ance by IL-10-treated dendritic cells. *J. Immunol.* **1997**. 159: 4772–4780.

73 Hoyne, G.F., Le, R., I, Corsin-Jimenez, M., Tan, K., Dunne, J., Forsyth, L.M., Dallman, M.J., Owen, M.J., Ish-Horowicz, D., Lamb, J.R., Serrate1-induced notch signalling regulates the decision between immunity and tolerance made by peripheral CD4(+) T cells. *Int. Immunol.* **2000**. 12: 177–185.

74 Yvon, E.S., Vigouroux, S., Rousseau, R.F., Biagi, E., Amrolia, P., Dotti, G., Wagner, H.J., Brenner, M.K., Over-expression of the Notch ligand, Jagged-1, induces alloantigen-specific human regulatory T cells. *Blood* **2003**. 102: 3815–3821.

75 Chang, C.C., Ciubotariu, R., Manavalan, J.S., Yuan, J., Colovai, A.I., Piazza, F., Lederman, S., Colonna, M., Cortesini, R., Dalla-Favera, R. et al., Tolerization of dendritic cells by T(S) cells: the crucial role of inhibitory receptors ILT3 and ILT4. *Nat. Immunol.* **2002**. 3: 237–243.

76 Manavalan, J.S., Rossi, P.C., Vlad, G., Piazza, F., Yarilina, A., Cortesini, R., Mancini, D., Suciu-Foca, N., High expression of ILT3 and ILT4 is a general

feature of tolerogenic dendritic cells. *Transpl. Immunol.* **2003**. 11: 245–258.

77 Khoury, S.J., Sayegh, M.H., The roles of the new negative T cell costimulatory pathways in regulating autoimmunity. *Immunity* **2004**. 20: 529–538.

78 Dong, H., Zhu, G., Tamada, K., Chen, L., B7-H1, a third member of the B7 family, co-stimulates T-cell proliferation and interleukin-10 secretion. *Nat. Med.* **1999**. 5: 1365–1369.

79 Yamazaki, T., Akiba, H., Iwai, H., Matsuda, H., Aoki, M., Tanno, Y., Shin, T., Tsuchiya, H., Pardoll, D.M., Okumura, K. et al., Expression of programmed death 1 ligands by murine T cells and APC. *J. Immunol.* **2002**. 169: 5538–5545.

80 Chapoval, A.I., Ni, J., Lau, J.S., Wilcox, R.A., Flies, D.B., Liu, D., Dong, H., Sica, G.L., Zhu, G., Tamada, K., Chen, L., B7-H3: a costimulatory molecule for T cell activation and IFN-gamma production. *Nat. Immunol.* **2001**. 2: 269–274.

81 Nishimura, H., Nose, M., Hiai, H., Minato, N., Honjo, T., Development of lupus-like autoimmune diseases by disruption of the PD-1 gene encoding an ITIM motif-carrying immunoreceptor. *Immunity* **1999**. 11: 141–151.

82 Tivol, E.A., Borriello, F., Schweitzer, A.N., Lynch, W.P., Bluestone, J.A., Sharpe, A.H., Loss of CTLA-4 leads to massive lymphoproliferation and fatal multiorgan tissue destruction, revealing a critical negative regulatory role of CTLA-4. *Immunity* **1995**. 3: 541–547.

83 Waterhouse, P., Penninger, J.M., Timms, E., Wakeham, A., Shahinian, A., Lee, K.P., Thompson, C.B., Griesser, H., Mak, T.W., Lymphoproliferative disorders with early lethality in mice deficient in Ctla-4. *Science* **1995**. 270: 985–988.

84 Suh, W.K., Gajewska, B.U., Okada, H., Gronski, M.A., Bertram, E.M., Dawicki, W., Duncan, G.S., Bukczynski, J., Plyte, S., Elia, A. et al., The B7 family member B7-H3 preferentially down-regulates T helper type 1-mediated immune responses. *Nat. Immunol.* **2003**. 4: 899–906.

85 Prasad, D.V., Richards, S., Mai, X.M., Dong, C., B7S1, a novel B7 family member that negatively regulates T cell activation. *Immunity* **2003**. 18: 863–873.

86 Takeuchi, O., Kawai, T., Muhlradt, P.F., Morr, M., Radolf, J.D., Zychlinsky, A., Takeda, K., Akira, S., Discrimination of bacterial lipoproteins by Toll-like receptor 6. *Int. Immunol.* **2001**. 13: 933–940.

87 Takeuchi, O., Sato, S., Horiuchi, T., Hoshino, K., Takeda, K., Dong, Z., Modlin, R.L., Akira, S., Cutting edge: role of Toll-like receptor 1 in mediating immune response to microbial lipoproteins. *J. Immunol.* **2002**. 169: 10–14.

88 Re, F., Strominger, J.L., Toll-like receptor 2 (TLR2) and TLR4 differentially activate human dendritic cells. *J. Biol Chem.* **2001**. 276: 37692–37699.

89 Yoshida, H., Hamano, S., Senaldi, G., Covey, T., Faggioni, R., Mu, S., Xia, M., Wakeham, A.C., Nishina, H., Potter, J., Saris, C.J., Mak, T.W., WSX-1 is required for the initiation of Th1 responses and resistance to L. major infection. *Immunity* **2001**. 15: 569–578.

90 Yoshimura, A., Lien, E., Ingalls, R.R., Tuomanen, E., Dziarski, R., Golenbock, D., Recognition of Gram-positive bacterial cell wall components by the innate immune system occurs via Toll-like receptor 2. *J. Immunol.* **1999**. 163: 1–5.

91 Underhill, D.M., Ozinsky, A., Hajjar, A.M., Stevens, A., Wilson, C.B., Bassetti, M., Aderem, A., The Toll-like receptor 2 is recruited to macrophage phagosomes and discriminates between pathogens. *Nature* **1999**. 401: 811–815.

92 Weigt, H., Muhlradt, P.F., Emmendorffer, A., Krug, N., Braun, A., Synthetic mycoplasma-derived lipopeptide MALP-2 induces maturation and function of dendritic cells. *Immunobiology* **2003**. 207: 223–233.

93 Qi, H., Denning, T.L., Soong, L., Differential induction of interleukin-10 and interleukin-12 in dendritic cells by microbial toll-like receptor activators and skewing of T-cell cytokine profiles. *Infect. Immun.* **2003**. 71: 3337–3342.

94 van der Kleij D., Latz, E., Brouwers, J.F., Kruize, Y.C., Schmitz, M., Kurt-Jones, E.A., Espevik, T., de Jong, E.C.,

Kapsenberg, M.L., Golenbock, D.T. et al., A novel host-parasite lipid cross-talk. Schistosomal lyso-phosphatidylserine activates Toll-like receptor 2 and affects immune polarization. *J. Biol. Chem.* **2002**. 277: 48122–48129.

95 Alexopoulou, L., Holt, A.C., Medzhitov, R., Flavell, R.A., Recognition of double-stranded RNA and activation of NF-kappaB by Toll- like receptor 3. *Nature* **2001**. 413: 732–738.

96 Aksoy, E., Zouain, C.S., Vanhoutte, F., Fontaine, J., Pavelka, N., Thieblemont, N., Willems, F., Ricciardi-Castagnoli, P., Goldman, M., Capron, M. et al., Double-stranded RNAs from the helminth parasite Schistosoma activate TLR3 in dendritic cells. *J. Biol. Chem.* **2005**. 280: 277–283.

97 Schulz, O., Diebold, S.S., Chen, M., Naslund, T.I., Nolte, M.A., Alexopoulou, L., Azuma, Y.T., Flavell, R.A., Liljestrom, P., Reis E Sousa, Toll-like receptor 3 promotes cross-priming to virus-infected cells. *Nature* **2005**. 433: 887–892.

98 Cella, M., Salio, M., Sakakibara, Y., Langen, H., Julkunen, I., Lanzavecchia, A., Maturation, activation, and protection of dendritic cells induced by double-stranded RNA. *J. Exp. Med.* **1999**. 189: 821–829.

99 Diebold, S.S., Montoya, M., Unger, H., Alexopoulou, L., Roy, P., Haswell, L.E., Al Shamkhani, A., Flavell, R., Borrow, P., Reis E Sousa, Viral infection switches non-plasmacytoid dendritic cells into high interferon producers. *Nature* **2003**. 424: 324–328.

100 Poltorak, A., He, X., Smirnova, I., Liu, M.Y., Huffel, C.V., Du, X., Birwell, D., Alejos, E., Silva, M., Galanos, C. et al., Defective LPS signalling in C3H/Hej and C57BL/10ScCR mice: mutations in Tlr4gene. *Science* **1998**. 282: 2085–2088.

101 Boonstra, A., Asselin-Paturel, C., Gilliet, M., Crain, C., Trinchieri, G., Liu, Y.-J., O'Garra, A., Felxibility of mouse classical and plasmacytoid-derived dendritic cells in directing T helper type 1 and 2 cell development: dependency on antigen dose and differential Toll-like receptor ligation. *J. Exp. Med.* **2003**. 197: 101–109.

102 Vieira, P.L., de Jong, E.C., Wierenga, E.A., Kapsenberg, M.L., Kalinski, P., Development of Th1-inducing capacity in myeloid dendritic cells requires environmental instruction. *J. Immunol.* **2000**. 164: 4507–4512.

103 Pulendran, B., Kumar, P., Cutler, C.W., Mohamadzadeh, M., Van Dyke, T., Banchereau, J., Lipopolysaccharides from distinct pathogens induce different classes of immune responses in vivo. *J. Immunol.* **2001**. 167: 5067–5076.

104 Goodridge, H.S., Marshall, F.A., Else, K.J., Houston, K.M., Egan, C., Al Riyami, L., Liew, F.Y., Harnett, W., Harnett, M.M., Immunomodulation via novel use of TLR4 by the filarial nematode phos-phorylcholine-containing secreted product, ES-62. *J. Immunol.* **2005**. 174: 284–293.

105 Whelan, M., Harnett, M.M., Houston, K.M., Patel, V., Harnett, W., Rigley, K.P., A filarial nematode-secreted product signals dendritic cells to acquire a pheno-type that drives development of Th2 cells. *J. Immunol.* **2000**. 164: 6453–6460.

106 Hayashi, F., Smith, K.D., Ozinsky, A., Hawn, T.R., Yi, E.C., Goodlett, D.R., Eng, J.K., Akira, S., Underhill, D.M., Aderem, A., The innate immune response to bacterial flagellin is mediated by Toll- like receptor 5. *Nature* **2001**. 410: 1099–1103.

107 Agrawal, S., Agrawal, A., Doughty, B., Gerwitz, A., Blenis, J., Van Dyke, T., Pulendran, B., Cutting edge: different Toll-like receptor agonists instruct dendritic cells to induce distinct Th responses via differential modulation of extracellular signal-regulated kinase-mitogen-activated protein kinase and c-Fos. *J. Immunol.* **2003**. 171: 4984–4989.

108 Means, T.K., Hayashi, F., Smith, K.D., Aderem, A., Luster, A.D., The Toll-like receptor 5 stimulus bacterial flagellin induces maturation and chemokine production in human dendritic cells. *J. Immunol.* **2003**. 170: 5165–5175.

109 Didierlaurent, A., Ferrero, I., Otten, L.A., Dubois, B., Reinhardt, M., Carlsen, H., Blomhoff, R., Akira, S., Kraehenbuhl, J.P., Sirard, J.C., Flagellin promotes myeloid differentiation factor 88-dependent development of Th2-type response. *J. Immunol.* **2004**. 172: 6922–6930.

110 Cunningham, A.F., Khan, M., Ball, J., Toellner, K.M., Serre, K., Mohr, E., MacLennan, I.C., Responses to the soluble flagellar protein FliC are Th2, while those to FliC on Salmonella are Th1. *Eur. J. Immunol.* **2004**. 34: 2986–2995.

111 Ito, T., Amakawa, R., Kaisho, T., Hemmi, H., Tajima, K., Uehira, K., Ozaki, Y., Tomizawa, H., Akira, S., Fukuhara, S., Interferon-a and interleukin-12 are induced differentially by toll-like receptor 7 ligands in human blood dendritic cell subsets. *J. Exp. Med.* **2002**. 195: 1507–1512.

112 Hemmi, H., Kaisho, T., Takeuchi, O., Sato, S., Sanjo, H., Hoshino, K., Horiuchi, T., Tomizawa, H., Takeda, K., Akira, S., Small anti-viral compounds activate immune cells via the TLR7 MyD88-dependent signaling pathway. *Nat. Immunol.* **2002**. 3: 196–200.

113 Jurk, M., Heil, F., Vollmer, J., Schetter, C., Krieg, A.M., Wagner, H., Lipford, G., Bauer, S., Human TLR7 or TLR8 independently confer responsiveness to the antiviral compound R-848. *Nat. Immunol.* **2002**. 3: 499.

114 Wagner, T.L., Ahonen, C.L., Couture, A.M., Gibson, S.J., Miller, R.L., Smith, R.M., Reiter, M.J., Vasilakos, J.P., Tomai, M.A., Modulation of TH1 and TH2 cytokine production with the immune response modifiers, R-848 and imiquimod. *Cell Immunol.* **1999**. 191: 10–19.

115 Brugnolo, F., Sampognaro, S., Liotta, F., Cosmi, L., Annunziato, F., Manuelli, C., Campi, P., Maggi, E., Romagnani, S., Parronchi, P., The novel synthetic immune response modifier R-848 (Resiquimod) shifts human allergen-specific CD4+ TH2 lymphocytes into IFN-gamma-producing cells. *J. Allergy Clin. Immunol.* **2003**. 111: 380–388.

116 Edwards, A.D., Diebold, S.S., Slack, E.M., Tomizawa, H., Hemmi, H., Kaisho, T., Akira, S., Reis E Sousa, Toll-like receptor expression in murine DC subsets: lack of TLR7 expression by CD8 alpha+ DC correlates with unresponsiveness to imidazoquinolines. *Eur. J. Immunol.* **2003**. 33: 827–833.

117 Gilliet, M., Boonstra, A., Paturel, C., Antonenko, S., Xu, X.L., Trinchieri, G., O'Garra, A., Liu, Y.J., The Development of Murine Plasmacytoid Dendritic Cell Precursors Is Differentially Regulated by FLT3-ligand and Granulocyte/Macrophage Colony-Stimulating Factor. *J. Exp. Med.* **2002**. 195: 953–958.

118 Krug, A., Towarowski, A., Britsch, S., Rothenfusser, S., Hornung, V., Bals, R., Giese, T., Engelmann, H., Endres, S., Krieg, A.M. et al., Toll-like receptor expression reveals CpG DNA as a unique microbial stimulus for plasmacytoid dendritic cells which synergizes with CD40 ligand to induce high amounts of IL-12. *Eur. J. Immunol.* **2001**. 31: 3026–3037.

119 Dzionek, A., Inagaki, Y., Okawa, K., Nagafune, J., Rock, J., Sohma, Y., Winkels, G., Zysk, M., Yamaguchi, Y., Schmitz, J., Plasmacytoid dendritic cells: from specific surface markers to specific cellular functions. *Hum. Immunol.* **2002**. 63: 1133–1148.

120 Asselin-Paturel, C., Boonstra, A., Dalod, M., Durand, I., Yessaad, N., Dezutter-Dambuyant, C., Vicari, A., O'Garra, A., Biron, C., Briere, F. et al., Mouse type I IFN-producing cells are immature APCs with plasmacytoid morphology. *Nat. Immunol.* **2001**. 2: 1144–1150.

121 Lazarus, R., Raby, B.A., Lange, C., Silverman, E.K., Kwiatkowski, D.J., Vercelli, D., Klimecki, W.J., Martinez, F.D., Weiss, S.T., TOLL-like receptor 10 genetic variation is associated with asthma in two independent samples. *Am. J. Respir. Crit Care Med.* **2004**. 170: 594–600.

122 Zhang, D., Zhang, G., Hayden, M.S., Greenblatt, M.B., Bussey, C., Flavell, R.A., Ghosh, S., A toll-like receptor that prevents infection by uropathogenic bacteria. *Science* **2004**. 303: 1522–1526.

123 Reis E Sousa, Yap, G., Schulz, O., Rogers, N., Schito, M., Aliberti, J., Hieny, S., Sher, A., Paralysis of dendritic cell IL-12 production by microbial products prevents infection-induced immuno-pathology. *Immunity* **1999**. 11: 637–647.

124 Park, A.Y., Hondowicz, B.D., Scott, P., IL-12 is required to maintain a Th1

response during Leishmania major infection. *J. Immunol.* **2000**. 165: 896–902.

125 de Jong, R., Altare, F., Haagen, I.A., Elferink, D.G., Boer, T., Breda Vriesman, P.J., Kabel, P.J., Draaisma, J.M., van Dissel, J.T., Kroon, F.P. et al., Severe mycobacterial and Salmonella infections in interleukin-12 receptor-deficient patients. *Science* **1998**. 280: 1435–1438.

126 Fieschi, C., Dupuis, S., Catherinot, E., Feinberg, J., Bustamante, J., Breiman, A., Altare, F., Baretto, R., Le Deist, F., Kayal, S. et al., Low penetrance, broad resistance, and favorable outcome of interleukin 12 receptor beta1 deficiency: medical and immunological implications. *J. Exp. Med.* **2003**. 197: 527–535.

127 Oxenius, A., Karrer, U., Zinkernagel, R.M., Hengartner, H., IL-12 is not required for induction of type 1 cytokine responses in viral infections. *J. Immunol.* **1999**. 162: 965–973.

128 Schijns, V.E., Haagmans, B.L., Wierda, C.M., Kruithof, B., Heijnen, IA, Alber, G., Horzinek, M.C., Mice lacking IL-12 develop polarized Th1 cells during viral infection. *J. Immunol.* **1998**. 160: 3958–3964.

129 Thomas, P.G., Carter, M.R., Atochina, O., Da'Dara, A.A., Piskorska, D., McGuire, E., Harn, D.A., Maturation of dendritic cell 2 phenotype by a helminth glycan uses a Toll-like receptor 4-dependent mechanism. *J. Immunol.* 2003. 171: 5837–5841.

130 Faveeuw, C., Mallevaey, T., Paschinger, K., Wilson, I.B., Fontaine, J., Mollicone, R., Oriol, R., Altmann, F., Lerouge, P., Capron, M. et al., Schistosome N-glycans containing core alpha 3-fucose and core beta 2-xylose epitopes are strong inducers of Th2 responses in mice. *Eur. J. Immunol.* **2003**. 33: 1271–1281.

131 Bergman, M.P., Engering, A., Smits, H.H., van Vliet, S.J., van Bodegraven, A.A., Wirth, H.P., Kapsenberg, M.L., Vandenbroucke-Grauls, C.M., van Kooyk, Y., Appelmelk, B.J., Helicobacter pylori modulates the T helper cell 1/T helper cell 2 balance through phase-variable interaction between lipopolysaccharide and DC-SIGN. *J. Exp. Med.* **2004**. 200: 979–990.

132 Akiba, H., Miyahira, Y., Atsuta, M., Takeda, K., Nohara, C., Futagawa, T., Matsuda, H., Aoki, T., Yagita, H., Okumura, K., Critical contribution of OX40 ligand to T helper cell type 2 differentiation in experimental leishmaniasis. *J. Exp. Med.* **2000**. 191: 375–380.

133 Steinman, R.M., Hawiger, D., Nussenzweig, M.C., Tolerogenic dendritic cells. *Annu. Rev. Immunol.* **2003**. 21: 685–711.

134 McGuirk, P., McCann, C., Mills, K.H., Pathogen-specific T regulatory 1 cells induced in the respiratory tract by a bacterial molecule that stimulates interleukin 10 production by dendritic cells: a novel strategy for evasion of protective T helper type 1 responses by Bordetella pertussis. *J. Exp. Med.* **2002**. 195: 221–231.

135 Urban, B.C., Ferguson, D.J., Pain, A., Willcox, N., Plebanski, M., Austyn, J.M., Roberts, D.J., Plasmodium falciparum-infected erythrocytes modulate the maturation of dendritic cells. *Nature* **1999**. 400: 73–77.

136 Mendez, S., Reckling, S.K., Piccirillo, C.A., Sacks, D., Belkaid, Y., Role for CD4(+) CD25(+) regulatory T cells in reactivation of persistent leishmaniasis and control of concomitant immunity. *J. Exp. Med.* **2004**. 200: 201–210.

137 Urban, B.C., Willcox, N., Roberts, D.J., A role for CD36 in the regulation of dendritic cell function. *Proc. Natl. Acad. Sci. U. S. A.* **2001**. 98: 8750–8755.

138 Dolganiuc, A., Kodys, K., Kopasz, A., Marshall, C., Do, T., Romics, L., Jr., Mandrekar, P., Zapp, M., Szabo, G., Hepatitis C virus core and nonstructural protein 3 proteins induce pro- and anti-inflammatory cytokines and inhibit dendritic cell differentiation. *J. Immunol.* 2003. 170: 5615–5624.

139 Kanto, T., Hayashi, N., Takehara, T., Tatsumi, T., Kuzushita, N., Ito, A., Sasaki, Y., Kasahara, A., Hori, M., Impaired allostimulatory capacity of peripheral blood dendritic cells recovered from hepatitis C virus-infected individuals. *J. Immunol.* **1999**. 162: 5584–5591.

140 MacDonald, A.J., Duffy, M., Brady, M.T., McKiernan, S., Hall, W., Hegarty, J., Curry, M., Mills, K.H., CD4 T helper type 1 and regulatory T cells induced against the same epitopes on the core protein in hepatitis C virus-infected persons. *J. Infect. Dis.* **2002**. 185: 720–727.

141 Mathieu, C., Adorini, L., The coming of age of 1, 25-dihydroxyvitamin D(3) analogs as immunomodulatory agents. *Trends Mol. Med.* **2002**. 8: 174–179.

142 Giacomini, E., Iona, E., Ferroni, L., Miettinen, M., Fattorini, L., Orefici, G., Julkunen, I., Coccia, E.M., Infection of human macrophages and dendritic cells with Mycobacterium tuberculosis induces a differential cytokine gene expression that modulates T cell response. *J. Immunol.* **2001**. 166: 7033–7041.

143 Atochina, O., Daly-Engel, T., Piskorska, D., McGuire, E., Harn, D.A., A schistosome-expressed immuno-modulatory glycoconjugate expands peritoneal Gr1(+) macrophages that suppress naive CD4(+) T cell proliferation via an IFN-gamma and nitric oxide-dependent mechanism. *J. Immunol.* **2001**. 167: 4293–4302.

144 Satoguina, J., Mempel, M., Larbi, J., Badusche, M., Loliger, C., Adjei, O., Gachelin, G., Fleischer, B., Hoerauf, A., Antigen-specific T regulatory-1 cells are associated with immunosuppression in a chronic helminth infection (onchocerciasis). *Microbes. Infect.* **2002**. 4: 1291–1300.

145 Hesse, M., Piccirillo, C.A., Belkaid, Y., Prufer, J., Mentink-Kane, M., Leusink, M., Cheever, A.W., Shevach, E.M., Wynn, T.A., The pathogenesis of schistosomiasis is controlled by cooperating IL-10-producing innate effector and regulatory T cells. *J. Immunol.* **2004**. 172: 3157–3166.

146 Ross, P.J., Lavelle, E.C., Mills, K.H., Boyd, A.P., Adenylate cyclase toxin from Bordetella pertussis synergizes with lipopolysaccharide to promote innate interleukin-10 production and enhances the induction of Th2 and regulatory T cells. *Infect. Immun.* **2004**. 72: 1568–1579.

147 Lavelle, E.C., McNeela, E., Armstrong, M.E., Leavy, O., Higgins, S.C., Mills, K.H., Cholera toxin promotes the induction of regulatory T cells specific for bystander antigens by modulating dendritic cell activation. *J. Immunol.* **2003**. 171: 2384–2392.

148 Lebre, M.C., Antons, J.C., Kalinski, P., Schuitemaker, J.H., van Capel, T.M., Kapsenberg, M.L., de Jong, E.C., Double-stranded RNA-exposed human keratino-cytes promote Th1 responses by induc-ing a Type-1 polarized phenotype in dendritic cells: role of keratinocyte-derived tumor necrosis factor alpha, type I interferons, and interleukin-18. *J. Invest Dermatol.* **2003**. 120: 990–997.

149 Pasare, C., Medzhitov, R., Toll-dependent control mechanisms of CD4 T cell activation. *Immunity* **2004**. 21: 733–741.

150 Sporri, R., Reis E Sousa, Inflammatory mediators are insufficient for full dendritic cell activation and promote expansion of CD4+ T cell populations lacking helper function. *Nat. Immunol.* **2005**. 6: 163–170.

VIII
CTL Priming and Crosspresentation

22
Crossprocessing and Crosspresentation

Mojca Škoberne and Nina Bhardwaj

22.1
Introduction

CD8$^+$ T cells are of prime importance for the recognition and elimination of tumor cells and cells that are infected by viruses or intracellular bacteria. They recognize short 8–10 AA residues bound to MHC class I molecules expressed on the surface of the majority of nucleated cells. MHC class I molecules display myriads of peptides that arise following the entry of cell-derived proteins into the endogenous (classical) antigen processing pathway. The initial activation of CD8$^+$ T cells is dependent upon professional antigen presenting cells (APCs) such as dendritic cells (DCs). APCs acquire antigens in the periphery and travel to secondary lymphoid organs where they stimulate T cells. However, as certain infectious agents may not target APCs, or APCs may rapidly succumb to infection, alternative mechanisms must be in place to induce protective CD8$^+$ T-cell responses. The same applies to antitumor immune responses where tumor cells by themselves are poorly functional as APCs.

It is now well established that APCs can internalize parts of- or whole infected cells and crosspresent the encoded antigens to T cells, a phenomenon termed "crosspriming", first coined in 1976 by Michael Bevan [1, 2]. Bevan injected mice with minor histocompatibility mismatched cells and observed that these cells prime host MHC I restricted cytotoxic T lymphocyte (CTL) responses to the exogenous antigen. These experiments were the first of several to establish the importance of crosspresentation in the induction of immune responses to viral and tumor antigens, as well as tolerance to self antigens.

Handbook of Dendritic Cells. Biology, Diseases, and Therapies.
Edited by M. B. Lutz, N. Romani, and A. Steinkasserer.
Copyright © 2006 WILEY-VCH Verlag GmbH & Co. KGaA, Weinheim
ISBN: 3-527-31109-2

22.2
Acquisition of Antigens for Crosspresentation

22.2.1
Cells that Crosspresent

The identity of the cells that crosspresent antigens is a matter of long and ongoing debate. Several cell types have been shown to crosspresent antigens including endothelial cells, B cells, macrophages and DCs, however only selected cells are involved in crosspriming *in vivo* as we discuss below [3–10].

Initial studies by Bevan et al. showed that T cells recognize minor histocompatibility antigens presented on the surface of splenocytes [1]. In the early 1990s Rock et al. showed that in spleens only certain cells can process and present extracellular antigens in the context of MHC class I molecules. These cells express MHC class II molecules and Fc receptors [11] and correspond to a macrophage/dendritic cell population. In the following years, as DCs were further characterized and their phagocytic capacity was confirmed [12, 13], their role in crosspresentation became more convincing. Shen et al. [14] addressed the dispute of whether DCs or contaminating macrophages in the DC cultures are responsible for crosspresentation by transducing GM-CSF into bone marrow cell cultures that were later immortalized. These cells expressed many DC characteristics and most importantly, they were able to present exogenous antigens to CTLs.

In more definitive studies, Kurts et al. [15] generated mice that selectively expressed MHC I molecules only in DCs, along with the ovalbumin (OVA) antigen under control of the rat insulin promoter (RIP), thereby restricting OVA expression to nonlymphoid tissue. When mice were injected with OVA-specific OT-I T cells, the T cells underwent substantial expansion in lymph nodes. These experiments indicated that DCs acquired OVA exogenously from the periphery, and crosspresented it to T cells within draining lymph nodes.

Ultimate confirmation of the primary role of DCs in crosspresentation (at least for the cell-associated antigens) came from Jung et al. [16]. The authors constructed a system where murine DCs were selectively rendered susceptible to killing by diptheria toxin. This approach enabled inducible *in vivo* depletion of DCs. Mice were adoptively transferred with OVA-specific OT-I cells, DCs were depleted and mice were injected with MHC I deficient splenocytes, loaded with OVA by osmotic shock. In such experimental conditions, but not in control mice where DCs were not depleted, stimulation of OT-I cells was greatly impaired.

In mice, however, DCs are not a uniform cell population. Based on the expression of CD8α, CD11b, CD4 and Gr-1, at least 6 different subsets have been described to date [17, 18]. The presentation of antigens by different DC subsets in mice seems complex and dependent on antigen form and route of inoculation. Soluble antigens injected intravenously (i.v.) can be presented to CD4+ T cells by CD8− DCs but only CD8+ DCs can crosspresent it to CD8+ T cells [19]. OVA-immune complexes (ICs) on the other hand can be presented to CD8+ T cells by both, CD8+ and CD8− DCs. However, only the CD8− DCs crosspresent ICs to CD4+ T cells [20].

The loss of all three Fc receptors impairs the presentation to CD8[+] T cells by CD8[−] DCs but not by CD8[+] DCs[−] perhaps because these DCs may acquire ICs via complement fixation and/or they use a mechanism similar to uptake of soluble antigen [20]. Crosspresentation of cell-associated antigens is also dependent upon distinct DC subsets. When β2m deficient OVA-loaded splenocytes were injected into mice, only the CD8[+] subset of DCs was capable of crosspresenting the antigen to CD8[+] T cells [21]. The DCs of the CD8α[+] subset are superior in comparison to the ones of CD8α[−] subsets in internalization of dead cells *in vitro* and *in vivo* [22, 23], but not latex or bacteria [22]. The mechanistic basis for this is not known, however. While the CD8α[+] DCs express slightly higher levels of apoptotic cell receptor CD36, the receptor appears to be dispensable for crosspresentation [24, 25].

The role of different DC subsets in cross-tolerance is equally complex. Initially it was shown that a bone marrow-derived cells were responsible for crosspresentation of pancreatic islet cell-associated OVA, and subsequent deletion of self-reactive CD8[+] T cells [26]. In the non-obese diabetic mice (NOD), diabetes is mediated through destruction of β islet cells by CD4[+] and CD8[+] T cells. When apoptosis of islet cells was deliberately induced, the CD11b[+] DC subset was shown to be responsible for increased presentation of the β islet antigens in draining lymph nodes. The treatment resulted in decreased CD4[+] T-cell responses to islet cells, development of regulatory T cells and ultimately, protection of mice to diabetes [27]. Belz et al. showed that expression of OVA or glycoprotein B under RIP, led to their presentation by CD8α[+] CD11b[−] DCs and deletional tolerance of CD8[+] T cells [28]. In a different model, gastric parietal cell-specific H[+]/K[+]-ATPase that is recognized by autoreactive T cells was constitutively processed by both CD8α high and CD8α low gastric DCs and transported to draining gastric lymph nodes [29]. Of note, the route of antigen inoculation can influence which DC subtype acquires antigens. Chung et al. demonstrated that intravenously injected OVA was preferentially internalized by CD8α[+] DCs, whereas intragastric injection of the protein resulted in uptake by CD8α[−]CD11b[+] DCs [30].

The identity of the crosspresenting cell is further complicated by observations that both immunity and tolerance may be induced after one DC subset migrates from the periphery and transfers the antigen to a second, lymph-node resident DC subset [31, 32].

How these differential roles of various mice DC subsets relate to humans is not known to date. In humans only two major DC subsets have been identified so far: myeloid (mDC) and plasmacytoid (pDC) DCs. Human mDCs were shown to efficiently take up and present apoptotic and necrotic tissues [33, 34] and to be involved in priming as well as tolerization [35–41]. On the contrary, human pDCs are primarily known as professional interferon α (IFNα) producing cells (IPC), and associated with innate antiviral responses. Recently, their role in adaptive immune responses is becoming better recognized [42–44], although their ability to crosspresent antigens is poorly explored. Studies indicate, that compared to mDCs they are inferior in the uptake of cellular material and may express different sets of antigen degrading enzymes such as cathepsins [44–47]. A recent study by Schnurr et al. showed that while CD1c[+] DCs efficiently crosspresent epitopes of full length NY-

ESO 1 protein in the context of class I MHC molecules, pDCs could only present MHC class II restricted epitopes [48]. Future studies will be needed to draw the final parallels between human and murine DC subtypes and to define their roles in the crosspresentation of antigens.

22.2.2
Sources of Antigens and Receptors involved in Crosspresentation

Phagocytes may acquire antigens for crosspresentation by macropinocytosis, phagocytosis or receptor-mediated endocytosis, however not all of them are equally efficient [49]. Sources of antigens that enter the crosspresentation pathway are equally diverse and include soluble antigens, immune complexes, antigens coupled to latex beads, heat-shock protein (HSP) bound peptides, cell-associated antigens, exosomes or even synthetic structures such as microspheres [50, 51]. (Antigen sources and receptors involved in their recognition are summarized in Table 22.1).

Below we discuss the relevance and mechanisms underlying the uptake of several of the above.

22.2.2.1 Apoptotic Cells
Apoptosis is a physiological form of cell death during embryogenesis, tissue turnover and following infection. Apoptotic cells are probably one of the most notorious "natural" sources of antigens for crosspresentation. The antigens packed within the cells result in crosspresentation that is several hundred-fold more efficient in comparison to soluble antigens [49], presumably because apoptotic cells are recognized by macrophages and DC via an array of receptors [52]. These include type I and type II integrins, phosphatidyl serine receptor (PSR), scavenger receptor CD36 and Lox-1, to name only a few. Macrophages and DCs however may not play equivalent roles in the uptake of apoptotic cells. Macrophages primarily function as apoptotic cell scavengers and thus prevent their secondary necrosis, release of dying cell contents and inflammation that could lead to induction of autoimmune responses [53–58]. On the contrary, DCs are less efficient in clearing the apoptotic cells [33, 59] but as shown *in vitro* and *in vivo*, they process and present the antigens encoded within apoptotic cells to lymphocytes and may do so following ligation of chosen apoptotic cell receptors. Initial studies in this field come from Albert et al. [34], who showed that virus-infected apoptotic cells are an excellent source of antigen for crosspresentation by DCs. DCs utilize scavenger receptor CD36 and two αv-integrin receptors (αvβ3 and αvβ5) to endocytose the apoptotic cells [33, 60]. However, the αv-integrins and CD36 were subsequently shown to be dispensable for crosspresentation, at least in mice [24, 25], indicating a redundancy amongst receptors in apoptotic cell phagocytosis. Indeed, Lox-1, PSR, complement receptors 3 and 4 and FcR were subsequently shown not only to be expressed on DCs, but also to bind apoptotic cells (e.g. after opsonization with iC3b or antibodies). Interestingly, phagocytes use these receptors not only to dock and internalize the apoptotic cells but also to localize the apoptotic cells. Phagocytic cells reach sites of exten-

Tab. 22.1   Selected sources of antigens for crosspresentation and their receptors.

	Receptors involved in recognition	Reference
Apoptotic cells	CD36 (may via TSP-1)	24, 25, 33
	$\alpha v \beta 5$	24, 33
	$\alpha v \beta 3$	24
	SR-A	170
	Lox-1	171
	PSR	172, 173
	MER	174
	CD14	175
	CD91	176
	CR3	41, 68, 69, 177
	CR4	41, 68, 69, 177
	ABC	178
	CD31	179
	G2A	63
Necrotic cells	Lox-1	171
(and components	CD91	86
encoded within)	TLR 3 (RNA)	180
	TLR 9 (DNA)	181
	TLR 2 (HSPs, HMGB-1)	182–185
	TLR 4 (HSPs, HMGB-1)	182–185
	RAGE (HMGB-1)	186
Exosomes	$\alpha v \beta 3$	117
	CD11a	117
	CD54	117
Nibbling of live cells	SR-A	108
Immune complexes	Fcγ receptors	20
	CR3, CR4 ?	20
	nonspecific uptake similar to soluble antigens	20
Heat shock proteins	CD91	86, 176, 187, 188
	CD14	189, 190
	CD40	93, 191, 192
	CD36	193
	Lox-1	171
	TLR2	183–185
	TLR4	183–185
	SR-A	194

sive apoptotic cell death by chemotaxis to lysophosphatidyl choline (LPC) [61], a molecule exposed on the surface of apoptotic cells in a PLA2-dependent manner [62] and then released by apoptotic cells as a chemoattractant. LPC is recognized by G2A, a recently identified receptor on phagocytes [63].

Uptake of infected apoptotic cells by DCs initiates effector T-cell responses. In contrast, the crosspresentation of uninfected counterparts in steady-state conditions may result in cross-tolerance [23, 27, 64]. Dying cells provide an excellent source of self antigens and can be carried to draining lymph nodes in the absence of inflammation [29, 65] presumably by DCs [29]. Uninfected apoptotic cells generally do not induce DC maturation [41, 66–69] and can even interfere with the response of APCs to inflammatory stimuli [41, 52]. Therefore, upon apoptotic cell uptake, DCs remain immature. Immature DCs have been shown to induce regulatory T cells – either *in vitro*, by repetitive stimulation of naïve T cells by allogenic immature DCs [70] or *in vivo*, following injection of immature DCs loaded with influenza antigen that induced regulatory CD8+ T cells in immunized individuals [35, 36]. Taken together, this has led to the hypothesis that uptake of apoptotic cells in the absence of inflammation will tolerize the responding T cells. Such a point of view has gained support through several recent publications [26, 27, 41, 66–69, 71, 72].

Mechanistically, the inhibitory effects of apoptotic cells are most likely linked to stimulation of discrete receptors. Binding of CD36, CD51 or CR3 by antibodies or apoptotic cells prevented DC maturation upon exposure to inflammatory signals such as LPS or CD40L [41, 66–69, 73, 74]. Until recently, it was not known whether modulation occurs at the transcriptional or posttranscriptional level, nor what the primary molecular targets are. Cvetanovic et al. [75] showed that binding of apoptotic cells to macrophages modulates NFκB and AP-1-dependent gene transcription. The authors suggest that a common transcriptional co-activator, such as CBP or its paralog p300, could be the target of apoptotic modulation [75], but experimental confirmation is still pending.

22.2.2.2 Necrotic Cells

Necrosis is a physical disruption of cells that occurs mostly in pathological situations or may follow apoptotic cell death when cells are not removed efficiently. In contrast to apoptotic cells where antigens remain confined in membrane-bound form, necrotic cells release their contents which are not only a source of antigens for crosspresentation [76–78], but simultaneously provide endogenous factors that induce activation of DCs [79–84]. Stimulatory properties of cell lysates are primarily attributed to various factors for example endogenous HSPs [85, 86], HMGB-1 or uric acid crystals [84, 87], to name a few. Given that antigens and maturation signals are delivered simultaneously it is not surprising that necrotic cells can elicit protective virus and tumor-specific T-cell responses [77, 78]. The relative role and contribution of apoptotic versus necrotic cells towards crosspresentation by DCs remains controversial and the issue is complicated by the use of various antigens (tumor versus infectious) and model systems.

22.2.2.3 Heat-shock Proteins

HSPs are highly conserved and abundant chaperones. They represent up to 5% of total intracellular proteins and upon exposure to stress, their levels can reach as high as 15% [88]. Aside from a number of housekeeping functions, HSPs assist in chaperoning non-covalently bound peptide antigens from the endosome or cytosol to MHC molecules [83]. When released into the extracellular environment, HSP-peptide complexes are recognized by APC by a myriad of receptors including CD91, CD14, CD40, CD36, Lox-1, TLR 2 and 4. HSPs induce APC activation or maturation [88, 89] and simultaneously deliver bound antigens for crosspresentation to T cells [86]. HSPs are recognized by DCs and channeled via receptor mediated endocytosis to the endocytic compartment [90]. A fraction of HSP-peptide complexes escapes destruction in phago-lysosomes, enters the cytosol and is processed in a TAP and proteasome-dependent manner [90].

Recognition of the stimulatory role of HSPs in the immune system can be attributed to the pioneering work of Srivastava [88] and Matzinger, who proposed the "danger hypothesis" [79]. To date numerous HSPs including gp96, HSP60, HSP70 and HSP90 have been associated with the induction of immune responses in mice and in humans [88]. Cytoplasmic HSP 70 and 90 and ER gp96 derived from virus-infected cells or tumor cells can be crosspresented, prime CD8$^+$ T cells and even induce protective immunity *in vivo* [77, 78, 91, 92]. Further support for their immunogenicity comes from the autoimmunity field. Millar et al. used a transgenic mouse model, where LCMV-GP is expressed in pancreatic islet cells, under the control of RIP. Normally, when the immunodominant epitope gp33 is administered to mice it tolerizes the responding T cells. However, when gp33 was co-administered with HSP70, the epitope specific T cells underwent autoimmune activation rather than anergy [93].

Shen et al. recently addressed the physiological relevance of HSPs in crosspresentation compared to other sources of antigens [94]. They used fibroblasts stably expressing OVA constructs in either a membrane bound, secreted or cytosolic form. After cell lysis, they exposed APCs to the three fractions and evaluated their ability to crossprime cytotoxic T lymphocytes to membrane-bound and cytosolic OVA. Whereas the cytosolic fraction could also include HSP-associated antigens the authors ascribe the primary role to native proteins. By depleting the cytosolic fraction of OVA with specific antibody, they abrogated the priming capacity of the cytosolic extract. The immune complexed OVA did not contain HSP70, HSP90, and grp94, suggesting that native protein vs HSP-associated peptides were a major source of antigen for crosspresentation. However, the contribution of other forms of HSPs cannot be formally excluded.

22.2.2.4 Immune Complexes

Fc receptors (FcRs) are a family of membrane glycoproteins that recognize and bind Fc portion of antibodies and include FcγR I (CD64), FcγR II (CD32) and FcγR III (CD64). Usually, the binding affinity and signaling potential of FcRs is greatly enchanced upon their oligomerization, thus in circumstances when immune complexes (ICs) of antibodies and antigens are present [95, 96]. FcRs are expressed on variety of hematopoietic cells, including mouse and human DCs. In mice, ICs are recognized by CD8$^+$ and CD8$^-$ DC subsets [20] and in humans immature DCs efficiently take up ICs [97, 98]. Among human DC subtypes cancer/testis antigen NY-

ESO-1 ICs were efficiently presented by monocyte-derived but not by CD34⁺ DCs [99] and efficacy of presentation was maturation-dependent. Accordingly, Langerhans cells, that are sensitive to maturation by IFN-γ, acquired crosspresentation capacity only upon stimulation with the cytokine [100]. Endocytosis of ICs via FcRs is followed by delivery to cytosol and entry into endogenous antigen processing and presentation pathway [101]. Formation of immune complexes and recognition by FcRs is associated with increase in crosspresentation of antigens [102, 103]. FcRs not only provide antigen for crosspresentation but may also deliver a DC activation signal in certain circumstances [98, 102]. However, because mice and humans express activatory as well as inhibitory FcR isoforms, the final outcome is dependent on cumulative signals provided through the two [104]. In the latter study, tumor cells opsonized by antibodies were more efficiently crosspresented when the inhibitory FcR was blocked. The delivery of ICs to APCs appears to have a significant role *in vivo*, resulting in the crosspriming of tumor specific T-cell responses [105, 106].

22.2.2.5 Nibbling from Live Cells

Although cell death is a very efficient mechanism for delivery into crosspresentation pathway it is not always required. Several reports show that DCs and macrophages can acquire the antigens by "nibbling" from live cells [10, 107, 108]. The process involves membrane transfer from donor to acceptor cell in a receptor-dependent manner, involving at least scavenger receptor A [108]. Next to nibbling, transfer of peptides via gap junctions is another mechanism of antigen transfer among living cells as we describe later [109].

22.2.2.6 Exosomes

Exsosomes are small vesicles (50–100 nm) of endocytic origin that are formed by reverse budding of late endosome membrane [110]. Exosome formation was initially attributed to neoplastic cells but was then extended to a variety of hematopietic cells, including DCs [111–115]. Although the function of exosomes is not yet fully explored they might be a means of intercellular communication.

Exosomes express MHC molecules on their surface and when DC-derived exosomes were pulsed with peptides they could elicit effector T-cell responses in tumor bearing mice [115]. A similar protective effect was seen when DCs were pulsed with microbial antigens [116]. Later on it was shown that tumors themselves constitutively secrete exosomes that are loaded with tumor antigens. When such exosomes are taken up by DCs they elicited T-cell responses through crosspresentation [112]. Interestingly, when the uptake of exosomes was studied *in vivo*, it was noted that in mouse blood-derived exosomes are taken up primarily by immature CD11c⁺ DCs, MOMA-1⁺ and ER-TR9⁺ macrophages. DCs internalized exosomes by means of αvβ3, CD11a and CD54 receptors in a process that did not induce their maturation [117]. Taken together, this suggests that exosomes can behave similarly to apoptotic cells. It is not inconceivable, therefore, that, depending on the conditions, exosomes could shift priming to effector or tolerogenic responses, however experimental support for this is absent.

22.2.2.7 TLR and MyD88 involvement in Crosspresentation

The importance of TLR signaling in crosspresentation was first suggested from observations that mice injected with conjugates of TLR ligands and antigens, could make efficient CTL responses [118, 119]. Moreover, the development of these responses occurred in the absence of CD4 T-cell help [119]. Interestingly, using bone-marrow derived DCs, it was shown that only selected TLR agonists potentiated the crosspresentation of OVA protein by DCs, namely TLR3 and TLR9 ligands, poly I:C and CpG DNA, respectively [120]. This could partially be explained by observations from West et al, who showed that triggering of TLRs initially mobilizes the actin cytoskeleton, stimulates macropinocytosis and thus enhances the presentation of antigens in the context of MHC class I and II molecules [121]. However, the authors observed similar effect with several TLR ligands, including, CpG, poly I:C, LPS and PGN [121].

Involvement of TLRs is even more intriguing in the case of apoptotic cells. Apoptotic cells are involved in the induction of tolerance, and in the settting of infection, the stimulation of effector T-cell responses. The nature and sequence of events that shifts the balance to either tolerance or immunity has long been speculated. Schulz et al. in their elegant studies offer a rather simple explanation [122]. Using a model system consisting of apoptotic cells loaded with TLR3 ligand (dsRNA), they showed initially that ligation of TLR3 is critical for initiating cross-priming. To verify that this applies during actual virus infection, they make use of Semiliki Forest Virus (SFV) which has very limited ability to infect the $CD8\alpha^+$ DCs in vitro, but can efficiently infect Vero cells. Vero cells were infected with genetically engineered OVA-expressing SFV that can undergo only a single round of infection (to prevent the possibility of indirect DC infection), induced to undergo apoptosis, and offered to DCs as a source of antigen. Only wild type but not $TLR3^{-/-}$ DCs were able to prime T-cell effectors *in vitro*. Similarly, *in vivo* priming to SFV-OVA was impaired in $TLR3^{-/-}$ mice, although not completely blocked. The effect could be mimicked by using OVA/poly I:C loaded Vero cells [122]. Activation of TLR3 stimulates DC maturation and secretion of variety of pro-inflammatory cytokines including type I IFNs that are known to enhance crosspriming [123]. The evaluation of crosspriming in TLR/IFNα/βR double KO mice will be necessary to define the role of type I IFNs in this model.

Chen et al. also used SFV-OVA virus and showed that directly infected DCs presented virus-encoded antigens to CD8+ T cells in a MyD88-independent manner, whereas crosspresentation and OT-I OVA-specific CD8+ T-cell stimulation was MyD88-dependent [124]. MyD88 is a signal adaptor molecule that is shared by most TLRs [125] and disruption of this molecule diminishes the expression of NFκB-dependant genes that participate in DC maturation. As TLR3 also signals in a MyD88-independent manner, participation of other MyD88-associated receptors in crosspriming to this virus can not be formally excluded. dsRNA can also trigger

DC maturation and type I IFN production through MyD88-independent, PKR-dependent or -independent mechanisms [125]. Thus, viruses which have dsRNA intermediates and which induce apoptosis, could promote crosspresentation by DCs through a number of mechanisms.

It will be interesting to see how dependence on TLRs and MyD88 correlates with observations published by Blander et al. [126] who show that phagosome maturation is regulated by the nature of its cargo. They studied uptake of bacteria and apoptotic cells by macrophages. Interestingly, bacteria but not apoptotic cells were rapidly localized in phago-lysosomes in a TLR and MyD88-dependent manner. When bacteria and apoptotic cells were incubated with macrophages simultaneously, co-localization of the two was not observed. Not surprisingly, and in support of observations made previously by Lucas et al. that simple addition of TLR agonists cannot modulate signals induced by apoptotic cells [127], the addition of LPS to apoptotic cells did not influence the rate of phago-lysosomes formation. The physiological relevance of this biased routing is unclear. Perhaps rapid phago-lysosomal formation after ingestion of bacteria ensures their rapid destruction while in the case of apoptotic cells the formation of late endosomes is delayed in order to allow antigenic peptides to enter the crosspresentation pathway. Trombetta et al., have shown recently that DCs express low levels of proteases in their phagosomes, but once stimulated to mature they up regulate expression and processing of internalized antigens [128]. Receptors involved in apoptotic cell uptake (at least uninfected apoptotic cells) may divert the dead cells into vesicles that have a paucity of TLRs. Another apparent question that remains to be answered in light of studies by Schulz [122] and Chen [124], is the fate of infected apoptotic cells: do they behave like their uninfected counterparts or do they undergo rapid uptake similar to the one observed after internalization of bacteria?

22.3
Mechanisms of Crossprocessing and Crosspresentation

What are the mechanisms that allow access of exogenous antigens into the presentation pathway that is normally restricted to antigens produced within the cell? Initially, investigation focused on entry of exogenous antigens into the classical pathway; however there is data suggesting that processing of exogenous antigens may be carried out in a self-sufficient phagosome-derived compartment, although this requires further confirmation. Here we list possible cellular mechanisms of crosspresentation that have been described over the last decade. (An overview of cellular mechanisms in crosspresentation is shown in Fig. 22.1).

22.3.1
Entry into the Classical Endocytic Pathway

Classical presentation of endogenous self- or pathogen-derived antigens involves their processing into peptides by cytosolic proteases, mainly proteasomes, followed

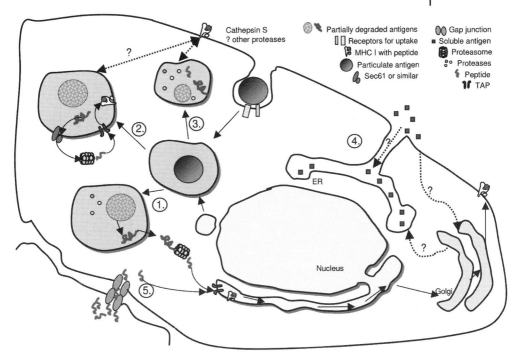

Fig. 22.1 Mechanisms of crosspresentation. Particulate antigens enter the antigen presenting cells most efficiently by receptor mediated endocytosis. Endosomes containing the ingested antigen then acquire selected endoplasmatic reticulum (ER)-derived elements and, after initial processing within the endosome, the antigens exit the late endosome and enter the cytoplasm. They then enter the classical pathway and are processed in a similar manner as endogenous antigens (1). Alternatively, the antigens may remain in the endosomes that acquire selected characteristics of ER and are fully processed in close viscinity of the phago- endosome. They re-enter the phago- endosomes and are there loaded onto MHC class I molecules (2). Whereas in the former cases the antigen processing involves trans- porter of antigen processing (TAP), this is not always required. Albeit less efficiently, the vacuolar processing pathway (3) may rely on vacuolar proteases, especially cathepsin S, to generate peptides that are loaded onto MHC class I molecules within the vacuole. A special pathway may be in place for soluble antigens (4) which by yet unidentified mechanisms directly access the ER. Gap junctions may permit the transfer of already processed peptides from donor to acceptor cells (5).

by transport via TAP molecules (transporter associated proteins) into the lumen of the endoplasmic reticulum (ER), where peptides bind to newly synthesized MHC class I molecules. MHC molecules bearing peptides are finally transported via the Golgi apparatus to the plasma membrane. Exogenous antigens such as IC, soluble proteins, apoptotic or necrotic cells and antigen-coated beads can also access this pathway from phagolysosomes and endosomes, following transfer into the cytosol [8, 49, 129, 130]. In human DCs, dying cells are captured into endocytic vesicles. After acidification and maturation of the vesicles, the antigens within the ingested apoptotic cells undergo proteolysis by cathepsins and are transported into the cytosol for further processing [130]. Cathepsin D has been identified as one such protease which participates in the crosspresentation of influenza infected apoptotic cells.

Crosspresentation is crucially dependent on actin cytoskeleton rearrangements [130, 131], a process that is extensively controlled by Rho GTPases [132] such as Rac1 or Cdc42. Transgenic mice expressing a dominant negative form of Rac1 under the CD11c promoter have reduced numbers of CD8α$^+$ DCs. The remaining DCs were substantially impaired in the *in vivo* uptake of exogenously delivered OVA loaded β2m$^{-/-}$ splenocytes. A consequence was decreased crosspresentation and crosspriming of OVA-specific OT I cells [131]. In accordance, it was shown previously that Rac 1 is activated upon binding of apoptotic cells to the integrin receptor αvβ5 [133]. Murine DCs that were transfected with vaccinia encoding constitutively active Rho GTPases increased the uptake of OVA protein and its presentation to CD4$^+$ T cells. The process was abrogated by toxin B from *Clostridium difficile* which inactivates Rho, Rac and Cdc42 [134]. As attention is being turned to the role of the cytoskeleton and its regulatory proteins in crosspresenting mechanisms, we will learn more of its involvement in the differential trafficking of internalized antigens.

22.3.2
Phagosome–endosome Compartment

An alternative pathway of processing exogenous antigens in the context of MHC class I presentation has recently been proposed by several groups. Phagosomes become enriched with components of ER whereby the resulting phagosome–endosome compartment with adjunct proteasomes forms a self-sufficient compartment for loading of MHC class I molecules with peptides ("one stop processing").

The first studies indicating that ER is involved at early phases of phagocytosis of crosspresentation came from the group of Desjardins. Gagnon and co-authors [135] showed that in macrophages, but not neutrophils, the ER donates the membrane to the forming phagosome, which in the process of maturation continues to incorporate elements of ER. This mechanism is involved in the uptake of inert particles, as well as microorganisms such as *Salmonella*, a bacterium that triggers crosspresentation [136]. Not surprisingly, in the next year three publications followed that showed involvement of ER membranes in phagosomes containing material to be crosspresented. The group of Desjardins [137] extended their previous studies by showing that in mouse macrophages, proteasomes are closely associated with phagosomal-ER compartment at its cytoplasmic side. Additionally they showed that the membranes of the compartment are enriched with TAP, another essential component for antigen presentation. When OVA containing latex beads were engulfed into phagosomes, the CD8$^+$ T cell-restricted epitope SIINFEKL could be detected with a specific antibody, with maximal expression 1–2 h after phagocytosis. In an independent study, Guermonprez et al. identified a similar process [138]. The authors used a complex procedure to isolate and purify phagosomes from phagocytosing murine DCs. Selected ER residential proteins such as

Sec61/Sec62, calnexin and TAP were recruited soon after formation of the phagosome. Additionally, components of MHC loading machinery: tapasin, ERp57, calreticulin and heavy chains of MHC class I molecules were detected in the phagosome. Exposure of OVA-latex beads to DCs resulted in functional OVA-MHC class I complexes and stimulation of CD8 T cell-restricted eptiope specific OT I T cells.

A similar mechanism may take place in human DCs. Ackerman et al. [139] showed that either primary human DCs or DC-like cell line KG-1 phagosomes as well as micropinosomes, all contain elements of the MHC loading machinery.

Lizée et al. [140] have identified a conserved sequence in the cytoplasmic domain of MHC class I heavy chain that apparently directs the molecules into the phagolysosomeal compartment. In mouse CD11c$^+$ DCs, deletion of a tyrosine residue in this domain alters routing of MHC class I molecules and consequently prevents crosspresentation of VSV and Sendai virus antigens and activation of specific CTLs.

Taken together, these studies imply that phagosomes enriched by elements donated by the ER could be a self-sufficient compartment for crosspresentation. According to the model, Sec61 would translocate protein segments derived from exogenous antigens into the cytoplasm where they would be processed into peptides by the closely adjacent proteasomes and retrotransported by TAP into the lumen of the phagosome. On the luminal side of phagosomes, the translocated peptides are loaded onto MHC molecules. The last step, the route by which peptide-MHC accesses the cell surface, has not been resolved, but may involve recycling MHC class I molecules.

However, the drawback of these studies is that they have relied on the use of artificial forms of antigen (latex beads), and therefore it remains unclear whether such ER-phagosomes are involved in the actual processing of apoptotic cells. Further studies will be essential to confirm the existence of this pathway and to determine whether it applies to physiological sources of antigens.

22.3.3
A Special Mechanism for Soluble Antigens?

A detour from the self-sufficient phagosome based mechanism of crosspresentation was recently proposed by the Creswell group, who made the observation that soluble antigens can gain access to the perinuclear ER compartment [141]. They used protease-resistant human β2m as a source of exogenous soluble antigen. They first showed that the β2m co-localized with the ER-marker calnexin but not with the lysosomal marker LAMP-1 or Golgi marker GM130. MHC class I molecules are not folded correctly in the absence of β2m, and consequently are eventually degraded in the ER. When exogenous β2m was added to DCs derived from the bone marrow (BM) of β2m-deficient mice, the expression of correctly folded MHC class I molecules on the cell surface was restored. The MHC molecules correctly folded with the help of exogenous β2m were even capable of stimulating OVA specific T cells when DCs were infected with vaccinia virus encoding OVA. Taken together, these data indicate, that the exogenous antigen accessed the ER to attain functionality in the described experimental system.

With this pathway, several questions remain to be addressed, including the mechanisms that deliver the soluble antigens to the ER: whether this involves the transient luminal continuities between pinosomes and ER [135] and/or retrograde transport through the Golgi [142].

22.3.4
Tap Dependence and Endocytic Exchange Mechanism (Vacuolar Pathway)

As discussed, exogenous antigens can be internalized into endocytic vesicles [8, 101, 143], and crosspresented onto MHC I molecules via different pathways. One involves the transfer of antigens into the cytosol from the phagosome, followed by further degradation in the cytosol by proteosomes and then transport via TAP into the ER. In addition, vacuolar processing (possibly in an ER-phagosome structure associated with TAP molecules), may permit "one-stop" processing [137, 138]. Finally, pre-processing in vacuoles may take place before access to the cytosol or the ER through an endocytic exchange mechanism. In most but not all of these cases, crosspresentation onto MHC class I molecules would be significantly TAP-dependent. However, there are now several examples of TAP-independent pathways. Below we review some of these data, how they relate to TAP-dependent pathways and how they might be integrated into the current models of crosspresentation.

Using a bone marrow chimera system, Huang et al., first convincingly demonstrated a role for TAP-dependent crosspriming *in vivo*. CT26 tumor lines (H-2d) expressing the influenza NP protein were delivered to lethally irradiated F1 strains (H-2 bxd), which had been reconstituted with wild type or TAP$^{-/-}$ (H-2b) bone marrow. As expected, crosspriming occurred in the TAP$^{+/+}$ reconstituted F1 mice, which developed a NP-specific, H-2Db restricted CTL response. However, no crosspriming was observed when chimeras were reconstituted with TAP$^{-/-}$ bone marrow. These studies supported a dominant TAP-dependent pathway for *in vivo* crosspriming to tumor cells [144]. Studies from other investigators confirmed that tumor and other cell associated antigens are also crosspresented, at least partly, by the TAP-dependent pathway [49, 71].

Subsequent studies by Sigal and Rock showed that TAP$^{+/+}$ bone marrow derived APCs were also essential for crosspriming to vaccinia and polio virus [145, 146]. The initiation of responses to LCMV and influenza virus also required bone marrow derived DCs, but unlike polio virus, both TAP-dependent and -independent pathways participated. Interestingly, the authors found that the TAP-independent pathway was much less efficient than the TAP-dependent route, on the order of 10–300-fold less [146]. Although the pathway by which the TAP-independent crosspriming takes place was not described, it was speculated that viral epitopes in exogenously acquired viral antigens were processed and presented to MHC class I molecules in endosomes. It is possible that the exogenous antigens are accessed as virus infected apoptotic cells. Consistent with this concept, Albert et al., have described both TAP-dependent and -independent pathways for the crosspresentation of influenza virus infected apoptotic cells by murine DCs. In their model they suggest that processed antigens are needed for efficient crosspresentation by the TAP-independent pathway [147], as crosspresentation was inhibited when apoptotic cells were pre-exposed to lactacystin (a proteosome inhibitor) before delivery to TAP$^{-/-}$ DCs.

Recent studies by Shen et al., have demonstrated a TAP independence of the vacuolar pathway of cross presentation, depending upon the physical form of the antigens. OVA incorporated into microspheres is apparently crosspresented by both TAP-dependent and -independent pathways. The authors showed that in the presence of TAP$^{-/-}$ DCs, this form of OVA, was inhibited from being cross presented by leupeptin (an inhibitor of cysteine proteases), This was further supported by the finding that cathepsin S, but not cathepsin L or B deficient DCs were deficient in the crosspresentation of microsphere complexed OVA. Furthermore, double TAP$^{-/-}$ and cathepsin S$^{-/-}$ mice completely failed to cross present the microsphere complexed OVA, indicating a critical role for cathepsin S in TAP-independent processing pathways. Cathepsin S also contributes to the crosspriming of cell-associated antigens as TAP$^{-/-}$ cathepsin S$^{-/-}$ animals failed to crosspresent OVA expressed in cells to OVA-specific T cells. Finally, a role for cathepsin S was identified in the crosspresentation of influenza viral epitopes. Compared to the overall response, however, the vacuolar pathway seemed less significant than the TAP-dependent pathways [148]. Larsson et al., have also reported on a partial requirement for cathepsins in the crosspresentation of virus infected apoptotic cells by human DCs. Blocking of aspartic proteases with pepstatin, led to inhibition of crosspresentation. Since proteosome inhibitors also partly blocked presentation, it is conceivable that some of the influenza antigens are first partially processed in the phagosome, before access to the cytosol. However, in this study it was not possible to rule out a TAP-independent, but cathepsin-dependent pathway [130]. These studies point to the possibility that different cathepsins may be involved in generating relevant peptides from various antigens.

TAP-dependent and -independent pathways have also been described using other forms of antigens for example HSP complexes containing antigenic peptides [149, 150]. In addition to being necessary for peptide transport from the cytosol to the ER, or retrograde transport from the cytosol to ER-phagosomes, TAP dependence may rely on the fact that MHC molecules in TAP$^{-/-}$ cells have reduced stability at phagolysosomal pH. Chefalo et al., showed that both tapasin and TAP were necessary for MHC-I to bind ER derived stabilizing peptides so that they could achieve the stability needed for peptide exchange in acidic vacuolar processing compartments [151]. These studies indicate that the endocytic vacuolar processing pathway requires the transfer of appropriately assembled MHC class I molecules from the ER.

22.3.5
Transfer of Peptides via Gap Junctions

A recently described pathway of crosspresentation involves the transfer of peptides from infected to neighboring cells by virtue of gap junctions [109]. Gap junctions are channels composed of connexin molecules in cell membranes that connect the neighboring cells for the purpose of ion, messenger and nutrient exchange. Apart from the limit in maximal size of transferred molecules (MW<1K) the mechanism seems to be unspecific. Neijssen et al. [109] show that a similar mechanism is used to also transfer peptides directly from the cytoplasm of one cell to the cytoplasm of a neighboring cell. Initially they used a cell line which does not express gap junctions. They then stably transfected the line with connexin 43 (Cx43). By comparing the control and Cx43 transfected lines and with the use of a chemical inhibitor of gap junction transfer, they were able to show that transfer of fluorescently labeled peptides from one to another cell depends exclusively on the use of gap junctions. Subsequently, a similar approach was used to show the transfer of immunogenic peptides and acquisition of antigen presentation ability of the acceptor cells.

The physiological relevance of this pathway remains unclear. As the cytosol is full of peptidases, only a small fraction of peptides would persist in the cytosol long enough to be transferred to the neighboring cells. Further destruction in the acceptor cell would ensure rapid dilution of the peptides, giving gap junction transferred peptides a very limited spatial spread. It has been proposed that peptides arising from infection of cells, could be transferred to adjacent non-infected cells and recognized by T cells. Such selective transfer could form a "buffer zone" that would contribute to self-limitation of infection. It is not clear whether the transfer of peptides via gap junctions plays any role in crosspriming. DCs do express proteins involved in gap junction formation [152], however, given the likely inefficiency of gap junction transfer, a major role in crosspriming seems unlikely.

22.4
Physiological Relevance of Crosspresentation

Let us finish by briefly tackling the subject that is more thoroughly discussed elsewhere in this publication. Why did the complicated mechanism of crosspresentation evolve? What is the physiological relevance of crosspriming? Whereas crosspresentation and crosspriming can take place in response to tumors [115, 153, 154], viruses [76, 77, 155–164] and bacteria [136, 165–167], the relevance in comparison to direct priming has been questioned [168, 169]. However, several circumstances call for the existence of alternative mechanisms of priming, including priming to tumor antigens which themselves are poor stimulators of T cells. Perhaps an even more apparent example is that of microbes that have developed immune escape mechanisms, for example abrogation of presentation on MHC class I molecules on infected cells but where nevertheless efficient long term immunity is still established (e.g. to CMV). These examples support the notion that mechanisms such as crosspresentation are immunologically relevant and sometimes indispensable [139].

Acknowledgements

We acknowledge support from the NIH an NCI, the Mary Kirkland Foundation, the Cancer Research Institute, the Burroughs Wellcome Foundation (NB). NB is a Doris Duke Distinguished Scientist and an Elisabeth Glaser Scientist.

References

1 Bevan, M. J., Minor H antigens introduced on H-2 different stimulating cells cross-react at the cytotoxic T cell level during in vivo priming. *J Immunol* **1976**. *117:* 2233–2238.

2 Bevan, M. J., Cross-priming for a secondary cytotoxic response to minor H antigens with H-2 congenic cells which do not cross-react in the cytotoxic assay. *J Exp Med* **1976**. *143:* 1283–1288.

3 Limmer, A., Ohl, J., Kurts, C., Ljunggren, H. G., Reiss, Y., Groettrup, M., Momburg, F., Arnold, B., Knolle, P. A., Efficient presentation of exogenous antigen by liver endothelial cells to CD8$^+$ T cells results in antigen-specific T-cell tolerance. *Nat Med* **2000**. *6:* 1348–1354.

4 Savinov, A. Y., Wong, F. S., Stonebraker, A. C., Chervonsky, A. V., Presentation of antigen by endothelial cells and chemoattraction are required for homing of insulin-specific CD8$^+$ T cells. *J Exp Med* **2003**. *197:* 643–656.

5 Ke, Y., Kapp, J. A., Exogenous antigens gain access to the major histocompatibility complex class I processing pathway in B cells by receptor-mediated uptake. *J Exp Med* **1996**. *184:* 1179–1184.

6 Kovacsovics-Bankowski, M., Clark, K., Benacerraf, B., Rock, K. L., Efficient major histocompatibility complex class I presentation of exogenous antigen upon phagocytosis by macrophages. *Proc Natl Acad Sci USA* **1993**. *90:* 4942–4946.

7 Kovacsovics-Bankowski, M., Rock, K. L., Presentation of exogenous antigens by macrophages: analysis of major histocompatibility complex class I and II presentation and regulation by cytokines. *Eur J Immunol* **1994**. *24:* 2421–2428.

8 Kovacsovics-Bankowski, M., Rock, K. L., A phagosome-to-cytosol pathway for exogenous antigens presented on MHC class I molecules. *Science* **1995**. *267:* 243–246.

9 Rock, K. L., Rothstein, L., Gamble, S., Fleischacker, C., Characterization of antigen-presenting cells that present exogenous antigens in association with class I MHC molecules. *J Immunol* **1993**. *150:* 438–446.

10 Ramirez, M. C., Sigal, L. J., Macrophages and dendritic cells use the cytosolic pathway to rapidly cross-present antigen from live, vaccinia-infected cells. *J Immunol* **2002**. *169:* 6733–6742.

11 Rock, K. L., Gamble, S., Rothstein, L., Presentation of exogenous antigen with class I major histocompatibility complex molecules. *Science* **1990**. *249:* 918–921.

12 Pancholi, P., Mirza, A., Schauf, V., Steinman, R. M., Bhardwaj, N., Presentation of mycobacterial antigens by human dendritic cells: lack of transfer from infected macrophages. *Infect Immun* **1993**. *61:* 5326–5332.

13 Inaba, K., Inaba, M., Naito, M., Steinman, R. M., Dendritic cell progenitors phagocytose particulates, including bacillus Calmette-Guerin organisms, and sensitize mice to mycobacterial antigens in vivo. *J Exp Med* **1993**. *178:* 479–488.

14 Shen, Z., Reznikoff, G., Dranoff, G., Rock, K. L., Cloned dendritic cells can present exogenous antigens on both MHC class I and class II molecules. *J Immunol* **1997**. *158:* 2723–2730.

15 Kurts, C., Cannarile, M., Klebba, I., Brocker, T., Dendritic cells are sufficient to cross-present self-antigens to CD8 T cells in vivo. *J Immunol* **2001**. *166:* 1439–1442.

16 Jung, S., Unutmaz, D., Wong, P., Sano, G., De los Santos, K., Sparwasser, T., Wu, S., Vuthoori, S., Ko, K., Zavala, F., Pamer, E. G., Littman, D. R., Lang, R. A., In vivo depletion of CD11c(+) dendritic cells abrogates priming of CD8(+) T cells by exogenous cell-associated antigens. *Immunity* **2002**. *17:* 211–220.

17 Pulendran, B., Modulating vaccine responses with dendritic cells and Toll-like receptors. *Immunol Rev* **2004**. *199:* 227–250.

18 Heath, W. R., Belz, G. T., Behrens, G. M., Smith, C. M., Forehan, S. P., Parish, I. A., Davey, G. M., Wilson, N. S., Carbone, F. R., Villadangos, J. A., Cross-presentation, dendritic cell subsets, and the generation of immunity to cellular antigens. *Immunol Rev* **2004**. *199:* 9–26.

19 Pooley, J. L., Heath, W. R., Shortman, K., Cutting edge: intravenous soluble antigen is presented to CD4 T cells by CD8⁻ dendritic cells, but cross-presented to CD8 T cells by CD8⁺ dendritic cells. *J Immunol* **2001**. *166:* 5327–5330.

20 den Haan, J. M., Bevan, M. J., Constitutive versus activation-dependent cross-presentation of immune complexes by CD8(⁺) and CD8(⁻) dendritic cells in vivo. *J Exp Med* **2002**. *196:* 817–827.

21 den Haan, J., Lehar, S., Bevan, M., CD8⁺ but not CD8⁻ dendritic cells cross-prime cytotoxic T cells in vivo. *J Exp Med* **2000**. *192:* 1685–1696.

22 Schulz, O., Reis e Sousa, C., Cross-presentation of cell-associated antigens by CD8alpha⁺ dendritic cells is attributable to their ability to internalize dead cells. *Immunology* **2002**. *107:* 183–189.

23 Iyoda, T., Shimoyama, S., Liu, K., Omatsu, Y., Akiyama, Y., Maeda, Y., Takahara, K., Steinman, R. M., Inaba, K., The CD8⁺ dendritic cell subset selectively endocytoses dying cells in culture and in vivo. *J Exp Med* **2002**. *195:* 1289–1302.

24 Schulz, O., Pennington, D. J., Hodivala-Dilke, K., Febbraio, M., Reis e Sousa, C., CD36 or alphavbeta3 and alphavbeta5 integrins are not essential for MHC class I cross-presentation of cell-associated antigen by CD8 alpha⁺ murine dendritic cells. *J Immunol* **2002**. *168:* 6057–6065.

25 Belz, G. T., Vremec, D., Febbraio, M., Corcoran, L., Shortman, K., Carbone, F. R., Heath, W. R., CD36 is differentially expressed by CD8⁺ splenic dendritic cells but is not required for cross-presentation in vivo. *J Immunol* **2002**. *168:* 6066–6070.

26 Kurts, C., Kosaka, H., Carbone, F. R., Miller, J. F., Heath, W. R., Class I-restricted cross-presentation of exogenous self-antigens leads to deletion of autoreactive CD8(⁺) T cells. *J Exp Med* **1997**. *186:* 239–245.

27 Hugues, S., Mougneau, E., Ferlin, W., Jeske, D., Hofman, P., Homann, D., Beaudoin, L., Schrike, C., Von Herrath, M., Lehuen, A., Glaichenhaus, N., Tolerance to islet antigens and prevention from diabetes induced by limited apoptosis of pancreatic beta cells. *Immunity* **2002**. 16: *169–181.*

28 Belz, G. T., Behrens, G. M., Smith, C. M., Miller, J. F., Jones, C., Lejon, K., Fathman, C. G., Mueller, S. N., Shortman, K., Carbone, F. R., Heath, W. R., The CD8alpha(⁺) dendritic cell is responsible for inducing peripheral self-tolerance to tissue-associated antigens. *J Exp Med* **2002**. *196:* 1099–1104.

29 Scheinecker, C., McHugh, R., Shevach, E. M., Germain, R. N., Constitutive presentation of a natural tissue auto-antigen exclusively by dendritic cells in the draining lymph node. *J Exp Med* **2002**. *196:* 1079–1090.

30 Chung, Y., Chang, J. H., Kweon, M. N., Rennert, P. D., Kang, C. Y., A CD8{alpha}-11b⁺ dendritic cells but not CD8{alpha}⁺ dendritic cells mediate cross-tolerance toward intestinal antigens. *Blood* **2005**. *106:* 201–206.

31 Belz, G. T., Smith, C. M., Kleinert, L., Reading, P., Brooks, A., Shortman, K., Carbone, F. R., Heath, W. R., Distinct migrating and nonmigrating dendritic cell populations are involved in MHC class I-restricted antigen presentation after lung infection with virus. *Proc Natl Acad Sci USA* **2004**. *101:* 8670–8675.

32 Carbone, F. R., Belz, G. T., Heath, W. R., Transfer of antigen between migrating and lymph node-resident DCs in peripheral T-cell tolerance and immunity. *Trends Immunol* **2004**. *25:* 655–658.

33 Albert, M. L., Pearce, S. F., Francisco, L. M., Sauter, B., Roy, P., Silverstein, R. L., Bhardwaj, N., Immature dendritic cells phagocytose apoptotic cells via alphavbeta5 and CD36, and cross-present antigens to cytotoxic T lymphocytes. *J Exp Med* **1998**. *188:* 1359–1368.

34 Albert, M. L., Sauter, B., Bhardwaj, N., Dendritic cells acquire antigen from apoptotic cells and induce class I-restricted CTLs. *Nature* **1998**. *392:* 86–89.

35 Dhodapkar, M. V., Young, J. W., Chapman, P. B., Cox, W. I., Fonteneau, J. F., Amigorena, S., Houghton, A. N., Steinman, R. M., Bhardwaj, N., Paucity of functional T-cell memory to melanoma antigens in healthy donors and melanoma patients. *Clin Cancer Res* **2000**. 6: 4831–4838.

36 Dhodapkar, M. V., Steinman, R. M., Krasovsky, J., Munz, C., Bhardwaj, N., Antigen-specific inhibition of effector T cell function in humans after injection of immature dendritic cells. *J Exp Med* **2001**. *193*: 233–238.

37 Sillanpaa, N., Magureanu, C. G., Murumagi, A., Reinikainen, A., West, A., Manninen, A., Lahti, M., Ranki, A., Saksela, K., Krohn, K., Lahesmaa, R., Peterson, P., Autoimmune regulator induced changes in the gene expression profile of human monocyte-dendritic cell-lineage. *Mol Immunol* **2004**. *41*: 1185–1198.

38 Levings, M. K., Gregori, S., Tresoldi, E., Cazzaniga, S., Bonini, C., Roncarolo, M. G., Differentiation of Tr1 cells by immature dendritic cells requires IL-10 but not CD25+CD4+ Tr cells. *Blood* **2005**. *105*: 1162–1169.

39 Ip, W. K., Lau, Y. L., Distinct maturation of, but not migration between, human monocyte-derived dendritic cells upon ingestion of apoptotic cells of early or late phases. *J Immunol* **2004**. *173*: 189–196.

40 Clayton, A. R., Prue, R. L., Harper, L., Drayson, M. T., Savage, C. O., Dendritic cell uptake of human apoptotic and necrotic neutrophils inhibits CD40, CD80, and CD86 expression and reduces allogeneic T cell responses: relevance to systemic vasculitis. *Arthritis Rheum* **2003**. *48*: 2362–2374.

41 Verbovetski, I., Bychkov, H., Trahtemberg, U., Shapira, I., Hareuveni, M., Ben-Tal, O., Kutikov, I., Gill, O., Mevorach, D., Opsonization of apoptotic cells by autologous iC3b facilitates clearance by immature dendritic cells, down-regulates DR and CD86, and up-regulates CC chemokine receptor 7. *J Exp Med* **2002**. *196*: 1553–1561.

42 Colonna, M., Trinchieri, G., Liu, Y. J., Plasmacytoid dendritic cells in immunity. *Nat Immunol* **2004**. *5*: 1219–1226.

43 McKenna, K., Beignon, A. S., Bhardwaj, N., Plasmacytoid dendritic cells: linking innate and adaptive immunity. *J Virol* **2005**. *79*: 17–27.

44 Fonteneau, J. F., Gilliet, M., Larsson, M., Dasilva, I., Munz, C., Liu, Y. J., Bhardwaj, N., Activation of influenza virus-specific CD4+ and CD8+ T cells: a new role for plasmacytoid dendritic cells in adaptive immunity. *Blood* **2003**. *101*: 3520–3526.

45 Grouard, G., Rissoan, M. C., Filgueira, L., Durand, I., Banchereau, J., Liu, Y. J., The enigmatic plasmacytoid T cells develop into dendritic cells with interleukin (IL)-3 and CD40-ligand. *J Exp Med* **1997**. *185*: 1101–1111.

46 Stent, G., Reece, J. C., Baylis, D. C., Ivinson, K., Paukovics, G., Thomson, M., Cameron, P. U., Heterogeneity of freshly isolated human tonsil dendritic cells demonstrated by intracellular markers, phagocytosis, and membrane dye transfer. *Cytometry* **2002**. *48*: 167–176.

47 Fiebiger, E., Meraner, P., Weber, E., Fang, I. F., Stingl, G., Ploegh, H., Maurer, D., Cytokines regulate proteolysis in major histocompatibility complex class II-dependent antigen presentation by dendritic cells. *J Exp Med* **2001**. *193*: 881–892.

48 Schnurr, M., Chen, Q., Shin, A., Chen, W., Toy, T., Jenderek, C., Green, S., Miloradovic, L., Drane, D., Davis, I. D., Villadangos, J., Shortman, K., Maraskovsky, E., Cebon, J., Tumor antigen processing and presentation depend critically on dendritic cell type and the mode of antigen delivery. *Blood* **2005**. *105*: 2465–2472.

49 Li, M., Davey, G. M., Sutherland, R. M., Kurts, C., Lew, A. M., Hirst, C., Carbone, F. R., Heath, W. R., Cell-associated ovalbumin is cross-presented much more efficiently than soluble ovalbumin in vivo. *J Immunol* **2001**. *166*: 6099–6103.

50 Larsson, M., Fonteneau, J. F., Bhardwaj, N., Cross-presentation of cell-associated antigens by dendritic cells. *Curr Top Microbiol Immunol* **2003**. *276*: 261–275.

51 Boisgerault, F., Rueda, P., Sun, C. M., Hervas-Stubbs, S., Rojas, M., Leclerc, C., Cross-priming of T cell responses by synthetic microspheres carrying a CD8+ T cell epitope requires an adjuvant signal. *J Immunol* **2005**. *174*: 3432–3439.

52 Savill, J., Dransfield, I., Gregory, C., Haslett, C., A blast from the past: clearance of apoptotic cells regulates immune responses. *Nat Rev Immunol* **2002**. *2*: 965–975.

53 Cohen, P. L., Caricchio, R., Abraham, V., Camenisch, T. D., Jennette, J. C., Roubey, R. A., Earp, H. S., Matsushima, G., Reap, E. A., Delayed apoptotic cell clearance and lupus-like autoimmunity in mice lacking the c-mer membrane tyrosine kinase. *J Exp Med* **2002**. *196:* 135–140.

54 Li, M. O., Sarkisian, M. R., Mehal, W. Z., Rakic, P., Flavell, R. A., Phosphatidyl-serine receptor is required for clearance of apoptotic cells. *Science* **2003**. *302:* 1560–1563.

55 Le, L. Q., Kabarowski, J. H., Weng, Z., Satterthwaite, A. B., Harvill, E. T., Jensen, E. R., Miller, J. F., Witte, O. N., Mice lacking the orphan G protein-coupled receptor G2A develop a late-onset autoimmune syndrome. *Immunity* **2001**. *14:* 561–571.

56 Taylor, P. R., Carugati, A., Fadok, V. A., Cook, H. T., rews, M., Carroll, M. C., Savill, J. S., Henson, P. M., Botto, M., Walport, M. J., A hierarchical role for classical pathway complement proteins in the clearance of apoptotic cells in vivo. *J Exp Med* **2000**. *192:* 359–366.

57 Boes, M., Role of natural and immune IgM antibodies in immune responses. *Mol Immunol* **2000**. *37:* 1141–1149.

58 Ehrenstein, M. R., Cook, H. T., Neuberger, M. S., Deficiency in serum immunoglobulin (Ig)M predisposes to development of IgG autoantibodies. *J Exp Med* **2000**. *191:* 1253–1258.

59 Bondanza, A., Zimmermann, V. S., Rovere-Querini, P., Turnay, J., Dumitriu, I. E., Stach, C. M., Voll, R. E., Gaipl, U. S., Bertling, W., Poschl, E., Kalden, J. R., Manfredi, A. A., Herrmann, M., Inhibition of phosphatidylserine recognition heightens the immunogenicity of irradiated lymphoma cells in vivo. *J Exp Med* **2004**. *200:* 1157–1165.

60 Rubartelli, A., Poggi, A., Zocchi, M. R., The selective engulfment of apoptotic bodies by dendritic cells is mediated by the alpha(v)beta3 integrin and requires intracellular and extracellular calcium. *Eur J Immunol* **1997**. *27:* 1893–1900.

61 Lauber, K., Bohn, E., Krober, S. M., Xiao, Y. J., Blumenthal, S. G., Lindemann, R. K., Marini, P., Wiedig, C., Zobywalski, A., Baksh, S., Xu, Y., Autenrieth, I. B., Schulze-Osthoff, K., Belka, C., Stuhler, G., Wesselborg, S., Apoptotic cells induce migration of phagocytes via caspase-3-mediated release of a lipid attraction signal. *Cell* **2003**. *113:* 717–730.

62 Kim, S. J., Gershov, D., Ma, X., Brot, N., Elkon, K. B., I-PLA(2) Activation during apoptosis promotes the exposure of membrane lysophosphatidylcholine leading to binding by natural immuno-globulin m antibodies and complement activation. *J Exp Med* **2002**. *196:* 655–665.

63 Kabarowski, J. H., Zhu, K., Le, L. Q., Witte, O. N., Xu, Y., Lysophosphatidyl-choline as a ligand for the immuno-regulatory receptor G2A. *Science* **2001**. *293:* 702–705.

64 Belz, G. T., Heath, W. R., Carbone, F. R., The role of dendritic cell subsets in selection between tolerance and immunity. *Immunol Cell Biol* **2002**. *80:* 463–468.

65 Huang, F. P., Platt, N., Wykes, M., Major, J. R., Powell, T. J., Jenkins, C. D., MacPherson, G. G., A discrete sub-population of dendritic cells transports apoptotic intestinal epithelial cells to T cell areas of mesenteric lymph nodes. *J Exp Med* **2000**. *191:* 435–444.

66 Stuart, L. M., Lucas, M., Simpson, C., Lamb, J., Savill, J., Lacy-Hulbert, A., Inhibitory effects of apoptotic cell ingestion upon endotoxin-driven myeloid dendritic cell maturation. *J Immunol* **2002**. *168:* 1627–1635.

67 Voll, R. E., Herrmann, M., Roth, E. A., Stach, C., Kalden, J. R., Girkontaite, I., Immunosuppressive effects of apoptotic cells. *Nature* **1997**. *390:* 350–351.

68 Morelli, A. E., Larregina, A. T., Shufesky, W. J., Zahorchak, A. F., Logar, A. J., Papworth, G. D., Wang, Z., Watkins, S. C., Falo, L. D., Jr., Thomson, A. W., Internalization of circulating apoptotic cells by splenic marginal zone dendritic cells: dependence on complement receptors and effect on cytokine production. *Blood* **2003**. *101:* 611–620.

69 Sohn, J. H., Bora, P. S., Suk, H. J., Molina, H., Kaplan, H. J., Bora, N. S., Tolerance is dependent on complement C3 fragment iC3b binding to antigen-presenting cells. *Nat Med* **2003**. *9:* 206–212.

70 Jonuleit, H., Schmitt, E., Schuler, G., Knop, J., Enk, A. H., Induction of interleukin 10-producing, nonproliferating CD4($^+$) T cells with regulatory properties by repetitive stimulation with allogeneic immature human dendritic cells. *J Exp Med* **2000**. *192:* 1213–1222.

71 Liu, K., Iyoda, T., Saternus, M., Kimura, Y., Inaba, K., Steinman, R. M., Immune tolerance after delivery of dying cells to dendritic cells in situ. *J Exp Med* **2002**. *196:* 1091–1097.

72 Steinman, R. M., Turley, S., Mellman, I., Inaba, K., The induction of tolerance by dendritic cells that have captured apoptotic cells. *J. Exp. Med.* **2000**. *191:* 411–416.

73 Urban, B. C., Ferguson, D. J., Pain, A., Willcox, N., Plebanski, M., Austyn, J. M., Roberts, D. J., Plasmodium falciparum-infected erythrocytes modulate the maturation of dendritic cells. *Nature* **1999**. *400:* 73–77.

74 Urban, B. C., Willcox, N., Roberts, D. J., A role for CD36 in the regulation of dendritic cell function. *Proc Natl Acad Sci USA* **2001**. *98:* 8750–8755.

75 Cvetanovic, M., Ucker, D. S., Innate immune discrimination of apoptotic cells: repression of proinflammatory macrophage transcription is coupled directly to specific recognition. *J Immunol* **2004**. *172:* 880–889.

76 Larsson, M., Fonteneau, J. F., Somersan, S., Sanders, C., Bickham, K., Thomas, E. K., Mahnke, K., Bhardwaj, N., Efficiency of cross presentation of vaccinia virus-derived antigens by human dendritic cells. *Eur J Immunol* **2001**. *31:* 3432–3442.

77 Herr, W., Ranieri, E., Olson, W., Zarour, H., Gesualdo, L., Storkus, W. J., Mature dendritic cells pulsed with freeze-thaw cell lysates define an effective in vitro vaccine designed to elicit EBV-specific CD4($^+$) and CD8($^+$) T lymphocyte responses. *Blood* **2000**. *96:* 1857–1864.

78 Plautz, G. E., Mukai, S., Cohen, P. A., Shu, S., Cross-presentation of tumor antigens to effector T cells is sufficient to mediate effective immunotherapy of established intracranial tumors. *J Immunol* **2000**. *165:* 3656–3662.

79 Matzinger, P., An innate sense of danger. *Ann N Y Acad Sci* **2002**. *961:* 341–342.

80 Gallucci, S., Lolkema, M., Matzinger, P., Natural adjuvants: endogenous activators of dendritic cells. *Nat. Med.* **1999**. *5:* 1249–1255.

81 Sauter, B., Albert, M. L., Francisco, L., Larsson, M., Somersan, S., Bhardwaj, N., Consequences of cell death: exposure to necrotic tumor cells, but not primary tissue cells or apoptotic cells, induces the maturation of immunostimulatory dendritic cells. *J Exp Med* **2000**. *191:* 423–434.

82 Skoberne, M., Beignon, A. S., Bhardwaj, N., Danger signals: a time and space continuum. *Trends Mol Med* **2004**. *10:* 251–257.

83 Li, Z., Menoret, A., Srivastava, P., Roles of heat-shock proteins in antigen presentation and cross-presentation. *Curr Opin Immunol* **2002**. *14:* 45–51.

84 Shi, Y., Evans, J. E., Rock, K. L., Molecular identification of a danger signal that alerts the immune system to dying cells. *Nature* **2003**. *425:* 516–521.

85 Somersan, S., Larsson, M., Fonteneau, J. F., Basu, S., Srivastava, P., Bhardwaj, N., Primary tumor tissue lysates are enriched in heat shock proteins and induce the maturation of human dendritic cells. *J Immunol* **2001**. *167:* 4844–4852.

86 Basu, S., Binder, R. J., Ramalingam, T., Srivastava, P. K., CD91 is a common receptor for heat shock proteins gp96, hsp90, hsp70, and calreticulin. *Immunity* **2001**. *14:* 303–313.

87 Messmer, D., Yang, H., Telusma, G., Knoll, F., Li, J., Messmer, B., Tracey, K. J., Chiorazzi, N., High mobility group box protein 1: an endogenous signal for dendritic cell maturation and Th1 polarization. *J Immunol* **2004**. *173:* 307–313.

88 Srivastava, P., Roles of heat-shock proteins in innate and adaptive immunity. *Nat Rev Immunol* **2002**. *2:* 185–194.

89 Binder, R. J., Vatner, R., Srivastava, P., The heat-shock protein receptors: some answers and more questions. *Tissue Antigens* **2004**. *64:* 442–451.

90 Palliser, D., Guillen, E., Ju, M., Eisen, H. N., Multiple intracellular routes in the cross-presentation of a soluble protein by murine dendritic cells. *J Immunol* **2005**. *174:* 1879–1887.

91 SenGupta, D., Norris, P. J., Suscovich, T. J., Hassan-Zahraee, M., Moffett, H. F., Trocha, A., Draenert, R., Goulder, P. J., Binder, R. J., Levey, D. L., Walker, B. D., Srivastava, P. K., Brander, C., Heat shock protein-mediated cross-presentation of exogenous HIV antigen on HLA class I and class II. *J Immunol* **2004**. *173:* 1987–1993.

92 Srivastava, P. K., Menoret, A., Basu, S., Binder, R. J., McQuade, K. L., Heat shock proteins come of age: primitive functions acquire new roles in an adaptive world. *Immunity* **1998**. *8:* 657–665.

93 Millar, D. G., Garza, K. M., Odermatt, B., Elford, A. R., Ono, N., Li, Z., Ohashi, P. S., Hsp70 promotes antigen-presenting cell function and converts T-cell tolerance to autoimmunity in vivo. *Nat Med* **2003**. *9:* 1469–1476.

94 Shen, L., Rock, K. L., Cellular protein is the source of cross-priming antigen in vivo. *Proc Natl Acad Sci USA* **2004**. *101:* 3035–3040.

95 Amigorena, S., Bonnerot, C., Fc receptor signaling and trafficking: a connection for antigen processing. *Immunol Rev* **1999**. *172:* 279–284.

96 Amigorena, S., Bonnerot, C., Fc receptors for IgG and antigen presentation on MHC class I and class II molecules. *Semin Immunol* **1999**. *11:* 385–390.

97 Heystek, H. C., Moulon, C., Woltman, A. M., Garonne, P., van Kooten, C., Human immature dendritic cells efficiently bind and take up secretory IgA without the induction of maturation. *J Immunol* **2002**. *168:* 102–107.

98 Geissmann, F., Launay, P., Pasquier, B., Lepelletier, Y., Leborgne, M., Lehuen, A., Brousse, N., Monteiro, R. C., A subset of human dendritic cells expresses IgA Fc receptor (CD89), which mediates internalization and activation upon cross-linking by IgA complexes. *J Immunol* **2001**. *166:* 346–352.

99 Nagata, Y., Ono, S., Matsuo, M., Gnjatic, S., Valmori, D., Ritter, G., Garrett, W., Old, L. J., Mellman, I., Differential presentation of a soluble exogenous tumor antigen, NY-ESO-1, by distinct human dendritic cell populations. *Proc Natl Acad Sci USA* **2002**. *99:* 10629–10634.

100 Matsuo, M., Nagata, Y., Sato, E., Atanackovic, D., Valmori, D., Chen, Y. T., Ritter, G., Mellman, I., Old, L. J., Gnjatic, S., IFN-gamma enables cross-presentation of exogenous protein antigen in human Langerhans cells by potentiating maturation. *Proc Natl Acad Sci USA* **2004**. *101:* 14467–14472.

101 Rodriguez, A., Regnault, A., Kleijmeer, M., Ricciardi-Castagnoli, P., Amigorena, S., Selective transport of internalized antigens to the cytosol for MHC class I presentation in dendritic cells. *Nat Cell Biol* **1999**. *1:* 362–368.

102 Regnault, A., Lankar, D., Lacabanne, V., Rodriguez, A., Thery, C., Rescigno, M., Saito, T., Verbeek, S., Bonnerot, C., Ricciardi-Castagnoli, P., Amigorena, S., Fcgamma receptor-mediated induction of dendritic cell maturation and major histocompatibility complex class I-restricted antigen presentation after immune complex internalization. *J Exp Med* **1999**. *189:* 371–380.

103 Machy, P., Serre, K., Leserman, L., Class I-restricted presentation of exogenous antigen acquired by Fcgamma receptor-mediated endocytosis is regulated in dendritic cells. *Eur J Immunol* **2000**. *30:* 848–857.

104 Dhodapkar, K. M., Kaufman, J. L., Ehlers, M., Banerjee, D. K., Bonvini, E., Koenig, S., Steinman, R. M., Ravetch, J. V., Dhodapkar, M. V., Selective blockade of inhibitory Fcgamma receptor enables human dendritic cell maturation with IL-12p70 production and immunity to antibody-coated tumor cells. *Proc Natl Acad Sci USA* **2005**. *102:* 2910–2915.

105 Rafiq, K., Bergtold, A., Clynes, R., Immune complex-mediated antigen presentation induces tumor immunity. *J Clin Invest* **2002**. *110:* 71–79.

106 Akiyama, K., Ebihara, S., Yada, A., Matsumura, K., Aiba, S., Nukiwa, T.,

Takai, T., Targeting apoptotic tumor cells to Fc gamma R provides efficient and versatile vaccination against tumors by dendritic cells. *J Immunol* **2003**. *170:* 1641–1648.

107 Harshyne, L. A., Watkins, S. C., Gambotto, A., Barratt-Boyes, S. M., Dendritic cells acquire antigens from live cells for cross-presentation to CTL. *J Immunol* **2001**. *166:* 3717–3723.

108 Harshyne, L. A., Zimmer, M. I., Watkins, S. C., Barratt-Boyes, S. M., A role for class A scavenger receptor in dendritic cell nibbling from live cells. *J Immunol* **2003**. *170:* 2302–2309.

109 Neijssen, J., Herberts, C., Drijfhout, J. W., Reits, E., Janssen, L., Neefjes, J., Cross-presentation by intercellular peptide transfer through gap junctions. *Nature* **2005**. *434:* 83–88.

110 Thery, C., Zitvogel, L., Amigorena, S., Exosomes: composition, biogenesis and function. *Nat Rev Immunol* **2002**. *2:* 569–579.

111 Trams, E. G., Lauter, C. J., Salem, N., Jr., Heine, U., Exfoliation of membrane ecto-enzymes in the form of micro-vesicles. *Biochim Biophys Acta* **1981**. *645:* 63–70.

112 Wolfers, J., Lozier, A., Raposo, G., Regnault, A., Thery, C., Masurier, C., Flament, C., Pouzieux, S., Faure, F., Tursz, T., Angevin, E., Amigorena, S., Zitvogel, L., Tumor-derived exosomes are a source of shared tumor rejection antigens for CTL cross-priming. *Nat Med* **2001**. *7:* 297–303.

113 Harding, C., Heuser, J., Stahl, P., Receptor-mediated endocytosis of transferrin and recycling of the transferrin receptor in rat reticulocytes. *J Cell Biol* **1983**. *97:* 329–339.

114 Raposo, G., Nijman, H. W., Stoorvogel, W., Liejendekker, R., Harding, C. V., Melief, C. J., Geuze, H. J., B lymphocytes secrete antigen-presenting vesicles. *J Exp Med* **1996**. *183:* 1161–1172.

115 Zitvogel, L., Regnault, A., Lozier, A., Wolfers, J., Flament, C., Tenza, D., Ricciardi-Castagnoli, P., Raposo, G., Amigorena, S., Eradication of established murine tumors using a novel cell-free vaccine: dendritic cell-derived exosomes. *Nat Med* **1998**. *4:* 594–600.

116 Aline, F., Bout, D., Amigorena, S., Roingeard, P., Dimier-Poisson, I., Toxoplasma gondii antigen-pulsed-dendritic cell-derived exosomes induce a protective immune response against T. gondii infection. *Infect Immun* **2004**. *72:* 4127–4137.

117 Morelli, A. E., Larregina, A. T., Shufesky, W. J., Sullivan, M. L., Stolz, D. B., Papworth, G. D., Zahorchak, A. F., Logar, A. J., Wang, Z., Watkins, S. C., Falo, L. D., Jr., Thomson, A. W., Endocytosis, intracellular sorting, and processing of exosomes by dendritic cells. *Blood* **2004**. *104:* 3257–3266.

118 Horner, A. A., Datta, S. K., Takabayashi, K., Belyakov, I. M., Hayashi, T., Cinman, N., Nguyen, M. D., Van Uden, J. H., Berzofsky, J. A., Richman, D. D., Raz, E., Immunostimulatory DNA-based vaccines elicit multifaceted immune responses against HIV at systemic and mucosal sites. *J Immunol* **2001**. *167:* 1584–1591.

119 Cho, H. J., Takabayashi, K., Cheng, P. M., Nguyen, M. D., Corr, M., Tuck, S., Raz, E., Immunostimulatory DNA-based vaccines induce cytotoxic lymphocyte activity by a T-helper cell-independent mechanism. *Nat Biotechnol* **2000**. *18:* 509–514.

120 Datta, S. K., Redecke, V., Prilliman, K. R., Takabayashi, K., Corr, M., Tallant, T., DiDonato, J., Dziarski, R., Akira, S., Schoenberger, S. P., Raz, E., A subset of Toll-like receptor ligands induces cross-presentation by bone marrow-derived dendritic cells. *J Immunol* **2003**. *170:* 4102–4110.

121 West, M. A., Wallin, R. P., Matthews, S. P., Svensson, H. G., Zaru, R., Ljunggren, H. G., Prescott, A. R., Watts, C., Enhanced dendritic cell antigen capture via toll-like receptor-induced actin remodeling. *Science* **2004**. *305:* 1153–1157.

122 Schulz, O., Diebold, S. S., Chen, M., Naslund, T. I., Nolte, M. A., Alexopoulou, L., Azuma, Y. T., Flavell, R. A., Liljestrom, P., Reis e Sousa, C., Toll-like receptor 3 promotes cross-priming to virus-infected cells. *Nature* **2005**. *433:* 887–892.

123 Le Bon, A., Etchart, N., Rossmann, C., Ashton, M., Hou, S., Gewert, D., Borrow, P., Tough, D. F., Cross-priming of CD8+ T cells stimulated by virus-induced type I interferon. *Nat Immunol* **2003**. *4:* 1009–1015.

124 Chen, M., Barnfield, C., Naslund, T. I., Fleeton, M. N., Liljestrom, P., MyD88 expression is required for efficient cross-presentation of viral antigens from infected cells. *J Virol* **2005**. *79:* 2964–2972.

125 Takeda, K., Akira, S., TLR signaling pathways. *Semin Immunol* **2004**. 16: 3–9.

126 Blander, J. M., Medzhitov, R., Regulation of phagosome maturation by signals from toll-like receptors. *Science* **2004**. *304:* 1014–1018.

127 Lucas, M., Stuart, L. M., Savill, J., Lacy-Hulbert, A., Apoptotic cells and innate immune stimuli combine to regulate macrophage cytokine secretion. *J Immunol* **2003**. *171:* 2610–2615.

128 Delamarre, L., Pack, M., Chang, H., Mellman, I., Trombetta, E. S., Differential lysosomal proteolysis in antigen-presenting cells determines antigen fate. *Science* **2005**. *307:* 1630–1634.

129 Reis e Sousa, C., Germain, R. N., Major histocompatibility complex class I presentation of peptides derived from soluble exogenous antigen by a subset of cells engaged in phagocytosis. *J Exp Med* **1995**. *182:* 841–851.

130 Fonteneau, J. F., Kavanagh, D. G., Lirvall, M., Sanders, C., Cover, T. L., Bhardwaj, N., Larsson, M., Characterization of the MHC class I cross-presentation pathway for cell-associated antigens by human dendritic cells. *Blood* **2003**. *102:* 4448–4455.

131 Kerksiek, K. M., Niedergang, F., Chavrier, P., Busch, D. H., Brocker, T., Selective Rac1 inhibition in dendritic cells diminishes apoptotic cell uptake and cross-presentation in vivo. *Blood* **2005**. *105:* 742–749.

132 Schwartz, M., Rho signalling at a glance. *J Cell Sci* **2004**. 117: 5457–5458.

133 Albert, M. L., Kim, J. I., Birge, R. B., alphavbeta5 integrin recruits the CrkII-Dock180-rac1 complex for phagocytosis of apoptotic cells. *Nat Cell Biol* **2000**. *2:* 899–905.

134 Aktories, K., Rho proteins: targets for bacterial toxins. *Trends Microbiol* **1997**. *5:* 282–288.

135 Gagnon, E., Duclos, S., Rondeau, C., Chevet, E., Cameron, P. H., Steele-Mortimer, O., Paiement, J., Bergeron, J. J., Desjardins, M., Endoplasmic reticulum-mediated phagocytosis is a mechanism of entry into macrophages. *Cell* **2002**. *110:* 119–131.

136 Pfeifer, J. D., Wick, M. J., Roberts, R. L., Findlay, K., Normark, S. J., Harding, C. V., Phagocytic processing of bacterial antigens for class I MHC presentation to T cells. *Nature* **1993**. *361:* 359–362.

137 Houde, M., Bertholet, S., Gagnon, E., Brunet, S., Goyette, G., Laplante, A., Princiotta, M. F., Thibault, P., Sacks, D., Desjardins, M., Phagosomes are competent organelles for antigen cross-presentation. *Nature* **2003**. *425:* 402–406.

138 Guermonprez, P., Saveanu, L., Kleijmeer, M., Davoust, J., Van Endert, P., Amigorena, S., ER-phagosome fusion defines an MHC class I cross-presentation compartment in dendritic cells. *Nature* **2003**. *425:* 397–402.

139 Ackerman, A. L., Kyritsis, C., Tampe, R., Cresswell, P., Early phagosomes in dendritic cells form a cellular compartment sufficient for cross presentation of exogenous antigens. *Proc Natl Acad Sci USA* **2003**. *100:* 12889–12894.

140 Lizee, G., Basha, G., Tiong, J., Julien, J. P., Tian, M., Biron, K. E., Jefferies, W. A., Control of dendritic cell cross-presentation by the major histocompatibility complex class I cytoplasmic domain. *Nat Immunol* **2003**. *4:* 1065–1073.

141 Ackerman, A. L., Kyritsis, C., Tampe, R., Cresswell, P., Access of soluble antigens to the endoplasmic reticulum can explain cross-presentation by dendritic cells. *Nat Immunol* **2005**. *6:* 107–113.

142 Sandvig, K., van Deurs, B., Transport of protein toxins into cells: pathways used by ricin, cholera toxin and Shiga toxin. *FEBS Lett* **2002**. *529:* 49–53.

143 Norbury, C. C., Hewlett, L. J., Prescott, A. R., Shastri, N., Watts, C., Class I MHC presentation of exogenous soluble antigen via macropinocytosis in bone marrow macrophages. *Immunity* 1995. *3:* 783–791.

144 Huang, A. Y., Bruce, A. T., Pardoll, D. M., Levitsky, H. I., In vivo cross-priming of MHC class I-restricted antigens requires TAP transporter. *Immunity* 1996. *4:* 349–355.

145 Sigal, L. J., Crotty, S., ino, R., Rock, K. L., Cytotoxic T-cell immunity to virus-infected non-haematopoietic cells requires presentation of exogenous antigen. *Nature* 1999. *398:* 77–80.

146 Sigal, L. J., Rock, K. L., Bone marrow-derived antigen-presenting cells are required for the generation of cytotoxic T lymphocyte responses to viruses and use transporter associated with antigen presentation (TAP)-dependent and -independent pathways of antigen presentation. *J Exp Med* 2000. *192:* 1143–1150.

147 Blachere, N. E., Darnell, R. B., Albert, M. L., Apoptotic cells deliver processed antigen to dendritic cells for cross-presentation. *PLoS Biol* 2005. *3:* e185.

148 Shen, L., Sigal, L. J., Boes, M., Rock, K. L., Important role of cathepsin S in generating peptides for TAP-independent MHC class I crosspresentation in vivo. *Immunity* 2004. *21:* 155–165.

149 Castellino, F., Boucher, P. E., Eichelberg, K., Mayhew, M., Rothman, J. E., Houghton, A. N., Germain, R. N., Receptor-mediated uptake of antigen/heat shock protein complexes results in major histocompatibility complex class I antigen presentation via two distinct processing pathways. *J Exp Med* 2000. *191:* 1957–1964.

150 Snyder, H. L., Bacik, I., Bennink, J. R., Kearns, G., Behrens, T. W., Bachi, T., Orlowski, M., Yewdell, J. W., Two novel routes of transporter associated with antigen processing (TAP)-independent major histocompatibility complex class I antigen processing. *J Exp Med* 1997. *186:* 1087–1098.

151 Chefalo, P. J., Grandea, A. G., 3rd, Van Kaer, L., Harding, C. V., Tapasin$^{-/-}$ and TAP1$^{-/-}$ macrophages are deficient in vacuolar alternate class I MHC (MHC-I) processing due to decreased MHC-I stability at phagolysosomal pH. *J Immunol* 2003. *170:* 5825–5833.

152 Oviedo-Orta, E., Howard Evans, W., Gap junctions and connexin-mediated communication in the immune system. *Biochim Biophys Acta* 2004. *1662:* 102–112.

153 Ochsenbein, A. F., Sierro, S., Odermatt, B., Pericin, M., Karrer, U., Hermans, J., Hemmi, S., Hengartner, H., Zinkernagel, R. M., Roles of tumour localization, second signals and cross priming in cytotoxic T-cell induction. *Nature* 2001. *411:* 1058–1064.

154 Huang, A. Y., Golumbek, P., Ahmadzadeh, M., Jaffee, E., Pardoll, D., Levitsky, H., Role of bone marrow-derived cells in presenting MHC class I-restricted tumor antigens. *Science* 1994. *264:* 961–965.

155 Arrode, G., Boccaccio, C., Lule, J., Allart, S., Moinard, N., Abastado, J. P., Alam, A., Davrinche, C., Incoming human cytomegalovirus pp65 (UL83) contained in apoptotic infected fibroblasts is cross-presented to CD8($^+$) T cells by dendritic cells. *J Virol* 2000. *74:* 10018–10024.

156 Arrode, G., Boccaccio, C., Abastado, J. P., Davrinche, C., Cross-presentation of human cytomegalovirus pp65 (UL83) to CD8$^+$ T cells is regulated by virus-induced, soluble-mediator-dependent maturation of dendritic cells. *J Virol* 2002. *76:* 142–150.

157 Tabi, Z., Moutaftsi, M., Borysiewicz, L. K., Human cytomegalovirus pp65– and immediate early 1 antigen-specific HLA class I-restricted cytotoxic T cell responses induced by cross-presentation of viral antigens. *J Immunol* 2001. *166:* 5695–5703.

158 Subklewe, M., Paludan, C., Tsang, M. L., Mahnke, K., Steinman, R. M., Munz, C., Dendritic cells cross-present latency gene products from Epstein-Barr virus-transformed B cells and expand tumor-reactive CD8($^+$) killer T cells. *J Exp Med* 2001. *193:* 405–411.

159 Larsson, M., Fonteneau, J. F., Lirvall, M., Haslett, P., Lifson, J. D., Bhardwaj, N., Activation of HIV-1 specific CD4 and CD8 T cells by human dendritic cells: roles for cross-presentation and non-infectious HIV-1 virus. *Aids* 2002. *16:* 1319–1329.

160 Andrieu, M., Desoutter, J. F., Loing, E., Gaston, J., Hanau, D., Guillet, J. G., Hosmalin, A., Two human immunodeficiency virus vaccinal lipopeptides follow different cross-presentation pathways in human dendritic cells. *J Virol* **2003**. *77:* 1564–1570.

161 Ignatius, R., Marovich, M., Mehlhop, E., Villamide, L., Mahnke, K., Cox, W. I., Isdell, F., Frankel, S. S., Mascola, J. R., Steinman, R. M., Pope, M., Canarypox virus-induced maturation of dendritic cells is mediated by apoptotic cell death and tumor necrosis factor alpha secretion. *J Virol* **2000**. *74:* 11329–11338.

162 Pollara, G., Speidel, K., Samady, L., Rajpopat, M., McGrath, Y., Ledermann, J., Coffin, R. S., Katz, D. R., Chain, B., Herpes simplex virus infection of dendritic cells: balance among activation, inhibition, and immunity. *J Infect Dis* **2003**. *187:* 165–178.

163 Schneider-Schaulies, S., ter Meulen, V., Triggering of and interference with immune activation: interactions of measles virus with monocytes and dendritic cells. *Viral Immunol* **2002**. *15:* 417–428.

164 Servet-Delprat, C., Vidalain, P. O., Valentin, H., Rabourdin-Combe, C., Measles virus and dendritic cell functions: how specific response cohabits with immunosuppression. *Curr Top Microbiol Immunol* **2003**. *276:* 103–123.

165 Janda, J., Schoneberger, P., Skoberne, M., Messerle, M., Russmann, H., Geginat, G., Cross-presentation of Listeria-derived CD8 T cell epitopes requires unstable bacterial translation products. *J Immunol* **2004**. *173:* 5644–5651.

166 Tvinnereim, A. R., Hamilton, S. E., Harty, J. T., Neutrophil involvement in cross-priming CD8+ T cell responses to bacterial antigens. *J Immunol* **2004**. *173:* 1994–2002.

167 Schaible, U. E., Winau, F., Sieling, P. A., Fischer, K., Collins, H. L., Hagens, K., Modlin, R. L., Brinkmann, V., Kaufmann, S. H., Apoptosis facilitates antigen presentation to T lymphocytes through MHC-I and CD1 in tuberculosis. *Nat Med* **2003**. *9:* 1039–1046.

168 Zinkernagel, R. M., On cross-priming of MHC class I-specific CTL: rule or exception? *Eur J Immunol* **2002**. *32:* 2385–2392.

169 Freigang, S., Egger, D., Bienz, K., Hengartner, H., Zinkernagel, R. M., Endogenous neosynthesis vs. cross-presentation of viral antigens for cytotoxic T cell priming. *Proc Natl Acad Sci USA* **2003**. *100:* 13477–13482.

170 Platt, N., Suzuki, H., Kodama, T., Gordon, S., Apoptotic thymocyte clearance in scavenger receptor class A-deficient mice is apparently normal. *J Immunol* **2000**. *164:* 4861–4867.

171 Delneste, Y., Magistrelli, G., Gauchat, J., Haeuw, J., Aubry, J., Nakamura, K., Kawakami-Honda, N., Goetsch, L., Sawamura, T., Bonnefoy, J., Jeannin, P., Involvement of LOX-1 in dendritic cell-mediated antigen cross-presentation. *Immunity* **2002**. *17:* 353–362.

172 Hoffmann, P. R., Kench, J. A., Vondracek, A., Kruk, E., Daleke, D. L., Jordan, M., Marrack, P., Henson, P. M., Fadok, V. A., Interaction between phosphatidylserine and the phosphatidylserine receptor inhibits immune responses in vivo. *J Immunol* **2005**. *174:* 1393–1404.

173 Fadok, V. A., Voelker, D. R., Campbell, P. A., Cohen, J. J., Bratton, D. L., Henson, P. M., Exposure of phosphatidylserine on the surface of apoptotic lymphocytes triggers specific recognition and removal by macrophages. *J Immunol* **1992**. *148:* 2207–2216.

174 Scott, R. S., McMahon, E. J., Pop, S. M., Reap, E. A., Caricchio, R., Cohen, P. L., Earp, H. S., Matsushima, G. K., Phagocytosis and clearance of apoptotic cells is mediated by MER. *Nature* **2001**. *411:* 207–211.

175 Savill, J., Apoptosis. Phagocytic docking without shocking. *Nature* **1998**. *392:* 442–443.

176 Ogden, C. A., deCathelineau, A., Hoffmann, P. R., Bratton, D., Ghebrehiwet, B., Fadok, V. A., Henson, P. M., C1q and mannose binding lectin engagement of cell surface calreticulin and CD91 initiates macropinocytosis and uptake of apoptotic cells. *J Exp Med* **2001**. 194: 781–795.

177 Wright, S. D., Silverstein, S. C., Receptors for C3b and C3bi promote phagocytosis but not the release of toxic oxygen from human phagocytes. *J Exp Med* **1983**. *158:* 2016–2023.

178 Marguet, D., Luciani, M. F., Moynault, A., Williamson, P., Chimini, G., Engulfment of apoptotic cells involves the redistribution of membrane phosphatidylserine on phagocyte and prey. *Nat Cell Biol* **1999**. *1:* 454–456.

179 Brown, S., Heinisch, I., Ross, E., Shaw, K., Buckley, C. D., Savill, J., Apoptosis disables CD31-mediated cell detachment from phagocytes promoting binding and engulfment. *Nature* **2002**. *418:* 200–203.

180 Kariko, K., Ni, H., Capodici, J., Lamphier, M., Weissman, D., mRNA is an endogenous ligand for toll-like receptor 3. *J Biol Chem* **2004**. *279:* 12542–12550.

181 Elias, F., Flo, J., Lopez, R. A., Zorzopulos, J., Montaner, A., Rodriguez, J. M., Strong cytosine-guanosine-independent immunostimulation in humans and other primates by synthetic oligodeoxynucleotides with PyNTTTTGT motifs. *J Immunol* **2003**. *171:* 3697–3704.

182 Park, J. S., Svetkauskaite, D., He, Q., Kim, J. Y., Strassheim, D., Ishizaka, A., Abraham, E., Involvement of Toll-like Receptors 2 and 4 in Cellular Activation by High Mobility Group Box 1 Protein. *J Biol Chem* **2004**. *279:* 7370–7377.

183 Ohashi, K., Burkart, V., Flohe, S., Kolb, H., Cutting edge: heat shock protein 60 is a putative endogenous ligand of the toll-like receptor-4 complex. *J Immunol* **2000**. *164:* 558–561.

184 Vabulas, R. M., Ahmad-Nejad, P., da Costa, C., Miethke, T., Kirschning, C. J., Hacker, H., Wagner, H., Endocytosed HSP60s use toll-like receptor 2 (TLR2) and TLR4 to activate the toll/interleukin-1 receptor signaling pathway in innate immune cells. *J Biol Chem* **2001**. *276:* 31332–31339.

185 Vabulas, R. M., Braedel, S., Hilf, N., Singh-Jasuja, H., Herter, S., Ahmad-Nejad, P., Kirschning, C. J., Da Costa, C., Rammensee, H. G., Wagner, H., Schild, H., The endoplasmic reticulum-resident heat shock protein Gp96 activates dendritic cells via the Toll-like receptor 2/4 pathway. *J Biol Chem* **2002**. *277:* 20847–20853.

186 Scaffidi, P., Misteli, T., Bianchi, M. E., Release of chromatin protein HMGB1 by necrotic cells triggers inflammation. *Nature* 2002. *418:* 191–195.

187 Binder, R. J., Han, D. K., Srivastava, P. K., CD91: a receptor for heat shock protein gp96. *Nat Immunol* 2000. *1:* 151–155.

188 Binder, R. J., Srivastava, P. K., Essential role of CD91 in re-presentation of gp96-chaperoned peptides. *Proc Natl Acad Sci USA* 2004. *101:* 6128–6133.

189 Kol, A., Lichtman, A. H., Finberg, R. W., Libby, P., Kurt-Jones, E. A., Cutting edge: heat shock protein (HSP) 60 activates the innate immune response: CD14 is an essential receptor for HSP60 activation of mononuclear cells. *J Immunol* 2000. *164:* 13–17.

190 Asea, A., Kraeft, S. K., Kurt-Jones, E. A., Stevenson, M. A., Chen, L. B., Finberg, R. W., Koo, G. C., Calderwood, S. K., HSP70 stimulates cytokine production through a CD14-dependant pathway, demonstrating its dual role as a chaperone and cytokine. *Nat Med* 2000. *6:* 435–442.

191 Wang, Y., Kelly, C. G., Karttunen, J. T., Whittall, T., Lehner, P. J., Duncan, L., MacAry, P., Younson, J. S., Singh, M., Oehlmann, W., Cheng, G., Bergmeier, L., Lehner, T., CD40 is a cellular receptor mediating mycobacterial heat shock protein 70 stimulation of CC-chemokines. *Immunity* 2001. *15:* 971–983.

192 Becker, T., Hartl, F. U., Wieland, F., CD40, an extracellular receptor for binding and uptake of Hsp70-peptide complexes. *J Cell Biol* 2002. *158:* 1277–1285.

193 van Kooten, C., Bancherau, J., CD40-CD40 ligand. *J Leukoc Biol* 2000. *67:* 2–17.

194 Berwin, B., Hart, J. P., Rice, S., Gass, C., Pizzo, S. V., Post, S. R., Nicchitta, C. V., Scavenger receptor-A mediates gp96/GRP94 and calreticulin internalization by antigen-presenting cells. *Embo J* 2003. *22:* 6127–6136.

23

A Systems Biologist's View of Dendritic Cell–Cytotoxic T Lymphocyte Interaction

Burkhard Ludewig and Gennady Bocharov

23.1
Introduction

CD8$^+$ cytotoxic T lymphocyte (CTL) responses represent a major immune mechanism of protection against many viral infections and tumors through the killing of altered (infected or malignant) cells [1]. Viruses generally provide signals to the immune system, either on the surface of infected cells or on antigen-presenting cells, that permit CD8$^+$ T cells to recognize virus-encoded peptides through their T-cell receptor (TCR). Viruses that induce an early and efficient CD8$^+$ T-cell response, such as the lymphocytic choriomeningitis virus (LCMV) in mice [2, 3] or the human immunodeficiency virus (HIV) [4, 5] tend to infect professional antigen-presenting cells such as DCs and macrophages. Importantly, DCs can be infected by viruses independently of a specific receptor [6], a finding that supports the notion that the general susceptibility of DCs to infection with a wide range of viruses may represent an important prerequisite for efficient CTL priming against viruses with more restricted tissue tropism [7].

Virus localization and transport, governing the kinetics of appearance of the viral antigens in different compartments of the body, determine whether CTL reactivity is induced, maintained, or aborted. Primary virus infection normally leads to CTL dynamics that have three distinctive phases: clonal expansion, contraction, and persistence of memory cells. Both the magnitude and the duration of the CTL response are tightly regulated at the single cell and the CTL population level. The early steps of T-cell proliferation/differentiation are still poorly characterized. For example, it still remains to be resolved how fast the effector function is acquired, or whether the activated CTLs follow "sequential" or "parallel" differentiation routes to become effector or memory CTL. Nonetheless, a successful/protective CTL response depends critically on the availability of specific antigens, efficient delivery of these antigens and their optimal presentation to the T cells within secondary lymphoid organs.

DCs are known for their remarkable potential to induce specific T-cell responses [8]. The extraordinary efficacy of DC to prime immune responses is shown by

Handbook of Dendritic Cells. Biology, Diseases, and Therapies.
Edited by M. B. Lutz, N. Romani, and A. Steinkasserer.

the fact that only 10^2 to 10^3 antigen-presenting DCs in the spleen are sufficient to elicit protective levels of antiviral CTL activation in mice [9]. Furthermore, a series of experimental studies in mice demonstrated that tumor-specific CTL can be induced by adoptive transfer of antigen-presenting DCs [10, 11]. However, the success of DC-based immunotherapeutical treatment of human cancer is still limited [12], most probably because the complexity of the regulation of the DC–CTL interaction is far beyond simple dose-effect type regulation schemes and its sensitivity to various factors is not yet completely understood. For example, DCs positively regulate immune responses by mediating antigen influx and antigen presentation in secondary lymphoid organs. However, they are rapidly lost during the cognate interaction with fully activated effector T cells [13].

The ability of DCs to induce protective CTL responses to a broad spectrum of pathogens (viruses and intracellular bacteria) under a wide range of conditions, implies that the interaction of DCs and CTLs is a *robust* biological system. This interaction is regulated at the single-cell level by a complex signal transduction and gene activation machinery and at the cell-population level by various modes of communication (e.g. cytokines, cognate interaction between MHC and TCR molecules, and recognition of particular surface molecules by cognate receptors) in different spatial compartments. The interaction of DC and CTL therefore is based on a highly regulated set of spatiotemporal interactions with the outcome depending on a large number of intra- and inter-cellular physical and biochemical parameters. Because of the inherent complexity of these processes, we follow here a "*systems biology*" analysis of the structure, dynamics and the operating principles that permit DCs to induce protective CTL responses, with a particular focus on the robustness of this phenomenon. We will (i) consider the components and structures that are central in the DC–CTL interaction; (ii) discuss the mechanisms that underlie specific DC–CTL interaction outcomes; and (iii) describe the sensitivity of the major patterns of CTL responses to variations in the interacting components or environment conditions. In our view, this conceptual framework might help to provide a consistent theoretical basis for transforming "disordered networks" of factors that potentially influence the interaction of DCs and CTLs into an "ordered hierarchy" of relevant parameters ranked according to their quantitative effects on the production of reliable and robust CTL responses.

23.2
Deciphering the Systems Biologist's Approach

"Systems biology" represents a framework for analysis and interpretation of the structure and dynamics of complex biological phenomena. Although it has a long history [14, 15], it is because of the recent advances in technologies for high-throughput and quantitative measurement of the interactions at the level of genes, proteins, cells, tissues, organs and systems that the interest in system-level approaches to biological phenomena has been renewed [16]. In theoretical immunology, the need for a systems approach was appreciated in late 1970s [17]. In experimental immunology, however, it has not become a universally accepted principle in the interpretation of immune phenomena yet, despite the fact that this approach has productively been used to address the issues of (i) the protection unit – the module of humoral immunity selected by evolution [18]; (ii) immune system recognition of antigen [19]; (iii) regulation of the immune responses [20]; and (iv) principles involved in effective operation of the immune system [21].

Systems biology focuses on the analysis of the structure, dynamics, design principles and control methods of biological systems in order to understand how robustness is achieved. Robustness means the ability to maintain stable functioning despite various perturbations [22]. It is recognized as an essential feature of biological systems and a conserved organizing principle in biology. Indeed, in a variety of biological systems ranging from genetic switches to physiological reactions, the robustness of their state and function to external and internal perturbations has been established [23]. The systems theory suggests that robustness is achieved through modularity, feedback, redundancy and structural stability. It is important that acquiring robustness against a certain set of perturbations is associated with fragility to other perturbations, which are outside the conventional set. This aspect of complex systems performance is considered to be related to the conservation of the fragility principle, and this issue has been discussed in detail in recent reviews by Kitano [23] and Csete and Doyle [24].

The ability of DCs to induce protective CTL responses to a broad spectrum of pathogens under a wide range of conditions with minimal immunopathological damage is critical in assuring the survival of the host. This implies that this physiological reaction of the immune system, namely the interaction of DCs with CTLs, represents a robust biological system. Therefore, "a systems biology" analysis of the various parameters of DC-mediated induction of CTL responses represents a coherent approach to deal with the complexity of the issue. This requires, however, a change in the type of questions that are being addressed before reviewing the current knowledge about DC–CTL interaction. The following section therefore introduces the key elements of the systems biology analysis of immune responses which govern the intrinsic robustness of DC–CTL interaction.

23.2.1
Modularity and Protocols

Modules are the components, parts or subsystems that contain the following features: (i) they have identifiable interfaces to other modules; (ii) they can be modified somewhat independently; (iii) they maintain some identity when isolated or rearranged; yet (iv) they derive additional identity from the rest of the system [24]. The modular approach greatly facilitates the conceptual understanding of the structure of complex systems. The meaning of modules varies depending on the level (from proteins and genes to pathways and networks and finally to tissues and organs) and the type (physical or logical/functional structure) of analysis. For example, the organs in the immune system, or the leukocyte subsets (DCs, T- or B-cells) represent two straightforward examples of physical modules. On the other hand, the antigen-processing pathways for peptide loading of MHC class I or class II molecules can be viewed as examples of two functional modules.

The modular structure of complex systems is thought to enhance their robustness by preventing spread or amplification of local perturbations [25]. In the case of the DC–CTL interaction, modules can be specified at various levels. For the purpose of our analysis we distinguish three basic modules: the thymus as "educational" module, the secondary lymphoid organs as "activation" module and the peripheral organs as "surveillance" module. Fig. 23.1 shows the modularity feature of the DC-dependent CTL maturation from a progenitor cell into a functional effector CTL.

Modular organization implies the need to understand the rules that prescribe interfaces or protocols between modules, permitting system functions that could not be achieved by isolated modules [24]. From a structural perspective, every cell in our body presents on the cell surface almost every cellular protein in complexes with MHC class I molecules providing information to the immune system about the cellular interior. This *MHC-restricted presentation* gives an example of a protocol, which is used in the DC–CTL interaction at the single cell level. For the TCR-mediated signaling event, there are additional molecules that must be integrated into consideration, such as CD28-CD80/CD86 or CLTA-4-CD80/CD80 interactions which represent second order parameters in the induction of CTL responses.

At the immune system level, it is the DC that provides the interface between the surveillance and activation modules (see Fig. 23.1). Indeed, the antigen, which is outside organized lymphoid tissues, is ignored unless it gets trapped by the antigen-presenting cells and brought into the secondary lymphoid organs [26]. There are some other general rules for the induction of immune responses, e.g. that the immune system responds to a *rapid perturbation* in its homeostasis – a term that includes an increase in antigenic stimulation and the requirement for specialized tissue milieus [20].

From the functional perspective, the ability to generate a biphasic signal response is an essential design constraint on the modular implementation of complex systems in biology [27]. The well-known low- and high-zone tolerance [28] or recently shown bell-shaped relationship between the peak CTL expansion and virus growth rate in LCMV system [29], confirm that the immune system is adjusted to respond to a certain optimal level in the activation signal, whereas the response to signals outside the optimal range is not induced.

23.2.2
Feedback Control

The concept of feedback control is central to the functioning of complex systems since it provides the robustness to external disturbances and variation of internal components [24]. The positive feedback acts to amplify small disturbances creating switches and breaking homogeneities, whereas the negative feedback serves to maintain patterns/states. It is noteworthy that in multilevel hierarchical systems, the task of the higher-level regulators is not to control, but rather to produce a coherent behavior/functioning of the lower level modules [30].

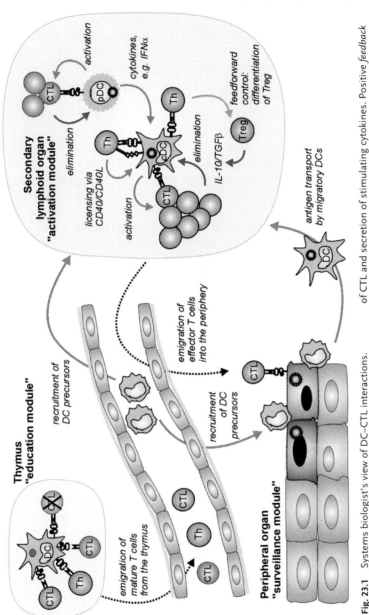

Fig. 23.1 Systems biologist's view of DC–CTL interactions. *Modularity* in the complex system of (direct and indirect) DC–CTL interactions is represented by different organizational modules: thymus ("educational module"), secondary lymphoid organs ("activation module"), and peripheral nonlymphoid organs ("surveillance module"). Different DC subsets, e.g. conventional DCs (cDC) versus plasmacytoid DCs (pDC) provide heterogenous redundancy and ensure homogenous outcome, that is, activation of CTL and secretion of stimulating cytokines. Positive *feedback* mechanisms (e.g. chemokine-driven recruitment of DC precursors into peripheral organs or secondary lymphoid organs, or CD40-mediated licensing of DCs) are indicated by green arrows. Negative feedback control loops such as elimination of DCs by effector CTL are depicted in red. Generation of regulatory Th cells (Treg) can be seen as feedforward control (blue arrow).

Immune processes have to be tightly regulated which is achieved by multiple interconnected control loops, implementing various types of feedback and feedforward controls. At the cellular level, the combination of negative (disturbances attenuating) and positive (disturbances amplifying) feedback loops is considered to be fundamental for generating the TCR-dependent signaling apparatus, which is both highly sensitive and discriminatory [21]. Individual cells tune and update their activation thresholds through changes in the co-stimulatory molecules expression, or intracellular kinetic competition between activating and deactivating signals [20].

The positive feedback is essential in all growth processes. in the immune system it serves to break up initial homogeneity of the lymphocyte repertoire and to adapt the repertoire to the pattern of antigenic perturbations experienced during the life of the host. In the context of the DC-mediated CTL priming the DC functions to amplify the antigen-specific CTL clone(s). It has been suggested [21] that mutually amplifying positive feedback interactions between CTL and DCs lead to a maximal response once an initial signal rises beyond a low minimal level (see Fig. 23.1 and Section 23.3.1). Conversely, there is growing evidence which suggests that CTL-mediated elimination of the antigen-presenting DC is probably a key process in the downregulation of adaptive immune responses so that the immune reaction does not become an immune overreaction [13, 31, 32]. The amplification effect of antigen-presenting DC on the CTL combined with the negative feedback from effector CTL on DC numbers implies that the cell population dynamics of the CTL–DC system *in vivo* reflects a predator–prey type of interaction [33]. This combination of positive and negative controls regulating the DC mediated CTL priming is depicted schematically in Fig. 23.1.

23.2.3
Redundancy

Redundancy means that multiple components/modules with equivalent or overlapping functions are present in the system and can substitute each other [23]. It allows the complex system to respond to a knockout by using alternative components ensuring robustness to component failure. The homogeneous redundancy, based on multiple copies of identical components, is considered to be effective when stimuli are targeted to a specific component. In general, the heterogeneous redundancy mediated by functionally close components is thought to ensure better protection against common mode failures. A combination of homogeneous outcome based upon heterogeneous implementation is an essential design feature of robust systems.

The immune system provides a rich source of examples of both homogeneous and heterogeneous type of redundancy at the systemic, cell population and molecular level; e.g. (i) its spatial structure represents an interconnected network of primary and secondary lymphoid organs with many similar yet distinct features maintaining a diverse repertoire of lymphocytes, (ii) $CD4^+$ and $CD8^+$ T cells display some similar functions, e.g. both can produce the effector cytokine interferon-gamma (IFNγ) or eliminate circulating target cells in a cognate interaction [34, 35], (iii) the chemokine network is a highly redundant system [36].

In the context of the DC-mediated induction of protective CTL responses, the redundancy is apparent in several ways. Two examples are presented in Fig. 23.1: (1) Redundancy on the CTL side is achieved in the form of multiple CTL clones which recognize the same epitope and are heterogeneous in the recognition features (binding affinity, on/off rates). The polyclonality and multispecificity of CTL responses are thought to be important in the control of viral infections, such as HBV [37] or HIV [38]. (2) Redundancy at the level of DC reveals itself in the existence of at least two distinct DC populations, known as conventional and plasmacytoid DCs, and a range of subsets with remarkable functional plasticity facilitating avid reactions against the pathogen.

23.2.4
Structural Stability

The system is called structurally stable if it preserves patterns of dynamics in spite of some perturbations in its parameters [39], this implies the existence of in-built resistance to particular perturbations. Concepts of stability are concerned with the effect on the system's behavior of perturbations in the system, e.g. kinetic parameters, cell numbers, environmental factors. The analyses of how systems behavior changes when perturbations are introduced into the system and which built-in mechanisms promote the stability are the focus of experimental immunology. Specifically, it has the aim to unravel the parameter variations, module knockouts, mutations, etc., which change the specific behavior. For example, the overall interaction of high-avidity CTLs with DCs displays an intrinsic structural stability once a certain threshold number of DCs presenting a particular peptide had entered a secondary lymphoid organ [33]. A further trait of the inherent structural stability in the interaction between DCs and CTLs is illustrated by the robust induction of antiviral CTL responses even in the absence of organized T-cell zones in secondary lymphoid organs. Mice lacking the chemokine receptor CCR7 [40] or its ligand chemokine CCL19 and CCL21 [41] display significant defects in the correct intralymphoid organs positioning of DCs and CTLs. Despite the lack of functional T-cell zones and a shift of DC–CTL interactions to the marginal zone, both CCR7-deficient [42] and *plt/plt* mice [43] exhibit only minor impairments in the induction of protective antiviral immune responses. These studies provide evidence for the relative insensitivity of DC–CTL interactions to particular perturbations.

23.3
From Systems Biology to DC–CTL Immunobiology

The frequency of specific CTL precursors (CTLp) for a distinct peptide/MHC complex is rather low, with approximately 1 in 100 000 in mice [44, 45] or humans [46]. In response to infection with LCMV, for example, these CTLp expand by a factor of 10^4–10^5, i.e. 100–300 CTL precursors (CTLp) per spleen expand to up to 10^7 effector CTLs specific for a single epitope [45, 47]. The doubling time during this proliferation period is around 6–8 h [44, 48]. Proliferation of CTLp is accompanied by differentiation into effector cells which possess the ability to secrete cytokine such as IFNγ [47], upregulate the expression of cytolytic granules and effector molecules such as perforin [49], and gain the ability to emigrate from secondary lymphoid organs into nonlymphoid organs through downregulation of CCR7 and CD62L [42]. The maximal expansion of CTL is followed by a contraction phase with death of >90% of the specific CTL effectors and the establishment of a rather stable memory population [50, 51].

It is important to note that the induction of CTL responses by vaccination with DCs expressing a high-affinity LCMV peptide follows similar kinetics as seen in the original viral infection. The maximal CTL expansion following priming with DCs is reached between days 6 and 8 [9, 33], the doubling time of the responding CTL population is 6–8 h, and the overall expansion scale is in the range of 10^4–10^5-fold [33]. CTLs elicited by DC-priming are able to perform *ex vivo* [9] and *in vivo* [11] cytolytic activity and protect mice from systemic and peripheral viral challenge [9, 52]. CTL frequencies remain elevated in the primed host, indicating establishment of a stable memory population [52]. These studies suggest that priming of antiviral CTLs is almost exclusively mediated by DCs presenting the immunodominant viral peptides. Indeed, recent studies support this view [53–55]. This section introduces the essential components involved in the interaction of DCs and CTLs, and discusses – in the light of a systems biologist's approach – the most important concepts describing DC-induced CTL responses.

23.3.1
Dynamics of CTL Activation and Differentiation

Antigenic stimulation of CTLs results in well-ordered changes in the transcriptional program of the cell defining naïve, effector and memory populations [56]. However, the differentiation pathways along which CTL effector and memory populations develop still remain controversial. Essentially, two different schemes, progressive versus programmed differentiation, have been put forward to explain the dynamics of CTL differentiation. Both views share the distinction between naïve (N, antigen unexperienced), effector (E, fully differentiated), effector memory (T_{EM}, CCR7- and CD62L-negative), and central memory (T_{CM}, CCR7- and CD62L-positive) T cells [51, 57]. The concept of programmed differentiation suggests that T cells differentiate along the following linear sequence: $N \rightarrow E \rightarrow T_{EM} \rightarrow T_{CM}$ [51]. Evidence for this view comes from studies showing that the brief antigen exposure (only 24 h) was sufficient to program activation, expansion and differentiation of naïve CTLs into effector CTLs [58, 59]. Importantly, the differentiation program proceeded in an antigen-independent manner towards the memory state [59]. Further analyses suggested that memory CTL (de)differentiate in an hierarchical manner from T_{EM} into T_{CM} [60, 61].

A contrasting model suggests that depending on the signal strength accumulated in the single T cell, hierarchical thresholds of proliferation and differentiation have to be passed [57]. Signal strength is defined here as the cumulative number of MHC–TCR interactions over time, and the overall co-stimulation received via surface molecules and soluble factors such as cytokines. On the basis of these considerations, a nonlinear, progressive differentiation of naïve T cells into either T_{CM},

T_{EM} or E following the interaction with DCs has been suggested [62]. Although this concept has been derived mainly on the basis of *in vitro* data [63–66], it is most helpful in explaining the large functional and phenotypical heterogeneity of both CD4$^+$ and CD8$^+$ T cells that is observed during viral infections [67, 68]. Progressive differentiation of T cells is most probably the basis for the observation that self-renewal of Th cells is inversely correlated with the differentiation status [69]. Likewise, CTL of the T_{CM}–type generated after LCMV infection have a greater capacity to proliferate after re-encounter with the antigen compared with T_{EM} CTL [60]. These findings and the concept of progressive differentiation are in accordance with the view that the balance between proliferation and differentiation depends on the strength and the quality of the antigenic impact [70, 71].

It is apparent that the partly conflicting models mentioned above might require re-evaluation, validation and/or refinement in order to come even closer to a realistic view of DC-induced CTL activation and differentiation. Intravital microscopy offers – to some extent – the possibility to "watch", for example, the interaction of CTLs with DCs in a higher resolution on a single cell level in the native anatomical context [54]. Results from a recent study indicate that there is an initial phase of the DC–CTL interaction where CTLs first scan multiple DCs which results in a reduction of their motility and an initial activation sequence with upregulation of distinct surface markers. Following the initial phase, rather stable DC–CTL clusters are formed for about 12 h, leading to the differentiation into cytokine-secreting effector CTLs [72]. CTLs then detach from the DCs and start to interact again shortly with antigen-bearing DCs, followed by CTL proliferation [72]. It is most likely that a rather undifferentiated subpopulation of DC-activated CTLs commences with a further proliferation–differentiation cycle whereas more differentiated CTLs emigrate into peripheral tissues and/or exert their effector function within secondary lymphoid organs. This view is supported by adoptive transfer studies showing that IFNγ producing Th cells failed to give rise to proliferating progenitors whereas non-IFNγ-producers exhibited self-renewal capacity [73].

23.3.2
Multiple Levels of Positive and Negative Feedback Control

Feedback control represents an important organizing principle of complex systems. The proper balance between positive and negative feedback loops is of utmost importance for the efficient response to perturbations and the maintenance of system stability. Regulation of antigen influx to secondary lymphoid organs via DC represents an important positive regulatory loop for the generation of CTL responses, whereas the elimination of DCs by effector CTLs can be seen as integrated negative feedback. Activation of DCs during the early interaction with CTLs or Th cells ("licensing") is part of a positive amplification mode. Negative feedforward control of excessive CTL responses, on the other hand, is most likely achieved through the concomitant induction of IL-10 producing regulatory Th cells.

23.3.2.1 Managing DC Recruitment and Antigen Translocation

DCs display two distinct functional phenotypes whereby immature DCs in non-lymphoid organs are specialized for uptake and processing of foreign antigens and DCs within secondary lymphoid organs represent more the mature phenotype with the ability to stimulate naïve T-cell responses. Immature DCs are localized in rather high densities in epithelial sites of the body surface, such as skin epidermis and the mucosae of the respiratory, gastrointestinal and urogenital tracts. Importantly, immature DCs are rare in "immunologically privileged" sites such as the central nervous system or the testis unless they are recruited to these sites by inflammatory stimuli. Immature DC display a distinct repertoire of inflammatory chemokine receptors (CCR1, CCR5, CCR6) [74] which bind a broad set of inflammatory chemokines, including RANTES, MCP-3, MIP-1α, MIP-1β, MIP-5. The ability of DC precursors and immature DCs to respond to inflammatory chemokines is probably relevant for their accumulation in peripheral tissues during the early phase of an inflammation. The localization of DCs within secondary lymphoid organs is controlled by the constitutive chemokines CCL19/ELC, CCL21/SLC, and CXCL13/BLC which are differentially expressed in the B and T-cell zones [75]. The chemokines CCL19 and CCL21 act via the CCR7 receptor and are capable of attracting naïve T cells and mature DCs [40, 76]. Importantly, pathogen-induced inflammation may accelerate both the recruitment of DC precursors into peripheral tissues [77, 78], and migration of DCs into secondary lymphoid organs [79, 80]. Thus, antigen sampling in the periphery and influx into secondary lymphoid organs is positively regulated through the distinct migration pattern of DCs: first, by increased recruitment of immature DCs to peripheral inflamed sites, and second, by promoting DC translocation to secondary lymphoid tissues.

23.3.2.2 Elimination of DCs by Effector CTL

The fact that DCs do not recirculate after they have homed to secondary lymphoid organs [81, 82] suggests that their life cycle is locally regulated during T/DC interactions. Whereas T helper cell-derived signals such as CD40/CD40L interaction activate DCs and/or maintain their viability [83, 84], other T cell-derived cytokines, such as IL-10 [85] or TGFβ [86] may inhibit DC function or induce apoptosis of mature DCs [83]. Antigen-presenting DCs may be rapidly lost during the cognate interaction with fully activated Th cells [87] or during massive nonspecific immune stimulation such as LPS treatment [88]. Furthermore, the cognate interaction with effector CTLs leads to the elimination of antigen-presenting DCs [31, 32]. The fact that loss of APCs due to killing by effector CTLs is a hallmark of infection with noncytopathic viruses such as LCMV [3, 89] suggests that elimination of APCs *in vivo* serves as a negative feedback mechanism to prevent exaggerated immune activation.

23.3.2.3 Rapid Amplification of Signals through Molecular "Ping–Pong" Interactions

Following encounter with an infectious agent, CTL responses are generated swiftly. Importantly, the induction of initial CTL responses against pathogens – with only few exceptions – is not dependent on the presence of Th cells [90, 91]. Such highly efficient activation is most likely set in motion through positive feedback loops with staggered back-and-forth signaling between DCs and the responding CTLs. Essentially, these processes start with cognate TCR–MHC interactions. Signaling via costimulatory molecules potentiates the APC function of DCs by upregulation of CD80 and CD86, which reciprocally enhance the activation of the T cells through triggering of CD28 [92, 93]. It is important to note that the function of these "second order" co-stimulatory signals depend on the "first order" signal delivered by the antigen.

The finding that CTL induction in case of limiting antigen availability [94–96] depends on the concomitant activation of Th cells indicates that the overall signal strength during the initial DC–CTL interaction (MHC–TCR contact plus co-stimulatory signals) has to reach a certain threshold. For example, amplification of the signal strength delivered to the CTLs by antigen-presenting DCs can be achieved through contact-dependent, CD40-mediated activation of the DCs by CD4$^+$ T cells, a process that has been termed "conditioning" [95] or "licensing" [97]. Since CD8$^+$ T cells can efficiently induce maturation of DCs, even in the absence of CD40 [98], it is conceivable that three-cell-type clusters of DCs, CTLs, and Th cells are necessary to generate efficient T-cell responses in cases of low antigenic impact [99].

23.3.2.4 Limiting the CTL "Overshoot" through Feedforward Control

The immune system reacts, as outlined above, to strong perturbations, caused by rapidly increasing amounts of antigen and concomitant inflammation. The initial delay in the control of the pathogen has to be seen as a result of an initially undeveloped specific immune response. However, the transiently overshooting pathogen load is, eventually, cleared from the tissues by an overshooting immune response [20, 71]. In order to minimize overshoot-associated immunopathological damage, the immune system employs potent feedforward control mechanisms both on the molecular and the cellular level.

Co-signaling molecules are important for the control and modulation of T-cell responses to antigen [93]. Whereas the co-stimulatory molecule CD28 is constitutively expressed on both CD4$^+$ and CD8$^+$ T cells, the co-inhibitory molecule CTLA-4 is only expressed on T cells following their activation. The engagement of the B7-family members CD80 and CD86 on the responding APCs exerts negative feedback on the proliferating T cells, e.g. by preventing IL-2 production and arresting cell cycle progression [100, 101]. "Programming" of the negative CLTA-4 signal during the CD28-dependent initial activation of T cells represents a typical feedforward control mechanism. Importantly, other members of the CD28 family such as PD-1 exert inhibitory functions during the activation of T cells [93]. Since the ligands of CD28, CTLA-4, and PD-1, i.e. CD80/CD86 and B7-H1, are expressed on DCs, it is reasonable to assume that it is the temporal sequence of the expression of co-stimulatory and co-inhibitory signals that regulates the magnitude and the differentiation pattern of T cells during their interaction with DCs.

Feedforward control on the cellular level is most likely a task of regulatory Th cells (Treg) which exert suppressive effects mainly via the co-inhibitory surface molecule CTLA-4 and the cytokine IL-10 [102]. Just like CTLA-4 expression, the production of IL-10 appears to be programmed early during the activation of Th cells [103]. Release of IL-10 is a hallmark of chronically activated Th1 and Th2 cells and thus, is important to avoid immunopathological damage in the host [104]. Taken together, the three-cell-type cluster interaction between DCs, CTLs and Th cells generates not only positive amplification loops during the initial T-cell proliferation via "licensing" of the DCs, but also implements feedforward control through CTLA-4 expressing and IL-10 secreting regulatory Th cells which contribute to counterbalance the CTL overshoot.

23.3.3
DC Subsets Provide Redundant Activating Signals

Robustness of complex systems does not only require sophisticated feedback mechanisms but also redundant modules and protocols to avoid disastrous system failure. We suggest that the existence of various DCs subtypes with distinct but largely overlapping functions secures redundancy for the swift activation of immune responses against pathogens and is essential for the induction of T-cell responses against tumors.

The exact differentiation pathway of DC precursors are still, to some extent, controversially discussed, however, it is now widely accepted that DCs arise from both multipotent myeloid and lymphoid bone-marrow precursors [105]. Furthermore, monocytes can give rise to immunostimulatory DCs *in vitro* under the influence of granulocyte-macrophage colony stimulating factor (GM-CSF) and IL-4 [106–109] or *in vivo* following their migration into secondary lymphoid organs [110, 111]. Langerhans cells (LCs), the prototype of immature DC, reside mainly within stratified squamous epithelia of the skin and mucous membranes were they closely attach to neighboring epithelial cells via an E-cadherin- and Ca^{2+}-dependent mechanism [112]. Separate precursors of LCs can be distinguished in bone-marrow cultures by their phenotype, by involvement of a recognizable monocyte stage and by their requirement for TGF-β [113]. Plasmacytoid DCs (pDCs) or interferon-producing cells can be differentiated from $CD4^+CD11c^-CD3^-$ precursors present in human blood or tonsillar lymphocyte preparations into DC-like cells using CD40 crosslinking [114]. The murine counterpart of the human pDC has been characterized recently by Trinchieri and colleagues [115]. Murine pDCs or interferon-alpha-producing cells (IPC) [115] and human pDCs [116, 117] secrete large amounts of type I interferon after viral stimulation and therefore most likely restrict initial viral replication and participate in the generation of antiviral Th1 responses [118]. The diverse differentiation pathways of DCs are indicative for a highly redundant system securing the generation of an array of immunostimulatory cells.

The existence of DCs that may co-develop with T cells from a distinct lymphoid progenitor cell subset in the thymus indicates that thymic CD8α⁺ DC may develop along a separate intrathymic differentiation pathway. Furthermore, both thymic and splenic DCs can differentiate from thymic "low CD4 precursors" [119], whereas bone marrow precursors produce both CD8α⁺ and CD8α⁻ DC populations in mouse spleen [120]. Initial *in vitro* experiments showed that CD8α⁺ or CD8α⁻ DC subsets did not differentiate into one or the other [121] which led to the hypothesis that CD8α⁺ and CD8α⁻ DC subsets represent separate DC lineages. However, recent studies indicate that CD8α⁻ LC can acquire CD8α upon maturation by CD40 ligation [122] and that CD8α⁻ DCs differentiate into CD8α⁺ DCs *in vivo* [123, 124] suggesting that CD8α expression is associated with the differentiation of DCs. Furthermore, the finding that monocytes can differentiate into both CD8α⁻ and CD8α⁺ DCs *in vivo* [110] makes it rather unlikely that CD8α is an exclusive marker for a distinct DC lineage.

The initial two-lineage-concept based on the phenotypical differences in CD8α expression led to the hypothesis that the two DC subsets may possess opposing functions, i.e. that "myeloid" CD8α⁻ DCs induce T cells responses, whereas (nowadays so-called) "lymphoid" CD8α⁺ DC in the periphery induce deletion of potentially self-reactive T cells [125], e.g. via Fas/FasL-mediated apoptosis during DC–T cell interactions [126]. This view is supported by findings of Ferguson et al. [127] and Belz et al. [128] that CD8α⁺ DCs mediate the induction of cross-tolerance. However, studies showing that both CD8α⁻ and CD8α⁺ DC induce vigorous T-cell responses after adoptive transfer into naive mice [129, 130], that CD8α⁺ DC are responsible for priming of immune responses against exogenous antigen [131], and that CD8α⁺ DC are the major cells involved in priming of antiviral CTL responses [132–134] suggest that induction of immune reactivity is the default pathway after DCs have encountered sufficient amounts of antigen.

The DC-lineage concept had been complemented by studies suggesting that plasmacytoid DC precursors favor exclusively the development of Th2-type responses whereas "myeloid" DCs derived from monocytic precursors determine Th1-type immune responses [135]. However, subsequent studies have shown that pDCs respond swiftly with secretion of large amounts of type I interferon following encounter with a broad range of viruses [136] which fosters the generation of antiviral Th1 cell responses [118]. Furthermore, peptide-pulsed pDC possess the ability to directly elicit effector CTL secreting IFNγ [137] indicating that, following appropriate stimulation, pDCs are capable of stimulating proliferation and differentiation of protective CTL responses.

The concept that distinct DC subsets or committed DC lineages may be endowed with distinct functions [138], e.g. induction versus tolerance or Th1- versus Th2-induction, is further challenged by recent studies that provide evidence for the high plasticity among DC subsets [139, 140]. It appears that the overall effect of antigen dose together with the quality of the inflammatory stimulus, e.g. stimulation via different Toll-like receptors (TLRs), determines whether pDCs and conventional DCs drive either a Th1 or and Th2 response [139]. Furthermore, LCMV infection elicits an apparent conversion of CD11c$^+$B220$^+$CD11b$^-$ pDC precursors into CD11c$^+$B220$^-$CD11b$^+$ "conventional" DCs with the ability to efficiently present antigen to T cells and produce IL-12 in response to TLR-4 ligation [140]. Taken together, these studies support the view that the different DC subsets with their interchangeable functions and their high plasticity represent an adaptable sentinel cell system that promotes effective and swift responses against invading pathogens.

23.3.4
Tuning of Dendritic Cell Activation

Activation of T cells depends on the kinetics of antigen appearance in secondary lymphoid organs, the "first order" parameter, and "second order" parameters such as stimulation via co-stimulatory surface molecules or inflammatory cytokines. Hence, an acute systemic perturbation with a rapid assembly of DCs delivering large amounts of antigen and T cells in the T-cell zone of lymph nodes or the spleen results in immune activation. Importantly, the immune system does not discriminate whether the antigen is "self" or "foreign" as long as sufficient amounts of antigen are presented for a minimal period of time by mature DCs [141, 142]. In case of low level perturbation, i.e. constant presentation of small amounts of antigen, T cells may be activated transiently, but without the development of a burst-like response [143]. This transient activation without extensive proliferation and differentiation into effector T cells results, however, in the adjustment of the activation thresholds of the individual cell [70]. An elevated activation threshold through repeated low-level antigenic stimulation will essentially result in the "ignorance" of the antigenic stimulus, as long as the stimulus remains at subthreshold levels. This model of "activation threshold tuning" implies that the excitation levels of T cells constantly adapt to the overall stimulation supplied by the antigen-presenting cells, mainly DCs, in secondary lymphoid organs [20, 70, 143]. Whereas tuning of T-cell activation threshold results in a constantly changing excitability, it appears that the initial activation threshold for DCs remains constantly at low levels, i.e. despite variations in the mode of activation due to context-dependency, it is an intrinsic feature of DC physiology to respond rapidly and profoundly to subtle perturbations in their environment.

23.3.4.1 Excitement through Pattern Recognition

Viruses and other pathogens display particular molecular signatures ("pathogen-associated molecular pattern") that are recognized by specific "pattern recognition receptors" [144]. DCs express a vast array of such receptors, including TLRs and C-type lectins. Whereas C-type lectins facilitate mainly binding of pathogens to and their ingestion by DCs [145], it is the triggering of intra- and extracellular TLRs that leads to rapid activation and maturation of DCs [146]. Maturation of DCs under the influence of TLR stimulation includes upregulation of co-stimulatory molecules, increased loading of MHC class II molecules, improved crosspresentation of exogenous antigens, and increased production of immune-stimulatory cytokines such as type I interferons and IL-12 [147]. Consequently, TLR-mediated signaling to pDCs and conventional DCs greatly enhances their ability to prime CTL responses [134, 137, 148], partially via concomitant licensing of Th cells [97]. Importantly, both human and mouse DC subsets express nearly the complete array of all known TLRs [146]. Thus, stimulation of DCs through TLRs and other pattern recognition receptors generates rapid pathogen-specific innate immune responses and conditions DCs for efficient stimulation of adaptive T-cell responses.

Two distinct features of TLR-mediated stimulation of DCs are particularly noteworthy. Namely that the activation of DC following contact with a TLR ligand *in vivo* occurs within only a few hours, and that this activation eventually diminishes the lifespan of the DCs [88, 149], the rapid apoptosis induced by TLR ligation can be – to some extent – counterbalanced during the early cognate interaction with T cells [149]. Thus, DCs represent a tunable cellular switch that integrates a large array of positive and negative signals both from the innate and the adaptive immune system.

23.3.4.2 DC Tuning and Tolerance to Self-antigens

It is apparent that the DC system with its different, functionally redundant subsets that all respond avidly to pathogen-associated innate stimuli, is highly efficient in activating T-cell responses. Infection-associated immunopathological damage is minimized by the above-mentioned negative feedback and feedforward control mechanisms. An important question, however, remains to be discussed, namely whether and how DCs contribute to establishment of self-tolerance. The most prominent view – at the moment – is that constitutive expression of self-antigens by immature DCs is associated with tolerance to self-antigens [150]. Indeed, tissue-derived self-antigens can be found in DCs [151, 152]. Furthermore, targeting of antigens to DCs in the absence of pro-inflammatory stimuli results in tolerance [153, 154]. However, presentation of significant amounts of self-antigen by DCs in secondary lymphoid organs – even in the absence of inflammatory signals – neither results in deletion of autoreactive T cells nor does it lead to autoimmunity [155]. Furthermore, *in vitro* and *in vivo* approaches show that even mature DCs can tolerize T cells [156, 157]. To resolve this discrepancy, a further refinement of the DC maturation scheme has been suggested with DCs being either semi-mature [157] or mature, but not licensed, to drive full differentiation of effector T cells [158]. Following this line of reasoning, the decision whether self-reactivity is induced or not depends on rather minuscule changes in the phenotype of a particular DC subset.

The demand for robustness in the interaction of DCs with T cells, i.e. efficient activation of pathogen-specific responses with minimal "collateral" damage, makes it difficult to envisage that, for example, partial upregulation of CD80 or CD86 or slight increases in IL-12 production by DCs would be the decisive switch that determines immune reactivity against self-antigens. We prefer to follow a more global approach with the "spatiotemporal" view of the regulation and homeostasis of immune responses [20, 21, 26]. Accordingly, peripheral self-antigens that are expressed below a certain threshold remain immunologically ignored because they do not reach secondary lymphoid organs [26]. Tissue antigens that reach secondary lymphoid organs in a steady state may – to some extent – induce T-cell activation with few autoreactive effector cells which, however, will not elicit autoimmune disease. Steady state influx of antigen to secondary lymphoid organs and presentation by DCs should eventually lead to activation threshold tuning in the responding T-cell population that keeps autoimmunity in check [20]. These rather simple rules explain why Th-cell responses against even abundantly expressed self-antigens such as collagen, cardiac myosin, or myelin basic protein can be easily induced once the antigen is delivered at supra-threshold levels with appropriate inflammatory stimuli, e.g. via subcutaneous application of a few hundred micrograms of purified antigen in complete Freund's adjuvant. Furthermore, in DC-based immunotherapy of tumors with DCs being loaded with MHC class I peptides, efficient priming of CTL responses against rather abundant self determinants (i.e. tumor antigens) is achieved if appropriate amounts (supra-threshold levels) of antigen are delivered in the right kinetics.

23.4
Conclusions

The induction of protective CTL responses by DCs represents a robust biological system as it ensures protection of individuals against the vast majority of pathogens and helps to maintain immune surveillance of tumors. Novel vaccination approaches for the improvement of CTL responses against infectious agents and tumors should be based upon the systems biology analysis of the robustness of DC–CTL interaction as well as the fragility of the pathological steady states. The framework presented should help to provide an economic conceptualization of *in vitro* and *in vivo* observations to avoid unnecessary complexities (e.g. divergent DC subsets or intermediate maturation stages). Thus, structure, dynamics, design and control principles of DC-induced CTL responses represent important issues that require the attention of further research efforts.

Acknowledgments

We thank Zvi Grossman for helpful discussions, and Tobias Junt for critical reading of the manuscript. This work was supported by the Kanton of St. Gallen, the Swiss National Science Foundation, the Wellcome Trust, the Alexander von Humboldt Foundation, the Leverhulme Trust, and the Russian Foundation for Basic Research.

References

1 Zinkernagel, R. M., Immunology taught by viruses. *Science* **1996**. *271*: 173–178.

2 Sevilla, N., Kunz, S., Holz, A., Lewicki, H., Homann, D., Yamada, H., Campbell, K. P., de la Torre, J. C., Oldstone, M. B., Immunosuppression and resultant viral persistence by specific viral targeting of dendritic cells. *J. Exp. Med.* **2000**. *192*: 1249–1260.

3 Borrow, P., Evans, C. F., Oldstone, M. B., Virus-induced immunosuppression: immune system-mediated destruction of virus-infected dendritic cells results in generalized immune suppression. *J. Virol.* **1995**. *69*: 1059–1070.

4 Kawamura, T., Gulden, F. O., Sugaya, M., McNamara, D. T., Borris, D. L., Lederman, M. M., Orenstein, J. M., Zimmerman, P. A., Blauvelt, A., R5 HIV productively infects Langerhans cells, and infection levels are regulated by compound CCR5 polymorphisms. *Proc. Natl. Acad. Sci. USA* **2003**. *100*: 8401–8406.

5 Turville, S. G., Santos, J. J., Frank, I., Cameron, P. U., Wilkinson, J., Miranda-Saksena, M., Dable, J., Stossel, H., Romani, N., Piatak, M., Jr., Lifson, J. D., Pope, M., Cunningham, A. L., Immunodeficiency virus uptake, turnover, and 2-phase transfer in human dendritic cells. *Blood* **2004**. *103*: 2170–2179.

6 Freigang, S., Egger, D., Bienz, K., Hengartner, H., Zinkernagel, R. M., Endogenous neosynthesis vs. cross-presentation of viral antigens for cytotoxic T cell priming. *Proc. Natl. Acad. Sci. USA* **2003**. *100*: 13477–13482.

7 Freigang, S., Probst, H. C., van den Broek, M., DC infection promotes antiviral CTL priming: the 'Winkelried' strategy. *Trends Immunol.* **2005**. *26*: 13–18.

8 Steinman, R. M., Pope, M., Exploiting dendritic cells to improve vaccine efficacy. *J. Clin. Invest* **2002**. *109*: 1519–1526.

9 Ludewig, B., Ehl, S., Karrer, U., Odermatt, B., Hengartner, H., Zinkernagel, R. M., Dendritic cells efficiently induce protective antiviral immunity. *J. Virol.* **1998**. *72*: 3812–3818.

10 Ochsenbein, A. F., Klenerman, P., Karrer, U., Ludewig, B., Pericin, M., Hengartner, H., Zinkernagel, R. M., Immune surveillance against a solid tumor fails because of immunological ignorance. *Proc. Natl. Acad. Sci. USA* **1999**. *96*: 2233–2238.

11 Ludewig, B., Ochsenbein, A. F., Odermatt, B., Paulin, D., Hengartner, H., Zinkernagel, R. M., Immunotherapy with dendritic cells directed against tumor antigens shared with normal host cells results in severe autoimmune disease. *J. Exp. Med.* **2000**. *191*: 795–804.

12 Rosenberg, S. A., Yang, J. C., Restifo, N. P., Cancer immunotherapy: moving beyond current vaccines. *Nat. Med.* **2004**. *10*: 909–915.

13 Ronchese, F., Hermans, I. F., Killing of dendritic cells: a life cut short or a purposeful death? *J. Exp. Med.* **2001**. *194*: F23–F26.

14 von Bertalanfy, L., *Modern theories of development: An introduction to theoretical biology.* Oxford University Press, New York **1933**.

15 Mesarovic, M. D., Systems theory and biology – view of a theoretician. In Mesarovic, M. D. (Ed.) *Systems theory and biology.* Springer, New York **1968**, pp 59–87.

16 Kitano, H., Systems biology: a brief overview. *Science* **2002**. *295*: 1662–1664.

17 Bell, G., Perelson, A. S., Pimbley, G., *Theoretical Immunology.* Marcel Dekker, New York **1978**.

18 Cohn, M., Langman, R. E., The protection: the unit of humoral immunity selected by evolution. *Immunol. Rev.* **1990**. *115*: 11–147.

19 Perelson, A. S., Wiegel, F. W., Some design principles for immune system recognition. *Complexity* **1999**. *4*: 29–37.

20 Grossman, Z., Paul, W. E., Self-tolerance: context dependent tuning of T cell antigen recognition. *Semin. Immunol.* **2000**. *12*: 197–203.

21 Germain, R. N., The art of the probable: system control in the adaptive immune system. *Science* **2001**. *293*: 240–245.

22 Kitano, H., Computational systems biology. *Nature* **2002**. *420*: 206–210.

23 Kitano, H., Cancer as a robust system: implications for anticancer therapy. *Nat. Rev. Cancer* **2004**. *4*: 227–235.

24 Csete, M. E., Doyle, J. C., Reverse engineering of biological complexity. *Science* 2002. *295:* 1664–1669.

25 Kitano, H., Biological robustness. *Nat. Rev. Genet.* 2004. *5:* 826–837.

26 Zinkernagel, R. M., Localization dose and time of antigens determine immune reactivity. *Semin. Immunol.* 2000. *12:* 163–171.

27 Levchenko, A., Bruck, J., Sternberg, P. W., Regulatory modules that generate biphasic signal response in biological systems. *Syst. Biol.* 2004. *1:* 139–148.

28 Mitchison, N. A., Induction of immuno-logical paralysis in two zones of dosage. *Proc. Roy. Soc. Lond B Biol. Sci.* 1964. *161:* 275–292.

29 Bocharov, G., Ludewig, B., Bertoletti, A., Klenerman, P., Junt, T., Krebs, P., Luzyanina, T., Fraser, C., Anderson, R. M., Underwhelming the immune response: effect of slow virus growth on CD8+-T-lymphocyte responses. *J. Virol.* 2004. *78:* 2247–2254.

30 Mesarovic, M. D., Sreenath, S. N., Keene, J. D., Search for organizing principles: understanding in systems biology. *Syst. Biol.* 2004. *1:* 19–27.

31 Hermans, I. F., Ritchie, D. S., Yang, J., Roberts, J. M., Ronchese, F., CD8+ T cell-dependent elimination of dendritic cells in vivo limits the induction of anti-tumor immunity. *J. Immunol.* 2000. *164:* 3095–3101.

32 Ludewig, B., Bonilla, W. V., Dumrese, T., Odermatt, B., Zinkernagel, R. M., Hengartner, H., Perforin-independent regulation of dendritic cell homeostasis by CD8(+) T cells in vivo: implications for adaptive immunotherapy. *Eur. J. Immunol.* 2001. *31:* 1772–1779.

33 Ludewig, B., Krebs, P., Junt, T., Metters, H., Ford, N. J., Anderson, R. M., Bocharov, G., Determining control parameters for dendritic cell-cytotoxic T lymphocyte interaction. *Eur. J. Immunol.* 2004. *34:* 2407–2418.

34 Barchet, W., Oehen, S., Klenerman, P., Wodarz, D., Bocharov, G., Lloyd, A. L., Nowak, M. A., Hengartner, H., Zinkernagel, R. M., Ehl, S., Direct quantitation of rapid elimination of viral antigen-positive lymphocytes by antiviral CD8(+) T cells in vivo. *Eur. J. Immunol.* 2000. *30:* 1356–1363.

35 Jellison, E. R., Kim, S. K., Welsh, R. M., Cutting edge: MHC class II-restricted killing in vivo during viral infection. *J. Immunol.* 2005. *174:* 614–618.

36 Rot, A., von Andrian, U. H., Chemokines in innate and adaptive host defense: basic chemokinese grammar for immune cells. *Annu. Rev. Immunol.* 2004. *22:* 891–928.

37 Rehermann, B., Fowler, P., Sidney, J., Person, J., Redeker, A., Brown, M., Moss, B., Sette, A., Chisari, F. V., The cytotoxic T lymphocyte response to multiple hepatitis B virus polymerase epitopes during and after acute viral hepatitis. *J. Exp. Med.* 1995. *181:* 1047–1058.

38 Lubaki, N. M., Ray, S. C., Dhruva, B., Quinn, T. C., Siliciano, R. F., Bollinger, R. C., Characterization of a polyclonal cytolytic T lymphocyte response to human immunodeficiency virus in persons without clinical progression. *J. Infect. Dis.* 1997. *175:* 1360–1367.

39 Polderman, J. W., Willems, J. C., *Introduction to mathematical systems theory.* Springer, New York 1998.

40 Forster, R., Schubel, A., Breitfeld, D., Kremmer, E., Renner-Muller, I., Wolf, E., Lipp, M., CCR7 coordinates the primary immune response by establish-ing functional microenvironments in secondary lymphoid organs. *Cell* 1999. *99:* 23–33.

41 Nakano, H., Tamura, T., Yoshimoto, T., Yagita, H., Miyasaka, M., Butcher, E. C., Nariuchi, H., Kakiuchi, T., Matsuzawa, A., Genetic defect in T lymphocyte-specific homing into peripheral lymph nodes. *Eur. J. Immunol.* 1997. *27:* 215–221.

42 Junt, T., Scandella, E., Forster, R., Krebs, P., Krautwald, S., Lipp, M., Hengartner, H., Ludewig, B., Impact of CCR7 on priming and distribution of antiviral effector and memory CTL. *J. Immunol.* 2004. *173:* 6684–6693.

43 Junt, T., Nakano, H., Dumrese, T., Kakiuchi, T., Odermatt, B., Zinkernagel, R. M., Hengartner, H., Ludewig, B., Antiviral immune responses in the

absence of organized lymphoid T cell zones in plt/plt mice. *J. Immunol.* **2002.** *168:* 6032–6040.

44 Bocharov, G. A., Modelling the dynamics of LCMV infection in mice: conventional and exhaustive CTL responses. *J. Theor. Biol.* **1998.** *192:* 283–308.

45 Blattman, J. N., Antia, R., Sourdive, D. J., Wang, X., Kaech, S. M., Murali-Krishna, K., Altman, J. D., Ahmed, R., Estimating the precursor frequency of naive antigen-specific CD8 T cells. *J. Exp. Med.* **2002.** *195:* 657–664.

46 Arstila, T. P., Casrouge, A., Baron, V., Even, J., Kanellopoulos, J., Kourilsky, P., A direct estimate of the human alphabeta T cell receptor diversity. *Science* **1999.** *286:* 958–961.

47 Gallimore, A., Glithero, A., Godkin, A., Tissot, A. C., Pluckthun, A., Elliott, T., Hengartner, H., Zinkernagel, R. M., Induction and exhaustion of lymphocytic choriomeningitis virus- specific cytotoxic T lymphocytes visualized using soluble tetrameric major histocompatibility complex class I-peptide complexes. *J. Exp. Med.* **1998.** *187:* 1383–1393.

48 Murali-Krishna, K., Altman, J. D., Suresh, M., Sourdive, D. J., Zajac, A. J., Miller, J. D., Slansky, J., Ahmed, R., Counting antigen-specific CD8 T cells: a reevaluation of bystander activation during viral infection. *Immunity* **1998.** *8:* 177–187.

49 Kagi, D., Ledermann, B., Burki, K., Seiler, P., Odermatt, B., Olsen, K. J., Podack, E. R., Zinkernagel, R. M., Hengartner, H., Cytotoxicity mediated by T cells and natural killer cells is greatly impaired in perforin-deficient mice. *Nature* **1994.** *369:* 31–37.

50 Zinkernagel, R. M., On differences between immunity and immunological memory. *Curr. Opin. Immunol.* **2002.** *14:* 523–536.

51 Kaech, S. M., Wherry, E. J., Ahmed, R., Effector and memory T-cell differentiation: implications for vaccine development. *Nat. Rev. Immunol.* **2002.** *2:* 251–262.

52 Ludewig, B., Oehen, S., Barchiesi, F., Schwendener, R. A., Hengartner, H., Zinkernagel, R. M., Protective antiviral cytotoxic T cell memory is most efficiently maintained by restimulation via dendritic cells. *J. Immunol.* **1999.** *163:* 1839–1844.

53 Jung, S., Unutmaz, D., Wong, P., Sano, G., De los, S. K., Sparwasser, T., Wu, S., Vuthoori, S., Ko, K., Zavala, F., Pamer, E. G., Littman, D. R., Lang, R. A., In vivo depletion of CD11c(+) dendritic cells abrogates priming of CD8(+) T cells by exogenous cell-associated antigens. *Immunity.* **2002.** *17:* 211–220.

54 Norbury, C. C., Malide, D., Gibbs, J. S., Bennink, J. R., Yewdell, J. W., Visualizing priming of virus-specific CD8+ T cells by infected dendritic cells in vivo. *Nat. Immunol.* **2002.** *3:* 265–271.

55 Probst, H. C., van den, B. M., Priming of CTLs by Lymphocytic Choriomeningitis Virus Depends on Dendritic Cells. *J. Immunol.* **2005.** *174:* 3920–3924.

56 Kaech, S. M., Hemby, S., Kersh, E., Ahmed, R., Molecular and functional profiling of memory CD8 T cell differentiation. *Cell* **2002.** *111:* 837–851.

57 Lanzavecchia, A., Sallusto, F., Progressive differentiation and selection of the fittest in the immune response. *Nat. Rev. Immunol.* **2002.** *2:* 982–987.

58 van Stipdonk, M. J., Lemmens, E. E., Schoenberger, S. P., Naive CTLs require a single brief period of antigenic stimulation for clonal expansion and differentiation. *Nat. Immunol.* **2001.** *2:* 423–429.

59 Kaech, S. M., Ahmed, R., Memory CD8+ T cell differentiation: initial antigen encounter triggers a developmental program in naive cells. *Nat. Immunol.* **2001.** *2:* 415–422.

60 Wherry, E. J., Teichgraber, V., Becker, T. C., Masopust, D., Kaech, S. M., Antia, R., von Andrian, U. H., Ahmed, R., Lineage relationship and protective immunity of memory CD8 T cell subsets. *Nat. Immunol.* **2003.** *4:* 225–234.

61 Wherry, E. J., Blattman, J. N., Murali-Krishna, K., van der, M. R., Ahmed, R., Viral persistence alters CD8 T-cell immunodominance and tissue distribution and results in distinct stages of functional impairment. *J. Virol.* **2003.** *77:* 4911–4927.

62 Sallusto, F., Geginat, J., Lanzavecchia, A., Central memory and effector memory T cell subsets: function, generation, and

maintenance. *Annu. Rev. Immunol.* **2004.** *22:* 745–63.

63 Langenkamp, A., Casorati, G., Garavaglia, C., Dellabona, P., Lanzavecchia, A., Sallusto, F., T cell priming by dendritic cells: thresholds for proliferation, differentiation and death and intraclonal functional diversification. *Eur. J. Immunol.* **2002.** *32:* 2046–2054.

64 Gett, A. V., Sallusto, F., Lanzavecchia, A., Geginat, J., T cell fitness determined by signal strength. *Nat. Immunol.* **2003.** *4:* 355–360.

65 Langenkamp, A., Messi, M., Lanzavecchia, A., Sallusto, F., Kinetics of dendritic cell activation: impact on priming of TH1, TH2 and nonpolarized T cells. *Nat. Immunol.* **2000.** *1:* 311–316.

66 Geginat, J., Sallusto, F., Lanzavecchia, A., Cytokine-driven proliferation and differentiation of human naive, central memory, and effector memory CD4(+) T cells. *J. Exp. Med.* **2001.** *194:* 1711–1719.

67 van Lier, R. A., ten Berge, I. J., Gamadia, L. E., Human CD8(+) T-cell differentiation in response to viruses. *Nat. Rev. Immunol.* **2003.** *3:* 931–939.

68 Roman, E., Miller, E., Harmsen, A., Wiley, J., von Andrian, U. H., Huston, G., Swain, S. L., CD4 effector T cell subsets in the response to influenza: heterogeneity, migration, and function. *J. Exp. Med.* **2002.** *196:* 957–968.

69 Hayashi, N., Liu, D., Min, B., Ben Sasson, S. Z., Paul, W. E., Antigen challenge leads to in vivo activation and elimination of highly polarized TH1 memory T cells. *Proc. Natl. Acad. Sci. USA* **2002.** *99:* 6187–6191.

70 Grossman, Z., Cellular tolerance as a dynamic state of the adaptable lymphocyte. *Immunol. Rev.* **1993.** *133:* 45–73.

71 Grossman, Z., Min, B., Meier-Schellersheim, M., Paul, W. E., Concomitant regulation of T-cell activation and homeostasis. *Nat. Rev. Immunol.* **2004.** *4:* 7–15.

72 Mempel, T. R., Henrickson, S. E., von Andrian, U. H., T-cell priming by dendritic cells in lymph nodes occurs in three distinct phases. *Nature* **2004.** *427:* 154–159.

73 Wu, C. Y., Kirman, J. R., Rotte, M. J., Davey, D. F., Perfetto, S. P., Rhee, E. G.,

Freidag, B. L., Hill, B. J., Douek, D. C., Seder, R. A., Distinct lineages of T(H)1 cells have differential capacities for memory cell generation in vivo. *Nat. Immunol.* **2002.** *3:* 852–858.

74 Sozzani, S., Allavena, P., Vecchi, A., Mantovani, A., The role of chemokines in the regulation of dendritic cell trafficking. *J. Leukoc. Biol.* **1999.** *66:* 1–9.

75 Cyster, J. G., Chemokines and cell migration in secondary lymphoid organs. *Science* **1999.** 286: 2098–2102.

76 Sallusto, F., Schaerli, P., Loetscher, P., Schaniel, C., Lenig, D., Mackay, C. R., Qin, S., Lanzavecchia, A., Rapid and coordinated switch in chemokine receptor expression during dendritic cell maturation. *Eur. J. Immunol.* **1998.** *28:* 2760–2769.

77 McWilliam, A. S., Napoli, S., Marsh, A. M., Pemper, F. L., Nelson, D. J., Pimm, C. L., Stumbles, P. A., Wells, T. N., Holt, P. G., Dendritic cells are recruited into the airway epithelium during the inflammatory response to a broad spectrum of stimuli. *J. Exp. Med.* **1996.** *184:* 2429–2432.

78 Stumbles, P. A., Strickland, D. H., Pimm, C. L., Proksch, S. F., Marsh, A. M., McWilliam, A. S., Bosco, A., Tobagus, I., Thomas, J. A., Napoli, S., Proudfoot, A. E., Wells, T. N., Holt, P. G., Regulation of dendritic cell recruitment into resting and inflamed airway epithelium: use of alternative chemokine receptors as a function of inducing stimulus. *J. Immunol.* **2001.** *167:* 228–234.

79 Martin, P., Ruiz, S. R., del Hoyo, G. M., Anjuere, F., Vargas, H. H., Lopez-Bravo, M., Ardavin, C., Dramatic increase in lymph node dendritic cell number during infection by the mouse mammary tumor virus occurs by a CD62L-dependent blood-borne DC recruitment. *Blood* **2002.** *99:* 1282–1288.

80 Yoneyama, H., Matsuno, K., Zhang, Y., Nishiwaki, T., Kitabatake, M., Ueha, S., Narumi, S., Morikawa, S., Ezaki, T., Lu, B., Gerard, C., Ishikawa, S., Matsushima, K., Evidence for recruitment of plasmacytoid dendritic cell precursors to inflamed lymph nodes through high endothelial venules. *Int. Immunol.* **2004.** *16:* 915–928.

81 Fossum, S., Lymph-borne dendritic leucocytes do not recirculate, but enter the lymph node paracortex to become interdigitating cells. *Scand. J. Immunol.* **1988.** *27:* 97–105.

82 Kupiec Weglinski, J. W., Austyn, J. M., Morris, P. J., Migration patterns of dendritic cells in the mouse. Traffic from the blood, and T cell-dependent and -independent entry to lymphoid tissues. *J. Exp. Med.* **1988.** *167:* 632–645.

83 Ludewig, B., Graf, D., Gelderblom, H. R., Becker, Y., Kroczek, R. A., Pauli, G., Spontaneous apoptosis of dendritic cells is efficiently inhibited by TRAP (CD40-ligand) and TNF-alpha, but strongly enhanced by interleukin- 10. *Eur. J. Immunol.* **1995.** *25:* 1943–1950.

84 Caux, C., Massacrier, C., Vanbervliet, B., Dubois, B., van Kooten, C., Durand, I., Banchereau, J., Activation of human dendritic cells through CD40 cross-linking. *J. Exp. Med.* **1994.** *180:* 1263–1272.

85 Enk, A. H., Angeloni, V. L., Udey, M. C., Katz, S. I., Inhibition of Langerhans cell antigen-presenting function by IL-10. A role for IL-10 in induction of tolerance. *J. Immunol.* **1993.** *151:* 2390–2398.

86 Strobl, H., Knapp, W., TGF-beta1 regulation of dendritic cells. *Microbes. Infect.* **1999.** *1:* 1283–1290.

87 Ingulli, E., Mondino, A., Khoruts, A., Jenkins, M. K., In vivo detection of dendritic cell antigen presentation to CD4(+) T Cells. *J. Exp. Med.* **1997.** *185:* 2133–2141.

88 DeSmedt, T., Pajak, B., Muraille, E., Lespagnard, L., Heinen, E., DeBaetselier, P., Urbain, J., Leo, O., Moser, M., Regulation of dendritic cell numbers and maturation by lipopolysaccharide in vivo. *J. Exp. Med.* **1996.** *184:* 1413–1424.

89 Odermatt, B., Eppler, M., Leist, T. P., Hengartner, H., Zinkernagel, R. M., Virus-triggered acquired immunodeficiency by cytotoxic T-cell-dependent destruction of antigen-presenting cells and lymph follicle structure. *Proc. Natl. Acad. Sci. USA* **1991.** *88:* 8252–8256.

90 Buller, R. M., Holmes, K. L., Hugin, A., Frederickson, T. N., Morse, H. C., III, Induction of cytotoxic T-cell responses in vivo in the absence of CD4 helper cells. *Nature* **1987.** *328:* 77–79.

91 Rahemtulla, A., Fung Leung, W. P., Schilham, M. W., Kundig, T. M., Sambhara, S. R., Narendran, A., Arabian, A., Wakeham, A., Paige, C. J., Zinkernagel, R. M., et al, Normal development and function of CD8+ cells but markedly decreased helper cell activity in mice lacking CD4. *Nature* **1991.** *353:* 180–184.

92 van Kooten, C., Banchereau, J., CD40-CD40 ligand. *J. Leukoc. Biol.* **2000.** 67: 2–17.

93 Chen, L., Co-inhibitory molecules of the B7-CD28 family in the control of T-cell immunity. *Nat. Rev. Immunol.* **2004.** *4:* 336–347.

94 Bennett, S. R., Carbone, F. R., Karamalis, F., Miller, J. F., Heath, W. R., Induction of a CD8+ cytotoxic T lymphocyte response by cross- priming requires cognate CD4+ T cell help. *J. Exp. Med.* **1997.** *186:* 65–70.

95 Ridge, J. P., Di Rosa, F., Matzinger, P., A conditioned dendritic cell can be a temporal bridge between a CD4+ T-helper and a T-killer cell. *Nature* **1998.** *393:* 474–478.

96 Schoenberger, S. P., Toes, R. E., van der Voort, E. I., Offringa, R., Melief, C. J., T-cell help for cytotoxic T lymphocytes is mediated by CD40– CD40L interactions. *Nature* **1998.** *393:* 480–483.

97 Smith, C. M., Wilson, N. S., Waithman, J., Villadangos, J. A., Carbone, F. R., Heath, W. R., Belz, G. T., Cognate CD4(+) T cell licensing of dendritic cells in CD8(+) T cell immunity. *Nat. Immunol.* **2004.** *5:* 1143–1148.

98 Ruedl, C., Kopf, M., Bachmann, M. F., CD8(+) T cells mediate CD40-independent maturation of dendritic cells in vivo. *J. Exp. Med.* **1999.** *189:* 1875–1884.

99 Mitchison, N. A., O'Malley, C., Three-cell-type clusters of T cells with antigen-presenting cells best explain the epitope linkage and noncognate requirements of the in vivo cytolytic response. *Eur. J. Immunol.* **1987.** *17:* 1579–1583.

100 Walunas, T. L., Lenschow, D. J., Bakker, C. Y., Linsley, P. S., Freeman, G. J., Green, J. M., Thompson, C. B.,

Bluestone, J. A., CTLA-4 can function as a negative regulator of T cell activation. *Immunity* **1994**. *1:* 405–413.

101 Walunas, T. L., Bakker, C. Y., Bluestone, J. A., CTLA-4 ligation blocks CD28-dependent T cell activation. *J. Exp. Med.* 1996. *183:* 2541–2550.

102 Bluestone, J. A., Abbas, A. K., Natural versus adaptive regulatory T cells. *Nat. Rev. Immunol.* **2003**. *3:* 253–257.

103 Assenmacher, M., Schmitz, J., Radbruch, A., Flow cytometric determination of cytokines in activated murine T helper lymphocytes: expression of interleukin-10 in interferon-gamma and in inter-leukin-4-expressing cells. *Eur. J. Immunol.* **1994**. *24:* 1097–1101.

104 O'Garra, A., Vieira, P. L., Vieira, P., Goldfeld, A. E., IL-10-producing and naturally occurring CD4+ Tregs: limiting collateral damage. *J. Clin. Invest* **2004**. *114:* 1372–1378.

105 Ardavin, C., Origin, precursors and differentiation of mouse dendritic cells. *Nat. Rev. Immunol.* **2003**. *3:* 582–590.

106 Peters, J. H., Gieseler, R., Thiele, B., Steinbach, F., Dendritic cells: from ontogenetic orphans to myelomonocytic descendants. *Immunol. Today* **1996**. *17:* 273–278.

107 Romani, N., Gruner, S., Brang, D., Kampgen, E., Lenz, A., Trockenbacher, B., Konwalinka, G., Fritsch, P., Steinman, R. M., Schuler, G., Proliferating dendritic cell progenitors in human blood. *J. Exp. Med.* **1994**. *180:* 83–93.

108 Sallusto, F., Lanzavecchia, A., Efficient presentation of soluble antigen by cultured human dendritic cells is maintained by granulocyte/macrophage colony- stimulating factor plus inter-leukin 4 and downregulated by tumor necrosis factor alpha. *J. Exp. Med.* **1994**. *179:* 1109–1118.

109 Zhou, L. J., Tedder, T. F., CD14(+) blood monocytes can differentiate into functionally mature CD83(+) dendritic cells. *Proc. Natl. Acad. Sci. USA 1996.* 93: 2588–2592.

110 Leon, B., Martinez, d. H., Parrillas, V., Vargas, H. H., Sanchez-Mateos, P., Longo, N., Lopez-Bravo, M., Ardavin, C., Dendritic cell differentiation potential of mouse monocytes: monocytes represent immediate precursors of CD8– and CD8+ splenic dendritic cells. *Blood* **2004**. *103:* 2668–2676.

111 Llodra, J., Angeli, V., Liu, J., Trogan, E., Fisher, E. A., Randolph, G. J., Emigration of monocyte-derived cells from atherosclerotic lesions characterizes regressive, but not progressive, plaques. *Proc. Natl. Acad. Sci. USA* **2004**. *101:* 11779–11784.

112 Tang, A., Amagai, M., Granger, L. G., Stanley, J. R., Udey, M. C., Adhesion of epidermal Langerhans cells to keratino-cytes mediated by E-cadherin. *Nature* **1993**. *361:* 82–85.

113 Jaksits, S., Kriehuber, E., Charbonnier, A. S., Rappersberger, K., Stingl, G., Maurer, D., CD34+ cell-derived CD14+ precursor cells develop into Langerhans cells in a TGF-beta 1-dependent manner. *J. Immunol.* **1999**. *163:* 4869–4877.

114 Grouard, G., Rissoan, M. C., Filgueira, L., Durand, I., Banchereau, J., Liu, Y. J., The enigmatic plasmacytoid T cells develop into dendritic cells with inter-leukin (IL)-3 and CD40-ligand. *J. Exp. Med.* **1997**. *185:* 1101–1111.

115 Asselin-Paturel, C., Boonstra, A., Dalod, M., Durand, I., Yessaad, N., Dezutter-Dambuyant, C., Vicari, A., O'Garra, A., Biron, C., Briere, F., Trinchieri, G., Mouse type I IFN-producing cells are immature APCs with plasmacytoid morphology. *Nat. Immunol.* **2001**. *2:* 1144–1150.

116 Siegal, F. P., Kadowaki, N., Shodell, M., Fitzgerald-Bocarsly, P. A., Shah, K., Ho, S., Antonenko, S., Liu, Y. J., The nature of the principal type 1 interferon-producing cells in human blood. *Science* **1999**. *284:* 1835–1837.

117 Cella, M., Jarrossay, D., Facchetti, F., Alebardi, O., Nakajima, H., Lanzavecchia, A., Colonna, M., Plasmacytoid monocytes migrate to inflamed lymph nodes and produce large amounts of type I interferon. *Nat. Med.* **1999**. *5:* 919–923.

118 Cella, M., Facchetti, F., Lanzavecchia, A., Colonna, M., Plasmacytoid dendritic cells activated by influenza virus and CD40L drive a potent TH1 polarization. *Nat. Immunol.* **2000**. *1:* 305–310.

119 Ardavin, C., Wu, L., Li, C. L., Shortman, K., Thymic dendritic cells and T cells develop simultaneously in the thymus form a common precursor population. *Nature* **1993**. *362:* 761–763.

120 Wu, L., Li, C. L., Shortman, K., Thymic dendritic cell precursors: relationship to the t lymphocyte lineage and phenotype of the dendritic cell progeny. *J. Exp. Med.* **1996**. *184:* 903–911.

121 Vremec, D., Shortman, K., Dendritic cell subtypes in mouse lymphoid organs: cross- correlation of surface markers, changes with incubation, and differences among thymus, spleen, and lymph nodes. *J. Immunol.* **1997**. *159:* 565–573.

122 Anjuere, F., Martinez, d. H., Martin, P., Ardavin, C., Langerhans cells acquire a CD8+ dendritic cell phenotype on maturation by CD40 ligation. *J. Leukoc. Biol.* **2000**. *67:* 206–209.

123 Merad, M., Fong, L., Bogenberger, J., Engleman, E. G., Differentiation of myeloid dendritic cells into CD8alpha- positive dendritic cells in vivo. *Blood* **2000**. *96:* 1865–1872.

124 Martinez, d. H., Martin, P., Arias, C. F., Marin, A. R., Ardavin, C., CD8alpha+ dendritic cells originate from the CD8alpha- dendritic cell subset by a maturation process involving CD8alpha, DEC-205, and CD24 up- regulation. *Blood* **2002**. *99:* 999–1004.

125 de St Groth, B. F., The evolution of self-tolerance: a new cell arises to meet the challenge of self-reactivity. *Immunol. Today* **1998**. *19:* 448–454.

126 Suss, G., Shortman, K., A subclass of dendritic cells kills CD4 T cells via Fas Fas- ligand- induced apoptosis. *J. Exp. Med.* **1996**. *183:* 1789–1796.

127 Ferguson, T. A., Herndon, J., Elzey, B., Griffith, T. S., Schoenberger, S., Green, D. R., Uptake of apoptotic antigen-coupled cells by lymphoid dendritic cells and cross-priming of CD8(+) T cells produce active immune unresponsive-ness. *J. Immunol.* **2002**. *168:* 5589–5595.

128 Belz, G. T., Behrens, G. M., Smith, C. M., Miller, J. F., Jones, C., Lejon, K., Fathman, C. G., Mueller, S. N., Shortman, K., Carbone, F. R., Heath, W. R., The CD8alpha(+) dendritic cell is responsible for inducing peripheral self-tolerance to tissue-associated antigens. *J. Exp. Med.* **2002**. *196:* 1099–1104.

129 Maldonado-Lopez, R., De Smedt, T., Michel, P., Godfroid, J., Pajak, B., Heirman, C., Thielemans, K., Leo, O., Urbain, J., Moser, M., CD8alpha+ and CD8alpha- subclasses of dendritic cells direct the development of distinct T helper cells in vivo. *J. Exp. Med.* **1999**. *189:* 587–592.

130 Ruedl, C., Bachmann, M. F., CTL priming by CD8(+) and CD8(-) dendritic cells in vivo. *Eur. J. Immunol.* **1999**. *29:* 3762–3767.

131 den Haan, J. M., Lehar, S. M., Bevan, M. J., CD8(+) but not CD8(-) dendritic cells cross-prime cytotoxic T cells in vivo. *J. Exp. Med.* **2000**. *192:* 1685–1696.

132 Smith, C. M., Belz, G. T., Wilson, N. S., Villadangos, J. A., Shortman, K., Carbone, F. R., Heath, W. R., Cutting edge: conventional CD8 alpha+ dendritic cells are preferentially involved in CTL priming after footpad infection with herpes simplex virus-1. *J. Immunol.* **2003**. *170:* 4437–4440.

133 Belz, G. T., Smith, C. M., Eichner, D., Shortman, K., Karupiah, G., Carbone, F. R., Heath, W. R., Cutting edge: conven tional CD8 alpha+ dendritic cells are generally involved in priming CTL immunity to viruses. *J. Immunol.* **2004**. *172:* 1996–2000.

134 Schulz, O., Diebold, S. S., Chen, M., Naslund, T. I., Nolte, M. A., Alexopoulou, L., Azuma, Y. T., Flavell, R. A., Liljestrom, P., Reis e Sousa, C., Toll-like receptor 3 promotes cross-priming to virus-infected cells. *Nature* **2005**. *433:* 887–892.

135 Rissoan, M. C., Soumelis, V., Kadowaki, N., Grouard, G., Briere, F., de Waal, M., Liu, Y. J., Reciprocal control of T helper cell and dendritic cell differentiation. *Science* **1999**. *283:* 1183–1186.

136 Dalod, M., Salazar-Mather, T. P., Malmgaard, L., Lewis, C., Asselin-Paurel, C., Briere, F., Trinchieri, G., Biron, C. A., Interferon alpha/beta and interleukin 12 responses to viral infections: pathways regulating dendritic cell cytokine expression in vivo. *J. Exp. Med.* **2002**. *195:* 517–528.

137 Salio, M., Palmowski, M. J., Atzberger, A., Hermans, I. F., Cerundolo, V., CpG-matured murine plasmacytoid dendritic cells are capable of in vivo priming of functional CD8 T cell responses to endogenous but not exogenous antigens. *J. Exp. Med.* **2004.** *199:* 567–579.

138 Heath, W. R., Belz, G. T., Behrens, G. M., Smith, C. M., Forehan, S. P., Parish, I. A., Davey, G. M., Wilson, N. S., Carbone, F. R., Villadangos, J. A., Cross-presentation, dendritic cell subsets, and the generation of immunity to cellular antigens. *Immunol. Rev.* **2004.** *199:* 9–26.

139 Boonstra, A., Asselin-Paturel, C., Gilliet, M., Crain, C., Trinchieri, G., Liu, Y. J., O'Garra, A., Flexibility of mouse classical and plasmacytoid-derived dendritic cells in directing T helper type 1 and 2 cell development: dependency on antigen dose and differential toll-like receptor ligation. *J. Exp. Med.* **2003.** *197:* 101–109.

140 Zuniga, E. I., McGavern, D. B., Pruneda-Paz, J. L., Teng, C., Oldstone, M. B., Bone marrow plasmacytoid dendritic cells can differentiate into myeloid dendritic cells upon virus infection. *Nat. Immunol.* **2004.** *5:* 1227–1234.

141 Ludewig, B., Odermatt, B., Landmann, S., Hengartner, H., Zinkernagel, R. M., Dendritic cells induce autoimmune diabetes and maintain disease via de novo formation of local lymphoid tissue. *J. Exp. Med.* **1998.** *188:* 1493–1501.

142 Ludewig, B., Junt, T., Hengartner, H., Zinkernagel, R. M., Dendritic cells in autoimmune diseases. *Curr. Opin. Immunol.* **2001.** *13:* 657–662.

143 Grossman, Z., Paul, W. E., Autoreactivity, dynamic tuning and selectivity. *Curr. Opin. Immunol.* **2001.** *13:* 687–698.

144 Janeway, C. A., Approaching the asymptote? Evolution and revolution in immunobiology. *Cold Spring Harb. Symp. Quant. Biol* **1989.** *54:* 1–13.

145 Geijtenbeek, T. B., van Vliet, S. J., Engering, A., 't Hart, B. A., van Kooyk, Y., Self- and nonself-recognition by C-type lectins on dendritic cells. *Annu. Rev. Immunol.* **2004.** *22:* 33–54.

146 Reis e Sousa, Toll-like receptors and dendritic cells: for whom the bug tolls. *Semin. Immunol.* **2004.** *16:* 27–34.

147 Iwasaki, A., Medzhitov, R., Toll-like receptor control of the adaptive immune responses. *Nat. Immunol.* **2004.** *5:* 987–995.

148 van Mierlo, G. J., Boonman, Z. F., Dumortier, H. M., den Boer, A. T., Fransen, M. F., Nouta, J., van der Voort, E. I., Offringa, R., Toes, R. E., Melief, C. J., Activation of dendritic cells that cross-present tumor-derived antigen licenses CD8+ CTL to cause tumor eradication. *J. Immunol.* **2004.** *173:* 6753–6759.

149 De Smedt, T., Pajak, B., Klaus, G. G., Noelle, R. J., Urbain, J., Leo, O., Moser, M., Antigen-specific T lymphocytes regulate lipopolysaccharide-induced apoptosis of dendritic cells in vivo. *J. Immunol.* **1998.** *161:* 4476–4479.

150 Steinman, R. M., Hawiger, D., Nussenzweig, M. C., Tolerogenic dendritic cells. *Annu. Rev. Immunol.* **2003.** *21:* 685–711. Epub; **2001** Dec 19.

151 Huang, F. P., Platt, N., Wykes, M., Major, J. R., Powell, T. J., Jenkins, C. D., MacPherson, G. G., A discrete sub-population of dendritic cells transports apoptotic intestinal epithelial cells to T cell areas of mesenteric lymph nodes. *J. Exp. Med.* **2000.** *191:* 435–444.

152 de Vos, A. F., van Meurs, M., Brok, H. P., Boven, L. A., Hintzen, R. Q., van, d., V, Ravid, R., Rensing, S., Boon, L., 't Hart, B. A., Laman, J. D., Transfer of central nervous system autoantigens and presentation in secondary lymphoid organs. *J. Immunol.* **2002.** *169:* 5415–5423.

153 Hawiger, D., Inaba, K., Dorsett, Y., Guo, M., Mahnke, K., Rivera, M., Ravetch, J. V., Steinman, R. M., Nussenzweig, M. C., Dendritic cells induce peripheral T cell unresponsiveness under steady state conditions in vivo. *J. Exp. Med.* **2001.** *194:* 769–779.

154 Probst, H. C., Lagnel, J., Kollias, G., van den, B. M., Inducible transgenic mice reveal resting dendritic cells as potent inducers of CD8+ T cell tolerance. *Immunity.* **2003.** *18:* 713–720.

155 Scheinecker, C., McHugh, R., Shevach, E. M., Germain, R. N., Constitutive presentation of a natural tissue auto-antigen exclusively by dendritic cells in

the draining lymph node. *J. Exp. Med.* **2002**. *196:* 1079–1090.

156 Albert, M. L., Jegathesan, M., Darnell, R. B., Dendritic cell maturation is required for the cross-tolerization of CD8+ T cells. *Nat. Immunol.* **2001**. *2:* 1010–1017.

157 Menges, M., Rossner, S., Voigtlander, C., Schindler, H., Kukutsch, N. A., Bogdan, C., Erb, K., Schuler, G., Lutz, M. B., Repetitive injections of dendritic cells matured with tumor necrosis factor alpha induce antigen-specific protection of mice from autoimmunity. *J. Exp. Med.* **2002**. *195:* 15–21.

158 Sporri, R., Reis e Sousa, Inflammatory mediators are insufficient for full dendritic cell activation and promote expansion of CD4+ T cell populations lacking helper function. *Nat. Immunol.* **2005**. *6:* 163–170.

IX
Dendritic Cells Cross-talk with Other Cell Types

24
Dendritic Cells and Natural Killer Cells

Magali Terme and Laurence Zitvogel

In the last 30 years, dendritic cells (DC) have been studied with regard to antigen-specific immune response and adaptive immunity. Recently, a key role of the DC has been highlighted: their capacity to control innate immunity by triggering NK-cell activation and the role for this interaction in T-cell priming. In this chapter, we will address different questions: (1) which DC subset interacts with NK cells? (2) in which location? (3) how is DC/NK interaction counter-regulated? (4) what is the role of DC-mediated NK-cell activation for CD4$^+$ T and CTL priming? (5) What is the potential pathophysiological relevance of the DC/NK cross-talk? Since this book focuses on DC biology, a short introduction on NK cells is necessary to approach our topic.

24.1
Introduction on NK Cells

NK cells represent the major population of the effectors of the innate arm of immunity. NK cells are large granular lymphocytes characterized by the absence of CD3 and the expression of FcγRIII/CD16 and CD56 in humans and NK1.1 in mice [1]. These cells are key players in immunosurveillance and in the defense against viral infections and tumors. In humans, NK cells are found in lymph nodes and at lower frequency in blood [2]. Lymph nodes harbor 40% of total lymphocytes and NK cells represent 5% of total lymph node lymphocytes, while blood contains 2% of total lymphocytes and NK cells represent 5–15% of total blood lymphocytes. Therefore, lymph node NK cells are, in the absence of infection and inflammation, ten times more abundant than blood NK cells [2]. In mouse, NK cells are mainly found in spleen, but are also detected in lymph nodes and blood and in nonlymphoid organs such as liver, lung, placenta, bone marrow.

Handbook of Dendritic Cells. Biology, Diseases, and Therapies.
Edited by M. B. Lutz, N. Romani, and A. Steinkasserer.
Copyright © 2006 WILEY-VCH Verlag GmbH & Co. KGaA, Weinheim
ISBN: 3-527-31109-2

NK cells have been firstly described as "non-MHC-restricted" cells. However, MHC Class I molecules present on the surface of tumor cells play a critical role on the inhibition of the NK cells. Besides, an inverse correlation has been established between the expression of MHC class I molecules on target cells and the susceptibility to NK-cell lysis [3, 4]. This correlation has led to "the missing self hypothesis" which has been confirmed by the characterization of MHC Class I specific inhibitory receptors. But some MHC Class I$^+$ tumor cell lines are killed by NK cells because of NK-cell activating receptors, such as NKG2D, which allow NK cells to overcome inhibitory signals generated by MHC Class I molecules on tumor cells. NK-cell receptors have recently been reviewed [5–9]. Thus, NK-cell function is tightly regulated by the integration of activating and inhibitory signals initiated by the engagement of cell surface receptors.

Activated NK cells are able to lyse infected or tumor cells, or secrete cytokines. They can directly kill target cells or cells covered with antibodies by the antibody-dependent cellular cytotoxicity (ADCC). Perforin/granzyme exocytosis [10] and death receptor engagement [11–13] are the two major mechanisms used by NK cells to induce target cell apoptosis. Activated NK cells can also secrete immunoregulatory cytokines, such as IFNγ, TNFα, IL-13, GM-CSF [14]. Cytotoxicity and cytokine secretion have been shown to be dependent on different NK-cell subsets. Indeed, in humans, two different NK-cell subsets have been identified by cell surface density of CD56. The majority of blood NK cells (90%) are CD56^{dim} and express high levels of CD16, whereas CD56^{bright} CD16^{dim} NK cells are found at minor levels [15]. CD56^{dim} NK cells are more cytotoxic than CD56^{bright} NK cells [16], whereas CD56^{bright} NK cells are involved in cytokine and chemokine production [14].

Recently, a provocative article demonstrates the capacity of mature human NK cells expressing high levels of MHC class II and co-stimulatory molecules after IL-2 activation to uptake and present particulate antigens or immune complexes and to stimulate antigen specific T cells *in vitro* [17]. However, the relevance of this observation for *in vivo* priming of naïve T cells in the setting of a viral infection remains to be shown.

24.2
Activation of NK Cells by DC

The original description of the existence of the DC/NK-cell interaction has been provided by our laboratory. Fernandez et al. [18] showed that injection of Flt3-L, a DC growth factor, or adoptive transfer of DC induced eradication of AK7 mesothelioma tumors. This antitumor effect was dependent on NK cells. Furthermore, *in vitro* bone-marrow-derived dendritic cells or the D1 DC cell line stimulated with TNFα could stimulate NK-cell effector functions (cytotoxicity and IFNγ production). Thereafter, several groups were interested in the study of the interaction between DC and NK cells. The regulation of NK-cell activation by DC is being slowly unraveled in the mouse and human settings, for both cytolytic and secretory activities as a function of the stimuli received by DC.

24.2.1
NK-cell Activation and DC Subsets

Gerosa et al. [19] have shown that stimulation of peripheral blood mononuclear cells with some Toll-like receptor (TLR) ligands, such as Poly I:C (for TLR3) or R848 (for TLR7–8), results in IFNγ secretion. NK cells represent the majority of IFNγ producing cells, in association with NKT cells and γδT cells. But these cells do not produce IFNγ directly upon stimulation by TLR ligands, except for TLR2 and TLR3 ligands that have been described to directly activate NK-cell functions [20–22]. The presence of HLA-DR$^+$ antigen presenting cells, and particularly dendritic cells (DC), is required to stimulate NK-cell IFNγ production [19]. However, not all DC subsets are able to stimulate NK cells. Myeloid DC stimulated by TLR ligands are able to stimulate both IFNγ production and cytotoxicity. Several *in vitro* human studies have highlighted the capacity of mature myeloid DC (monocyte-derived DC (MD-DC) and CD34$^+$-derived DC) stimulated by LPS, a TLR4 ligand, to activate resting NK cells what results in the enhancement of cytotoxicity and IFNγ secretion functions, proliferation and upregulation of CD25 and CD69 on NK cells [23–28]. On the other hand, plasmacytoid DC (pDC) pretreated with viruses or TLR9 ligands trigger only NK-cell cytolytic function [19]. By contrast, Langerhans cells fail to induce NK-cell stimulation [29].

In mouse, as in humans, myeloid DC trigger both NK-cell effector functions. Indeed, bone-marrow-derived DC (BM-DC) or splenocytes-derived D1-cell line stimulated by LPS or TNFα are able to activate NK-cell IFNγ production and cytolytic function [18, 30–32]. In contrast with human pDC, plasmacytoid DC isolated from MCMV-infected mice can not only activate cytotoxicity but also IFNγ production by NK cells [33].

24.2.2
Molecular Mechanisms of the DC-mediated NK-cell Activation

Several studies have highlighted the fact that the maturation state of DC influence their capacity to activate NK cells. Immature DC require a maturation stimuli, such as LPS, TNFα, *Mycobacterium tuberculosis* or IFNα, to activate resting NK cells [23–26]. However, we will describe later in this chapter that in some circumstances pathophysiological stimuli can promote DC-mediated NK-cell activation without inducing complete DC maturation.

Both soluble factors and cell–cell contacts are involved in NK-cell triggering by DC [23–26, 28]. Mature DC are a major source of IL-12 [34] and other cytokines, such as IL-18, that are mainly implicated in the stimulation of NK-cell IFNγ production [19, 23, 35]. Myeloid DC or pDC stimulated respectively by TLR3 ligands and by viruses or TLR9 ligands secrete type I IFN that is mostly involved in activation of NK-cell cytotoxicity [19]. Cell-to-cell contacts and synapse formation between DC and NK cells are also needed [23, 35]. Formation of stimulatory synapses between DC and NK cells depends on lipid raft mobilization and remodeling of cytoskeleton and promotes the polarized secretion of preassembled stores of IL-12 by DCs towards the NK-cell [35]. The synaptic delivery of IL-12 by DC was found to be required for IFNγ secretion by NK cells, as assessed using inhibitors of cytoskeleton rearrangements and transwell experiments. Besides, Langerhans cells fail to induce NK-cell activation presumably because they do not produce IL-12 and express IL-15Rα [29].

While cytokines play an important role in mature DC-mediated NK-cell activation, cell surface receptors are also involved. Then, Jinushi et al. [36] have shown that IFNα treatment of MD-DC induce the upregulation of MHC-Class I related chain-A and B (MICA/B), two ligands for the NK-cell activating receptor NKG2D, that contribute to NK-cell triggering. IL-15-derived DC seems to be implicated in the upregulation of MICA/B on MD-DC after IFNα exposure [37].

In murine studies, most DC are derived from bone marrow and treated with TLR4 ligands such as LPS or *E. coli* bacteria. Following TLR4 triggering, BM-DC propagated in the absence of IL-4 transiently produce IL-2 [38, 39] and stimulate NK-cell cytotoxicity and IFNγ secretion by two different mechanisms. IL-2-derived DC is involved in the activation of IFNγ production while type I IFN produced by DC triggers NK-cell cytotoxicity as in humans [32]. By contrast, IL-12 and IL-18 do not have a key role in this setting. *In vivo*, inoculation of *E. coli* promotes IL-2 secretion by DC that contribute to NK-cell activation [32]. However, when IL-4-propagated BM-DC are exposed to LPS, they do not produce IL-2, because of the inhibitory role of IL-4 on IL-2 production [40]. By contrast, upon LPS treatment BM-DC generated in the presence of IL-4 upregulate IL-15Rα and produce IL-15 that contribute to the activation of NK-cell cytotoxicity [31]. In this setting, IL-12 acts in synergy with IL-15 and IL-15Rα molecules to induce NK-cell IFNγ secretion [31]. Using IL-12-deficient mice, Borg et al. [35] have confirmed that IL-12 have a pivotal role in the induction of IFNγ production by mature BM-DC.

In the absence of TLR4 stimulation, IL-4-propagated BM-DC are also able to activate NK-cell functions, unlike BM-DC generated in the absence of IL-4 [30]. The triggering receptor expressed on myeloid cells-2 (TREM2) associated with KARAP/DAP12 adaptor molecule is upregulated on BM-DC by IL-4 and involved in DC-mediated NK-cell activation [30].

24.3
Reciprocal Interaction of DC and NK Cells

Communication between DC and NK cells is not unidirectional but bi-directional and should really be considered as a dialog. Thus, not only do DC have the ability to activate NK cells, but activated NK cells are also able to stimulate maturation or lysis of DC. According to Piccioli et al. [25], the outcome of the NK-DC interactions is tightly regulated by the NK/DC cell ratio. Indeed, at a low NK/DC ratio DC maturation prevails, while at a high NK/DC ratio DC lysis occurs [25].

24.3.1
DC Maturation Induced by NK Cells

Once NK cells are activated, they can interact with immature MD-DC and induce, or at least augment, their maturation characterized by the upregulation of CCR7, MHC Class II molecules and co-stimulatory molecules, such as CD80, CD86, CD83, but also by enhanced antigen-presenting capacity, and production of IL-12 [23, 25]. This NK cell-dependent DC maturation requires both cell–cell contact and TNFα secretion [23, 25]. In the presence of suboptimal stimuli, e.g. CpG, activated NK cells favor pDC maturation leading to type I IFN and TNF production [19]. In a mouse model, we have highlighted that NK cells, that have been activated by IL-4-propagated BM-DC, can in turn promote maturation of BM-DC characterized by the upregulation of CD80 and CD86 molecules on DC surface. This maturation is in part due to the TREM2/KARAP/DAP12 pathway, at least for the upregulation of CD86 [30]. The relevance of this phenomenon *in vivo* has been demonstrated in a mouse model of MHC ClI^{lo} tumor. Indeed, ClI^{lo} tumors induce NK-cell activation that in cascade stimulate DC activation, characterized by IL-12 production [41].

Thus, activated NK cells can induce DC maturation either directly or in synergy with suboptimal levels of microbial signals. NK cell-induced DC activation is dependent on both TNFα/IFNγ secretion and a cell–cell contact. As shown by Vitale et al. in a human setting [42], the NK cell-mediated DC maturation involving the release of TNFα by NK cells depends on the triggering of the NK-cell receptor NKp30 and is counter-regulated by KIR and NKG2A inhibitory receptors. In a mouse tumor model, the group of Mocikat [43] has been able to show a critical role for IL-12 produced by endogenous DC following IFNγ secretion by activated NK cells, underscoring a link between innate immunity and cross priming of CTL [43] (as discussed below).

Therefore, NK help might be critical for optimal DC activation and subsequent induction of T-cell responses in conditions where inflammation is poor but where NK-cell activation could occur through direct recognition of target cells. In particular, this may be relevant to defense against cancer, as discussed below.

24.3.2
Lysis of DC by Activated NK Cells

The relevance of the NK cell-mediated DC lysis is more intriguing. Immature DC are preferentially killed by NK cells [24, 25, 44, 45], but after maturation by microbial stimulation or cytokines, DC become resistant to NK-cell killing [44–46]. This is due to the upregulation of MHC Class I molecules on DC upon maturation that protect DC from NK-cell lysis [46, 47]. Killing of immature DC is confined to a small NK-cell subset expressing the CD94/NKG2A inhibitory receptor [48]. Thus, DC that express high levels of HLA-E, the CD94/NKG2A ligand, are protected from NK-cell lysis. However, a small subset of NK cells expressing low levels of NKG2A is also able to lyse mature DC [48]. Activating receptors, such as NKp30 and maybe others, are also needed to kill immature DC [24]. A recent *in vivo* study performed in mice has shown that immature DC are eliminated by NK cells via a pathway dependent on the TNF-related apoptosis-inducing ligand (TRAIL) [49]. Depletion of NK cells or neutralization of TRAIL function during immunization with immature DC loaded with tumor antigens significantly enhanced cognate T-cell responses, suggesting that the lysis of DC by NK cells might limit vaccination efficacy [49].

24.4
Where do DC Meet NK Cells?

24.4.1
In Lymph Nodes

The presence of NK cells in lymph nodes (LN) has been controversial for a long time. But recently, some studies have highlighted the existence of a large amount of NK cells in inflamed and uninflamed lymph nodes [50, 51]. NK cells seem to be ten times more abundant in LN than in blood [51]. In inflamed LN, NK cells have been localized in the T-cell area, using immunohistochemistry and PCR [50]. In uninflamed LN, the presence of NK cells in the T-cell area has been confirmed, including clusters in the parafollicular regions of the T-cell zone [52]. In the T-cell area, DEC-205$^+$ DC have also been identified, suggesting that LN might be an essential site of DC/NK-cell encounter [52]. These LN-NK cells are mainly CD56^{bright} and are non cytolytic [51]. The CD56^{bright} NK-cell subset has been described as being the most reactive NK-cell subset to DC *in vitro* [52, 53].

However, a question remains: does DC interact with resident NK cells or with NK cells recruited from blood? Recent studies have shown that LPS-activated MD-DC can interact with resting CD56^{bright} NK cells isolated from LN within 6 h and induce NK-cell IFNγ production [52]. This study suggests that DC could interact with resident LN-NK cells and modulate cytokine secretion thereby regulating Th differentiation. Ferlazzo et al. have suggested that CD56^{bright} NK cells could be a precursor for CD56^{dim} NK cells, because when CD56^{bright} NK cells are cultured with IL-2 during one week they acquire a phenotype similar to CD56^{dim} NK cells [51]. Thus, DC could induce NK-cell to become cytolytic effectors in LN. Another hypothesis is that NK cells could be recruited by DC from blood through high endothelial venules (HEV). In blood, CD56^{bright} NK cells express high levels of CCR7 and CD62L, two molecules involved in the recruitment towards LN. However, we cannot exclude that CD56^{dim} NK cells, which express a lot of chemokine receptors such as the fractalkine receptor CX3CR1, or CXCR3 the receptor for IP-10 or I-TAC, could be recruited from blood through HEV by chemokines secreted by DC.

Different groups have tried to study the recruitment of NK cells by DC in LN in murine models. Our laboratory has been the first to show that when IL-4-propagated or LPS-matured BM-DC are inoculated into the footpad of Nude mice or immunocompetent C57Bl/6 mice, NK cells are activated and their number is enhanced in the draining LN [30]. Adoptive transfer of labeled NK cells showed that NK cells are recruited from blood through HEV upon mature DC or *Leishmania major* pro-mastigotes injection in footpad [54, 55]. In the same way, IL-4-propagated DC seem to provoke NK-cell recruitment from blood rather than resident NK-cell prolifera-tion, since resident NK cells do not incorporate BrdU (Terme et al., in preparation). Using intravital microscopy, NK-cell behavior has been characterized in LN in the steady state and upon *L. major* infection. At the steady state, NK cells are found in the LN outer paracortex in close vicinity with DC near HEV. But these cells are slow motile, unlike T cells which move rapidly. After *L. major* infection, NK cells rapidly accumulate in the outer paracortex where they can interact with DC and T lymphocytes and secrete IFNγ, but they do not acquire higher motility [55]. Foot-pad injection of LPS-matured DC trigger NK-cell recruitment in LN in a CXCR3-dependent manner, rather than in a CCR7-dependent manner [54]. These NK cells produce IFNγ that is necessary for Th1 polarization [54]. We will discuss in another chapter the role of DC/NK-cell cross-talk on T-cell polarization and priming.

24.4.2
In the Periphery

DC/NK-cell interaction can also happen at sites of inflammation. The CD56^{dim} CD16$^+$ NK-cell subset, that represents the prevalent NK-cell subset in the peripher-al blood expresses a chemokine receptor repertoire that allows them to migrate in inflamed tissues. They can migrate in tissues in response to different chemokines, e.g. CCL3, CXCL8, CX3CL1 or CCL22, that are produced by DC or other cells, such as macrophages, endothelial cells or neutrophiles [15, 56]. Thus, NK cells can pene-trate into the tissues, where they can encounter resident DC. Moreover, different arguments suggest that DC/NK cross-talk exists in periphery. NK cells have been detected in close contact with DC in Malassezia-induced atopic dermatitis lesion [57], and in imatinib mesylate-induced lichenoid dermatitis [58]. An aberrant PEN5$^+$ NK-cell infiltration has also been found in tissues infiltrated with malig-nant Langerhans histocytosis, suggesting that a dysregulation of the LC/NK cross-talk could participate in chronic inflammation (Borg C., unpublished observa-tions).

24.5
DC/NK Cross-talk and T Lymphocytes

24.5.1
Bridging Innate and Adaptive Immunity

As previously discussed (Section 24.3.1), activated NK cells can induce DC matura-tion. Thus, NK-cell activation could be crucial for T-cell responses. Mailliard et al. [59] have shown that in the presence of tumor targets and cytokines (IFNα), hu-man NK cells become activated. Those activated NK cells promote the maturation of DC that acquire the capacity to produce IL-12, thereby promoting Th1 polarization and a CD8$^+$ T-cell-specific response [59] *in vitro*. In a mouse model where the growth of MHC Class I^{lo} tumor targets induced NK-cell activation, IL-12 produced by DC after NK-cell dependent maturation induced a CD8$^+$ T-cell memory response [41].

Other studies have highlighted the fact that IFNγ produced by activated NK cells can influence T-cell responses. Indeed, resistance to *L. major* involves a Th1 response that is dependent on IFNγ produced by NK cells [60]. In the RMA-S-Rae1β tumor model, tumor rejection is mediated by NK cells via a NKG2D-dependent mechanism, but NK-cell activation is also responsible for a secondary T-cell-specific response [61].

A recent and elegant study has revealed that recruitment of NK cells in lymph nodes upon stimulation by LPS-matured OVA-pulsed DC or adjuvants such as Ri-

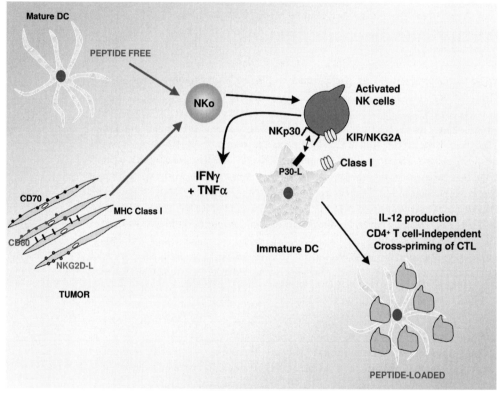

Fig. 24.1 A DC/NK/DC cross-talk leading to efficient cross-presentation of tumor antigens. Tumor cells over-expressing NKG2D ligands or CD70 or CD80 might directly trigger NK-cell activation. Alternatively, exogenous LPS-activated mature DC can turn on NK cells *in vivo*. DC-activated NK cells produce IFNγ which in turn, promotes endogenous DC activation and production of IL-12 and elicitation of CTL cross priming in the absence of CD4$^+$ T-cell help [43]. In the human setting, DC-activated NK cells or IL-2 activated NK cells will induce DC maturation through engagement of NKp30, in a TNF-α dependent-manner.

bi or R848 correlates with Th1 priming [54]. Depletion experiments have shown that NK cells provide IFNγ that is pivotal for Th1 response [54] (Fig. 24.1). However, the role of IFNγ on Th1 response is not really understood and it remains unclear at which level IFNγ can act: on naïve T cells or on host DC.

A recent study [43] assigns a new role for the DC/NK-cell cross-talk in immune responses. The authors demonstrate efficient CTL priming and long term CD8⁺ T-cell memory against A20 tumor antigens in the absence of CD4⁺ T-cell help using CD40$^{-/-}$ mature DC, unpulsed with tumor antigens but capable of NK-cell triggering. Their data clearly show that expression of NKG2D ligands on A20 are required for NK-cell triggering, and that IFNγ was necessary for (1) NK-cell activation, (2) activation of endogenous DC and (3) IL-12 production leading to CTL induction (Fig. 24.1).

24.5.2
Modulation of the DC/NK-cell Cross-talk by CD4+CD25+ Regulatory T Cells and Conventional T Cells

Recent advances in our understanding of the biology of regulatory T cells prompted us to determine which impact regulatory T cells (Treg) might have on NK cells and also on DC/NK-cell cross-talk. We have highlighted an inverse correlation between NK-cell activation and Treg in cancer-bearing patients (GIST and melanoma). In humans, Treg directly inhibit activation of NK-cell cytotoxicity and IL-12-mediated IFNγ production, and reduce NKG2D expression on NK cells. This NK-cell inhibition is due to membrane-bound TGFβ [80]. In the mouse model, only activated regulatory T cells (activation by anti-CD3 and anti-CD28) express membrane-bound TGFβ and have the ability to directly inhibit NK-cell cytotoxicity and induce a downregulation of NKG2D on NK cells upon adoptive transfer in Nude mice. Moreover, Treg control NK-cell homeostatic proliferation *in vivo* since Treg cells deficiency or depletion by cyclophosphamide (CTX) or anti-CD25 antibody results in an enhanced proliferation of NK cells [80]. In CTX-treated mice, adoptive transfer of IL-4-generated BM-DC induce an augmentation of NK-cell proliferation in the draining lymph node compared with the contralateral lymph node (Terme et al., in preparation). Moreover, in CTX-treated mice, NK-cell activation in draining lymph nodes upon DC injection is enhanced as assessed by CD69 expression on NK cells and IFNγ production. Thus, Treg inhibit DC-mediated NK-cell activation *in vivo. In vitro*, resting Treg inhibit DC-mediated induction of IFNγ production but also NK-cell-mediated DC maturation (Terme et al., in preparation).

Therefore, activated Treg directly inhibit NK-cell cytotoxicity in mouse and human in a TGFβ-dependent manner, whereas resting Treg restrict DC-mediated NK-cell activation. However, some subsets of T lymphocytes are also able to enhance NK-cell activation. In experiments where IL-4 propagated BM-DC were injected in Nude Mice, the number of DX5⁺ or NK1.1⁺ NK cells expressing CD69 was increased by 2-fold by 24 h, whereas in immunocompetent C57Bl/6 mice, the number of CD69⁺ NK cells was augmented by 3–4-fold [30] suggesting a potential role of IL-2 produced by CD4⁺ T cells in this NK-cell activation [50].

24.6
The DC/NK-cell Cross-talk in Physiopathology

24.6.1
In Infectious Diseases

24.6.1.1 Viral Infections

NK cells contribute to innate defense during certain viral infections, such as infections caused by *Herpes* viruses, hepatitis viruses or HIV [62, 63]. For example, NK cells have a key role in controling murine cytomegalovirus (MCMV) infection [64]. NK-cell cytotoxicity and cytokine or chemokine production are both involved in the defense against this virus, cytotoxicity is predominant in the spleen, whereas IFNγ production occurs in the liver [65]. Not all NK cells seem to be important in the elimination of MCMV. Indeed, the Ly49H activating NK-cell receptor which recognizes the MCMV m157 glycoprotein is restricted to a particular NK-cell subset [64, 66]. CpG motifs, which are ligands for TLR9, are abundant in the genomes of α and β *Herpes* viruses such as HSV and MCMV. DC and pDC have been shown to recognize MCMV through TLR9 [67]. Although DC lose their capacity to stimulate T cells upon MCMV infection, probably because of the loss of their IL-2 production capacity [68], Andrews et al. have highlighted the importance of the DC (especially CD8α⁺ DC subset) in the expansion of the Ly49H⁺ NK-cell subset via IL-12 and IL-18 [69]. Ly49H⁺ NK cells in turn favor the survival of CD8α⁺ DC [69], confirming the hypothesis that the interaction between DC and NK cells is bi-directional. pDC are also pivotal players in viral infections. During MCMV infection, pDC recognize MCMV through TLR9 and secrete high levels of type I IFN and IL-12 which induce NK-cell activation and promote viral clearance by Ly49H⁺ NK cells [33, 67]. Upon pDC depletion, type I IFN secretion is dramatically reduced, but other cell types compensate and secrete large amounts of IL-12 guaranteeing high levels of IFNγ and normal NK-cell responses to MCMV [67]. Thus, MCMV recognition via TLR9 promotes pDC, DC and also macrophages to secrete cytokines allowing NK-cell activation and MCMV clearance [67].

NK-cell functions are impaired early in HIV infection [70]. Poggi et al. have shown that HIV-1 Tat, an HIV product, inhibits DC-mediated NK-cell activation by blocking calcium influx and calcium-calmodulin kinase II activation elicited via LFA-1 [71]. Since HIV-1 Tat also inhibits IL-12 secretion by DC, it might be account for the inhibition of NK-cell activation in HIV [72, 73].

NK cells have also been involved in the control of Influenza and Sendai virus [74]. Since plasmacytoid DC secrete type I IFN in response to these viruses, a role for the DC/NK-cell interaction in the control of these infections can not be excluded [75, 76].

24.6.1.2 Bacterial Infections

NK cells play a role in the control of viral infections but also in the clearance of bacterial infection. As we previously discussed, myeloid DC are able to secrete IL-2 during the first hours of bacterial infection [39]. In a murine model of *E. coli*, DC-derived IL-2 promote NK-cell activation and especially IFNγ production which has an important antibacterial function [32]. The number of bacteria in the spleen of mice injected with IL-2-deficient BM-DC is significantly higher than in spleens of mice treated with WT DC, thus IL-2-deficient DC lead to a limited bacterial clearance compared with WT BM-DC [32].

24.6.2
In Cancer

NK cells have been involved in the control of different tumors, such as neuroblastoma, chronic myeloid leukemia (CML) and gastro-intestinal stromal tumors (GIST). GIST cells display typical features of NK-cell sensitivity: TAP-1 deficiency, over-expression of MIC and ULBP (the ligands for NKG2D receptor), loss of MHC class I molecules [58]. Moreover, they are lysed by NK cells from normal volunteers as efficiently as the NK-cell susceptible K562 targets [58].

Our pioneering studies demonstrated the relevance of the DC-mediated NK-cell activation in tumor rejection in a mesothelioma tumor model [18]. Gleevec/STI571, a specific inhibitor of BCR/ABL, c-kit and PDGF-R tyrosine kinases, is used successfully for the treatment of CML and GIST. We have shown that administration of STI571 promote NK-cell dependent antitumor effects in tumor transplantation models using tumor cells that are resistant to STI571 *in vitro* [58]. Antitumor effects are more pronounced when mice are treated with STI571 associated with the DC-growth factor Flt3-L. *In vitro* studies in which GM-CSF and IL-4-propagated BM-DC are incubated with increasing dosage of STI571 highlighted that nanomolar concentrations of STI571 are sufficient to endow DC with ability to stimulate NK cells *in vitro* but do not promote DC maturation. STI571-stimulated DC promote NK-cell IFNγ production independently of IL-12 but require cell–cell contacts. Since identical results are obtained using STI571-treated BM-DC and c-kit-deficient BM-DC, STI571 probably acts by inhibiting the c-kit pathway in DC. Importantly, in GIST-bearing patients Gleevec induce NK-cell activation in 49% of the cases. Gleevec-mediated NK-cell activation is correlated with the clinical outcome. None of the patients who display enhanced NK-cell functions exhibit progressive disease, while all ten patients with refractory GIST have poor NK-cell activity. Moreover, the time to progression is significantly longer in patients with NK-cell activation compared with those who did not show NK-cell activation after Gleevec treatment [58].

Ruggeri et al. [77, 78] have highlighted the role of allogeneic NK cells in graft-versus-leukemia effects (GvL) in acute myeloid leukemia (AML) patients after haploidentical bone-marrow transplantation and in engrafment. But the mechanisms involved in NK-cell activation remain obscure. HLA-C/KIR mismatches between recipient leukemic cells and donor NK cells might not fully account for NK-cell activation *in vivo*. Host DC might play a key role for NK-cell activation. Different hypothesis could be envisaged. Indeed, the cytokine storm induced by the myeloablative conditioning could promote DC maturation and NK-cell activation. But with mature DC the risk of graft-versus-host disease (GvHD) development is very high [79]. In chronic myeloid patients, the success of donor lymphocyte infusion in controlling the residual disease remains also poorly understood. Since DC from CML patients bear the BCR/ABL translocation, we hypothesized that BCR/ABL translocation confers to myeloid DC NK-cell stimulatory capacities. Immature monocyte-derived DC from CML patients selectively stimulate NK-cell effector functions unlike immature MD-DC derived from healthy volunteers [81]. Using a mouse CML model obtained by infection of mouse bone marrow by retroviruses bearing the BCR/ABL translocation, we further demonstrated that the BCR/ABL translocation is responsible for the selective NK-cell stimulatory capacities of the CML DC [81]. Surprisingly, although NK cells derived from CML bearing mice display impaired functions and do not respond to IL-2 stimulation, they produce IFNγ in response to BCR/ABL-DC. Since other genetic defects leading to myeloproliferative disorders (JunB- or ICSBP-deficient mice) did not promote NK-cell activation, the BCR/ABL translocation appears unique to confer NK-cell stimulatory capacities. BCR/ABL induce expression of ligands for the NKG2D activating NK-cell receptor on murine and human myeloid DC. STI571, which is a specific inhibitor of BCR/ABL, block the expression of NKG2D ligands on DC. CML-DC-induced killing activity is significantly hampered by anti-NKG2D antibody or STI571 pretreatment in human and mouse settings, confirming the role of NKG2D in NK-cell activation. Interestingly, BCR/ABL translocation does not induce DC maturation. BCR/ABL DC have poor allostimulatory capacities, suggesting a reduced risk for GvHD development [81]. Therefore, the clonal BCR/ABL DC displayed the unique and selective capacity to activate NK cells, suggesting that they may participate in the NK-cell control of the disease. However, while STI571 treatment is critical at the early stage of the disease for its direct antileukemic effects, it might be deleterious at later stages because of their effect on NKG2D ligands expression on DC.

Exosomes are small vesicules that are secreted from professional APC such as B cells or dendritic cells. Exosomes will be described in more detail in Chapter 25. DC-derived exosomes (DEX) harbor functional MHC Class I/peptides and MHC Class II/peptides that can induce specific T-cell responses. A phase I clinical trial has been completed in France. Fifteen stage IIIb or IV melanoma patients received four peptide-pulsed DEX vaccines. Since clinical regressions have been observed without any detection of T-cell response, innate effector functions have been assessed on these patients. In 7 out of 14 patients that could be evaluated, NK-cell activation has been detected. *In vitro* studies showed that DEX can enhance CD69 expression on NK-cell patients, and that NKG2D ligands can be found at the surface of DEX. In murine studies, inoculation of DEX derived from immature BM-DC in footpad induce an enhancement of NK-cell number accompanied by an activation of NK cells in draining lymph nodes compared with contralateral LN. The role of NKG2D in DEX-mediated NK-cell activation has been confirmed in this mouse model using anti-NKG2D neutralizing antibody. The recruitment or proliferation of NK cells is dependent on IL-2-derived T cells because IL-2 neutralizing antibody blocked NK-cell augmentation (Chaput et al., submitted).

24.7
Concluding Remarks

NK cells and DC were identified more than 20 years ago and have been studied almost independently until 1998. DC were then defined as sentinels of the immune system, capable of recognizing "danger" and specialized in the initiation of the adaptive response against danger-causing agents. In addition, DC-dependent activation of CD8⁺ T cells requires CD4⁺ T-cell help. Although still valid in many ways, this model has now to face a growing body of evidence showing that (1) not all receptors sensing "danger" or pathogenic agents are expressed by DC; (2) other cells than CD4⁺ T cells may provide help to DC maturation; (3) DC have the potential to influence both innate and adaptive immunity. In particular, NK cells make functional interactions with DC that can clearly influence induction/regulation of innate and adaptive immune responses. NK activation through NK-cell receptors can promote T-cell responses against tumors, by inducing DC activation. Reciprocally, DC have been found to induce potent NK-cell activation in antiviral responses. IL-12, IL-15, IL-18 and IFN-αβ production by DC as well as cell–cell contact are critical for the enhancement of NK-cell effector functions.

NK/DC interactions are critical in situations where receptors allowing the recognition of the pathogenic agent are only expressed by one of the two subsets. By extrapolation, one may hypothesize that the expression of receptors allowing early detection of all pathogenic agents is compartmentalized in (innate) immune cell types. In this model, activation of the whole immune system is thus absolutely dependent on cell–cell cross-talk that spreads activation among cells that compose it. At one extreme, DC would be the first cells to be activated by microbes, thanks to their expression of relevant innate sensors (TLR, NOD) and turn on the system. At another extreme, in the case of a tumor that does not cause overt inflammation but does express ligands for activating NK-cell receptors, NK cells would be the first cells to be activated and turn on the system.

It is also intriguing to note the close similarities between the two cell types. Indeed, (1) the development of NK cells and some DC subtypes is selectively dependent on Flt3-L; (2) Some NK-cell markers are expressed on DC (i.e. NKRP1) and vice versa (CD11c): leukemic cells expressing both DC and NK markers and CD11c⁺/DX5⁺ bitypic cells in lymphocytic choriomeningitis virus infection (LCMV) protecting mice against autoimmune diabetes have also been described; (3) DC may have cytotoxic properties or may acquire them in response to stimulation with type I IFN, whereas activated NK cells have recently been identified as potent antigen presenting cells.

Finally, most studies on NK-DC interactions have focused on conditions of microbial and tumoral challenge. It remains to dissect whether NK/DC cross-talk is also involved in autoimmune situations.

References

1 Rolstad B, Seaman WE. Natural killer cells and recognition of MHC class I molecules: new perspectives and challenges in immunology. *Scand. J. Immunol.* **1998**, *47*, 412–425.

2 Ferlazzo G, Munz C. NK cell compartments and their activation by dendritic cells. *J. Immunol.* **2004**, *172*, 1333–1339.

3 Karre K, Ljunggren HG, Piontek G, Kiessling R. Selective rejection of H-2-deficient lymphoma variants suggests alternative immune defence strategy. *Nature* **1986**, *319*, 675–678.

4 Ljunggren HG, Karre K. Host resistance directed selectively against H-2-deficient lymphoma variants. Analysis of the mechanism. *J. Exp. Med.* **1985**, *162*, 1745–1759.

5 Moretta A, et al. Activating receptors and coreceptors involved in human natural killer cell-mediated cytolysis. *Annu. Rev. Immunol.* **2001**, *19*, 197–223.

6 Diefenbach A, Raulet DH. Strategies for target cell recognition by natural killer cells. *Immunol. Rev.* **2001**, *181*, 170–184.

7 Diefenbach A, Raulet DH. Innate immune recognition by stimulatory immunoreceptors. *Curr. Opin. Immunol.* **2003**, *15*, 37–44.

8 Lanier LL. On guard–activating NK cell receptors. *Nat. Immunol.* **2001**, *2*, 23–27.

9 Cerwenka A, Lanier LL. Natural killer cells, viruses and cancer. *Nat. Rev. Immunol.* **2001**, *1*, 41–49.

10 Kagi D, et al. Cytotoxicity mediated by T cells and natural killer cells is greatly impaired in perforin-deficient mice. *Nature* **1994**, *369*, 31–37.

11 Smyth MJ, et al. Nature's TRAIL – on a path to cancer immunotherapy. *Immunity* **2003**, *18*, 1–6.

12 Smyth MJ, et al. Tumor necrosis factor-related apoptosis-inducing ligand (TRAIL) contributes to interferon gamma-dependent natural killer cell protection from tumor metastasis. *J. Exp. Med.* **2001**, *193*, 661–670.

13 Takeda K, et al. Involvement of tumor necrosis factor-related apoptosis-inducing ligand in surveillance of tumor metastasis by liver natural killer cells. *Nat. Med.* **2001**, *7*, 94–100.

14 Cooper MA, et al. Human natural killer cells: a unique innate immunoregulatory role for the CD56(bright) subset. *Blood* **2001**, *97*, 3146–3151.

15 Cooper MA, Fehniger TA, Caligiuri MA. The biology of human natural killer-cell subsets. *Trends Immunol.* **2001**, *22*, 633–640.

16 Nagler A, Lanier LL, Cwirla S, Phillips JH. Comparative studies of human FcRIII-positive and negative natural killer cells. *J. Immunol.* **1989**, *143*, 3183–3191.

17 Hanna J, et al. Novel APC-like properties of human NK cells directly regulate T cell activation. *J. Clin. Invest.* **2004**, *114*, 1612–1623.

18 Fernandez NC, et al. Dendritic cells directly trigger NK cell functions: cross-talk relevant in innate anti-tumor immune responses in vivo. *Nat. Med.* **1999**, *5*, 405–411.

19 Gerosa F, et al. The reciprocal interaction of NK cells with plasmacytoid or myeloid dendritic cells profoundly affects innate resistance functions. *J. Immunol.* **2005**, *174*, 727–734.

20 Chalifour A, et al. Direct bacterial protein PAMPs recognition by human NK cells involves TLRs and triggers {alpha}-defensin production. *Blood* **2004**.

21 Sivori S, et al. CpG and double-stranded RNA trigger human NK cells by Toll-like receptors: induction of cytokine release and cytotoxicity against tumors and dendritic cells. *Proc. Natl. Acad. Sci. USA* **2004**, *101*, 10 116–10 121.

22 Schmidt KN, et al. APC-independent activation of NK cells by the Toll-like receptor 3 agonist double-stranded RNA. *J. Immunol.* **2004**, *172*, 138–143.

23 Gerosa F, et al. Reciprocal activating interaction between natural killer cells and dendritic cells. *J. Exp. Med.* **2002**, *195*, 327–333.

24 Ferlazzo G, et al. Human dendritic cells activate resting natural killer (NK) cells and are recognized via the NKp30 receptor by activated NK cells. *J. Exp. Med.* **2002**, *195*, 343–351.

25 Piccioli D, Sbrana S, Melandri E, Valiante NM. Contact-dependent stimulation and inhibition of dendritic cells by natural killer cells. *J. Exp. Med.* **2002**, *195*, 335–341.

26 Fernandez NC, et al. Dendritic cells (DC) promote natural killer (NK) cell functions: dynamics of the human DC/NK cell cross talk. *Eur. Cytokine Netw.* **2002**, *13*, 17–27.

27 Yu Y, et al. Enhancement of human cord blood CD34+ cell-derived NK cell cytotoxicity by dendritic cells. *J. Immunol.* **2001**, *166*, 1590–1600.

28 Amakata Y, Fujiyama Y, Andoh A, Hodohara K, Bamba T. Mechanism of NK cell activation induced by coculture with dendritic cells derived from peripheral blood monocytes. *Clin. Exp. Immunol.* **2001**, *124*, 214–222.

29 Munz C, et al. Mature myeloid dendritic cell subsets have distinct roles for activation and viability of circulating human natural killer cells. *Blood* **2005**, *105*, 266–273.

30 Terme M, et al. IL-4 confers NK stimulatory capacity to murine dendritic cells: a signaling pathway involving KARAP/DAP12-triggering receptor expressed on myeloid cell 2 molecules. *J. Immunol.* **2004**, *172*, 5957–5966.

31 Koka R, et al. Cutting edge: murine dendritic cells require IL-15Ralpha to prime NK cells. *J. Immunol.* **2004**, *173*, 3594–3598.

32 Granucci F, et al. A contribution of mouse dendritic cell-derived IL-2 for NK cell activation. *J. Exp. Med.* **2004**, *200*, 287–295.

33 Dalod M, et al. Dendritic cell responses to early murine cytomegalovirus infection: subset functional specialization and differential regulation by interferon alpha/beta. *J. Exp. Med.* **2003**, *197*, 885–898.

34 Trinchieri G. Interleukin-12 and the regulation of innate resistance and adaptive immunity. *Nat. Rev. Immunol.* **2003**, *3*, 133–146.

35 Borg C, et al. NK cell activation by dendritic cells (DCs) requires the formation of a synapse leading to IL-12 polarization in DCs. *Blood* **2004**, *104*, 3267–3275.

36 Jinushi M, et al. Critical role of MHC class I-related chain A and B expression on IFN-alpha-stimulated dendritic cells in NK cell activation: impairment in chronic hepatitis C virus infection. *J. Immunol.* **2003**, *170*, 1249–1256.

37 Jinushi M, et al. Autocrine/paracrine IL-15 that is required for type I IFN-mediated dendritic cell expression of MHC class I-related chain A and B is impaired in hepatitis C virus infection. *J. Immunol.* **2003**, *171*, 5423–5429.

38 Granucci F, et al. Inducible IL-2 production by dendritic cells revealed by global gene expression analysis. *Nat. Immunol.* **2001**, *2*, 882–888.

39 Granucci F, Feau S, Angeli V, Trottein F, Ricciardi-Castagnoli P. Early IL-2 production by mouse dendritic cells is the result of microbial-induced priming. *J. Immunol.* **2003**, *170*, 5075–5081.

40 Sauma D, et al. Interleukin-4 selectively inhibits interleukin-2 secretion by lipopolysaccharide-activated dendritic cells. *Scand. J. Immunol.* **2004**, *59*, 183–189.

41 Mocikat R, et al. Natural killer cells activated by MHC class I(low) targets prime dendritic cells to induce protective CD8 T cell responses. *Immunity* **2003**, *19*, 561–569.

42 Vitale M, et al. NK-dependent DC maturation is mediated by TNF{alpha} and IFN{gamma} released upon engagement of the NKp30 triggering receptor. *Blood* **2005**.

43 Adam C, et al. DC-NK cell cross-talk as a novel CD4+ T cell-independent pathway for antitumor CTL induction. *Blood* **2005**.

44 Wilson JL, et al. Targeting of human dendritic cells by autologous NK cells. *J. Immunol.* **1999**, *163*, 6365–6370.

45 Carbone E, et al. Recognition of autologous dendritic cells by human NK cells. *Eur J. Immunol.* **1999**, *29*, 4022–4029.

46 Ferlazzo G, et al. The interaction between NK cells and dendritic cells in bacterial infections results in rapid induction of NK cell activation and in the lysis of uninfected dendritic cells. *Eur. J. Immunol.* **2003**, *33*, 306–313.

47 Ferlazzo G, Semino C, Melioli G. HLA class I molecule expression is up-regulated during maturation of dendritic cells, protecting them from natural killer cell-mediated lysis. *Immunol. Lett.* **2001**, *76*, 37–41.

48 Chiesa MD, et al. The natural killer cell-mediated killing of autologous dendritic cells is confined to a cell subset expressing CD94/NKG2A, but lacking inhibitory killer Ig-like receptors. *Eur. J. Immunol.* **2003**, *33*, 1657–1666.

49 Hayakawa Y, et al. NK cell TRAIL eliminates immature dendritic cells in vivo and limits dendritic cell vaccination efficacy. *J. Immunol.* **2004**, *172*, 123–129.

50 Fehniger TA, et al. CD56bright natural killer cells are present in human lymph nodes and are activated by T cell-derived IL-2: a potential new link between adaptive and innate immunity. *Blood* **2003**, *101*, 3052–3057.

51 Ferlazzo G, et al. The abundant NK cells in human secondary lymphoid tissues require activation to express killer cell Ig-like receptors and become cytolytic. *J. Immunol.* **2004**, *172*, 1455–1462.

52 Ferlazzo G, et al. Distinct roles of IL-12 and IL-15 in human natural killer cell activation by dendritic cells from secondary lymphoid organs. *Proc. Natl. Acad. Sci. USA* **2004**, *101*, 16606–16611.

53 Vitale M, et al. The small subset of CD56brightCD16– natural killer cells is selectively responsible for both cell proliferation and interferon-gamma production upon interaction with dendritic cells. *Eur. J. Immunol.* **2004**, *34*, 1715–1722.

54 Martin-Fontecha A, et al. Induced recruitment of NK cells to lymph nodes provides IFN-gamma for T(H)1 priming. *Nat. Immunol.* **2004**, *5*, 1260–1265.

55 Bajenoff M, et al. NK cells are slow motile cells localized in the lymph node outerparacortex where they make contacts with dendritic cells. Pasteur, I., 24–28 nov. Paris.

56 Robertson MJ. Role of chemokines in the biology of natural killer cells. *J. Leukoc. Biol.* **2002**, *71*, 173–183.

57 Buentke E, et al. Natural killer and dendritic cell contact in lesional atopic dermatitis skin–Malassezia-influenced cell interaction. *J. Invest. Dermatol.* **2002**, *119*, 850–857.

58 Borg C, et al. Novel mode of action of c-kit tyrosine kinase inhibitors leading to NK cell-dependent antitumor effects. *J. Clin. Invest.* **2004**, *114*, 379–388.

59 Mailliard RB, et al. Dendritic cells mediate NK cell help for Th1 and CTL responses: two-signal requirement for the induction of NK cell helper function. *J. Immunol.* **2003**, *171*, 2366–2373.

60 Scharton TM, Scott P. Natural killer cells are a source of interferon gamma that drives differentiation of CD4+ T cell subsets and induces early resistance to Leishmania major in mice. *J. Exp. Med.* **1993**, *178*, 567–577.

61 Westwood JA, et al. Cutting edge: novel priming of tumor-specific immunity by NKG2D-triggered NK cell-mediated tumor rejection and Th1-independent CD4+ T cell pathway. *J. Immunol.* **2004**, *172*, 757–761.

62 Biron CA, Byron KS, Sullivan JL. Severe herpesvirus infections in an adolescent without natural killer cells. *N. Engl. J. Med.* **1989**, *320*, 1731–1735.

63 Biron CA, Nguyen KB, Pien GC, Cousens LP, Salazar-Mather TP. Natural killer cells in antiviral defense: function and regulation by innate cytokines. *Annu. Rev. Immunol.* **1999**, *17*, 189–220.

64 Arase H, Mocarski ES, Campbell AE, Hill AB, Lanier LL. Direct recognition of cytomegalovirus by activating and inhibitory NK cell receptors. *Science* **2002**, *296*, 1323–1326.

65 Tay CH, Welsh RM. Distinct organ-dependent mechanisms for the control of murine cytomegalovirus infection by natural killer cells. *J. Virol.* **1997**, *71*, 267–275.

66 Smith HR, et al. Recognition of a virus-encoded ligand by a natural killer cell activation receptor. *Proc. Natl. Acad. Sci. USA* **2002**, *99*, 8826–8831.

67 Krug A, et al. TLR9-dependent recognition of MCMV by IPC and DC generates coordinated cytokine responses that activate antiviral NK cell function. *Immunity* **2004**, *21*, 107–119.

68 Andrews DM, Andoniou CE, Granucci F, Ricciardi-Castagnoli P, Degli-Esposti MA. Infection of dendritic cells by murine cytomegalovirus induces functional paralysis. *Nat. Immunol.* **2001**, *2*, 1077–1084.

69 Andrews DM, Scalzo AA, Yokoyama WM, Smyth MJ, Degli-Esposti MA. Functional interactions between dendritic cells and NK cells during viral infection. *Nat. Immunol.* **2003**, *4*, 175–181.

70 Sirianni MC, Tagliaferri F, Aiuti F. Pathogenesis of the natural killer cell deficiency in AIDS. *Immunol. Today* **1990**, *11*, 81–82.

71 Poggi A, et al. NK cell activation by dendritic cells is dependent on LFA-1-mediated induction of calcium-calmodulin kinase II: inhibition by HIV-1 Tat C-terminal domain. *J. Immunol.* **2002**, *168*, 95–101.

72 Poggi A, Rubartelli A, Zocchi MR. Involvement of dihydropyridine-sensitive calcium channels in human dendritic cell function. Competition by HIV-1 Tat. *J. Biol. Chem.* **1998**, *273*, 7205–7209.

73 Rubartelli A, Poggi A, Sitia R, Zocchi MR. HIV-I Tat: a polypeptide for all seasons. *Immunol. Today* **1998**, *19*, 543–545.

74 Mandelboim O, et al. Recognition of haemagglutinins on virus-infected cells by NKp46 activates lysis by human NK cells. *Nature* **2001**, *409*, 1055–1060.

75 Fonteneau JF, et al. Activation of influenza virus-specific CD4+ and CD8+ T cells: a new role for plasmacytoid dendritic cells in adaptive immunity. *Blood* **2003**, 101, 3520–3526.

76 Siegal FP, et al. The nature of the principal type 1 interferon-producing cells in human blood. *Science* **1999**, *284*, 1835–1837.

77 Ruggeri L, et al. Role of natural killer cell alloreactivity in HLA-mismatched hematopoietic stem cell transplantation. *Blood* **1999**, *94*, 333–339.

78 Ruggeri L, et al. Effectiveness of donor natural killer cell alloreactivity in mismatched hematopoietic transplants. *Science* **2002**, *295*, 2097–2100.

79 Shlomchik WD, et al. Prevention of graft versus host disease by inactivation of host antigen-presenting cells. *Science* **1999**, *285*, 412–415.

80 Ghiringhelli F., et al. CD4⁺ CD25⁺ regulatory T cells inhibit natural killer cell functions in a transforming growth factor-β-dependent manner. *J. Exp. Med.* **2005**, *202*, 1075–1085.

81 Terme M., et al. BCR/ABL promotes dendritic cell-mediated natural killer cell activation. *Cancer Res.* **2005**, *65*, 6409–6417.

25
Intercellular Communication via Protein Transfer

Marca H.M. Wauben

25.1
What are Exosomes, and Where do they Come From?

A mediator involved in protein transfer between cells is the exosome. Exosomes are small membrane vesicles formed within multivesicular endosomes, also called multivesicular bodies (MVB), and released in the extracellular space upon fusion of MVB with the cell surface membrane [1–3]. They can be purified from cell-culture supernatant by a series of centrifugation or filtration steps followed by ultra-centrifugation to pellet the exosomes. Exosomes are typically between 30–100 nm in diameter (reviewed in [4, 5]). They are secreted by a multitude of cell types, including reticulocytes [6], platelets [7], mast cells [8, 9], B lymphocytes [10, 11], T lymphocytes [12–15], dendritic cells (DC) [11, 16–20], tumor cells [21–24], epithelial cells [25, 26], and trophoblasts [27]. Ample evidence for the presence of exosomes *in vivo* has been documented. They have been demonstrated in various bodily fluids such as serum, urine, malignant effusions, and broncho-alveolar lavage, and at the surface of follicular DC in germinal centers [reviewed in 4]. Furthermore, two lines of research recently converged, as it has been demonstrated that the so-called tumor-derived membrane vesicles, already reported in the early 1980s [Reviewed in 28], are identical with the described tumor-derived exosomes [24].

In general the protein composition of exosomes clearly differs from that of total cell lysates, plasma membrane or apoptotic vesicles, indicating selective sorting events for protein inclusion in exosomes [5, 10, 11, 13, 29]. The common exosome-specific constituents are molecules involved in antigen presentation and co-stimulation, for example major histocompatibility complex (MHC) class I/II and CD86, tetraspanins, annexins, heat shock proteins (hsp), cytoskeletal proteins, raft-associated proteins and glycolipids [Reviewed in 4, 5]. Analysis of exosomes with a different origin revealed also the presence of cell-type specific proteins [8–10, 11, 13, 16, 20, 21, 24, 25, 29–31]. Besides these origin-dependent differences, the maturation or activation state of the exosome-producing cell can influence both the quantity and quality of the exosomes [15, 20]. In several cell types, for example primary DC, epithelial cells, EBV-transformed B cells and tumor cells, the fusion of endo-

Handbook of Dendritic Cells. Biology, Diseases, and Therapies.
Edited by M. B. Lutz, N. Romani, and A. Steinkasserer.
Copyright © 2006 WILEY-VCH Verlag GmbH & Co. KGaA, Weinheim
ISBN: 3-527-31109-2

cytic compartments with the plasma membrane, and thus the secretion of exosomes, is considered to be constitutive [10, 20, 21, 24, 25, 32, 33]. However *in vitro* analysis of exosome secretion by immature monocyte-derived DC, indicated that monocytes are poor producers of exosomes during their differentiation process, while immature DC readily secrete exosomes [11, 20]. Further progression from immature DC to mature DC seems to coincide with a reduction in exosome secretion, as LPS-treated mature DC produce less exosomes than immature DC [20, 32]. This indicates that exosome secretion can be regulated during differentiation. Indeed the first description of maturation-dependent exosome secretion was in reticulocytes during the late stages of erythrocyte differentiation [1, 2, 6].

Besides maturation dependent exosome secretion, activation-dependent exosome secretion has been reported. For T cells it has been shown that upon activation via the T-cell receptor (TCR) exosomes containing adhesion molecules, MHC class I and II molecules, and phosphorylated zeta and CD3/TCR, are released more abundantly [13]. Also in mast cells the fusion of late endocytic vesicles with the cell surface is regulated and occurs only after activation [8, 33]. Interestingly, it is not only the quantity of secreted exosomes that differs between immature and mature DC, but also qualitative differences have been observed, as analyzed by protein composition [20]. It is important to note that human and mouse DC-derived exosomes are not completely identical. Exosomes from monocyte-derived human DC, expressing ICAM-1 and CD86 [11], seem to display a phenotype between immature and mature murine DC-derived exosomes [20]. These maturation/activation state-dependent qualitative differences are not only present in DC-derived exosomes, but are also shown in exosome-like vesicles derived from a T-cell clone, which was either anergized or activated *in vitro* [15].

The qualitative differences, represented by the presence of cell-type or activation/maturation state-dependent specific proteins, and the quantitative differences, resulting from maturation/activation state-dependent secretion of exosomes, may define the specific targeting of the exosomes and their biological effect.

25.2
Which Cells are Targets for Exosomes, and how do Exosomes Interact with these Cells?

The physical interaction of exosomes and target cells is an essential step in their ability to perform their physiological function. Exosomes derived from DC, epithelial cells, mast cells, tumor cells, B lymphocytes, and T lymphocytes have been shown to transfer proteins to DC [9, 15, 17–20, 23, 31, 34], B cells [9, 15, 20] or follicular DC [35]. What exactly defines the cellular target of particular exosomes, as well as the interaction mechanism between exosomes and their target, is largely unclear. Different modes of interaction can be envisioned for different cell types, and may relate to specific physiological functions. Exosomes may either merely attach to the cell surface of target cells or fuse with the plasma membrane. Alternatively they can be endocytosed, as also occurs with vesicles derived from apoptotic cells [36–39]. The best characterized example of exosome-mediated protein transfer is the targeting of B-lymphocyte-derived exosomes to the surface of follicular DC in germinal centers [35]. These MHC class II positive B-cell-derived exosomes are present in abundance on the surface of follicular DC, a cell type that neither secretes exosomes nor synthesizes MHC class II molecules [35, 40]. The functionality of adhesion molecules on B-cell-derived exosomes has also been demonstrated in the anchoring of these exosomes to extracellular matrix components, and tumor necrosis factor (TNF) alpha activated fibroblasts [41]. However, until now there is no evidence that exosomes cluster with the plasma membrane of myeloid DC in a manner similar to that described for follicular DC. The mechanism of exosomal interaction with myeloid DC is far less clear. There is ample evidence that exosomes derived from various cellular sources can transfer antigens and functional MHC-peptide complexes to myeloid DC [9, 15–21, 31, 34, 42]. Although it cannot be excluded that exosomes transfer the peptide-MHC complexes directly to the cell surface of myeloid DC, it has been shown that DC have the capacity to internalize exosomes efficiently [9, 21, 22, 43]. The effectiveness of antigen-delivery via exosomes largely depends on the exosomal expression of specific receptors or ligands needed for DC docking [9, 32, 43]. In this respect the presence of hsp in exosomes, and the common receptor for hsp (CD91) on DC could be a pathway for efficient internalization of exosomes [9, 32, 43]. Although it has been suggested that the soluble molecule MFG-E8/lactadherin is an opsonin involved in docking exosomes to target cells [32, 43], this molecule seems not to be essential for addressing exosomes to DC [20]. Other surface molecules such as externalized phosphatidyl serine, CD11a, CD54, and the tetraspanins CD9 and CD81 have also been implicated to play a role in attachment and/or uptake of exosomes by DC [43]. The fact that the ability of bone-marrow-derived DC to capture exosomes decreases with the maturation state, and thereby the endocytic capacity of DC, further supports a role for the endocytic route of exosomes after DC binding [43]. Once internalized and sorted into the endosomal pathway, the DC can process and present antigens derived from exosomes [43]. Such targeting to the endosomal route seems to conflict with the transfer of functional peptide-MHC complexes from exosomes to the plasma membrane of the DC. However, retrograde MHC class II transport from late endosomes to the cell surface, during which endogenous exosomes bearing MHC class II back-fuse with the membrane of MVB has been described [44, 45], and could possibly be a mechanism by which internalized exosomes deliver peptide-MHC complexes to the plasma membrane of the DC.

25.3
What is the Consequence of Exosome Binding or Uptake for the Target Cell?

In studies using DC-derived exosomes or tumor-cell-derived exosomes no evidence was found for DC maturation as a consequence of exosome binding or uptake [21, 34, 46].

Also binding of B-cell-derived exosomes to immature DC did not induce DC maturation [9], while adhesion of these exosomes to fibroblasts readily trigger Ca++ signals [41]. In contrast, mast-cell-derived exosomes induced phenotypical and functional maturation of immature DC, as defined by the upregulation of MHC class II and co-stimulatory molecules and the enhancement of the T-cell stimulatory capacity [9]. Interestingly, although hsp are described to be a common constituent of exosomes, hsp60 and hsc70 are selectively enriched in mast-cell-derived exosomes [9, 21, 29]. The presence of CD91, a common receptor for hsp, on the surface of DC could play a role in the internalization of these exosomes [9], and the functional outcome of this process.

In general B lymphocytes have a poor capacity to activate naïve T cells. Interestingly, exosomes derived from mature DC endow B lymphocytes with the ability to prime naïve T cells, while immature-DC-derived exosomes lack this capacity [20]. Also in the presence of DC, the T-cell activating capacity of mature DC-derived exosomes appeared to be much stronger than of immature DC-derived exosomes [20].

Thus the differential composition of exosomes released by distinct cell types, or under dissimilar conditions, determines the effect on the target cell.

25.4
What is the Physiological Role of Exosomes in the Immune System?

Originally exosomes were thought to function as waste-bins for the elimination of obsolete proteins. This view was based on the fact that during the maturation phase of reticulocytes, secretion of exosomes allows the elimination of proteins that are no longer necessary for the function of differentiated red blood cells [6]. However it remained unclear why, besides the intracellular degradation pathway of fusion of MVB with lysosomes [47], an alternative route of plasma membrane fusion exists [6].

In the immune system activation-dependent, polarized vesicle secretion is a feature shared by a number of different immune cells. NK cells and cytotoxic T cells direct their lytic granules with exquisite timing and precision to the plasma membrane of the target cells [12, 14, 48]. In this process exosomes have an active biological role as messengers to deliver the kiss of death [48]. Activation-dependent secretion of exosomes has also been described for CD4$^+$ T cells and mast cells [8, 13, 15, 33]. The physiological role of these non-lytic vesicles secreted after cell activation, however is less well understood. The same holds true for the more constitutive exosomal secretion by APC, endothelial cells and tumor cells [10, 21, 25, 28, 32, 33].

Exosomes present in the extracellular fluids and circulation can be regarded as an alternative antigen source. It has been reported that high concentrations of exosomes, expressing MHC-peptide complexes and co-stimulation, can activate *in vitro* T-cell clones and lines very weakly [10, 42], but fail to stimulate naïve T cells [18, 22]. However, in the presence of DC the T-cell stimulatory capacity of these exosomes was highly increased [18–20, 34, 42]. This led to the hypothesis that exosomes transfer proteins to APC in a functional manner [9, 17–20, 23, 31, 34, 42, 43]. Indeed tumor-peptide loaded DC-derived exosomes can strongly stimulate tumor-specific cytotoxic T-cell responses after *in vivo* administration [16]. These promising findings have already resulted in a phase I clinical trial in which metastatic melanoma patients were immunized with autologous DC-derived exosomes pulsed with MAGE 3 peptides [49]. The physiological role of exosomal transfer of MHC-peptide complexes between DC could be the amplification of the T-cell response as a result of the increase of the number of antigen presenting cells (APC) bearing the specific peptide-MHC complexes [16, 18, 34]. Alternatively, exosomes could be a vector for the transfer of antigens from migrating DC, such as Langerhans cells emigrating from infected skin to draining lymph nodes, to resident lymph node DC, such as CD8+ DC [50]. Not only DC-derived exosomes are transferred to other DC, also exosomes derived from mast cells, epithelial cells, tumor cells and T cells have been shown to target DC [9, 15, 22, 26]. It is possible that exosomes play a role in the immune surveillance by transferring cell type-, activation/maturation state-, and micro-environmental specific snapshots to DC. As such, exosomes are not exclusively linked to T-cell stimulatory properties but in fact can also induce tolerance. Several studies have shown that anergic regulatory T cells can modulate the APC function by inducing a tolerogenic phenotype after cell–cell contact [15, 51–54]. Recently, it has been found that both activated anergic and effector T cells donate MHC-peptide complexes to APC [15]. However, only APC that acquired MHC-peptide complexes from regulatory-T-cell-derived exosomes became tolerogenic, while effector-T-cell-derived vesicles endowed the APC with T-cell stimulatory properties [15]. This led to the hypothesis that the exosomal transfer of an as yet not defined anergic-T-cell-derived molecule endowed the APC with tolerogenic properties. In this respect the identification of molecules associated with immune regulation in exosomal preparations of several cell types is a major future challenge.

For tumor-cell-derived exosomes it has been found that transfer of tumor antigens to DC can result in crosspresentation and specific activation of cytotoxic T cells [21, 22]. However, tumor cells have also developed strategies to circumvent immune recognition, such as the ectopic expression of the human leukocyte antigen (HLA) class I molecule HLA-G, interacting with inhibitory receptors on immune cells [23]. In melanoma-derived exosomes the presence of HLA-G could be demonstrated, which may result in the inhibition rather than the enhancement of the T-cell stimulatory capacity of targeted DC, thereby allowing the tumor to escape immune surveillance [23]. Another example is the presence of bioactive FasL in exosomes secreted by tumor cells [24, 55], or by T-cell leukemia and normal human T-cell blasts after mitogenic stimulation [14]. FasL-containing exosomes could be involved in the induction of autocrine or paracrine cell death during immune regulation. Such immune regulatory function of exosome-associated FasL has also been suggested to play a role in the tolerance for paternal antigens during pregnancy, as trophoblasts appeared to secrete FasL-contaning exosomes [27]. Tumor cells may exploit this mechanism and may shed FasL-containing exosomes to subvert the immune system, and thereby escape immune surveillance [24, 56].

Tab. 25.1 Exosomal transfer of membrane proteins between (immune) cells.

Source of exosomes	Target cell	Reference
DC	DC	16–20, 34, 43, 46
Tumor cell	DC	21, 22
Epithelial cell	DC	26, 31
Mast cell	DC, B lymphocyte	9
CD4 T lymphocyte	DC, B lymphocyte	15
DC	B lymphocyte	20
B lymphocyte	Follicular DC	35

Other examples of tolerance induction after administration of exosomes have been reported in a heart transplantation model in which injection of donor DC-derived exosomes, but not recipient DC-derived exosomes, before transplantation resulted in tolerance rather than immunity [17]. Also injection of exosomes released by intestinal epithelial cells of antigen fed rat resulted in antigen-specific tolerance in naïve recipients [26]. Although most studies focused on the exosomal protein transfer to DC, exosomes also target other cell types, for example T cells [57]. Since T cell–T cell presentation has been shown to be very efficient in anergy induction [reviewed in 58], exosomal protein transfer to T cells may also be involved in tolerance induction.

In summary, exosomes can serve as intercellular messengers for the transfer of proteins between a wide variety of cells (Table 25.1). The outcome of the immune response induced by exosomal protein transfer largely depends on the cellular origin of the exosomes and their target cell.

25.5
Cell–Cell Contact-dependent Transfer of Membrane Proteins

Besides exosome-mediated protein transfer, a cell–cell contact-dependent manner of intercellular transfer has been observed as well. The transfer of membrane fragments across immunological synapses [59] emerges as a common route of communication between immune cells. For a long time protein acquisition by immune cells at the immunological synapse has been regarded as a unidirectional process from target cell or APC to effector cell. For T cells it has been shown that they acquire not only MHC class I and II peptide complexes and lipids from APC during T cell-APC interaction, but also incorporate other membrane associated proteins such as, Ig, CD80, CD86, OX40L, CD4, IL2-R, and ICAM-1 [60–69]. After specific receptor–ligand interactions have established cell–cell contact, such as CD28-CD80/CD86 or TCR-peptide-MHC, a wide variety of surface molecules from APC can be absorbed by antigen-experienced and naïve CD8 and CD4 T cells [60, 61, 65, 70]. Both resting and activated T cells can absorb molecules from APC, albeit activated T cells are more efficient [60, 69]. Importantly, this phenomenon is not an *in vitro* artefact since it also occurs *in vivo* [60, 69]. Once the T cell is activated, TCR or CD28 ligation is no longer required for the direct capture of MHC-peptide complexes [64, 69, 71]. Activated T cells acquire MHC molecules not only from syngeneic APC, but also from allogeneic and xenogeneic APC [64, 71, 72]. Furthermore, activated, but not resting, CD4 T cells can acquire a variety of molecules in an antigen or ligand nonspecific manner after cell–cell contact with endothelial cells, DC, and monocytes [66, 71, 73]. This suggests that TCR ligation is required for T-cell activation, but is thereafter not strictly needed for acquisition of molecules.

Besides the acquisition of APC-derived molecules several immune cells have been shown to absorb molecules from their target cells. Cytotoxic T cells can rapidly capture membrane fragments from target cells in a TCR signaling-dependent fashion [62, 63]. Since target cells did not acquire CD8 from cytotoxic T cells, it has been assumed that this is a unidirectional process [63]. Also B cells can absorb membrane proteins from target cells. After immune synapse formation, the B cell receptor mediates the efficient capture of integral membrane proteins expressed on target cells [74]. NK cells rapidly acquire MHC class I molecules and viral receptors from target cells after formation of an inhibitory immune synapse between NK cells and target cells [75–77]. However, in the inhibitory immune synapse cell surface proteins are not only transferred from target cell to NK cell, but also vice versa from NK cell to target cell [78]. Although in many reports it is assumed that protein transfer in the APC-T cell immunological synapse is a unidirectional process from APC to T cell, more and more evidence is accumulating that protein transfer is in fact bidirectional [15, 71, 79, 80]. In this respect the description of homotypic synapses between leukemia cells, and the bidirectional transfer of membrane fragments across these synapses is also worth mentioning [81].

Besides the fact that DC can readily capture antigen from dead and apoptotic cells for presentation to MHC class I restricted CTL, they can also acquire cell surface molecules from viable endothelial cells, macrophages, B cells, activated T cells, and other both mature and immature DC in a cell–cell contact-dependent manner [15, 79, 80, 82–85]. Whether the maturation state of the DC influences the efficiency of membrane molecule absorption from other cells is still a controversial issue [79, 80, 82]. Notably macrophages, well known for their excellent phagocytic capacity, appeared to be rather inefficient in acquiring membrane molecules from other cells [80].

It has been assumed that resting immune cells do not acquire membrane fragments from cells transiently encountered along nonproductive interactions. However, it appeared that organized immunological synapses are also formed between mature DC and naïve T cells in the absence of antigen, and even in the absence of MHC molecules, and that these antigen-independent synapses allow the transmission of signals [86, 87]. Whether in these antigen-independent synapses, intercellular protein transfer also ensues, remains to be seen.

Although the precise requirements for intercellular protein exchange, and the transfer direction are not fully understood, the phenomenon of swapping membrane molecules has been acknowledged to occur between a wide variety of immune cells (Table 25.2).

Tab. 25.2   Protein transfer between (immune) cells during cell–cell contact.

Protein donor	Protein acceptor	Reference
DC	CD8 T lymphocyte	60, 68, 69
"Model"APC[1]	CD8 T lymphocyte	60–62, 69
DC	CD4 T lymphocyte	60, 65, 69
"model"APC	CD4 T lymphocyte	60, 65, 66, 69
Target cell	CD8 T lymphocyte	63
Endothelial cell	CD4 T lymphocyte	66, 73
Endothelial cell	DC	82
T lymphocyte	DC	79
Melanoma cell-line	DC	79
Adenocarcinoma cell-line	DC	85
DC	DC	82, 85
Target cell	NK cell	75–78
NK cell	Target cell	78
Leukemia cell	Leukemia cell	81

[1] Model APC; Drosophila cells [60, 61, 69], DC cell line [61], RMA-S [61], Transfected RMA and 3T3 [62], Fibroblast cell lines [65], COS-1 cell lines [65, 66], A20 [65], MC38 cell lines [65], Jurkat transfectants [66], SVT2 transfectants [66].

25.6
How are Membrane Proteins Transferred Between Immune Cells, and What is their Fate?

The mechanism of direct cell–cell contact-dependent transfer of membrane molecules is largely unknown. In several studies it has been shown that both proteins and lipid markers are transferred, suggesting that membrane fractions rather than soluble proteins are exchanged [15, 62–64, 73, 79, 80]. Besides transfer via exosomes, it has been suggested that other vesicles, for example those formed from extracted plasma membrane fragments, are involved in the synaptic protein transfer process [74, 80]. Alternatively, membrane fusion leading to the formation of membrane bridges between cytotoxic T lymphocytes and target cells or between APC and T cells [63, 70], and the formation of membrane nanotubes connecting multiple cells [63, 88], have been demonstrated and implicated as means for protein transfer. Overall, intercellular protein transfer is a complex process during which several different mechanisms may run in parallel (Fig. 25.1).

The process of active protein transfer triggered by receptor signaling, which occurs rapidly after conjugate formation between cells, has been called "trogocytosis" (Troxis is Greek noun derivative of 'trogo' the ancient Greek verb meaning 'to eat, chew, gnaw') [89]. Trogocytosis (troxis necrosis) originally referred to the gradual disappearance of hepatocytes as a result of lymphocyte–hepatocyte binding and internalization of liver surface molecules by the lymphocyte during hepatitis [90]. Several reports describe APC-derived MHC molecules, which are rapidly absorbed by T cells, being co-internalized with the TCR within several hours through endo-

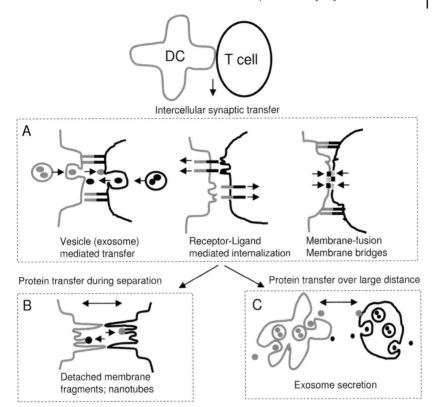

Fig. 25.1 Intercellular transfer of membrane proteins: A give-and-take relationship. Schematic representation of possible mechanisms involved in intercellular protein transfer. (A) During the intimate contact between a DC and T cell, protein exchange can occur in the immunological synapse. (B) During the separation of the DC and T-cell membrane fragments can be exchanged as a consequence of mechanical traction. C) After activation, the DC and T cell secrete exosomes, which can act as vector for protein transfer over a larger distance.

cytosis, and being localized in endosomes and lysosomes [60, 61]. However, receptor-mediated internalization is probably only part of the story, as it has been demonstrated that after immune synapse formation an enormous quantity and a wide variety of molecules can be absorbed [60, 62]. For resting T cells the requirement for cell–cell contact appeared to be more stringent than for activated T cells, which also show absorption of proteins in a trans-well system [69]. Once activated, T cells can absorb molecules in the absence of specific ligands [64, 66, 71]. Entire proteins are transferred and stabilized in the recipient cell membrane, often seen as discrete punctate formation, in a functional manner [60–63, 65, 66, 69, 70]. The role of the cytoskeleton in the absorption process also differs between resting and activated T cells, because resting T cells appeared to be more dependent on the actin cytoskeleton for both TCR- and CD28-mediated protein absorption [69]. Interestingly, this dependence was lost when the specific MHC-peptide concentration on the APC was increased [69]. These findings illustrate that the requirements for protein absorption by T cells are rather complex, and depend on the activation state (resting versus activated T cells), maturation state (naïve versus memory T cells), the molecules involved (TCR, CD28 etc.), and the concentration of ligand on the donor cell.

Similar to protein acquisition by T cells, protein absorption by DC can either result in internalization, processing and presentation of the acquired antigens, or in the functional integration of intact acquired protein complexes in the plasma membrane [79, 80, 82, 85]. Recently, a new route of intercellular peptide exchange via gap junction channels between cells has also been described, via which peptides are transferred from the cytoplasm of one cell into the cytoplasm of its neighbor [91]. Also monocyte-derived DC are capable of using this route for intercellular transfer of small peptides [91].

25.7
What is the Physiological Role of Membrane Protein Swapping in the Immune System?

Although the physiological role of intercellular protein exchange is largely unknown, several interesting hypotheses have been postulated. It has been speculated that absorption of membrane molecules in the immunological synapse is needed for the dissociation of the two cells forming the synapse. Indeed the formation of tight clusters between T cells and APC raises the question of how these cells can subsequently disengage. The rapid absorption and internalization of ligands at the contact site could play a role in T cell–APC dissociation, allowing the activated T cell to move from one APC to another [60].

Alternatively, the efficient MHC-peptide stripping of APC by high affinity T cells may provide a mechanism of T-cell competition for specific MHC-peptide ligands. This could play a role in affinity maturation of T-cell responses, giving the biggest advantage to high affinity T cells [68, 80]. The absorption and internalization of MHC-peptide complexes might also be involved in limiting the risk of over-stimulation of the antigen specific T-cell population.

Depending on the fate of the transferred molecules, intercellular protein exchange between immune cells may also result in the inheritance of functional molecules from conjugating cells, which are not transcribed by the recipient cell. This may influence not only the phenotype, but also the function of the recipient cell. In this respect trogocytosis should be regarded as a vector for intercellular communication. The functional acquisition of APC or target cell-derived MHC-peptide complexes by cytotoxic T cells may play an active role in immune regulation by the induction of fratricide. During this process T cells become sensitive to lysis by neighboring cytotoxic T cells specific for the same ligand [61–63]. This process could be involved in the elimination of cytotoxic T cells that have interacted with many targets, and as such play a role in the down-regulation of the immune response. Also functional MHC class II molecules can be acquired from professional APC by activated T cells [60, 61, 63, 64, 68]. Since MHC class II expression on T cells can result in the induction of anergy [Reviewed in 58], it can be envisaged that acquired MHC class II-peptide complexes presented on activated T cells may contribute to immune regulation [64, 65]. On the other hand T cells not only acquire MHC-peptide complexes from APC, but can also absorb co-stimulatory molecules, for example CD80 or OX40L [65, 66]. Such T cells may act as professional APC, and as such give rise to an enhanced immune response [65, 66].

DC can present exogenous antigens derived from the extracellular milieu or captured from neighboring cells via the MHC class I antigen presentation route, a process called cross-presentation [92]. The capacity of DC to acquire intact MHC-peptide complexes from viable cells, and the functional presentation of these complexes on the DC surface, can be regarded as a pathway for cross-presentation [15, 79, 80, 82–85].

Whether the cross-presentation ultimately results in T-cell tolerance or immunity depends on the context in which the protein transfer takes place, meaning the composition of transferred proteins and the resulting effect on DC maturation and activation. Both the quantity and quality of membrane protein swapping during T cell–APC contact is affected by the strength of the interaction, and the activation and maturation state of the T cell and APC. As such intercellular membrane protein transfer between T cells and APC seems to represent a cell–cell contact-dependent manner to regulate immune responses.

25.8
Concluding Remarks

It is challenging to postulate that intercellular protein transfer plays a significant role in the regulation of the immune response. The further dissection of immunogenic versus tolerogenic signals delivered by exosomes or membrane patches is an important future challenge. Insights into these different aspects of intercellular protein transfer may offer novel opportunities for the development of immunotherapeutic strategies not only for the treatment of cancer, but also for prevention of allograft rejection and autoimmune diseases.

Abbreviations

DC   dendritic cell
NK   natural killer cell
MVB   multivesicular body
MHC   major histocompatibility complex
hsp   heat shock protein
TCR   T-cell receptor
TNF   tumor necrosis factor
APC   antigen presenting cell
HLA   human leukocyte antigen

References

1 Harding, C., Heuser, J., Stahl, P., Endocytosis and intracellular processing of transferrin and colloidal-gold transferrin in rat reticulocytes: demonstration of a pathway for receptor shedding. *Eur. J. Cell. Biol.* **1984.** *35:* 256–263.

2 Pan, B. T., Teng, K., Wu, C., Adam, M., Johnstone, R. M., Electron microscopic evidence for externalization of the transferrin receptor in vesicular form in sheep reticulocytes. *J. Cell. Biol.* **1985.** *101:* 942–948.

3 Johnstone, R. M., Adam, M., Hammond, J. R., Orr, L., Turbide, C., Vesicle formation during reticulocyte maturation. Association of plasma membrane activities with released vesicles (exosomes). *J. Biol. Chem.* **1987.** *262:* 9412–9420.

4 Fevrier, B., Raposo, G., Exosomes: endosomal-derived vesicles shipping extracellular messages. *Curr. Opin. Cell. Biol.* **2004.** *16:* 415–421.

5 Thery, C., Zitvogel, L., Amigorena, S., Exosomes: composition, biogenesis and function. *Nat. Rev. Immunol.* **2002.** *2:* 569–579.

6 Johnstone, R. M., Mathew, A., Mason, A. B., Teng, K., Exosome formation during maturation of mammalian and avian reticulocytes: evidence that exosome release is a major route for externalization of obsolete membrane proteins. *J. Cell. Physiol.* **1991.** *147:* 27–36.

7 Heijnen, H. F., Schiel, A. E., Fijnheer, R., Geuze, H. J., Sixma, J. J., Activated platelets release two type of membrane vesicles: microvesicles by surface shedding and exosomes derived from exocytosis of multivesicular bodies and alpha-granules. *Blood* **1999.** *94:* 3791–3799.

8 Raposo, G., Tenza, D., Mecheri, S., Peronet, R., Bonnerot, C., Desaymard, C., Accumulation of major histocompatibility complex class II molecules in mast cell secretory granules and their release upon degranulation. *Mol. Biol. Cell.* **1997.** *8:* 2631–2645.

9 Skokos, D., Botros, H. G., Demeure, C., Morin, J., Peronet, R., Birkenmeier, G., Boudaly, S., Mecheri, S., Mast-cell-derived exosomes induce phenotypic and functional maturation of dendritic cells

and elicit specific immune responses *in vivo. J. Immunol.* **2003.** *170:* 3037–3045.

10 Raposo, G., Nijman, H. W., Stoorvogel, W., Liejendekker, R., Harding, C. V., Melief, C. J., Geuze, H. J., B lymphocytes secrete antigen-presenting vesicles. *J. Exp. Med.* **1996.** *183:* 1161–1172.

11 Clayton, A., Court, J., Navabi, H., Adams, M., Mason, M. D., Hobot, J. A., Newman, G. R., Jasani, B., Analysis of antigen presenting cell derived exosomes, based on immuno-magnetic isolation and flow cytometry. *J. Immunol. Methods* **2001.** *247:* 163–174.

12 Peters, P. J., Borst, J., Oorschot, V., Fukuda, M., Krahenbuhl, O., Tschopp, J., Slot, J. W., Geuze, H. J., Cytotoxic T lymphocyte granules are secretory lysosomes, containing both perforin and granzymes. *J. Exp. Med.* **1991.** *173:* 1099–1109.

13 Blanchard, N., Lankar, D., Faure, F., Regnault, A., Dumont, C., Raposo, G., Hivroz, C., TCR activation of human T cells induces the production of exosomes bearing the TCR/CD3/zeta complex. *J. Immunol.* **2002.** *168:* 3235–3241.

14 Martinez-Lorenzo, M. J., Anel, A., Gamen, S., Monle, N. I., Lasierra, P., Larrad, L., Piniero, A., Alava, M. A., Naval, J., Activated human T cells release bioactive Fas ligand and APO2 ligand in microvesicles. *J. Immunol.* **1999.** *163:* 1274–1281.

15 Nolte-'t Hoen, E. N. M., Wagenaar-Hilbers, J. P. A., Peters, P. J., Gadella, B. M., van Eden, W., Wauben M. H. M., Uptake of membrane molecules from T cells endows antigen-presenting cells with novel functional properties. *Eur. J. Immunol.* **2004.** *34:* 3115–3125.

16 Zitvogel, L., Regnault, A., Lozier, A., Wolfers, J., Flament, C., Tenza, D., Ricciardi-Castagnoli, P., Raposo, G., Amigorena, S., Eradication of established murine tumors using a novel cell-free vaccine: dendritic-cell-derived exosomes. *Nat. Med.* **1998.** *4:* 594–600.

17 Peche, H., Heslan, M., Usal, C., Amigorena, S., Cuturi, M. C., Presentation of donor major histocompatibility complex antigens by bone marrow dendritic-cell-derived exosomes

modulates allograft rejection. *Transplantation* **2003**. *76:* 1503–1510.

18 Thery, C., Duban, L., Segura, E., Veron, P., Lantz, O., Amigorena, S., Indirect activation of naive CD4(+) T cells by dendritic cell-derived exosomes. *Nat. Immunol.* **2002**. *3:* 1156–1162.

19 Gansuvd, B., Hagihara, M., Higuchi, A., Ueda, Y., Tazume, K., Tsuchiya, T., Hunkhtuvshih, N., Kato, S., Hotta, T., Umbilical cord blood dendritic cells are a rich source of soluble HLA-DR: synergistic effect of exosomes and dendritic cells on autologous or allogeneic T cell proliferation. *Hum. Immunol.* **2003**. *64:* 427–439.

20 Segura, E., Nicco, C., Lombard, B., Veron, P., Raposo, G., Batteux, F., Amigorena, S., Thery, C., ICAM-1 on exosomes from mature dendritic cells is critical for efficient naïve T cell priming. *Blood* **2005**. *106:* 216–223 (prepublished online March 24, 2005).

21 Wolfers, J., Lozier, A., Raposo, G., Regnault, A., Thery, C., Masurier, C., Flament, C., Pouzieux, S., Faure, F., Tursz, T., Angevin, E., Amigorena, S., Zitvogel, L., Tumor-derived exosomes are a source of shared tumor rejection antigens for CTL cross-priming. *Nat. Med.* **2001**. *7:* 297–303.

22 Andre, F., Schartz, N. E., Movassagh, M., Flament, C., Pautier, P., Morice, P., Pomel, C., Lhomme, C., Escudier, B., Le Chevalier, T., Tursz, T., Amigorena, S., Raposo, G., Angevin, E., Zitvogel, L., Malignant effusions and immunogenic tumour-derived exosomes. *Lancet* **2002**. *360:* 295–305.

23 Riteau, B., Faure, F., Menier, C., Viel, S., Carosella, E. D., Amigorena, S., Rouas-Freiss, N., Exosomes bearing HLA-G are released by melanoma cells. *Hum. Immunol.* **2003**. *64:* 1064–1072.

24 Taylor, D. D., Gercel-Taylor, C., Tumor-derived exosomes and their role in cancer-associated T-cell signaling defects. *Br. J. Cancer.* **2005**. *92:* 305–311.

25 van Niel, G., Raposo, G., Candalh, C., Boussac, M., Hershberg, R., Cerf-Bensussan, N., Heyman, M., Intestinal epithelial cells secrete exosome-like vesicles. *Gastroenterology* **2001**. *121:* 337–349.

26 Karlsson, M., Lundin, S., Dahlgren, U., Kahu, H., Pettersson, I., Telemo, E., "Tolerosomes" are produced by intestinal epithelial cells. *Eur. J. Immunol.* **2001**. *31:* 2892–2900.

27 Frangsmyr, L., Baranov, V., Nagaeva, O., Stendahl, U., Kjellberg, L., Mincheva-Nilsson, L., Cytoplasmic microvesicular form of Fas ligand in human early placenta: switching the tissue immune privilege hypothesis from cellular to vesicular level. *Mol. Hum. Reprod.* **2004**. *11:* 35–41.

28 Taylor, D. D., Taylor, C. G., Jiang, C. G., Black, P. H., Characterization of plasma membrane shedding from murine melanoma cells. *Int. J. Cancer* **1988**. *41:* 629–635.

29 Thery, C., Boussac, M., Veron, P., Ricciardi-Castagnoli, P., Raposo, G., Garin, J., Amigorena, S., Proteomic analysis of dendritic-cell-derived exosomes: a secreted subcellular compartment distinct from apoptotic vesicles. *J. Immunol.* **2001**. *166:* 7309–7318.

30 Wubbolts, R., Leckie, R. S., Veenhuizen, P. T., Schwarzmann, G., Mobius, W., Hoernschemeyer, J., Slot, J. W., Geuze, H. J., Stoorvogel, W., Proteomic and biochemical analyses of human B-cell-derived exosomes. Potential implications for their function and multivesicular body formation. *J. Biol. Chem.* **2003**. *278:* 10963–10972.

31 van Niel, G., Mallegol, J., Bevilacqua, C., Candalh, C., Brugiere, S., Tomaskovic-Crook, E., Heath, J. K., Cerf-Bensussan, N., Heyman, M., Intestinal epithelial exosomes carry MHC class II/peptides able to inform the immune system in mice. *Gut* **2003**. *52:* 1690–1697.

32 Thery, C., Regnault, A., Garin, J., Wolfers, J., Zitvogel, L., Ricciardi-Castagnoli, P., Raposo, G., Amigorena, S., Molecular characterization of dendritic-cell-derived exosomes: selective accumulation of the heat shock protein hsc73. *J. Cell. Biol.* **1999**. *147:* 599–610.

33 Blott, E. J., Griffiths, G. M., Secretory lysosomes. *Nat. Rev. Mol. Cell Biol.* **2002**. *3:* 122–131.

34 Andre, F., Chaput, N., Schartz, N. E., Flament, C., Aubert, N., Bernard, J., Lemonnier, F., Raposo, G., Escudier, B., Hsu, D. H., Tursz, T., Amigorena, S., Angevin, E., Zitvogel, L., Exosomes as potent cell-free peptide-based vaccine. I. Dendritic-cell-derived exosomes transfer functional MHC class I/peptide complexes to dendritic cells. *J. Immunol.* **2004**. *172:* 2126–2136.

35 Denzer, K., van Eijk, M., Kleijmeer, M. J., Jakobson, E., de Groot, C., Geuze, H. J., Follicular dendritic cells carry MHC class II-expressing microvesicles at their surface. *J. Immunol.* **2000**. *165:* 1259–1265.

36 Gallucci, S., Lolkema, M., Matzinger, P., Natural adjuvants : endogenous activators of dendritic cells. *Nat. Med.* **1999**. *5:* 1249–1255.

37 Albert, M. L., Pearce, S. F., Francisco, L. M., Sauter, B., Roy, P., Silverstein, R. L., Bhardwaj, N., Immature dendritic cells phagocytose apoptotic cells via alphavbeta5 and CD36, and cross-present antigens to cytotoxic T lymphocytes. *J. Exp. Med.* **1998**. *188:* 1359–1368.

38 Iyoda, T., Shimoyama, S., Liu, K., Omatsu, Y., Akiyama, Y., Meada, Y., Takahara, K., Steinman, R. M., Inaba, K., The CD8+ dendritic cell subset selectively endocytoses dying cells in culture and in vivo. *J. Exp. Med.* **2002**. *195:* 1289–1302.

39 Morelli, A. E., Larregina, A. T., Shufesky, W. J., Zahorchak, A. F., Logar, A. J., Papworth, G. D., Wang, Z., Watkins, S. C., Falo, L. D., Thomson, A. W., Internalization of circulating apoptotic cells by splenic marginal zone dendritic cells: dependence on complement receptors and effect on cytokine production. *Blood* **2003**. *101:* 611–620.

40 Gray, D., Kosco, M., Stockinger, B., Novel pathways of antigen presentation for the maintenance of memory. *Int. Immunol.* **1991**. *3:* 141–148.

41 Clayton, A., Turkes, A., Dewitt, S., Steadman, R., Mason, M. D., Hallett, M. B., Adhesion and signaling by B cell-derived exosomes: the role of integrins. *FASEB J.* **2004**. *18:* 977–979.

42 Vincent-Schneider, H., Stumptner-Cuvelette, P., Lankar, D., Pain, S., Raposo, G., Benaroch, P., Bonnerot, C., Exosomes bearing HLA-DR1 molecules need dendritic cells to efficiently stimulate specific T cells. *Int. Immunol.* **2002**. *14:* 713–722.

43 Morelli, A. E., Larregina, A. T., Shufesky, W. J., Sullivan, M. L. G., Beer-Stolz, D., Papworth, G. D., Zahorchak, A. F., Logar, A. J., Wang, Z., Watkins, S. C., Falo, L. D., Thomson, A. W., Endocytosis, intracellular sorting, and processing of exosomes by dendritic cells. *Blood* **2004**. *104:* 3257–3266.

44 Kleijmeer, M., Ramm, G., Schuurhuis, D., Griffith, J., Rescigno, M., Ricciardi-Castagnoli, P., Rudensky, A. Y., Ossendorp, F., Melief, C. J., Stoorvogel, W., Geuze, H. J., Reorganization of multivesicular bodies regulates MHC class II antigen presentation by dendritic cells. *J. Cell. Biol.* **2001**. *155:* 53–63.

45 Chow, A., Toomre, D., Garrett, W., Mellman, I., Dendritic cell maturation triggers retrograde MHC class II transport from lysosomes to the plasma membrane. *Nature* **2002**. *418:* 988–994.

46 Chaput, N., Schartz, N. E. C., Andre, F., Taieb, J., Novault, S., Bonnaventure, P., Aubert, N., Bernard, J., Lemonnier, F., Merad, M. Adema, G., Adams, M., Ferrantini, M., Carpentier, A. F., Escudier, B., Tursz, T., Angevin, E., Zitvogel, L., Exosomes as potent cell-free peptide-based vaccine. II. Exosomes in CpG adjuvants efficiently prime native Tc1 lymphocytes leading to tumor rejection. *J. Immunol.* **2004**. *172:* 2137–2146.

47 Futter, C. E., Pearse, A., Hewlett, L. J., Hopkins: Multivesicular endosomes containing internalized EGF-EGF receptor complexes mature and then fuse directly with lysosomes. *J. Cell. Biol.* **1996**. *132:* 1011–1023.

48 Trambas, C. M., Griffiths, G. M., Delivering the kiss of death. *Nat. Immunol.* **2003**. *4:* 399–403.

49 Escudier, B., Dorval, T., Chaput, N., Andre, F., Caby, M-P., Novault, S., Flament, C., Leboulaire, C., Borg, C., Amigorena, S. et al., Vaccination of metastatic melanoma patients with autologous dendritic cell (DC) derived-exosomes: results of the first phase I clinical trial. *J. Transl. Med.* **2005**. *3:* 1–13.

50 Allan, R. S., Smith, C. M., Belz, G. T., van Lint, A. L., Wakim, L. M., Heath, W. R., Carbone, F. R., Epidermal viral immunity induced by CD8alpha+ dendritic cells but not by Langerhans cells. *Science* **2003**. *301*: 1925–1928.

51 Taams, L. S., van Rensen, A. J., Poelen, M. C., van Els, C. A., Besseling, A. C., Wagenaar, J. P., van Eden, W., Wauben, M. H., Anergic T cells actively suppress T cell responses via the antigen-presenting cell. *Eur. J. Immunol.* **1998**. *28:* 2902–2912.

52 Taams, L. S., Boot, E. P., van Eden, W., Wauben, M. H., 'Anergic' T cells modulate the T-cell activating capacity of antigen-presenting cells. *J. Autoimmun.* **2000**. *14:* 335–341.

53 Vendetti, S., Chai, J. G., Dyson, J., Simpson, E., Lombardi, G., Lechler, R., Anergic T cells inhibit the antigen presenting function of dendritic cells. *J. Immunol.* **2000**. *165:* 1175–1181.

54 Frasca, L., Scotta, C., Lombardi, G., Piccolella, E., Human anergic CD4+ T cells can act as suppressor cells by affecting autologous dendritic cell conditioning and survival. *J. Immunol.* **2002**. *168:* 1060–1068.

55 Taylor, D. D., Gercel-Taylor, C., Lyons, K. S., Stanson, J., Whiteside, T. L., T-cell apoptosis and suppression of T-cell receptor/CD3-zeta by Fas ligand-containing membrane vesicles shed from ovarian tumors. *Clin. Cancer Res.* **2003**. *9:* 5113–5119.

56 Whiteside, T. L., Tumor-derived exosomes or microvesicles: another mechanism of tumor escape from the host immune system? *Br. J. Cancer* **2005**. *92:* 209–211.

57 Arnold, P. Y., Mannie, M. D., Vesicles bearing MHC class II molecules mediate transfer of antigen from antigen-presenting cells to CD4+ T cells. *Eur. J. Immunol.* **1999**. *29:* 1363–1373.

58 Taams, L. S., Wauben, M. H., Anergic T cells as active regulators of the immune response. *Hum. Immunol.* **2000**. *61:* 633–639.

59 Monks, C. R., Freiberg, B. A., Kupfer, H., Sciaky, N., Kupfer, A., Three-dimensional segregation of supramolecular activation clusters in T cells. *Nature* **1998**. 395: 82–86.

60 Hwang, I., Huang, J. F., Kishimoto, H., Brunmark, A., Peterson, P. A., Jackson, M. R., Surh, C. D., Cai, Z., Sprent, J., T cells can use either T cell receptor or CD28 receptors to absorb and internalize cell surface molecules derived from antigen-presenting cells. *J. Exp. Med.* **2000**. *191:* 1137–1148.

61 Huang, J. F., Yang, Y., Sepulveda, H., Shi, W., Hwang, I., Peterson, P. A., Jackson, M. R., Sprent, J., Cai, Z., TCR-Mediated internalization of peptide-MHC complexes acquired by T cells. *Science* **1999**. *286:* 952–954.

62 Hudrisier, D., Riond, J., Mazarguil, H., Gairin, J. E., Joly, E., Cutting edge: CTLs rapidly capture membrane fragments from target cells in a TCR signaling-dependent manner. *J. Immunol.* **2001**. *166:* 3645–3649.

63 Stinchcombe, J. C., Bossi, G., Booth, S., Griffiths, G. M., The immunological synapse of CTL contains a secretory domain and membrane bridges. *Immunity* **2001**. *15:* 751–761.

64 Patel, D. M., Arnold, P. Y., White, G. A., Nardella, J. P., Mannie, M. D., Class II MHC/peptide complexes are released from APC and are acquired by T cell responders during specific antigen recognition. *J. Immunol.* **1999**. *163:* 5201–5210.

65 Sabzevari, H., Kantor, J., Jaigirdar, A., Tagaya, Y., Naramura, M., Hodge, J., Bernon, J., Schlom, J., Acquisition of CD80 (b7-1) by T cells. *J. Immunol.* **2001**. *166:* 2505–2513.

66 Baba, E., Takahashi, Y., Lichtenfeld, J., Tanaka, R., Yoshida, A., Sugamura, K., Yamamoto, N., Tanaka, Y., Functional CD4 T cells after intercellular molecular transfer of 0X40 ligand. *J. Immunol.* **2001**. *167:* 875–883.

67 Hudrisier, D., Bongrand, P., Intercellular transfer of antigen-presenting cell determinants onto T cells: molecular mechanisms and biological significance. *FASEB J.* **2002**. *16:* 477–486.

68 Kedl, R. M., Schaefer, B. C., Kappler, J. W., Marrack, P., T cells down-modulate peptide-MHC complexes on APCs in vivo. *Nat. Immunol.* **2002**. *3:* 27–32.

69 Hwang, I., Sprent, J., Role of the actin cytoskeleton in T cell absorption and

internalization of ligands from APC. *J. Immunol.* **2001**. *166:* 5099–5107.

70 Wetzel, S. A., McKeithan, T. W., Parker, D. C., Peptide-specific intercellular transfer of MHC class II to CD4+ T cells directly from the immunological synapse upon cellular dissociation. *J. Immunol.* **2005**. *174:* 80–89.

71 Patel, D. M., Mannie, M. D., Intercellular exchange of class II major histocompatibility complex/peptide complexes is a conserved process that requires activation of T cells but is constitutive in other types of antigen presenting cell. *Cell. Immunol.* **2001**. *214:* 165–172.

72 Lorber, M. I., Loken, M. R., Stall, A. M., Fitch, F. W., I-A antigens on cloned alloreactive murine T lymphocytes are acquired passively. *J. Immunol.* **1982**. *128:* 2798–2803.

73 Brezinschek, R. I., Oppenheimer-Marks, N., Lipsky, P. E., Activated T cells acquire endothelial cell-surface determinants during transendothelial migration. *J. Immunol.* **1999**. *162:* 1677–1684.

74 Batista, F. D., Iber, D., Neuberger, M. S., B cells acquire antigen from target cells after synapse formation. *Nature* **2001**. *411:* 489–494.

75 Carlin, L. M., Eleme, K., McCann, F. E., Davis, D. M., Intercellular transfer and supramolecular organization of human leukocyte antigen c at inhibitory natural killer cell immune synapses. *J. Exp. Med.* **2001**. *194:* 1507–1517.

76 Sjostrom, A., Eriksson, M., Cerboni, C., Johansson, M. H., Sentman, C. L., Karre, K., Hoglund, P., Acquisition of external major histocompatibility complex class I molecules by natural killer cells expressing inhibitory Ly49 receptors. *J. Exp. Med.* **2001**. *194:* 1519–1530.

77 Tabiasco, J., Vercellone, A., Meggetto, F., Hudrisier, D., Brousset, P., Fournie, J-J., Acquisition of viral receptor by NK cells through immunological synapse. *J. Immunol.* **2003**. *170:* 5993–5998.

78 Vanherberghen, B., Andersson, K., Carlin, L. M., Nolte-'t Hoen, E. N. M., Williams, G. S., Hoglund, P., Davis, D. M., Human and murine inhibitory natural killer cell receptors transfer from natural killer cells to target cells. *Proc.*

Natl. Acad. Sci. USA **2004**. *101:* 16873–16878.

79 Russo, V., Zhou, D., Sartirana, C., Rovere, P., Villa, A., Rossini, S., Traversari, C., Bordignon, C., Acquisition of intact allogeneic human leukocyte antigen molecules by human dendritic cells. *Blood* **2000**. *95:* 3473–3477.

80 Harshyne, L. A., Watkins, S. C., Gambotto, A., Barratt-Boyes, S. M., Dendritic cells acquire antigens from live cells for cross-presentation to CTL. *J. Immunol.* **2001**. 166: 3717–3723.

81 Poupot, M., Fournie, J-J., Spontaneous membrane transfer through homotypic synapses between lymphoma cells. *J. Immunol.* **2003**. *171:* 2517–2523.

82 Herrera, O. B., Golshayan, D., Tibbott, R., Ochoa, F. S., James, M. J., Marelli-Berg, F. M., Lechler, R. I., A novel pathway of alloantigen presentation by dendritic cells. *J. Immunol.* **2004**. *173:* 4828–4837.

83 Bedford, P., Garner, K., Knight, S. C., MHC class II molecules transferred between allogeneic dendritic cells stimulate primary mixed leukocyte reactions. *Int. Immunol.* **1999**. *11:* 1739–1744.

84 Knight, S. C., Iqball, S., Roberts, M. S., Macatonia, S., Bedford, P.A., Transfer of antigen between dendritic cells in the stimulation of primary T-cell proliferation. *Eur. J. Immunol.* **1998**. *28:* 1636–1644.

85 Harshyne, L. A., Zimmer, M. I., Watkins, S. C., Barratt-Boyes, S. M., A role for class A scavenger receptor in dendritic cell nibbling from live cells. *J. Immunol.* **2003**. *170:* 2302–2309.

86 Revy, P., Sospedra, M., Barbour, B., Trautmann, A., Functional antigen-independent synapses formed between T cells and dendritic cells. *Nat. Immunol.* **2001**. *2:* 925–931.

87 Benvenuti, F., Lagaudriere-Gesbert, C., Grandjean, I., Jancic, C., Hivroz, C., Trautmann, A., Lantz, O., Amigorena, S., Dendritic cell maturation controls adhesion, synapse formation, and the duration of the interactions with naïve T lymphocytes. *J. Immunol.* **2004**. *172:* 292–301.

88 Onfelt, B., Nedvetzki, S., Yanagi, K., Davis, D. M., Cutting edge: Membrane nanotubes connect immune cells. *J. Immunol.* **2004**. *173:* 1511–1513.

89 Joly, E., Hudrisier, D., What is trogo-cytosis and what is its purpose ? *Nat. Immunol.* **2003**. *4:* 815.

90 Wang, M. X., Morgan, T., Lungo, W., Wang, L., Sze, G. Z., French, S. W., "Piecemeal" necrosis: renamed troxis necrosis. *Exp. Mol. Pathol.* **2001**. *71:* 137–146.

91 Neijssen, J., Herberts, C., Drijfhout, J. W., Reits, E., Janssen, L., Neefjes, J., Cross-presentation by intercellular peptide transfer through gap junctions. *Nature* **2005**. *434:* 83–88.

92 Heath, W. R., Carbone, F. R., Cross-presentation in viral immunity and self-tolerance. *Nat. Rev. Immunol.* **2001**. *1:* 126–134.

X
Tolerogenic Dendritic Cells

26
Differentiation Stages and Subsets of Tolerogenic Dendritic Cells

Manfred B. Lutz

26.1
Introductory Remarks

Dendritic cells (DC) represent sentinel cells integrated in or residing close to epithelia and thus standing first in line to face invading pathogens. They are major players of the immune system to distinguish self from non-self [1, 2]. Equipped with pathogen recognition receptors, such as the Toll-like receptors (TLR), DC can recognize evolutionarily conserved microbial structures and transmit "danger" signals which lead to DC maturation and migration to the draining lymph node [3, 4]. Although DC are located in almost all body tissues, their frequency is higher in epithelia and also in secondary lymphoid organs such as the thymus, spleen and lymph nodes. However, the high frequency of DC in the secondary lymphoid organs during the steady state may point to their permanent role in tolerance induction, besides their central role for adaptive immune responses after activation. While it was accepted quite early that antigen presentation by DC in the thymus is involved in central tolerance, there is recently accumulating evidence that DC of the spleen and lymph nodes regulate peripheral tolerance. There is considerable hope that such tolerogenic DC might enter clinical application in transplantation, autoimmunity and allergy/asthma [5–11]. In this chapter the tolerogenic capacity and mechanisms of different DC subsets and differentiation stages of DC will be reviewed (Fig. 26.1).

Handbook of Dendritic Cells. Biology, Diseases, and Therapies.
Edited by M. B. Lutz, N. Romani, and A. Steinkasserer.
Copyright © 2006 WILEY-VCH Verlag GmbH & Co. KGaA, Weinheim
ISBN: 3-527-31109-2

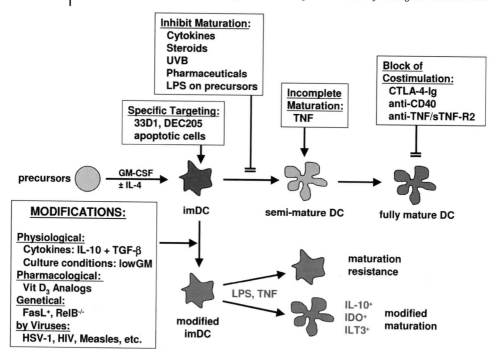

Fig. 26.1 Methods of *in vitro* generation or *in vivo* targeting of tolerogenic DC. Many methods of generating tolerogenic DC are focused around immature DC. They can be targeted specifically *in vivo* by delivery of apoptotic material or targeting to specific surface receptors. Spontaneous maturation of *in vitro* generated DC can be achieved by addition of maturation inhibitors. Inhibition of *in vivo* maturation is attempted by inhibiting co-stimulation of already matured DC. Incompletely matured DC (semi-mature DC) can still be tolerogenic. Due to the problem of maintaining stable immature DC, DC modifications are desired that result in maturation-resistant immature DC or modified/alternatively matured DC expressing inhibitory surface receptors (IDO, ILT-3) or cytokines (IL-10).

26.2
Mechanisms of T-cell Tolerance Induction

The mechanisms of T-cell tolerance have been categorized into five distinct phenomena: ignorance, anergy, deletion, immune deviation, and regulation/suppression. They have been observed by using DC *in vitro* and in mouse models *in vivo* [12] and may help to explain the "behavior" of T cells when stimulated under highly controlled conditions but, as demonstrated below, the maintenance of tolerance *in vivo* appears more complex and may not be attributed exclusively to the one exclusive mechanism.

26.2.1
Ignorance

When antigens expressed in peripheral organs do not reach the blood or secondary lymphoid organs to be presented or crosspresented on MHC I and II molecules, T cells specific for these antigens remain ignorant. Conversely, circulating naïve T cells will not be able to extravasate and survey the tissues. Although ignorance has been postulated as a theoretical mechanism to maintain tolerance, it is difficult to demonstrate *in vivo* whether an antigen is not presented at all or presented at very low levels so that specific effects of T cells are below the detection limit. For obvious reasons there is no role for DC in ignorance as non-presenting cells.

26.2.2
Anergy

T-cell anergy, or hyporesponsiveness, was first demonstrated with T-cell clones *in vitro* [13]. The theory of clonal T-cell anergy induction proposes that a TCR signal (signal 1) is provided in the absence of a co-stimulatory signal through the CD28 molecule (signal 2). However, both signals are required for IL-2 gene activation and clonal T-cell expansion [14].

The TCR signal alone induces first, a Ca^{2+} influx resulting in NF-AT activation, which then trigger a series of signaling events, including GRAIL activation which inhibits IL-2 production by the anergic cells [15–17] and second, CTLA-4 upregulation. A second encounter of such partially preactivated T cells with APC competent in B7-1/B7-2 co-stimulation will now preferentially trigger the CTLA-4 molecule because its affinity is 10-fold higher than CD28 [18, 19]. CTLA-4 signals activate the cell cycle inhibitors $p27^{Kip1}$ and $p21^{Cip1}$ [20, 21] and potentially other factors [22], and thus prevents cell cycle entry and IL-2 production. Thus, partial T-cell stimulation leads to their suboptimal activation, resulting in anergy, characterized by TCR hyporesponsiveness, block of cell cycle entry and impaired IL-2 production [23]. Nevertheless, in most anergy settings the T cells remain responsive to high doses of IL-2 alone without further stimulation of the TCR and CD28 which is in contrast to reg T (see below).

Many of the T cells engaged in this way might die from apoptosis while the surviving T cells persist alive for at least one month in mice remaining in an anergic state. During this time period anergic T cells may acquire regulatory functions on other effector T cells and DC [24, 25].

26.2.3
Deletion

The most clear-cut form of T-cell tolerance is the complete deletion of T cells with unwanted specificity. This form of T-cell elimination has been described in different settings. Deletion of T cells can occur in the absence of co-stimulation [26, 27], by the lack of growth factors [28] or space [29], i.e. as a result of suboptimal stimulation. In contrast, deletion commonly occurs as a result of optimal T-cell priming by activation-induced cell death (AICD) in order to limit or terminate T-cell responses of activated T cells through CD95/CD95L dependent mechanisms [30–33].

Especially for CD8$^+$ CTL responses, the deletional pathway is well documented [34–36]. Presentation of exogenous antigens on MHC I molecules occurs via cross-presentation and can also lead to deletional tolerance [37].

26.2.4
Immune Deviation

T helper cell differentiation can occur two major pathways. The Th1 differentiation gives rise to predominantly IFN-γ producing T cells while Th2 cells produce IL-4, IL-5 and IL-13. As the Th1 cytokines can promote CD8$^+$ CTL responses, this arm of the adaptive T helper cell activity has been called "cellular immune response". In contrast, the Th2 cytokine IL-4 was the first to be identified as promoting the isotype switch of B cells to IgE, and therefore the Th2 help has been called a "humoral immune response". This dichotomy is not rigidly applied since Th1 cytokines can also support certain isotype switches (IFN-γ to IgG2a in the mouse) [38] and the activity of macrophages and NK cells partially depends on antibody binding to Fc receptors for microbial recognition or antibody-dependent cellular cytotoxicity (ADCC).

The phenomenon of immune deviation is based on the observation that T helper cell priming can be polarized towards Th1 in the presence of IFN-γ and blocking of IL-4, while Th2 polarization is driven by IL-4 and blocking of IL-12 and/or IFN-γ [39]. If an immune response against a microbe requires Th1 immunity to resolve the infection, such as *Leishmania major* in C57BL/6 mice, the immune response is dramatically impaired if it is polarized towards the "wrong" Th2 arm as observed in BALB/c mice. As a fatal result BALB/c mice succumb to the infection [40]. This indicates that a given pool of naïve T cells with specificity for a certain antigen can be differentiated in either the Th1 or the Th2 direction, and all subsequent immunizations for the same antigen will then also follow the preset type of memory immune response.

26.2.5
The Concept of "Immune Balance"

If a harmless antigen enters the body, it is conceivable that a nonfatal type of immune reactivity may be induced. This may be especially applicable for allergies. For example, in nonallergic individuals an immunoreactivity to the Bet v 1 allergen can be observed, but the responsive T-cell clones produce predominantly IFN-γ and little IL-4 and the serum contains substantial IgG but no detectable IgE. In this way an involvement of mast cells in the immune response is avoided [41]. Thus, only if the "wrong" type of immune response is initiated, may diseases such as allergies and asthma develop. This concept of permanent immune deviation to harmless antigens/allergens has also been called "immune balance" [42, 43].

26.2.6
Regulation/suppression

Regulatory T cells (reg T) represent specialized effector T cells which are able to downregulate immune responses. All reg T cells have in common that they require TCR stimulation to activate their suppressive mechanisms, which then, however, can act antigen-unspecifically and also provide bystander suppression. Two distinct classes can be distinguished: (a) the thymus-derived naturally-occurring CD4$^+$ CD25$^+$ Foxp3$^+$ reg T cells, which constantly express the IL-2Rα chain (CD25) and mainly regulate through cell contact-dependent mechanisms [44, 45]; (b) several types of peripherally induced reg T such as the Tr1 which mainly regulate through secretion of IL-10 [46]. Also Th3 cells have been described which mainly secrete TGF-β and thereby act suppressively [47]. In addition there are also CD4$^-$CD8$^-$ reg T [48, 49] and CD8$^+$CD28$^-$ reg T subsets [50] with dramatic suppressor functions.

The role of reg T during the steady state *in vivo* is unclear. The induction of both Tr1 and CD4$^+$ CD25$^+$ reg T requires co-stimulation through CD28 [51, 52]. This would indicate that immature DC in the steady state cannot be responsible for their induction because they are largely incompetent for co-stimulation. In fact, in healthy individuals all CD4$^+$CD25$^+$ reg T cells are in a resting state, i.e. negative for activation markers such as CD69. Also experiments comparing immature and mature DC in mice clearly showed that mature DC were superior in the activation and expansion of CD4$^+$CD25$^+$ reg T cells [53, 54].

However, human immature MoDC have been shown to stimulate the occurrence of IL-10-producing Tr1 *in vitro* [55]. Immature MoDC already express certain amounts of co-stimulatory molecules CD80/CD86. Together, this might indicate that low levels of TCR stimulation plus low levels of co-stimulation might be sufficient for reg T activation.

26.2.7
Combinations

When the fate of antigen-specific T cells is followed *in vivo* and the analysis is not restricted to the determination as to whether a certain tolerance mechanism occurs or not, a combination of several tolerance mechanisms mostly seems to occur.

Anergy and cytokine-mediated suppression can occur simultaneously in models of tolerance against superantigens in mice [56] and anergic T cells can turn into suppressing/regulatory T cells *in vivo* [57, 58] and as shown for human T cells *in vitro* [59].

Many models of *in vivo* tolerance induction observed an initial expansion of the antigen-specific T cells after antigen application followed by their deletion after a couple of days and finally the few surviving T cells showed signs of anergy. This has been studied for the nonimmunogenic i.v. injection of superantigens [58, 60] or OVA protein [61–63].

Together, the kinetics of tolerization or immunization of T cells *in vivo* seem quite similar but occur at different levels of intensity (Fig. 26.2). Parameters direct-

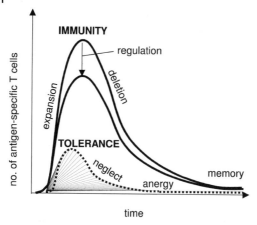

Fig. 26.2 Quantitative and kinetic view of T-cell immunity and tolerance. Induction of T-cell immunity, e.g. by vaccination, is accompanied by the clonal expansion of the antigen-specific T cells, followed by the decline of cell numbers (deletion) and resulting in the maintenance of an antigen-specific memory T-cell pool. All these processes can be suppressed, e.g. by regulatory T cells. If tolerance is induced, e.g. by systemic application of an antigen, the series of events remains similar to immunization, but with lower numbers due to the suboptimal activation levels of the individual T cells. The expanded T cells decline here rather due to death by neglect and result in long-lived anergic T cells.

ing tolerance or immunity might therefore rather depend on the dose, duration and timing of antigen presentation [29, 64–66]. Many of these parameters can be directed by different types of APC but also by different maturation stages of DC.

26.3
Tolerogenic DC Subsets *in vivo*

One major question regarding tolerogenic DC is whether certain subsets of DC exist with an intrinsic tolerogenic potential as opposed to the sensor-DC that upon activation become immunogenic DC. The current literature suggests that environmental signals can either inhibit DC maturation or modify DC so that they convert to tolerogenic APC.

26.3.1
Thymic DC

DC resident within the thymus are involved in thymocyte selection by presenting antigens from the circulation [67]. Thymic DC present these antigens for longer time periods as compared with DC of other organs [68]. A closer analysis of thymic DC revealed that they are shortlived and predominantly of the MHC II$^+$, CD11c$^+$, CD205$^+$, CD8α^+ subtype [69, 70]. So the tolerogenic nature of antigen presentation by thymic DC does not seem to be intrinsically different from that of the spleen. Rather, the responses of immature thymocytes and mature T cells seem to determine different outcomes [71]. Intrathymic injection of isolated thymic or splenic DC has been shown to induce T-cell anergy to Mls antigens [72].

26.3.2
DC in Lymph Nodes and Spleen

DC reside as interdigitating cells in the spleen and lymph nodes even in the steady state and not only after infections. There they can be characterized either by their immature or partially mature state [73] or as phenotypically distinct and functionally specialized subsets (see Chapter 12).

DC within spleen and lymph nodes abundantly present self-antigens in the steady state as measured by the self-peptide/MHC II complex recognizing antibody Y-Ae [74]. DC from Peyer's patches constitutively produce IL-10 and after isolation induce IL-4 and IL-10 releasing T cells [75]. When CD8 T cells were followed in inducible transgenic mice where both DC and antigen-reactive T cells appear at a physiological frequency, strong tolerance is induced against a *de novo* induced antigen [76].

Similarly to the DC1/DC2 paradigm for the induction of respective Th1 or Th2 immunity, it is now a matter of debate whether certain intrinsically tolerogenic DC subsets exist or whether all known DC subsets can exert tolerogenic function, depending on their environmental instruction. One source of environmental signals for DC might be derived from low level and locally restricted apoptotic cells which occur continuously during tissue remodeling. In the mouse spleen two major DC subsets have been characterized quite well. One is the classical myeloid CD8α^- DC and the other was initially termed "lymphoid" DC characterized by their CD8α expression. *In vitro* and *in vivo* apoptotic cell uptake could be observed by CD8α^+ DC which crosspresented the apoptotic material and induced initial CD8$^+$ T-cell expansion followed by deletion [12, 77] or induced CD8$^+$ T-cell unresponsiveness [78]. As the CD8α expression largely overlaps with the CD205 expression, the targeting of antigens to this specific molecule may also be attributed to this DC subset [79, 80]. It is of note that in most of the experiments used to analyze lymph node DC, resident versus immigrating DC from the draining tissues have not been analyzed separately.

26.3.3
Migratory DC from Peripheral Organs

Peripheral lymph nodes draining the skin also contain mature DC during the steady state. These DC are MHC II^{high}, B7-2^{high}, CD40^{high} DC and clearly represent epidermal LC and dermal DC which have migrated to the lymph node [73]. Their steady-state migration is dependent on CCR7 expression [81]. Such migratory DC are not induced by any pathogenic or normal skin flora as they can be also detected in germ-free mice (Kanazawa and Lutz, unpublished observations).

In peripheral lymph nodes draining various organs, the transport of apoptotic material during the steady state has been demonstrated. The earliest reports show the transport of melanocytes by epidermal LC [82] which has later been confirmed using a transgenic melanocytosis model [83]. Cannulation of the pseudo-afferent lymphatics in the rat revealed two populations of migrating DC, with one transporting self-antigens from the intestine to the mesenteric lymph node [84]. Similarly, inhaled antigen is transported by DC to the lung-draining lymph node and there induced Tr1 cells [52]. This mechanism might be exploited by microorganisms for immune escape as demonstrated for *Bordetella pertussis* [85]. Two more groups show the presentation of cell-associated self-antigens in the lymph nodes draining the stomach [86] or pancreas [87]. Although this presentation might indicate the self-antigen transport by migratory DC, this has not been demonstrated.

26.3.4
Plasmacytoid DC

Initially the plasmacytoid DC (pDC) have been termed DC2 for their capacity to induce immune deviation by promoting Th2 responses and thereby inhibiting Th1 cells [88, 89]. However in the mouse, the contrary direction of deviation has been demonstrated by pDC. Murine Th2 cell mediated asthma could be prevented by pDC but not myeloid BM-DC [90]. *In vivo*, B220$^+$ pDC can be detected in the thymus, lymph nodes and the spleen in an immature/resting stage where they might induce CD4$^+$ T cells with regulatory capacity [91], T-cell anergy [92] or CD8$^+$ IL-10-producing reg T [93]. A distinct B220$^-$ Gr-1$^-$ CD11c^{low} CD45RB^{hi} DC population with plasmacytoid morphology was found in murine spleens to be potent producers of IL-10 [94]. Such IL-10 producing pDC might develop under tolerogenic conditions *in vivo* by the influence of factors derived from splenic stroma cells [95]. Triggering of the CD200 receptor on pDC induces the release of indoleamine-2,3-dioxigenase (IDO) and initiates tryptophan catabolism which impairs T-cell functions [96].

26.4
DC Precursors

Besides the tolerogenic capacity of differentiated DC, there is accumulating evidence that DC precursors might also bear tolerogenic capacity. During the culture period of 8–10 days for BM-DC generation with GM-CSF according to a standard protocol [97], the DC precursors go through a stage around day 3–4 where they transiently acquire suppressive capacities (Table 26.1) [98]. Such DC precursors have been described as myeloid suppressor cells (MSC) with major influence on Th1 cells in *Candida* infection [99] and CD8$^+$ CTL responses in murine tumor models [100–102]. Human MSC seem to influence antitumor immune responses in patients with head-and-neck cancer [103–106] but they are poorly characterized.

Tab. 26.1 Characteristics of murine suppressive DC precursors
(i.e. Myeloid Suppressor Cells, MSC).

Morphology	non-adherent, irregular round cell shape, ring-shaped nucleus
Antigen uptake	CD14$^-$, TLR4$^-$, FcγR II&III$^+$, CD205$^{+/-}$,
Antigen presentation	MHC I$^+$, MHC II$^-$, CD1d$^+$
Co-stimulation	CD80$^+$, CD86$^{+/-}$, CD40$^-$, CD54$^+$
Development	CD34$^-$, CD31$^+$, ER-MP58$^+$, CD13$^-$, Gr-1^{low} (Ly-6C$^+$, Ly-6G^{low}), CD11c–
Lineage	CD4$^-$, CD8α$^-$, B220$^-$, NK1.1$^-$, DX5$^-$, asialoGM1$^+$, F4/80$^+$, **CD11b$^+$**, MOMA1$^-$, 33D1$^-$, CD169$^-$
Homing	CD62L$^+$, CCR7$^-$, CCR5$^-$
Activation requirement	IFN-γ
Suppressive mechanisms	cell-contact required, NO release
Disease implication	suppression of antitumor CTL immunity, NOD

26.5
Immature DC

26.5.1
Tissue Resident DC

The prototype of an immature DC *in vivo* is the epidermal Langerhans cell in a healthy individual. In the steady state LC show a stellate morphology and together form a meshwork structure at a high cell density (1000 LC mm^{-2}) to sense for pathogens. LC express very little MHC II on their surface and B7 molecules are absent [107]. In this respect they are even less mature than the immature DC residing in the spleen and lymph nodes [108]. Also all other organs contain about 1% tissue-resident, immature DC but the percentages are highest in epithelial and secondary lymphoid tissues (2–4%) [109].

26.5.2
Induction of T-cell Anergy by Immature DC

Immature DC are characterized by their high capacity to endocytose antigens but low levels of MHC I and II and co-stimulatory molecule expression on the cell surface [110] and therby fulfill the criteria for anergy induction.

Intracellularly, immature DC accumulate large amounts of MHC II molecules within different types of endosomal compartments where antigen loading can take place [111]. Immature DC also continuously endocytose antigens or apoptotic material from their environment [112, 113] so that self-antigens then could appear on MHC I and/or II molecules also on the surface of DC in the steady state *in vivo*. Although the formation of MHC II/peptide complexes seems not to occur in immature DC [114]. The question remains however, why immature DC express MHC II molecules on their surface and what peptides are presented?

In conjunction with the low levels of surface MHC I and II expression, immature DC express only limited amounts of co-stimulatory molecules on their surface [115] and do not release pro-inflammatory cytokines [8, 116]. As indicated above, some B7 co-stimulation might be required to trigger CTLA-4 for induction [18] or maintenance of anergy [117].

26.5.3
Maturation Inhibitors

After isolation of DC from peripheral [118] or secondary lymphoid tissues [115] but also during the generation of murine BM-DC cultures [97, 119] DC mature "spontaneously", presumably due to the release of pro-inflammatory cytokines by different cell types of the disrupted tissue or macrophages developing in culture [120]. Thereby DC upregulate MHC and co-stimulatory molecules and acquire immunogenic properties. For this reason numerous methods have been developed to inhibit DC maturation and arrest the DC in a tolerogenic state (Table 26.2).

26.5.4
Maturation Resistance

Although the list of factors inhibiting DC maturation is constantly increasing, the stability of the immature phenotype has not been addressed by many studies. The induction of a stably immature phenotype, however, is necessary to maintain or improve the tolerogenic potential of the DC after injection. What this means is illustrated in Fig. 26.3, but has also been observed after injection of immature DC which then upregulated B7 molecules when detected in secondary lymphoid organs [121].

Maturation resistance has only been demonstrated by a few methods so far. The treatment of murine BM-DC cultures with only low doses of GM-CSF (5 U ml^{-1}) rendered the cells insensitive to maturation with TNF, LPS or anti-CD40 antibodies [122]. With these immature maturation-resistant DC, the mean heart allograft rejection time of 24 days that was reached with immature maturation-sensitive DC [121] could be extended to >120 days. Potent inhibitors of DC maturation in conjunction with maturation resistance found so far are: vitamin D3 analogs [123, 124]; the combined treatment of DC with IL-10 and TGF-β [125, 126]; and dexamethasone with human Mo-DC [127].

Tab. 26.2 Inhibitors of DC maturation.

Treatment or condition	Maturation resistance	T-cell tolerance mechanism	References
α-MSH			143
Aspirin		DTH inhibition	144, 145
Bordetella pertussis		IL-10⁺ reg T	85
CD47		n.d.	146–148
Diverse corticosteroids		anergy IL-10⁺ DC	149–155
Dexamethasone	yes		127
Cyclosporin A			151, 156–161
Early LPS		anergy	162
E-cadherin		n.d.	163
Fumaric acid esters		n.d.	164
Haptoglobin		n.d.	165
IL-10		anergy suppression	166–173
IL-10 plus TGF-β	yes		125, 126
Low GM-CSF, no IL-4	yes	anergy	122
Magnesium ions		n.d.	174
Malaria (*P. falciparum*)		dysregulation	175
Mycobacterium tuberculosis			176
Mycophenolate mofetil		n.d.	177
N-acetyl-L-cysteine		n.d.	178
Rapamycin			179
Sanglifehrin A			180
Soluble CD83		n.d.	181
Substance P		n.d.	182
TGF-β		deviation	120, 183–185
Trypanosoma cruzi		n.d.	186
UV-B light		anergy IL-12 p40 by DC	187–189
Vascular endothelial growth factor (VEGF)		n.d.	190
Viruses		diverse	191–198
Vitamin D3 analogs	yes	anergy	151, 199–202

n.d. = not determined

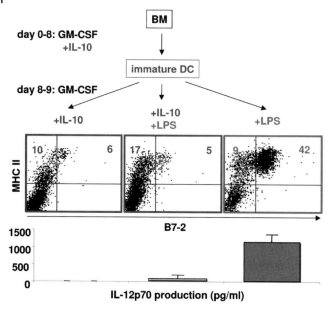

Fig. 26.3 Importance of maturation resistance by immature DC for their tolerogenicity. BM-DC were cultured for 8 days with GM-CSF (200 U ml^{-1}) plus IL-10 (10 ng ml^{-1}) to inhibit spontaneous maturation. Then the cells were washed and replated with GM-CSF and IL-10, or IL-10 plus LPS or LPS for another 24 h. FACS analysis was performed with the cells for surface MHC II (M5/114-PE) and B7-2 (FITC) and the IL-12p70 production measured by ELISA. The data show that in the absence of the inhibitory IL-10 signal, the immature DC rapidly mature on LPS.

26.6
Semi-mature DC

The term semi-mature was introduced to describe a stage of activation/maturation of DC that shows distinct characteristics of partial maturation. Semi-mature DC show an intermediate phenotype between immature DC, representing the resting tissue type, and completely mature DC, representing the migratory, T-cell priming type [8]. DC that have been suboptimally matured with TNF downregulated their endocytosis capacity and expressed high levels of DC maturation markers (CD25, CD205, 2A1), MHC II and co-stimulatory molecules, acquired lymph node homing potential (CCR7), but were unable to secrete cytokines or soluble mediators such as IL-1β, IL-6, IL-10, IL-12p40, IL-12p70, NO and the chemokine MCP-1, in contrast to LPS or LPS plus anti-CD40 fully matured DC. Functionally, we found that such TNF/DC can induce peptide-specific tolerance in the Th1-mediated autoimmune model experimental autoimmune encephalomyelitis (EAE) [128]. The incomplete DC maturation we observed by TNF *in vitro*, has also been demonstrated by endogenous inflammatory mediators *in vivo*. Although CD4$^+$ T cells were activated, they could not induce any isotype switch in B cells [129]. The TNF/DC can be frozen and thawed for practical reasons. Intravenous injection is superior to the intraperitoneal route and subcutaneous injection is not protective because the TNF/DC are co-localized and exposed to tissue injury signals counteracting their tolerogenicity (Fig. 26.4) [130].

More recent data indicate that semi-mature DC can be protective only in Th1 models such as EAE [128] or collagen-induced arthritis (CIA) [131] but not in Th2 models such as allergies [132], asthma (Erb/Lutz unpublished observations and [133]) or experimental autoimmune myocarditis (EAM) [134]. On the other hand LPS-matured BM-DC can ameliorate the Th2 model asthma [135] and TNF/DC can promote Th2 immunity in *L. major* infections in Th2-biased susceptible BALB/c mice (Fig. 26.4) (Wiethe/Gessner/Lutz manuscript in preparation). The latter point indicates that TNF/DC are also inducing Th2 immunity and might thereby immunedeviate the Th1 cells in the EAE model. This is even further supported by the fact that TNF/DC require co-activation of IL-4 and IL-13 secreting NKT cells through CD1d molecules on the DC surface for EAE protection [136]. Thus, TNF/DC induce a Th2/NKT2 shift which as a prerequisite to establish the protective Th2/Tr1 phenotype as observed upon repetitive administration of TNF/DC.

Other features of semi-mature DC generated by TNF treatment *in vitro* further underscore their relative rather than absolute tolerogenic potential. First, semi-mature DC are not terminally differentiated, since subsequent stimulation *in vitro* and

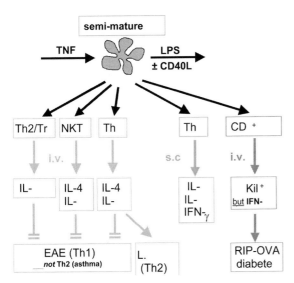

Fig. 26.4 Tolerogenic and immunogenic properties of *in vitro* generated semi-mature TNF/DC. Semi-mature DC are generated by stimulation with TNF and can further mature with LPS or LPS/anti-CD40 treatment. Repetitive i.v. injections of peptide-loaded DC will induce IL-10[+] IL-4[+] IL-13[+] CD4[+] T cells and IL-4[+] IL-13[+] NKT cells which in conjunction suppress Th1-mediated EAE but not Th2-mediated asthma. Injection of DC via the s.c. route abrogates protection by inducing an unpolarized Th0 response. Intravenous injection also promotes Th2 immunity in the *Leishmania* model and CD8[+] CTL activity in the RIP-OVA diabetes model despite application via the i.v. route. Thus semi-mature DC are only semi-tolerogenic depending on the type of T-cell response to be influenced. There however, they are highly effective and can suppress peptide-specifically.

in vivo by LPS or LPS plus anti-CD40 can convert them into fully mature DC, producing cytokines and abrogating their tolerogenic potential [130]. Second, by loading semi-mature DC simultaneously with peptides on MHC class I and II molecules we could show that these cells then tolerize the Th1 CD4$^+$ T-cell compartment for EAE induction, but immunize the CD8$^+$ T-cell compartment to induce CTL activity and induce diabetes in the transgenic RIP-OVA model. This demonstrates that CD4$^+$ and CD8$^+$ T cells have different co-stimulatory requirements for their activation (Fig. 26.4) [137].

26.6.1
Steady-state Migratory DC

Besides the *in vitro* generation of semi-mature DC the question remains as to whether such a differentiation stage of DC might exist *in vivo* and whether this correlate would then be tolerogenic as well. We outlined before that there is evidence that steady-state migratory DC in normal healthy mice represent the *in vivo* counterpart of *in vitro* generated semi-mature DC [8]. However our recent data indicate that semi-mature DC generated *in vitro* by TNF treatment induce TH2/NKT2 immune deviation, while the semi-mature steady state migratory DC *in vivo* might induce other mechanisms of CD4 T cell tolerance such as anergy and deletion (Kanazawa, Azukizawa, Lutz, in preparation).

In peripheral lymph nodes of healthy mice different lineage subsets of DC (CD4$^-$8$^-$, CD4$^+$8$^-$, CD4$^-$8$^+$, plasmacytoid) can be identified, but also different stages of maturation, i.e. immature CD40^{low} and mature CD40^{high} DC. The CD40^{high} DC also express high levels of MHC II and B7 molecules and are migratory DC from the skin, thus representing Langerhans cells and dermal DC [73]. We extended these findings and found that the CD40^{high} DC did not produce cytokines/chemokines (IL-6, IL-12p40, IL-10, MCP-1) apart from little TNF [138]. Because these CD40^{high} DC very much resemble our *in vitro*-generated TNF/DC, we would argue that they are at a semi-mature stage.

The transport of tissue antigens or apoptotic material during the steady state has been demonstrated from the intestine [84], stomach [86], skin [83]. There is further evidence that such DC-mediated transport, processing and presentation of harmless antigens applied into the lung or expressed as a neo-self antigen will lead to T-cell tolerance by inducing CD4$^+$ Tr1 cells or CD8$^+$ deletion [52, 76, 87]. Such DC presenting self-antigens in the steady state on MHC I and II can appear mature as they express high levels of MHC I and II and co-stimulatory molecules but nevertheless act as if tolerogenic [74].

Thus, uptake of apoptotic material might partially activate the immature tissue resident DC for antigen processing, enhanced presentation (MHC II^{high}) and migration (CCR7^{high}) [139–141]. The resulting semi-mature, steady state migratory DC will present the transported self-antigens in a tolerogenic manner after arriving in the draining lymph node [8]. The mechanisms of T-cell tolerance might include induction of T-cell anergy/hyporesponsiveness [138], IL-10$^+$ reg T [52] and deletion [87].

26.7
Fully Mature DC

An interesting question is whether even fully mature DC might be able to induce certain forms of T-cell tolerance, despite the fact that their major task is the priming of adaptive immune responses.

Immunization studies showed that repetition of the priming process either with the help of adjuvants or by DC will boost the response, as can be measured by the percentages of specific T cells or antibody titers. Independent of the vaccination method, a maximum will be reached for each parameter. The intraclonal competition of specific T cells for antigen-presenting cells, including DC, which display the relevant peptide for restimulation, or simply the competition for space within the lymph node, might contribute to limit the amount of T-cell clonal expansion [29]. However, this is not a tolerogenic function of the antigen-priming DC.

Another aspect for the limitation of immune reactivity is the constitutive presence of the subset of $CD4^+ CD25^+$ T cells released from the thymus and detectable at 3–10% in the circulation and secondary lymphoid organs of mice and humans. The question remains as to which would be the type of DC able to activate $CD4^+ CD25^+$ T cells regulatory capacity *in vivo*. A proportion of $CD4^+CD25^+$ reg T permanently seems to express activation markers and divide in the steady state [75]. Adoptive transfer of OVA-specific $CD4^+ CD25^+$ TCR-transgenic DO11.10 T cells showed that only LPS-matured DC were able to expand the transferred reg T but not immature DC [54, 142]. These data indicate that $CD4^+ CD25^+$ reg T are co-induced with effector T cells to limit the immune response against foreign antigens. How the balance is regulated between priming and regulation remains elusive.

Acknowledgements

Many thanks to Gerold Schuler and Alexander Steinkasserer for their continuous support and Jonathan Austyn, Christian Bogdan, Thomas Brocker, Jean Davoust, Klaus Erb, André Gessner, Manfred Kopf, Pieter Leenen, Niki Romani, Michael Sixt, Patrizia Stoitzner for exciting collaborations and discussions. The author was supported by the Deutsche Forschungsgemeinschaft DFG (SFB466, SFB642, LU851/1-1, LU851/2-1, GK592), the Interdisciplinary Center for Clinical Research (IZKF) Erlangen, and both the ELAN Fonds and the Marohn Foundation of the University Hospital Erlangen.

Abbreviations

APC	antigen presenting cell	i.v.	intravenous
BM	bone marrow	pDC	plasmacytoid DC
DC	dendritic cells	reg T	regulatory T cells
LC	Langerhans cells	s.c.	subcutaneous

References

1 Burnet, F. M., *The clonal selection theory of acquired immunity*. Vanderbilt University Press, Nashville, TN: **1959**.

2 Janeway, C. A., Jr., The immune system evolved to discriminate infectious nonself from noninfectious self. *Immunol Today* **1992**. *13:* 11–16.

3 Matzinger, P., Tolerance, danger, and the extended family. *Annu Rev Immunol* **1994**. *12:* 991–1045.

4 Janeway, C. A., Jr., How the immune system protects the host from infection. *Microbes Infect* **2001**. *3:* 1167–1171.

5 Fairchild, P. J., Waldmann, H., Dendritic cells and prospects for transplantation tolerance. *Curr Opin Immunol* **2000**. *12:* 528–535.

6 Jonuleit, H., Schmitt, E., Steinbrink, K., Enk, A. H., Dendritic cells as a tool to induce anergic and regulatory T cells. *Trends Immunol* **2001**. *22:* 394–400.

7 Lambrecht, B. N., Hoogsteden, H. C., Pauwels, R. A., Dendritic cells as regulators of the immune response to inhaled allergen: recent findings in animal models of asthma. *Int Arch Allergy Immunol* **2001**. *124:* 432–446.

8 Lutz, M. B., Schuler, G., Immature, semi-mature, and fully mature dendritic cells: Which signals induce tolerance or immunity? *Trends Immunol* **2002**. *23:* 445–449.

9 Morelli, A. E., Thomson, A. W., Dendritic cells: regulators of alloimmunity and opportunities for tolerance induction. *Immunol Rev* **2003**. *196:* 125–146.

10 Turley, S. J., Dendritic cells: inciting and inhibiting autoimmunity. *Curr Opin Immunol* **2002**. *14:* 765–770.

11 Steinman, R. M., Hawiger, D., Nussenzweig, M. C., Tolerogenic dendritic cells. *Annu Rev Immunol* **2003**. *21:* 685–711.

12 Liu, K., Iyoda, T., Saternus, M., Kimura, Y., Inaba, K., Steinman, R. M., Immune tolerance after delivery of dying cells to dendritic cells in situ. *J Exp Med* **2002**. *196:* 1091–1097.

13 Lamb, J. R., Skidmore, B. J., Green, N., Chiller, J. M., Feldmann, M., Induction of tolerance in influenza virus-immune T lymphocyte clones with synthetic peptides of influenza hemagglutinin. *J Exp Med* **1983**. *157:* 1434–1447.

14 Schwartz, R. H., T cell anergy. *Annu Rev Immunol* **2003**. *21:* 305–334.

15 Macian, F., Garcia-Cozar, F., Im, S. H., Horton, H. F., Byrne, M. C., Rao, A., Transcriptional mechanisms underlying lymphocyte tolerance. *Cell* **2002**. *109:* 719–731.

16 Heissmeyer, V., Macian, F., Im, S. H., Varma, R., Feske, S., Venuprasad, K., Gu, H., Liu, Y. C., Dustin, M. L., Rao, A., Calcineurin imposes T cell unresponsiveness through targeted proteolysis of signaling proteins. *Nat Immunol* **2004**. *5:* 255–265.

17 Mueller, D. L., E3 ubiquitin ligases as T cell anergy factors. *Nat Immunol* **2004**. *5:* 883–890.

18 Greenwald, R. J., Boussiotis, V. A., Lorsbach, R. B., Abbas, A. K., Sharpe, A. H., CTLA-4 regulates induction of anergy in vivo. *Immunity* **2001**. *14:* 145–155.

19 Salomon, B., Bluestone, J. A., Complexities of cd28/b7: ctla-4 costimulatory pathways in autoimmunity and transplantation. *Annu Rev Immunol* **2001**. *19:* 225–252.

20 Boussiotis, V. A., Freeman, G. J., Taylor, P. A., Berezovskaya, A., Grass, I., Blazar, B. R., Nadler, L. M., p27kip1 functions as an anergy factor inhibiting interleukin 2 transcription and clonal expansion of alloreactive human and mouse helper T lymphocytes. *Nat Med* **2000**. *6:* 290–297.

21 Jackson, S. K., DeLoose, A., Gilbert, K. M., Induction of anergy in Th1 cells associated with increased levels of cyclin-dependent kinase inhibitors p21Cip1 and p27Kip1. *J Immunol* **2001**. *166:* 952–958.

22 Verdoodt, B., Blazek, T., Rauch, P., Schuler, G., Steinkasserer, A., Lutz, M. B., Funk, J. O., The cyclin-dependent kinase inhibitors p27Kip1 and p21Cip1 are not essential in T cell anergy. *Eur J Immunol* **2003**. *33:* 3154–3163.

23 Macian, F., Im, S. H., Garcia-Cozar, F. J., Rao, A., T-cell anergy. *Curr Opin Immunol* **2004**. *16:* 209–216.

24 Lechler, R., Chai, J. G., Marelli-Berg, F., Lombardi, G., The contributions of T-cell anergy to peripheral T-cell tolerance. *Immunology* **2001**. *103:* 262–269.

25 Appleman, L. J., Boussiotis, V. A., T cell anergy and costimulation. *Immunol Rev* **2003**. *192:* 161–180.

26 Ferber, I., Schonrich, G., Schenkel, J., Mellor, A. L., Hammerling, G. J., Arnold, B., Levels of peripheral T cell tolerance induced by different doses of tolerogen. *Science* **1994**. *263:* 674–676.

27 Critchfield, J. M., Racke, M. K., Zuniga-Pflucker, J. C., Cannella, B., Raine, C. S., Goverman, J., Lenardo, M. J., T cell deletion in high antigen dose therapy of autoimmune encephalomyelitis. *Science* **1994**. *263:* 1139–1143.

28 Förster, I., Hirose, R., Arbeit, J. M., Clausen, B. E., Hanahan, D., Limited capacity for tolerization of CD4+ T cells specific for a pancreatic beta cell neo-antigen. *Immunity* **1995**. *2:* 573–585.

29 Rachmilewitz, J., Lanzavecchia, A., A temporal and spatial summation model for T-cell activation: signal integration and antigen decoding. *Trends Immunol* **2002**. *23:* 592–595.

30 Dhein, J., Walczak, H., Baumler, C., Debatin, K. M., Krammer, P. H., Auto-crine T-cell suicide mediated by APO-1/(Fas/CD95). *Nature* **1995**. *373:* 438–441.

31 Brunner, T., Mogil, R. J., LaFace, D., Yoo, N. J., Mahboubi, A., Echeverri, F., Martin, S. J., Force, W. R., Lynch, D. H., Ware, C. F., et al., Cell-autonomous Fas (CD95)/Fas-ligand interaction mediates activation-induced apoptosis in T-cell hybridomas. *Nature* **1995**. *373:* 441–444.

32 Ju, S. T., Panka, D. J., Cui, H., Ettinger, R., el-Khatib, M., Sherr, D. H., Stanger, B. Z., Marshak-Rothstein, A., Fas(CD95)/FasL interactions required for programmed cell death after T-cell activation. *Nature* **1995**. *373:* 444–448.

33 Green, D. R., Droin, N., Pinkoski, M., Activation-induced cell death in T cells. *Immunol Rev* **2003**. *193:* 70–81.

34 Zinkernagel, R. M., Moskophidis, D., Kundig, T., Oehen, S., Pircher, H., Hengartner, H., Effector T-cell induction and T-cell memory versus peripheral deletion of T cells. *Immunol Rev* **1993**. *133:* 199–223.

35 Wells, A. D., Li, X. C., Strom, T. B., Turka, L. A., The role of peripheral T-cell deletion in transplantation tolerance. *Philos Trans R Soc Lond B Biol Sci* **2001**. *356:* 617–623.

36 Lechler, R. I., Garden, O. A., Turka, L. A., The complementary roles of deletion and regulation in transplantation tolerance. *Nat Rev Immunol* **2003**. *3:* 147–158.

37 Heath, W. R., Belz, G. T., Behrens, G. M., Smith, C. M., Forehan, S. P., Parish, I. A., Davey, G. M., Wilson, N. S., Carbone, F. R., Villadangos, J. A., Cross-presentation, dendritic cell subsets, and the generation of immunity to cellular antigens. *Immunol Rev* **2004**. *199:* 9–26.

38 Stavnezer, J., Molecular processes that regulate class switching. *Curr Top Microbiol Immunol* **2000**. *245:* 127–168.

39 Mosmann, T. R., Coffman, R. L., TH1 and TH2 cells: different patterns of lymphokine secretion lead to different functional properties. *Annu Rev Immunol* **1989**. *7:* 145–173.

40 Gumy, A., Louis, J. A., Launois, P., The murine model of infection with Leishmania major and its importance for the deciphering of mechanisms under-lying differences in Th cell differentia-tion in mice from different genetic backgrounds. *Int J Parasitol* **2004**. *34:* 433–444.

41 Ebner, C., Schenk, S., Najafian, N., Siemann, U., Steiner, R., Fischer, G. W., Hoffmann, K., Szepfalusi, Z., Scheiner, O., Kraft, D., Nonallergic individuals recognize the same T cell epitopes of Bet v 1, the major birch pollen allergen, as atopic patients. *J Immunol* **1995**. *154:* 1932–1940.

42 Akdis, M., Verhagen, J., Taylor, A., Karamloo, F., Karagiannidis, C., Crameri, R., Thunberg, S., Deniz, G., Valenta, R., Fiebig, H., Kegel, C., Disch, R., Schmidt-Weber, C. B., Blaser, K., Akdis, C. A., Immune responses in healthy and allergic individuals are characterized by a fine balance between allergen-specific T regulatory 1 and T helper 2 cells. *J Exp Med* **2004**. *199:* 1567–1575.

43 Romagnani, S., Immunologic influences on allergy and the TH1/TH2 balance.

J Allergy Clin Immunol **2004.** *113:* 395–400.

44 Hori, S., Takahashi, T., Sakaguchi, S., Control of autoimmunity by naturally arising regulatory CD4+ T cells. *Adv Immunol* **2003.** *81:* 331–371.

45 Piccirillo, C. A., Shevach, E. M., Naturally-occurring CD4+CD25+ immunoregulatory T cells: central players in the arena of peripheral tolerance. *Semin Immunol* **2004.** *16:* 81–88.

46 Roncarolo, M. G., Bacchetta, R., Bordignon, C., Narula, S., Levings, M. K., Type 1 T regulatory cells. *Immunol Rev* **2001.** *182:* 68–79.

47 Weiner, H. L., Induction and mechanism of action of transforming growth factor-beta-secreting Th3 regulatory cells. *Immunol Rev* **2001.** *182:* 207–214.

48 Zhang, Z. X., Young, K., Zhang, L., CD3+CD4–CD8– alphabeta-TCR+ T cell as immune regulatory cell. *J Mol Med* **2001.** *79:* 419–427.

49 Fischer, K., Voelkl, S., Heymann, J., Przybylski, G. K., Mondal, K., Laumer, M., Kunz-Schughart, L., Schmidt, C. A., Andreesen, R., Mackensen, A., Isolation and characterization of human antigen-specific TCR{alpha}{beta}+ CD4–CD8– double negative regulatory T cells. *Blood* **2005.** *105:* 2828–2835.

50 Filaci, G., Suciu-Foca, N., CD8+ T suppressor cells are back to the game: are they players in autoimmunity? *Autoimmun Rev* **2002.** *1:* 279–283.

51 Salomon, B., Lenschow, D. J., Rhee, L., Ashourian, N., Singh, B., Sharpe, A., Bluestone, J. A., B7/CD28 costimulation is essential for the homeostasis of the CD4+CD25+ immunoregulatory T cells that control autoimmune diabetes. *Immunity* **2000.** *12:* 431–440.

52 Akbari, O., DeKruyff, R. H., Umetsu, D. T., Pulmonary dendritic cells producing IL-10 mediate tolerance induced by respiratory exposure to antigen. *Nat Immunol* **2001.** *2:* 725–731.

53 Yamazaki, S., Iyoda, T., Tarbell, K., Olson, K., Velinzon, K., Inaba, K., Steinman, R. M., Direct expansion of functional CD25+ CD4+ regulatory T cells by antigen-processing dendritic cells. *J Exp Med* **2003.** *198:* 235–247.

54 Oldenhove, G., de Heusch, M., Urbain-Vansanten, G., Urbain, J., Maliszewski, C., Leo, O., Moser, M., CD4+ CD25+ regulatory T cells control T helper cell type 1 responses to foreign antigens induced by mature dendritic cells in vivo. *J Exp Med* **2003.** *198:* 259–266.

55 Jonuleit, H., Schmitt, E., Schuler, G., Knop, J., Enk, A. H., Induction of interleukin 10-producing, nonproliferating CD4(+) T cells with regulatory properties by repetitive stimulation with allogeneic immature human dendritic cells. *J. Exp. Med.* **2000.** *192:* 1213–1222.

56 Miller, C., Ragheb, J. A., Schwartz, R. H., Anergy and cytokine-mediated suppression as distinct superantigen-induced tolerance mechanisms in vivo. *J. Exp. Med.* **1999.** *190:* 53–64.

57 Rammensee, H. G., Kroschewski, R., Frangoulis, B., Clonal anergy induced in mature V beta 6+ T lymphocytes on immunizing Mls-1b mice with Mls-1a expressing cells. *Nature* **1989.** *339:* 541–544.

58 Kawabe, Y., Ochi, A., Selective anergy of V beta 8+,CD4+ T cells in Staphylococcus enterotoxin B-primed mice. *J. Exp. Med.* **1990.** *172:* 1065–1070.

59 Lombardi, G., Sidhu, S., Batchelor, R., Lechler, R., Anergic T cells as suppressor cells in vitro. *Science* **1994.** *264:* 1587–1589.

60 Webb, S., Morris, C., Sprent, J., Extrathymic tolerance of mature T cells: clonal elimination as a consequence of immunity. *Cell* **1990.** *63:* 1249–1256.

61 Kearney, E. R., Pape, K. A., Loh, D. Y., Jenkins, M. K., Visualization of peptide-specific T cell immunity and peripheral tolerance induction in vivo. *Immunity* **1994.** *1:* 327–339.

62 Pape, K. A., Merica, R., Mondino, A., Khoruts, A., Jenkins, M. K., Direct evidence that functionally impaired CD4+ T cells persist in vivo following induction of peripheral tolerance. *J Immunol* **1998.** *160:* 4719–4729.

63 Reinhardt, R. L., Khoruts, A., Merica, R., Zell, T., Jenkins, M. K., Visualizing the generation of memory CD4 T cells in the whole body. *Nature* **2001.** *410:* 101–105.

64 Zinkernagel, R. M., Localization dose and time of antigens determine immune

reactivity. *Semin. Immunol.* **2000.** *12:* 163–171; discussion 257–344.

65 Gett, A. V., Sallusto, F., Lanzavecchia, A., Geginat, J., T cell fitness determined by signal strength. *Nat Immunol* **2003.** *4:* 355–360.

66 Mempel, T. R., Henrickson, S. E., Von Andrian, U. H., T-cell priming by dendritic cells in lymph nodes occurs in three distinct phases. *Nature* **2004.** *427:* 154–159.

67 Kyewski, B. A., Fathman, C. G., Rouse, R. V., Intrathymic presentation of circulating non-MHC antigens by medullary dendritic cells. An antigen-dependent microenvironment for T cell differentiation. *J Exp Med* **1986.** *163:* 231–246.

68 Muller, K. P., Schumacher, J., Kyewski, B. A., Half-life of antigen/major histo-compatibility complex class II complexes in vivo: intra- and interorgan variations. *Eur J Immunol* **1993.** *23:* 3203–3207.

69 Ardavin, C., Shortman, K., Cell surface marker analysis of mouse thymic dendritic cells. *Eur J Immunol* **1992.** *22:* 859–862.

70 Wu, L., Vremec, D., Ardavin, C., Winkel, K., Suss, G., Georgiou, H., Maraskovsky, E., Cook, W., Shortman, K., Mouse thymus dendritic cells: kinetics of development and changes in surface markers during maturation. *Eur J Immunol* **1995.** *25:* 418–425.

71 Matzinger, P., Guerder, S., Does T-cell tolerance require a dedicated antigen-presenting cell? *Nature* **1989.** *338:* 74–76.

72 Inaba, M., Inaba, K., Hosono, M., Kumamoto, T., Ishida, T., Muramatsu, S., Masuda, T., Ikehara, S., Distinct mechanisms of neonatal tolerance induced by dendritic cells and thymic B cells. *J Exp Med* **1991.** *173:* 549–559.

73 Ruedl, C., Koebel, P., Bachmann, M., Hess, M., Karjalainen, K., Anatomical origin of dendritic cells determines their life span in peripheral lymph nodes. *J. Immunol* **2000.** *165:* 4910–4916.

74 Inaba, K., Pack, M., Inaba, M., Sakuta, H., Isdell, F., Steinman, R. M., High levels of a major histocompatibility complex II-self peptide complex on dendritic cells from the T cell areas of lymph nodes. *J Exp Med* **1997.** *186:* 665–672.

75 Fisson, S., Darrasse-Jeze, G., Litvinova, E., Septier, F., Klatzmann, D., Liblau, R., Salomon, B. L., Continuous activation of autoreactive CD4+ CD25+ regulatory T cells in the steady state. *J Exp Med* **2003.** *198:* 737–746.

76 Probst, H. C., Lagnel, J., Kollias, G., van den Broek, M., Inducible transgenic mice reveal resting dendritic cells as potent inducers of CD8+ T cell tolerance. *Immunity* **2003.** *18:* 713–720.

77 Iyoda, T., Shimoyama, S., Liu, K., Omatsu, Y., Akiyama, Y., Maeda, Y., Takahara, K., Steinman, R. M., Inaba, K., The CD8+ dendritic cell subset selec-tively endocytoses dying cells in culture and in vivo. *J Exp Med* **2002.** *195:* 1289–1302.

78 Ferguson, T. A., Herndon, J., Elzey, B., Griffith, T. S., Schoenberger, S., Green, D. R., Uptake of apoptotic antigen-coupled cells by lymphoid dendritic cells and cross-priming of CD8(+) T cells produce active immune unrespon-siveness. *J Immunol* **2002.** *168:* 5589–5595.

79 Hawiger, D., Inaba, K., Dorsett, Y., Guo, M., Mahnke, K., Rivera, M., Ravetch, J. V., Steinman, R. M., Nussenzweig, M. C., Dendritic cells induce peripheral t cell unresponsiveness under steady state conditions in vivo. *J Exp Med* **2001.** *194:* 769–780.

80 Mahnke, K., Qian, Y., Knop, J., Enk, A. H., Induction of CD4+/CD25+ regulatory T cells by targeting of antigens to immature dendritic cells. *Blood* **2003.** *101:* 4862–4869.

81 Ohl, L., Mohaupt, M., Czeloth, N., Hintzen, G., Kiafard, Z., Zwirner, J., Blankenstein, T., Henning, G., Forster, R., CCR7 governs skin dendritic cell migration under inflammatory and steady-state conditions. *Immunity* **2004.** *21:* 279–288.

82 Mishima, Y., Melanosomes in phagocytic vacuoles in Langerhans cells. *J. Cell. Biol.* **1966.** *30:* 417–423.

83 Hemmi, H., Yoshino, M., Yamazaki, H., Naito, M., Iyoda, T., Omatsu, Y., Shimoyama, S., Letterio, J. J.,

Nakabayashi, T., Tagaya, H., Yamane, T., Ogawa, M., Nishikawa, S., Ryoke, K., Inaba, K., Hayashi, S., Kunisada, T., Skin antigens in the steady state are trafficked to regional lymph nodes by transforming growth factor-beta1-dependent cells. *Int Immunol* 2001. *13:* 695–704.

84 Huang, F. P., Platt, N., Wykes, M., Major, J. R., Powell, T. J., Jenkins, C. D., MacPherson, G. G., A discrete sub-population of dendritic cells transports apoptotic intestinal epithelial cells to T cell areas of mesenteric lymph nodes. *J. Exp. Med.* 2000. *191:* 435–444.

85 McGuirk, P., McCann, C., Mills, K. H., Pathogen-specific T Regulatory 1 Cells Induced in the Respiratory Tract by a Bacterial Molecule that Stimulates Interleukin 10 Production by Dendritic Cells: A Novel Strategy for Evasion of Protective T Helper Type 1 Responses by Bordetella pertussis. *J Exp Med* 2002. *195:* 221–231.

86 Scheinecker, C., McHugh, R., Shevach, E. M., Germain, R. N., Constitutive presentation of a natural tissue auto-antigen exclusively by dendritic cells in the draining lymph node. *J Exp Med* 2002. *196:* 1079–1090.

87 Belz, G. T., Behrens, G. M., Smith, C. M., Miller, J. F., Jones, C., Lejon, K., Fathman, C. G., Mueller, S. N., Shortman, K., Carbone, F. R., Heath, W. R., The CD8alpha(+) dendritic cell is responsible for inducing peripheral self-tolerance to tissue-associated antigens. *J Exp Med* 2002. *196:* 1099–1104.

88 Rissoan, M.-C., Soumelis, V., Kadowaki, N., Grouard, G., Briere, F., de Waal Malefyt, R., Liu, Y.-J., Reciprocal control of T helper cell and dendritic cell differentiation. *Science* 1999. *283:* 1183–1186.

89 Liu, Y. J., Kadowaki, N., Rissoan, M. C., Soumelis, V., T cell activation and polarization by DC1 and DC2. *Curr Top Microbiol Immunol* 2000. *251:* 149–159.

90 De Heer, H. J., Hammad, H., Soullie, T., Hijdra, D., Vos, N., Willart, M. A., Hoogsteden, H. C., Lambrecht, B. N., Essential role of lung plasmacytoid dendritic cells in preventing asthmatic reactions to harmless inhaled antigen. *J Exp Med* 2004. *200:* 89–98.

91 Martin, P., Del Hoyo, G. M., Anjuere, F., Arias, C. F., Vargas, H. H., Fernandez, L. A., Parrillas, V., Ardavin, C., Character-ization of a new subpopulation of mouse CD8alpha+ B220+ dendritic cells endowed with type 1 interferon production capacity and tolerogenic potential. *Blood* 2002. *100:* 383–390.

92 Kuwana, M., Kaburaki, J., Wright, T. M., Kawakami, Y., Ikeda, Y., Induction of antigen-specific human CD4(+) T cell anergy by peripheral blood DC2 precursors. *Eur J Immunol* 2001. *31:* 2547–2557.

93 Gilliet, M., Liu, Y. J., Generation of human CD8 T regulatory cells by CD40 ligand-activated plasmacytoid dendritic cells. *J Exp Med* 2002. *195:* 695–704.

94 Wakkach, A., Fournier, N., Brun, V., Breittmayer, J. P., Cottrez, F., Groux, H., Characterization of dendritic cells that induce tolerance and T regulatory 1 cell differentiation in vivo. *Immunity* 2003. *18:* 605–617.

95 Zhang, M., Tang, H., Guo, Z., An, H., Zhu, X., Song, W., Guo, J., Huang, X., Chen, T., Wang, J., Cao, X., Splenic stroma drives mature dendritic cells to differentiate into regulatory dendritic cells. *Nat Immunol* 2004. *5:* 1124–1133.

96 Fallarino, F., Asselin-Paturel, C., Vacca, C., Bianchi, R., Gizzi, S., Fioretti, M. C., Trinchieri, G., Grohmann, U., Puccetti, P., Murine plasmacytoid dendritic cells initiate the immunosuppressive pathway of tryptophan catabolism in response to CD200 receptor engagement. *J Immunol* 2004. *173:* 3748–3754.

97 Lutz, M. B., Kukutsch, N., Ogilvie, A. L., Rößner, S., Koch, F., Romani, N., Schuler, G., An advanced culture method for generating large quantities of highly pure dendritic cells from mouse bone marrow. *J Immunol Methods* 1999. *223:* 77–92.

98 Rößner, S., Wiethe, C., Hänig, J., Seifarth, C., Lutz, M. B., Myeloid dendrit-ic cell precursors generated from bone marrow suppress T cell responses via cell contact and nitric oxide production in vitro. *Eur J Immunol* 2005: (in press).

99 Mencacci, A., Montagnoli, C., Bacci, A., Cenci, E., Pitzurra, L., Spreca, A., Kopf, M., Sharpe, A. H., Romani, L.,

CD80+Gr-1+ myeloid cells inhibit development of antifungal Th1 immunity in mice with candidiasis. *J Immunol* **2002**. *169:* 3180–3190.

100 Bronte, V., Wang, M., Overwijk, W. W., Surman, D. R., Pericle, F., Rosenberg, S. A., Restifo, N. P., Apoptotic death of CD8+ T lymphocytes after immunization: induction of a suppressive population of Mac-1+/Gr-1+ cells. *J. Immunol* **1998**. *161:* 5313–5320.

101 Bronte, V., Chappell, D. B., Apolloni, E., Cabrelle, A., Wang, M., Hwu, P., Restifo, N. P., Unopposed production of granulocyte-macrophage colony-stimulating factor by tumors inhibits CD8+ T cell responses by dysregulating antigen-presenting cell maturation. *J. Immunol* **1999**. *162:* 5728–5737.

102 Bronte, V., Apolloni, E., Cabrelle, A., Ronca, R., Serafini, P., Zamboni, P., Restifo, N. P., Zanovello, P., Identification of a CD11b(+)/Gr-1(+)/CD31(+) myeloid progenitor capable of activating or suppressing CD8(+) T cells. *Blood* **2000**. *96:* 3838–3846.

103 Young, M. R., Wright, M. A., Lozano, Y., Prechel, M. M., Benefield, J., Leonetti, J. P., Collins, S. L., Petruzzelli, G. J., Increased recurrence and metastasis in patients whose primary head and neck squamous cell carcinomas secreted granulocyte-macrophage colony-stimulating factor and contained CD34+ natural suppressor cells. *Int J Cancer* **1997**. *74:* 69–74.

104 Almand, B., Resser, J. R., Lindman, B., Nadaf, S., Clark, J. I., Kwon, E. D., Carbone, D. P., Gabrilovich, D. I., Clinical significance of defective dendritic cell differentiation in cancer. *Clin Cancer Res* **2000**. *6:* 1755–1766.

105 Almand, B., Clark, J. I., Nikitina, E., van Beynen, J., English, N. R., Knight, S. C., Carbone, D. P., Gabrilovich, D. I., Increased production of immature myeloid cells in cancer patients: a mechanism of immunosuppression in cancer. *J Immunol* **2001**. *166:* 678–689.

106 Lathers, D. M., Achille, N., Kolesiak, K., Hulett, K., Sparano, A., Petruzzelli, G. J., Young, M. R., Increased levels of immune inhibitory CD34+ progenitor cells in the peripheral blood of patients with node positive head and neck squamous cell carcinomas and the ability of these CD34+ cells to differentiate into immune stimulatory dendritic cells. *Otolaryngol Head Neck Surg* **2001**. *125:* 205–212.

107 Romani, N., Holzmann, S., Tripp, C. H., Koch, F., Stoitzner, P., Langerhans cells – dendritic cells of the epidermis. *Apmis* **2003**. *111:* 725–740.

108 Wilson, N. S., El-Sukkari, D., Belz, G. T., Smith, C. M., Steptoe, R. J., Heath, W. R., Shortman, K., Villadangos, J. A., Most lymphoid organ dendritic cell types are phenotypically and functionally immature. *Blood* **2003**. *102:* 2187–2194.

109 Austyn, J. M., Dendritic cells. *Cur Opin Hematol* **1998**. *5:* 3–15.

110 Sallusto, F., Cella, M., Danieli, C., Lanzavecchia, A., Dendritic cells use macropinocytosis and the mannose receptor to concentrate macromolecules in the major histocompatibility complex class II compartment: downregulation by cytokines and bacterial products. *J Exp Med* **1995**. *182:* 389–400.

111 Mellman, I., Steinman, R. M., Dendritic cells: specialized and regulated antigen processing machines. *Cell* **2001**. *106:* 255–258.

112 Sallusto, F., Lanzavecchia, A., Efficient presentation of soluble antigen by cultured human dendritic cells is maintained by granulocyte/macrophage colony-stimulating factor plus interleukin 4 and downregulated by tumor necrosis factor alpha. *J Exp Med* **1994**. *179:* 1109–1118.

113 Inaba, K., Turley, S., Yamaide, F., Iyoda, T., Mahnke, K., Inaba, M., Pack, M., Subklewe, M., Sauter, B., Sheff, D., Albert, M., Bhardwaj, N., Mellman, I., Steinman, R. M., Efficient presentation of phagocytosed cellular fragments on the major histocompatibility complex class II products of dendritic cells. *J. Exp. Med.* **1998**. *188:* 2163–2173.

114 Inaba, K., Turley, S., Iyoda, T., Yamaide, F., Shimoyama, S., Reis e Sousa, C., Germain, R. N., Mellman, I., Steinman, R. M., The formation of immunogenic major histocompatibility complex II-peptide ligands in lysosomal compartments of dendritic cells is regulated by

inflammatory stimuli. *J Exp Med* **2000**. *191:* 927–936.

115 Inaba, K., Witmer, P. M., Inaba, M., Hathcock, K. S., Sakuta, H., Azuma, M., Yagita, H., Okumura, K., Linsley, P. S., Ikehara, S., et, a. l., The tissue distribution of the B7-2 costimulator in mice: abundant expression on dendritic cells in situ and during maturation in vitro. *J. Exp. Med.* **1994**. *180:* 1849–1860.

116 Romani, N., Kampgen, E., Koch, F., Heufler, C., Schuler, G., Dendritic cell production of cytokines and responses to cytokines. *Int Rev Immunol* **1990**. *6:* 151–161.

117 Lohr, J., Knoechel, B., Kahn, E. C., Abbas, A. K., Role of B7 in T cell tolerance. *J Immunol* **2004**. 173: 5028–5035.

118 Schuler, G., Steinman, R. M., Murine epidermal Langerhans cells mature into potent immunostimulatory dendritic cells in vitro. *J Exp Med* **1985**. *161:* 526–546.

119 Inaba, K., Inaba, M., Romani, N., Aya, H., Deguchi, M., Ikehara, S., Muramatsu, S., Steinman, R. M., Generation of large numbers of dendritic cells from mouse bone marrow cultures supplemented with granulocyte/macrophage colony-stimulating factor. *J Exp Med* **1992**. *176:* 1693–1702.

120 Yamaguchi, Y., Tsumura, H., Miwa, M., Inaba, K., Contrasting effects of TGF-beta 1 and TNF-alpha on the development of dendritic cells from progenitors in mouse bone marrow. *Stem Cells* **1997**. *15:* 144–153.

121 Fu, F., Li, Y., Qian, S., Lu, L., Chambers, F., Starzl, T. E., Fung, J. J., Thomson, A. W., Costimulatory molecule-deficient dendritic cell progenitors (MHC class II+, CD80dim, CD86–) prolong cardiac allograft survival in nonimmunosuppressed recipients. *Transplantation* **1996**. *62:* 659–665.

122 Lutz, M. B., Suri, R. M., Niimi, M., Ogilvie, A. L. J., Kukutsch, N. A., Rößner, S., Schuler, G., Austyn, J. M., Immature dendritic cells generated with low doses of GM-CSF in the absence of IL-4 are maturation-resistant and prolong allograft survival in vivo. *Eur J Immunol* **2000**. *30:* 1813–1822.

123 Griffin, M. D., Lutz, W., Phan, V. A., Bachman, L. A., McKean, D. J., Kumar, R., Dendritic cell modulation by 1alpha,25 dihydroxyvitamin D3 and its analogs: a vitamin D receptor-dependent pathway that promotes a persistent state of immaturity in vitro and in vivo. *Proc Natl Acad Sci USA* **2001**. *98:* 6800–6805.

124 Adorini, L., Giarratana, N., Penna, G., Pharmacological induction of tolerogenic dendritic cells and regulatory T cells. *Semin Immunol* **2004**. *16:* 127–134.

125 Sato, K., Yamashita, N., Baba, M., Matsuyama, T., Regulatory dendritic cells protect mice from murine acute graft-versus-host disease and leukemia relapse. *Immunity* **2003**. *18:* 367–379.

126 Sato, K., Yamashita, N., Baba, M., Matsuyama, T., Modified myeloid dendritic cells act as regulatory dendritic cells to induce anergic and regulatory T cells. *Blood* **2003**. *101:* 3581–3589.

127 Woltman, A. M., de Fijter, J. W., Kamerling, S. W., Paul, L. C., Daha, M. R., van Kooten, C., The effect of calcineurin inhibitors and corticosteroids on the differentiation of human dendritic cells. *Eur J Immunol* **2000**. *30:* 1807–1812.

128 Menges, M., Rößner, S., Voigtländer, C., Schindler, H., Kukutsch, N. A., Bogdan, C., Erb, K., Schuler, G., Lutz, M. B., Repetitive injections of dendritic cells matured with tumor necrosis factor-α induce antigen-specific protection of mice from autoimmunity. *J Exp Med* **2002**. *195:* 15–21.

129 Sporri, R., Reis e Sousa, C., Inflammatory mediators are insufficient for full dendritic cell activation and promote expansion of CD4+ T cell populations lacking helper function. *Nat Immunol* **2005**. *6:* 163–170.

130 Voigtländer, C., Rößner, S., Cierpka, C., Theiner, G., Hänig, J., Wiethe, C., Menges, M., Schuler, G., Lutz, M. B., Dendritic cells matured with TNF can be further activated in vitro and in vivo thereby converting their tolerogenicity into immunogenicity **2005**: (submitted).

131 van Duivenvoorde, L. M., Louis-Plence, P., Apparailly, F., van der Voort, E. I., Huizinga, T. W., Jorgensen, C., Toes, R.

E., Antigen-specific immunomodulation of collagen-induced arthritis with tumor necrosis factor-stimulated dendritic cells. *Arthritis Rheum* **2004**. *50:* 3354–3364.

132 Hammad, H., Lambrecht, B. N., Pochard, P., Gosset, P., Marquillies, P., Tonnel, A. B., Pestel, J., Monocyte-derived dendritic cells induce a house dust mite-specific Th2 allergic inflammation in the lung of humanized SCID mice: involvement of CCR7. *J Immunol* **2002**. *169:* 1524–1534.

133 Lambrecht, B. N., De Veerman, M., Coyle, A. J., Gutierrez-Ramos, J. C., Thielemans, K., Pauwels, R. A., Myeloid dendritic cells induce Th2 responses to inhaled antigen, leading to eosinophilic airway inflammation. *J Clin Invest* **2000**. *106:* 551–559.

134 Eriksson, U., Ricci, R., Hunziker, L., Kurrer, M. O., Oudit, G. Y., Watts, T. H., Sonderegger, I., Bachmaier, K., Kopf, M., Penninger, J. M., Dendritic cell-induced autoimmune heart failure requires cooperation between adaptive and innate immunity. *Nat Med* **2003**. *9:* 1484–1490.

135 Kuipers, H., Hijdra, D., De Vries, V. C., Hammad, H., Prins, J. B., Coyle, A. J., Hoogsteden, H. C., Lambrecht, B. N., Lipopolysaccharide-induced suppression of airway Th2 responses does not require IL-12 production by dendritic cells. *J Immunol* **2003**. *171:* 3645–3654.

136 Wiethe, C., Schuler, G., Kopf, M., Lutz, M. B., Tolerance induction by semi-mature dendritic cells requires cooperation of antigen-specific CD4+ T cells with CD1d-restricted NKT cells **2005**: (submitted).

137 Kleindienst, P., Wiethe, C., Lutz, M. B., Brocker, T., Simultaneous induction of CD4 T cell tolerance and CD8 T cell immunity by semimature dendritic cells. *J Immunol* **2005**. *174:* 3941–3947.

138 Kanazawa, N., Azukizawa, H., Lutz, M. B. et al., Steady state migratory DC are semi-mature and induce CD4+ T cell tolerance in the draining lymph node **2005**: (in preparation).

139 Steinman, R. M., Turley, S., Mellman, I., Inaba, K., The induction of tolerance by dendritic cells that have captured apoptotic cells. *J Exp Med* **2000**. *191:* 411–416.

140 Verbovetski, I., Bychkov, H., Trahtemberg, U., Shapira, I., Hareuveni, M., Ben-Tal, O., Kutikov, I., Gill, O., Mevorach, D., Opsonization of apoptotic cells by autologous iC3b facilitates clearance by immature dendritic cells, down-regulates DR and CD86, and up-regulates CC chemokine receptor 7. *J Exp Med* **2002**. *196:* 1553–1561.

141 Bouchon, A., Hernandez-Munain, C., Cella, M., Colonna, M., A DAP12-mediated pathway regulates expression of CC chemokine receptor 7 and maturation of human dendritic cells. *J Exp Med* **2001**. *194:* 1111–1122.

142 Tarbell, K. V., Yamazaki, S., Olson, K., Toy, P., Steinman, R. M., CD25+ CD4+ T cells, expanded with dendritic cells presenting a single autoantigenic peptide, suppress autoimmune diabetes. *J Exp Med* **2004**. *199:* 1467–1477.

143 Luger, T. A., Kalden, D., Scholzen, T. E., Brzoska, T., alpha-melanocyte-stimulating hormone as a mediator of tolerance induction. *Pathobiology* **1999**. 67: 318–321.

144 Matasic, R., Dietz, A. B., Vuk-Pavlovic, S., Cyclooxygenase-independent inhibition of dendritic cell maturation by aspirin. *Immunology* **2000**. *101:* 53–60.

145 Hackstein, H., Morelli, A. E., Larregina, A. T., Ganster, R. W., Papworth, G. D., Logar, A. J., Watkins, S. C., Falo, L. D., Thomson, A. W., Aspirin inhibits in vitro maturation and in vivo immunostimulatory function of murine myeloid dendritic cells. *J Immunol* **2001**. *166:* 7053–7062.

146 Demeure, C. E., Tanaka, H., Mateo, V., Rubio, M., Delespesse, G., Sarfati, M., CD47 engagement inhibits cytokine production and maturation of human dendritic cells. *J Immunol* **2000**. *164:* 2193–2199.

147 Avice, M. N., Rubio, M., Sergerie, M., Delespesse, G., Sarfati, M., Role of cd47 in the induction of human naive t cell anergy. *J Immunol* **2001**. *167:* 2459–2468.

148 Latour, S., Tanaka, H., Demeure, C., Mateo, V., Rubio, M., Brown, E. J., Maliszewski, C., Lindberg, F. P., Oldenborg, A., Ullrich, A., Delespesse, G., Sarfati, M., Bidirectional negative regulation of human T and dendritic cells by CD47 and its cognate receptor signal-regulator protein-alpha: down-regulation of IL-12 responsiveness and inhibition of dendritic cell activation. *J Immunol* **2001**. *167*: 2547–2554.

149 Moser, M., De Smedt, T., Sornasse, T., Tielemans, F., Chentoufi, A. A., Muraille, E., Van Mechelen, M., Urbain, J., Leo, O., Glucocorticoids down-regulate dendritic cell function in vitro and in vivo. *Eur. J. Immunol.* **1995**. *25*: 2818–2824.

150 Vieira, P. L., Kalinski, P., Wierenga, E. A., Kapsenberg, M. L., de, J. E., Glucocorticoids inhibit bioactive IL-12p70 production by in vitro- generated human dendritic cells without affecting their T cell stimulatory potential [In Process Citation]. *J. Immunol.* **1998**. *161*: 5245–5251.

151 Singh, S., Aiba, S., Manome, H., Tagami, H., The effects of dexamethasone, cyclosporine, and vitamin D(3) on the activation of dendritic cells stimulated by haptens. *Arch Dermatol Res* **1999**. *291*: 548–554.

152 de Jong, E. C., Vieira, P. L., Kalinski, P., Kapsenberg, M. L., Corticosteroids inhibit the production of inflammatory mediators in immature monocyte-derived DC and induce the development of tolerogenic DC3. *J Leukoc Biol* **1999**. *66*: 201–204.

153 Matasic, R., Dietz, A. B., Vuk-Pavlovic, S., Dexamethasone inhibits dendritic cell maturation by redirecting differentiation of a subset of cells. *J Leukoc Biol* **1999**. *66*: 909–914.

154 Vanderheyde, N., Verhasselt, V., Goldman, M., Willems, F., Inhibition of human dendritic cell functions by methylprednisolone. *Transplantation* **1999**. *67*: 1342–1347 (http://www.biomednet.com/db/medline/99287333).

155 Rea, D., van Kooten, C., van Meijgaarden, K. E., Ottenhoff, T. H., Melief, C. J., Offringa, R., Gluco-corticoids transform CD40-triggering of dendritic cells into an alternative activation pathway resulting in antigen-presenting cells that secrete IL-10. *Blood* **2000**. *95*: 3162–3167.

156 Knight, S. C., Balfour, B., O'Brien, J., Buttifant, L., Sensitivity of veiled (dendritic) cells to cyclosporine. *Transplantation* **1986**. *41*: 96–100.

157 Knight, S. C., Bedford, P. A., Effect of cyclosporine and retinoic acid on dendritic cell function. *Transplant Proc* **1987**. *19*: 320.

158 Knight, S. C., Roberts, M., Macatonia, S. E., Edwards, A. J., Blocking of acquisition and presentation of antigen by dendritic cells with cyclosporine. Studies with fluorescein isothiocyanate. *Transplantation* **1988**. *46*: 48S-53S.

159 Horrocks, C., Duncan, J. I., Sewell, H. F., Ormerod, A. D., Thomson, A. W., Differential effects of cyclosporine A on Langerhans cells and regulatory T-cell populations in severe psoriasis: an immunohistochemical and flow cytometric analysis. *J Autoimmun* **1990**. *3*: 559–570.

160 Roberts, M. S., Knight, S. C., Low-dose immunosuppression by cyclosporine operating via antigen-presenting dendritic cells. *Transplantation* **1990**. *50*: 91–95.

161 Lee, J., Ganster, R., Geller, D., Burckart, G., Thomson, A., Lu, L., Cyclosporine A inhibits the expression of costimulatory molecules on in vitro-generated dendritic cells: association with reduced nuclear translocation of nuclear factor kappa B. *Transplantation* **1999**. *68*: 1255–1263.

162 Lutz, M. B., Kukutsch, N. A., Menges, M., Rößner, S., Schuler, G., Culture of bone marrow cells in GM-CSF plus high doses of lipopolysaccharide generates exclusively immature dendritic cells which induce alloantigen-specific CD4 T cell anergy in vitro. *Eur J Immunol* **2000**. *30*: 1048–1052.

163 Riedl, E., Stockl, J., Majdic, O., Scheinecker, C., Knapp, W., Strobl, H., Ligation of E-cadherin on in vitro-generated immature Langerhans-type dendritic cells inhibits their maturation. *Blood* **2000**. *96*: 4276–4284.

164 Zhu, K., Mrowietz, U., Inhibition of dendritic cell differentiation by fumaric acid esters. *J Invest Dermatol* **2001**. *116*: 203–208.

165 Xie, Y., Li, Y., Zhang, Q., Stiller, M. J., Wang, C. L., Streilein, J. W., Haptoglobin is a natural regulator of Langerhans cell function in the skin. *J Dermatol Sci* **2000**. *24*: 25–37.

166 Macatonia, S. E., Doherty, T. M., Knight, S. C., O'Garra, A., Differential effect of IL-10 on dendritic cell-induced T cell proliferation and IFN-gamma production. *J Immunol* **1993**. *150:* 3755–3765.

167 Enk, A. H., Angeloni, V. L., Udey, M. C., Katz, S. I., Inhibition of Langerhans cell antigen-presenting function by IL-10. A role for IL-10 in induction of tolerance. *J Immunol* **1993**. *151:* 2390–2398.

168 Caux, C., Massacrier, C., Vanbervliet, B., Barthelemy, C., Liu, Y. J., Bancherau, J., Interleukin 10 inhibits T cell alloreaction induced by human dendritic cells. *Int Immunol* **1994**. *6:* 1177–1185.

169 Peguet-Navarro, J., Moulon, C., Caux, C., Dalbiez, G. C., Bancherau, J., Schmitt, D., Interleukin-10 inhibits the primary allogeneic T cell response to human epidermal Langerhans cells. *Eur J Immunol* **1994**. *24:* 884–891.

170 Buelens, C., Verhasselt, V., De Groote, D., Thielemans, K., Goldman, M., Willems, F., Human dendritic cell responses to lipopolysaccharide and CD40 ligation are differentially regulated by interleukin-10. *Eur J Immunol* **1997**. *27:* 1848–1852.

171 Buelens, C., Verhasselt, V., De Groote, D., Thielemans, K., Goldman, M., Willems, F., Interleukin-10 prevents the generation of dendritic cells from human peripheral blood mononuclear cells cultured with interleukin-4 and granulocyte/macrophage-colony-stimulating factor. *Eur J Immunol* **1997**. *27:* 756–762.

172 Sato, K., Yamashita, N., Matsuyama, T., Human peripheral blood monocyte-derived interleukin-10-induced semi-mature dendritic cells induce anergic CD4(+) and CD8(+) T cells via presentation of the internalized soluble antigen and cross-presentation of the phagocytosed necrotic cellular fragments. *Cell Immunol* **2002**. *215:* 186–194.

173 Steinbrink, K., Graulich, E., Kubsch, S., Knop, J., Enk, A. H., CD4(+) and CD8(+) anergic T cells induced by interleukin-10-treated human dendritic cells display antigen-specific suppressor activity. *Blood* **2002**. *99:* 2468–2476.

174 Schempp, C. M., Dittmar, H. C., Hummler, D., Simon-Haarhaus, B., Schulte-Monting, J., Schopf, E., Simon, J. C., Magnesium ions inhibit the antigen-presenting function of human epidermal Langerhans cells in vivo and in vitro. Involvement of ATPase, HLA-DR, B7 molecules, and cytokines. *J Invest Dermatol* **2000**. *115:* 680–686.

175 Urban, B. C., Ferguson, D. J., Pain, A., Willcox, N., Plebanski, M., Austyn, J. M., Roberts, D. J., Plasmodium falciparum-infected erythrocytes modulate the maturation of dendritic cells. *Nature* **1999**. *400:* 73–77.

176 Geijtenbeek, T. B., Van Vliet, S. J., Koppel, E. A., Sanchez-Hernandez, M., Vandenbroucke-Grauls, C. M., Appelmelk, B., Van Kooyk, Y., Mycobacteria target DC-SIGN to suppress dendritic cell function. *J Exp Med* **2003**. *197:* 7–17.

177 Mehling, A., Grabbe, S., Voskort, M., Schwarz, T., Luger, T. A., Beissert, S., Mycophenolate mofetil impairs the maturation and function of murine dendritic cells. *J Immunol* **2000**. *165:* 2374–2381.

178 Verhasselt, V., Vanden Berghe, W., Vanderheyde, N., Willems, F., Haegeman, G., Goldman, M., N-acetyl-L-cysteine inhibits primary human T cell responses at the dendritic cell level: association with NF-kappaB inhibition. *J Immunol* **1999**. *162:* 2569–2574.

179 Hackstein, H., Taner, T., Zahorchak, A. F., Morelli, A. E., Logar, A. J., Gessner, A., Thomson, A. W., Rapamycin inhibits IL-4—induced dendritic cell maturation in vitro and dendritic cell mobilization and function in vivo. *Blood* **2003**. *101:* 4457–4463.

180 Steinschulte, C., Taner, T., Thomson, A. W., Bein, G., Hackstein, H., Cutting edge: sanglifehrin A, a novel cyclophilin-binding immunosuppressant blocks bioactive IL-12 production by human dendritic cells. *J Immunol* **2003**. *171:* 542–546.

181 Lechmann, M., Krooshoop, D. J., Dudziak, D., Kremmer, E., Kuhnt, C., Figdor, C. G., Schuler, G., Steinkasserer,

A., The extracellular domain of CD83 inhibits dendritic cell-mediated T cell stimulation and binds to a ligand on dendritic cells. *J Exp Med* **2001**. *194:* 1813–1821.

182 Staniek, V., Misery, L., Peguet-Navarro, J., Abello, J., Dutremepuich, J.-D., Claudy, A., Schmitt, D., Binding and in vitro modulation of human epidermal Langerhans cell functions by substance P. *Arch Derm Res* **1997**. *289:* 285–291.

183 Bonham, C. A., Lu, L., Banas, R. A., Fontes, P., Rao, A. S., Starzl, T. E., Zeevi, A., Thomson, A. W., TGF-beta 1 pre-treatment impairs the allostimulatory function of human bone marrow-derived antigen-presenting cells for both naive and primed T cells. *Transpl Immunol* **1996**. *4:* 186–191.

184 Takeuchi, M., Alard, P., Streilein, J. W., TGF-beta promotes immune deviation by altering accessory signals of antigen-presenting cells. *J Immunol* **1998**. *160:* 1589–1597.

185 Geissmann, F., Revy, P., Regnault, A., Lepelletier, Y., Dy, M., Brousse, N., Amigorena, S., Hermine, O., Durandy, A., TGF-beta 1 prevents the noncognate maturation of human dendritic Langerhans cells. *J Immunol* **1999**. *162:* 4567–4575.

186 Van Overtvelt, L., Vanderheyde, N., Verhasselt, V., Ismaili, J., De Vos, L., Goldman, M., Willems, F., Vray, B., Trypanosoma cruzi infects human dendritic cells and prevents their maturation: inhibition of cytokines, HLA-DR, and costimulatory molecules. *Infect Immun* **1999**. *67:* 4033–4040.

187 Simon, J. C., Tigelaar, R. E., Bergstresser, P. R., Edelbaum, D., Cruz, P. D., Ultraviolet B radiation converts Langerhans cells from immunogenic to tolerogenic antigen-presenting cells . Induction of specific clonal anergy in CD4 T helper cells. *J. immunol.* **1991**. *146:* 485–491.

188 Schmitt, D. A., Ullrich, S. E., Exposure to ultraviolet radiation causes dendritic Cells/Macrophages to secrete immune-suppressive IL-12p40 homodimers. *J Immunol* **2000**. *165:* 3162–3167.

189 Denfeld, R. W., Hara, H., Tesmann, J. P., Martin, S., Simon, J. C., UVB-irradiated dendritic cells are impaired in their APC function and tolerize primed Th1 cells but not naive CD4+ T cells. *J Leukoc Biol* **2001**. *69:* 548–554.

190 Gabrilovich, D. I., Chen, H. L., Girgis, K. R., Cunningham, H. T., Meny, G. M., Nadaf, S., Kavanaugh, D., Carbone, D. P., Production of vascular endothelial growth factor by human tumors inhibits the functional maturation of dendritic cells. *Nat Med* **1996**. *2:* 1096–1103.

191 Macatonia, S. E., Gompels, M., Pinching, A. J., Patterson, S., Knight, S. C., Antigen-presentation by macrophages but not by dendritic cells in human immunodeficiency virus (HIV) infection. *Immunology* **1992**. *75:* 576–581.

192 Blauvelt, A., Clerici, M., Lucey, D. R., Steinberg, S. M., Yarchoan, R., Walker, R., Shearer, G. M., Katz, S. I., Functional studies of epidermal Langerhans cells and blood monocytes in HIV-infected persons. *J Immunol* **1995**. *154:* 3506–3515.

193 Grosjean, I., Caux, C., Bella, C., Berger, I., Wild, F., Banchereau, J., Kaiserlian, D., Measles virus infects human dendritic cells and blocks their allostimulatory properties for CD4+ T cells. *J Exp Med* **1997**. *186:* 801–812.

194 Fugier-Vivier I., Servet-Delprat C., Rivailler, P., Rissoan, M. C., Liu, Y. J., C., R.-C., Measles virus suppresses cell-mediated immunity by interfering with the survival and functions of dendritic and T cells. *J Exp Med* **1997**. *186:* 813–823.

195 Salio, M., Cella, M., Suter, M., Lanzavecchia, A., Inhibition of dendritic cell maturation by herpes simplex virus. *Eur J Immunol* **1999**. *29:* 3245–3253.

196 Kruse, M., Rosorius, O., Kratzer, F., Stelz, G., Kuhnt, C., Schuler, G., Hauber, J., Steinkasserer, A., Mature dendritic cells infected with herpes simplex virus type 1 exhibit inhibited T-cell stimulatory capacity. *J Virol* **2000**. *74:* 7127–7136.

197 Sevilla, N., Kunz, S., Holz, A., Lewicki, H., Homann, D., Yamada, H., Campbell, K. P., de La Torre, J. C., Oldstone, M. B., Immunosuppression and resultant viral persistence by specific viral targeting of dendritic cells. *J Exp Med* **2000**. *192:* 1249–1260.

198 Raftery, M. J., Schwab, M., Eibert, S. M., Samstag, Y., Walczak, H., Schonrich, G., Targeting the function of mature dendritic cells by human cytomegalovirus. A multilayered viral defense strategy. *Immunity* **2001**. *15:* 997–1009.

199 Penna, G., Adorini, L., 1 Alpha,25-dihydroxyvitamin D3 inhibits differentiation, maturation, activation, and survival of dendritic cells leading to impaired alloreactive T cell activation. *J Immunol* **2000**. *164:* 2405–2411.

200 Mattner, F., Smiroldo, S., Galbiati, F., Muller, M., Di Lucia, P., Poliani, P. L., Martino, G., Panina-Bordignon, P., Adorini, L., Inhibition of Th1 development and treatment of chronic-relapsing experimental allergic encephalomyelitis by a non-hypercalcemic analogue of 1,25-dihydroxyvitamin D(3). *Eur J Immunol* **2000**. *30:* 498–508.

201 Piemonti, L., Monti, P., Sironi, M., Fraticelli, P., Leone, B. E., Dal Cin, E., Allavena, P., Di Carlo, V., Vitamin D3 affects differentiation, maturation, and function of human monocyte-derived dendritic cells. *J Immunol* **2000**. *164:* 4443–4451.

202 Griffin, M. D., L utz, W., Phan, V. A., Bachman, L. A., McKean, D. J., Kumar, R., Dendritic cell modulation by 1alpha,25 dihydroxyvitamin D3 and its analogs: a vitamin D receptor-dependent pathway that promotes a persistent state of immaturity in vitro and in vivo. *Proc Natl Acad Sci USA* **2001**. *98:* 6800–6805.

27
Dendritic Cell Manipulation with Biological and Pharmacological Agents to Induce Regulatory T Cells

Luciano Adorini and Giuseppe Penna

27.1
Introduction

Dendritic cells (DCs), a highly specialized antigen-presenting cell (APC) system critical for the initiation of CD4$^+$ T-cell responses are present, in different stages of maturation, in the circulation as well as in lymphoid and nonlymphoid organs, where they exert a sentinel function. After antigen uptake, DCs migrate through the afferent lymph to T-dependent areas of secondary lymphoid organs where they can prime naïve T cells. During migration to lymphoid organs, DCs mature into potent APCs by increasing their immunostimulatory properties while decreasing antigen-capturing capacity [1].

It is now clear that DCs can be not only immunogenic but also tolerogenic, both intrathymically and in the periphery [2], and they can modulate T-cell development [3]. In particular, immature DCs have been found to have tolerogenic properties, and to induce T cells with suppressive activity [4]. However, the simplistic concept that immature DCs are intrinsically and uniquely able to induce suppressor T cells (Ts) has been dispelled by the observation that mature DCs can also be very efficient inducers of Ts cells [5], a property already noted for semi-mature DCs [6].

DCs are heterogeneous not only in terms of maturation state, but also of origin, morphology, phenotype and function [1, 7], and DC subsets have also been considered as specialized inducers of effector or suppressor T cells [8]. Two distinct DC subsets were originally defined in the human blood based on the expression of CD11c, and they have been subsequently characterized as belonging to the myeloid or lymphoid lineage. Although different denominations have been used, they can be defined as myeloid (M-DCs) and plasmacytoid (P-DCs) DCs [8, 9]. A cell population resembling human P-DCs has also been identified in the mouse [10]. M-DCs are characterized by a monocytic morphology; express myeloid markers like CD13 and CD33, the β2 integrin CD11c, the activatory receptor ILT1 and low levels of the IL-3 receptor α chain CD123. Conversely, P-DCs have a morphology resembling plasma cells, are devoid of myeloid markers, express high levels of CD4, CD62L and CD123. M-DCs produce high levels of IL-12, while P-DCs high

Handbook of Dendritic Cells. Biology, Diseases, and Therapies.
Edited by M. B. Lutz, N. Romani, and A. Steinkasserer.
Copyright © 2006 WILEY-VCH Verlag GmbH & Co. KGaA, Weinheim
ISBN: 3-527-31109-2

levels of IFN-α [9], cytokines with clearly distinct effects on T-cell activation and differentiation.

Interest in the role of regulatory/suppressor T cells (Ts) cells has recently resurged and, among the various populations of Ts cells described, naturally-occurring thymic and peripheral CD4$^+$ T cells that co-express CD25 are currently the most actively investigated [11]. CD4$^+$CD25$^+$ Ts cells prevent the activation and proliferation of potentially autoreactive T cells that have escaped thymic deletion [12]. They fail to proliferate and secrete cytokines in response to polyclonal or antigen-specific stimulation, and are not only anergic but also inhibit the activation of responsive T cells [11]. Although CD25, CD152, and glucocorticoid-induced TNF-related protein (GITR) are markers of CD4$^+$CD25$^+$ Ts cells, they are also expressed by activated T cells [11]. A more faithful marker distinguishing CD4$^+$CD25$^+$ Ts cells from recently activated CD4$^+$ T cells is Foxp3, a member forkhead family of transcription factors that is required for CD25$^+$Ts development and is sufficient for their suppressive function [13–15]. Foxp3$^+$ CD4$^+$CD25$^+$ Ts cells play an important role in preventing the induction of several autoimmune diseases, such as the autoimmune syndrome induced by day 3 thymectomy in genetically susceptible mice [12], inflammatory bowel disease [16], T1D in thymectomized rats [17] and in NOD mice [18, 19]. CD25$^+$Ts are reduced in NOD compared to other mouse strains, and this reduction could be a factor in their susceptibility to T1D [18, 20]. CD25$^+$Ts and effector T cells coexist within the pancreatic lesion before the onset of type 1 diabetes, and several factors, such as blockade of ICOS [21], can perturb this balance, precipitating autoimmunity. A defect in peripheral regulatory cells affecting both CD25$^+$Ts and NK cells has been described also in T1D patients [22], and autoreactive T cells in diabetics are skewed to a pro-inflammatory Th1 phenotype lacking the IL-10-secreting T cells found in nondiabetic, HLA-matched controls [23]. The clinical relevance of CD4$^+$CD25$^+$ Ts cells has also been shown in patients affected by rheumatoid arthritis [24] and multiple sclerosis [25].

Because DCs are pleiotropic modulators of T-cell activity, manipulation of DC function, to favor the induction of DCs with tolerogenic properties leading to the development of Ts cells, could be exploited to modulate immune responses. Considerable efforts are ongoing to translate this concept into clinical practice, also by rationalizing the tolerogenic effects exerted by immunosuppressive and immunomodulatory drugs currently used to control autoimmune diseases and graft rejection [26, 27].

27.2
Mechanisms Promoting Tolerogenic Dendritic Cells

Tolerogenic DCs are characterized by reduced expression of co-stimulatory molecules, in particular CD40, CD80, CD86. In addition, they show reduced IL-12 and increased IL-10 production, and often an early stage of maturation [2]. While these well-established phenotypic and functional properties of tolerogenic DCs can easily explain their propensity to induce regulatory rather than effector T cells, several other mechanisms may play a role in favoring Ts cell induction by tolerogenic DCs.

27.2.1
Indoleamine 2,3-dioxygenase

One mechanism by which DCs can regulate T-cell responses is via expression of indoleamine 2,3-dioxygenase (IDO), the rate-limiting enzyme of tryptophan catabolism. IDO-transgenic DCs decrease the concentration of tryptophan, increase the concentration of kynurenine, the main tryptophan metabolite, and suppress allogeneic T-cell proliferation *in vitro* due to T-cell death, because suppressive tryptophan catabolites exert a cytotoxic action preferentially on activated T cells [28]. Although the concept that cells expressing IDO can suppress T-cell responses and promote tolerance is a relatively new paradigm in immunology, accumulating evidence supports this possibility, including studies on maternal tolerance to the fetus, tumor resistance, chronic infections and autoimmune diseases [29]. In particular, IDO-expressing DCs contribute to the generation and maintenance of peripheral tolerance by depleting autorective T cells [29]. Interestingly, CD4$^+$CD25$^+$ Ts cells can condition DCs to express IDO functional activity and suppressive properties, a process that requires IFN-γ production by DCs [30].

A mechanistic link has also been established between the tolerance-inducing activity of CTLA-4-Ig and its capacity to induce IDO expression in DCs [31]. IDO-competent DC subsets acquire potent and dominant T-cell suppressive properties as a consequence of IDO upregulation, blocking the ability of T cells to respond to other stimulatory DCs in the same cultures [32]. Selective IDO upregulation in DCs does not inhibit T-cell activation, but prevents T-cell clonal expansion due to rapid death of activated T cells, suggesting that IDO-competent DCs provide a regulatory bridge, mediated by CTLA4-CD80/CD86 engagement, between regulatory and naïve responder T cells [32]. IDO expression is induced by CTLA-4-Ig in specific DC subsets, notably in CD8α^+ and plasmacytoid DCs, and provides a potential mechanistic explanation for their T-cell regulatory properties [33]. Mouse plasmacytoid DCs can also initiate the immunosuppressive pathway of tryptophan catabolism, via type I IFN receptor signaling, in response to CD200 receptor engagement [34].

Human DCs that express IDO also inhibit T-cell proliferation *in vitro* [35]. Interestingly, IDO-mediated suppressor activity is present in fully mature as well as immature human CD123$^+$ plasmacytoid DCs. IDO$^+$ DCs can be readily detected *in vivo*, suggesting that these cells may represent a regulatory subset of APCs in man [36]. Thus, inducible IDO expression appears to play an important role in making DC tolerogenic, and its modulation has potential therapeutic applications.

27.2.2
Immunoglobulin-like Transcripts

Immunoglobulin-like transcripts (ILTs) are receptors structurally and functionally related to killer cell inhibitory receptors (KIR) [37] that have been shown to be involved in immunoregulation [38]. ILT family members can be subdivided into two main types. One, comprising ILT1, ILT7, ILT8, and leukocyte Ig-like receptor 6, is characterized by a short cytoplasmic tail delivering an activating signal through the immunoreceptor tyrosine-based activatory motif (ITAM) of the associated common γ chain of the Fc receptor. Members of the second type, including ILT2, ILT3, ILT4, and ILT5, contain a cytoplasmic immunoreceptor tyrosine-based inhibitory motif (ITIM) transducing a negative signal [39]. When inhibitory ILTs are activated, their ITIM domains become phosphorylated, and recruit p56^{lck} and SH2-containing protein-tyrosine-phosphatase 1 (SHP-1), leading to downstream events and gene modulation [40]. The high homology between ILTs and KIRs suggests that ILTs can also interact with class I MHC molecules, but this has been so far confirmed only for ILT2 and ILT4 [41, 42].

Most cell types involved in innate or acquired immune responses, including myeloid, lymphoid and dendritic cells, express at least one member of the ILT family, which may play an important role in immunoregulation [40]. For example, the inhibitory receptor ILT3 has been shown to negatively regulate activation of antigen-presenting cells [43]. A connection between ILTs and tolerance induction has been established by the observation that CD8$^+$CD28$^-$ suppressor T cells upregulate ILT3 and ILT4 expression on DCs, rendering them tolerogenic [44]. Such tolerogenic DCs have been reported to anergize alloreactive CD4$^+$CD45RO$^+$CD25$^+$ T cells converting them into regulatory T cells which, in turn, continue the cascade of suppression by tolerizing other DCs [45]. Alloantigen specific CD8$^+$CD28$^-$Foxp3$^+$ T suppressor cells have also been shown to induce ILT3$^+$ ILT4$^+$ tolerogenic endothelial cells, inhibiting alloreactivity [46]. Consistent with these results, rat CD8$^+$Foxp3$^+$ Ts cells have been shown to induce PIR-B, an hortolog of inhibitory ILTs [47], in DCs and heart endothelial cells, and to mediate tolerance to allogeneic heart transplants [48].

27.3
Induction of Tolerogenic Dendritic Cells

Dendritic cells induce and regulate T-cell responses, and tolerogenic DCs can promote the development of regulatory T cells with suppressive activity. Thus, the possibility to manipulate DCs using different pharmacological or biological agents, enabling them to exert tolerogenic activities, could be exploited to better control a variety of chronic inflammatory conditions, from autoimmune diseases to allograft rejection. Both biological and pharmacological agents have been shown capable of inducing tolerogenic DCs [26, 27, 49]. Notably, several *in vitro* studies have demonstrated that human regulatory T cells can be induced by DCs manipulated to acquire and/or enhance tolerogenic properties, and *in vivo* data are also accumulating.

27.3.1
Biological Agents Promoting Tolerogenic Dendritic Cells

Cytokines represent the best known class of biological agents currently used to favor the induction of tolerogenic DCs. In particular, DCs differentiated in the presence of IL-10, TGF-β, TNF-α, or G-CSF can acquire phenotypic and functional properties characteristic of tolerogenic DCs.

27.3.1.1 **IL-10**
Although several biological agents can favor, directly or indirectly, induction of regulatory T cells, IL-10 is probably the best example of an immunomodulatory protein that promotes regulatory T-cell induction, at least in part, by targeting DCs.

APCs and T cells are primary targets of IL-10, a potent suppressor of several effector functions of macrophages, dendritic cells, T cells and NK cells. In addition, IL-10 contributes to regulate proliferation and differentiation of B cells, mast cells, and thymocytes [50]. An important property of IL-10, from an immunotherapeutic perspective, is its capacity to inhibit Th1 cells. The inhibition of the Th1 cell pathway by IL-10 is mediated by several mechanisms, including inhibition of IL-12 production by APCs, and blocking of IFN-γ synthesis by differentiated Th1 cells [50]. However, IL-10 targets more efficiently naïve rather than activated or memory T cells, possibly due to the downregulation of IL-10R α chain. In addition, IL-10 strongly inhibits production of pro-inflammatory monokines as IL-1, IL-6, IL-8, TNF-α and GM-CSF, as well as of reactive oxygen and nitrogen species following activation of human or mouse macrophages. The intrinsically strong anti-inflammatory properties of IL-10 are further enhanced by its capacity to induce regulatory T cells. In particular, the presence of IL-10 during differentiation of CD4⁺ T cells results in the development of a defined subset of regulatory T cells (Tr1 cells) characterized by low proliferation, absence of IL-2 production, and a specific cytokine profile characterized by IL-10 and IFN-γ but no IL-4 nor IL-5 production [51].

It is possible that these inhibitory functions of IL-10 and related cytokines can find therapeutic applications, and their activity in modulating APC functions and in promoting development of regulatory T cells suggests a possible use in the treatment of autoimmune diseases [52]. Supporting this hypothesis, severe colitis was abrogated in a model of inflammatory bowel disease by systemic administration of IL-10 but, interestingly, not of IL-4. In addition, IL-10 treatment can ameliorate mouse lupus via inhibition of pathogenic Th1 cytokines [50]. IL-10 treatment is currently being tested in multiple inflammatory conditions, including rheumatoid arthritis, inflammatory bowel disease, psoriasis, allograft rejection and chronic hepatitis C [53]. The results from clinical trials are heterogeneous but, so far, inferior to the expectations [53].

27.3.1.2 **TGF-β**

TGF-β, a member of a large family of evolutionary conserved proteins known for their pleiotropic activities, can promote or inhibit cell growth and function. TGF-β1 is produced by every leukocyte lineage, including lymphocytes, macrophages, and dendritic cells. It can modulate expression of adhesion molecules, provide a chemotactic gradient for leukocytes and other cells participating in an inflammatory response, and inhibit them once they have become activated [54].

The important role of TGF-β in autoimmune diseases is shown by the massive autoimmune inflammation affecting multiple organs in mice deficient in TGF-β or with induced disruption of the TGF-β type II receptor, as well as, for example, by the inhibition of EAE following TGF-β administration, and by enhancement of EAE upon its neutralization [54]. In addition, TGF-β is considered a major mediator in oral tolerance [55]. Although the disease-limiting properties of TGF-β in autoimmune diseases seem attractive, disruption of the balance between its opposing activities can contribute to aberrant development, malignancy, or pathogenic immune and inflammatory responses characterized by widespread tissue fibrosis and deposition of extracellular matrix, as shown by the reversible decline in the glomerular filtration rate observed in an open-label trial in patients with secondary progressive MS [56].

Several studies have shown that TGF-β1 inhibits *in vitro* activation and maturation of DCs, preventing the upregulation of critical T-cell co-stimulatory molecules on DC surface, inhibiting IL-12 production, and reducing their antigen-presenting capacity [57–59]. Thus, in addition to direct inhibitory effects of TGF-β1 on effector T lymphocytes [60], and to induction of the Ts transcription factor foxp3 [61], its inhibitory effects at the DC level may critically contribute to the immunosuppressive effects. In contrast to these negative regulatory effects of TGF-β1 on DC function and maturation, certain subpopulations of immature DCs in non lymphoid tissues are positively regulated by TGF-β1 signaling. In particular, epithelial-associated DC populations seem to critically require TGF-β1 stimulation for development and function, as shown for the development of epithelial Langerhans cells *in vitro* and *in vivo*, in which TGF-β1 seems to enhance also antigen processing and co-stimulatory functions [62].

27.3.1.3 **TNF-α**

Tumor necrosis factor (TNF)-α is a pro-inflammatory cytokine with interesting immunoregulatory properties [63], that can modulate DC development, phenotype and function [6]. Maturation by TNF-α induce high levels of MHC class II and co-stimulatory molecules on DCs, but they remain weak producers of pro-inflammatory cytokines, in particular IL-12 [6]. These incompletely matured DCs (semi-mature DCs) induce peptide-specific IL-10-producing T cells *in vivo* and prevent EAE [64]. DCs with a similar phenotype were previously found to inhibit, when injected 1 week before transplantation, haplotype-specific cardiac allograft rejection, with a marked increase in median graft survival time from 8 to >100 days [65]. Injection of semi-mature DCs could also protect mice from GVHD and induce the expansion of IL-10-producing CD4$^+$CD25$^+$ Ts cells [66], suggesting that semi-mature DCs may be beneficial in the treatment of several immune-mediated diseases.

27.3.1.4 G-CSF

Granulocyte colony-stimulating factor (G-CSF), the key hematopoietic growth factor of the myeloid lineage, has been recently found to possess marked immunoregulatory properties [67, 68].

CD4$^+$ T cells exposed *in vivo* to G-CSF acquire Tr1-type properties, once triggered *in vitro* through the T-cell receptor, including IL-10-dominated cytokine production profile, intrinsically low proliferation, and contact-independent suppression of antigen-driven proliferation [69]. The immunomodulatory effects of G-CSF might be mediated by DCs expressing high levels of co-stimulatory molecules and HLA-DR, but decreased IL-12p70 secretion and poor allostimulatory capacity [70], reminiscent of semi-mature DCs [6]. The ability of G-CSF to promote key tolerogenic interactions between DCs and regulatory T-cells has been recently demonstrated by the enhanced recruitment of TGF-β1-expressing CD4$^+$CD25$^+$ Ts cells after adoptive transfer of DCs isolated from G-CSF- compared to vehicle-treated mice into naïve NOD recipients [71].

27.3.2
Pharmacological Agents Promoting Tolerogenic Dendritic Cells

A variety of immunosuppressive agents are currently used to inhibit transplantation rejection and to treat autoimmune diseases (Table 27.1). Some of these pharmacological agents have been instrumental in the control of allograft rejection, giving a decisive impulse to clinical transplantation in the late 1970s. Interestingly, the mechanism of action of major immunosuppressive drugs, like the calcineurin inhibitors cyclosporine A and tacrolimus, has been only understood after almost 20 years of clinical use [72]. Thus, it is perhaps not surprising that a novel mechanism of action shared by many immunosuppressive and anti-inflammatory agents, based on the induction of DCs with tolerogenic properties, has only recently emerged [26, 27, 49, 73].

Indeed, several immunosuppressive agents currently used to treat allograft rejection and autoimmune diseases have been shown to induce DCs with tolerogenic phenotype and function (Table 27.1). Notable examples are glucocorticoids [74–77], mycophenolate mofetil (MMF) [78, 79], and sirolimus [80, 81]. These agents impair DC maturation and inhibit upregulation of co-stimulatory molecules, secretion of pro-inflammatory cytokines, in particular IL-12, and allostimulatory capacity. Sirolimus appears to be a very interesting agent, because it induces tolerogenic DCs [81], and sirolimus-treated alloantigen-pulsed DCs infused 1 week before transplantation inhibit antigen-specific T-cell responsiveness and prolong graft survival [82].

Conversely, controversial effects of calcineurin inhibitors, like cyclosporine A and tacrolimus, have been reported on DC maturation, although these drugs have a clear inhibitory effect on DC, decreasing their cytokine production and allostimulatory capacity [77, 83, 84]. Other immunosuppressive agents, like desoxyspergualin, also inhibit the allostimulatory capacity of DCs, impairing their maturation and IL-12 production as well [85–87]. Similar effects are exerted on DCs by anti-inflammatory agents, such as acetylsalicylic acid [88, 89], butyric acid [90] and *N*-acetyl-L-cysteine [91].

AGENT	EFFECTS ON DENDRITIC CELL						
	Differentiation	Maturation	Costimulatory molecules	IL-12 Production	IL-10 Production	Allostimulatory capacity	NF-kB Activation
BIOLOGICAL							
IL-10	↓	↓	↓	↓		↓	
TGF-β	↓	↓	↓	↓		↓	
TNF-α		↓	↓	↓		↓	
G-CSF		↓	↓	↓		↓	
PHARMACOLOGICAL							
Acetylsalicylic acid		↓	↓	↓	↕	↓	↓
Butyric acid	↓	↓	↓	↓			↓
Calcineurin inhibitors	↕	↓	↓	↓	↕	↓	↓
Deoxyspergualin	↓	↓	↓	↓		↓	↓
Glucocorticoids		↓	↓	↓	↕	↓	↓
N-acetyl-L-cysteine	↓	↓	↓	↓		↓	
Mycophenolate mofetil	↓	↓	↓	↓		↓	
Sirolimus	↓	↓	↓	↓		↓	
Vitamin D receptor agonists		↓	↓	↓	→	↓	↓

Tab. 27.1 Effects of pharmacological and biological agents with anti-inflammatory and immunosuppressive properties on dendritic cells. Compiled from references quoted in the text. Downwards arrows indicate inhibition, upwards arrows stimulation, and horizontal arrows no effect. Blanks indicate information not available.

Although the pro-tolerogenic effects of several pharmacological agents on DCs are well established, little is known about their capacity to induce regulatory T cells promoting transplantation tolerance. MMF was able, as a monotherapy, to induce some limited levels of transplantation tolerance even if no induction of tolerogenic DCs was observed *in vivo* [92]. Conversely, calcineurin inhibitors have been reported to prevent transplantation tolerance induced by co-stimulation blockers, although the issue is still unresolved [93], but successful establishment of alloantigen-specific hyporesponsiveness by NF-_B inhibitor-treated DCs was not inhibited by concomitant calcineurin inhibition [94]. In addition, the sirolimus derivative everolimus did not hamper *in vitro* the suppressive activity of CD4$^+$CD25$^+$ Ts cells, suggesting that these cells may still exert suppressive activity in transplant recipients treated with drugs interfering with IL-2 signaling [95].

Finally, as discussed in greater detail below, the activated form of vitamin D, 1,25(OH)$_2$D$_3$, and its analogues have also been found to inhibit DC maturation, leading to reduced expression of co-stimulatory molecules and IL-12 production. These tolerogenic DCs show decreased capacity to stimulate alloreactive T cells, and promote the differentiation of CD4$^+$CD25$^+$ Ts cells.

27.4
Induction of Tolerogenic Dendritic Cells by VDR Agonists

The activated form of vitamin D, 1,25(OH)$_2$D$_3$, is a secosteroid hormone that has, in addition to its central function in calcium and bone metabolism, important effects on the growth and differentiation of many cell types, and pronounced immunoregulatory properties [96–100]. The biological effects of 1,25(OH)$_2$D$_3$ are mediated by the vitamin D receptor (VDR), a member of the superfamily of nuclear hormone receptors functioning as a agonist-activated transcription factor that binds to specific DNA sequence elements, vitamin D responsive elements, in vitamin D responsive genes and ultimately influences their rate of RNA polymerase II-mediated transcription [101].

APCs, and notably DCs, express the VDR and are key targets of VDR agonists, both *in vitro* and *in vivo*. A number of studies, summarized in Table 27.2, has clearly demonstrated that 1,25(OH)$_2$D$_3$ and its analogues markedly modulate DC phenotype and function [102–107]. These studies, performed either on monocyte-derived DCs from human peripheral blood or on bone-marrow derived mouse DCs, have consistently shown that *in vitro* treatment of DCs with 1,25(OH)$_2$D$_3$ and its analogues leads to downregulated expression of the co-stimulatory molecules CD40, CD80, CD86, and to decreased IL-12 and enhanced IL-10 production, resulting in decreased T-cell activation (Table 27.2). The block of maturation, coupled with abrogation of IL-12 and strongly enhanced production of IL-10, highlight the important functional effects of 1,25(OH)$_2$D$_3$ and its analogues on DCs and are, at least in part, responsible for the induction of DCs with tolerogenic properties. The combination of these effects can explain the capacity of VDR agonists to induce DCs with tolerogenic properties that favor suppressor T-cell enhancement. DCs are able to synthesize $1,25(OH)_2D_3$ *in vitro* as a consequence of increased 1α-hydroxylase expression [108], and this could also contribute to promote regulatory T-cell induction. It is also possible that $1,25(OH)_2D_3$ may contribute to the physiological control of immune responses, and possibly be also involved in maintaining tolerance to self antigens, as suggested by the enlarged lymph nodes containing a higher frequency of mature DCs in VDR-deficient mice [109].

Tab. 27.2 Phenotypic and functional modifications induced by VDR ligands in human myeloid dendritic cells.

Phenotype	Effect
Maturation marker expression	
CD83	decreased
DC-LAMP	decreased
Antigen uptake	
Mannose receptor expression	increased
Co-stimulatory molecule expression	
CD40	decreased
CD80	decreased
CD86	decreased
Inhibitory molecule expression	
ILT3	increased
ILT4	unmodified
B7-H1	unmodified
Chemokine receptor expression	
CCR7	decreased

Function	Effect
Cytokine production	
IL-10	increased
IL-12	decreased
Chemokine production	
CCL2	increased
CCL17	decreased
CCL18	increased
CCL20	decreased
CCL22	increased
Apoptosis	
Maturation-induced	increased
T-cell activation	
Response to alloantigens	decreased

Compiled from refs. [102, 134] and from the author's unpublished data.

27.4.1
Tolerogenic Dendritic Cells Induced by VDR Agonists lead to enhancement of regulatory T cells

The prevention of DC differentiation and maturation as well as the modulation of their activation and survival leading to DCs with tolerogenic phenotype and function (Table 27.2) play an important role in the immunoregulatory activity of VDR agonists, and appear to be critical for the capacity of this hormones to induce CD4+CD25+ Ts cells that are able to control autoimmune responses and allograft rejection (Table 27.3).

VDR agonists enhance CD4+CD25+ Ts cells and promote tolerance induction in transplantation and autoimmune disease models. A short treatment with 1,25(OH)$_2$D$_3$ and mycophenolate mofetil, a selective inhibitor of T and B cell proliferation that also modulates APCs, induces tolerance to islet allografts associated with an increased frequency of CD4+CD25+ Ts cells able to adoptively transfer transplantation tolerance [92]. The induction of tolerogenic DCs could indeed represent a therapeutic strategy promoting tolerance to allografts [99] and the observation that immature myeloid DCs can induce T-cell tolerance to specific antigens in human volunteers represents an important proof of concept for this approach [110].

CD4+CD25+ Ts cells able to inhibit the T-cell response to a pancreatic autoantigen and to significantly delay disease transfer by pathogenic CD4+CD25⁻ T cells are also induced by treatment of adult nonobese diabetic (NOD) mice with the VDR agonist BXL-219 [111]. This treatment arrests insulitis, blocks the progression of Th1 cell infiltration into the pancreatic islets, and inhibits type 1 diabetes development at nonhypercalcemic doses [111]. Although the type 1 diabetes and islet transplantation models are quite different, in both cases administration of VDR

Tab. 27.3 VDR agonists foster the induction of regulatory T cells.

In dendritic cells

Inhibit IL-12
Enhance IL-10
Down-regulate CD40, CD80, CD86
Block maturation
Upregulate ILT3 expression

Leading to T cells characterized by

Reduced Th1 development
Hyporesponsiveness to auto and alloantigens
Increased CTLA-4 expression
Decreased CD40L expression
Enhanced CD4+CD25+ suppressor T cells

Compiled from references [102–107, 115] and from the author's unpublished results.

agonists doubles the number of CD4$^+$CD25$^+$ Ts cells, in the spleen and pancreatic lymph nodes, respectively.

However, tolerogenic DCs may not always be necessarily involved in the generation of Ts cells by VDR agonists. A combination of 1,25(OH)$_2$D$_3$ and dexamethasone has been shown to induce human and mouse naïve CD4$^+$ T cells to differentiate *in vitro* into Ts cells, even in the absence of APCs [112]. These Ts cells produced IL-10, but no IL-5 nor IFN-γ, thus distinguishing them from the previously described Tr1 cells [51]. Upon transfer, the IL-10-producing Ts cells could prevent central nervous system inflammation, indicating their capacity to exert a suppressive function *in vivo* [112]. Thus, although DCs appear to be primary targets for the immunomodulatory activities of VDR agonists, they can also act directly on T cells, as expected by VDR expression in both cell types and by the presence of common targets in their signal transduction pathways, such as the nuclear factor NF-κB that is downregulated in APCs and in T cells.

27.4.2
Upregulation of Inibitory Receptor Expression in Dendritic Cells by VDR agonists

To further characterize mechanisms accounting for the induction of DCs with tolerogenic properties by VDR agonists, we have examined the expression of immunoglobulin-like transcripts (ILT), receptors structurally and functionally related to killer cell inhibitory receptors (KIR) [43], by 1,25(OH)$_2$D$_3$-treated DCs. We have found that incubation of monocyte-derived human DCs, either immature or during maturation, with 1,25(OH)$_2$D$_3$ leads to a selective upregulation of ILT3 [27]. Analysis of DC subsets revealed a higher ILT3 expression on P-DCs compared to M-DCs [113, 114]. CD40 ligation reduced ILT3 expression on M-DCs but had little effect on P-DCs [115]. Maintaining high ILT3 expression on P-DCs matured via CD40 ligation is of interest, because this cell population has been shown to induce CD8$^+$ regulatory T cells able to suppress the proliferation of naïve CD8$^+$ cells through an IL-10-dependent pathway [116]. While incubation with 1,25(OH)$_2$D$_3$ did not affect the already high ILT3 expression by P-DCs, it increased its expression on M-DCs considerably [115]. The downregulation of ILT3 on M-DCs by T cell-dependent signals, and the upregulation of this inhibitory receptor by 1,25(OH)$_2$D$_3$ in DCs suggests a novel mechanism for the immunomodulatory properties of this hormone that could play a role in the control of T-cell responses.

As tolerogenic DCs induced by different pharmacological agents share several properties (Table 27.1), we analyzed upregulation of ILT3 expression in immature and mature DCs by selected immunomodulatory agents. 1,25(OH)$_2$D$_3$ markedly upregulates ILT3 expression on both immature and mature DCs, whereas IL-10 has a much less pronounced effect, and dexamethasone no observable activity. In the same experiment, all the three agents inhibited DC maturation, as shown by decreased CD83 expression [27]. An *in vivo* correlate could be established by the marked upregulation of ILT3 expression in DCs of psoriatic lesions treated with the VDR agonist calcipotriol, whereas no ILT3 expression was induced by topical treatment of psoriatic plaques with the glucocorticoid mometazone [135]. These results indicate that drug-induced ILT3 upregulation is not a general feature of tolerogenic DCs, as proposed by a recent study [117], and are consistent with the view that VDR agonists and glucocorticoids modulate DCs using distinctive pathways [118]. Although ILT3 expression by DCs is required for induction of regulatory T cells, DC pretreatment with $1,25(OH)_2D_3$ leads to induction of CD4$^+$Foxp3$^+$ cells with suppressive activity irrespective of the presence of neutralizing anti-ILT3 mAb, indicating that ILT3 expression is dispensable for the capacity of $1,25(OH)_2D_3$-treated DCs to induce regulatory T cells (ref. Penna, G., A. Roncari, S. Amuchastegui, K. C. Daniel, E. Berti, M. Colonna, and L. Adorini. 2005. Expression of the inhibitory receptor ILT3 on dendritic cells is dispensable for induction of CD4$^+$Foxp3$^+$ regulatory T cells by 1,25-dihydroxyvitamin D3. *Blood 106*: 3490–3497).

27.4.3
Modulation of Chemokine Production by VDR Agonists can affect Recruitment of Effector T cells and CD4$^+$CD25$^+$ Ts cells to Inflammatory Sites

In both islet transplantation and type 1 diabetes models, treatment with VDR agonists has a profound effect on the migration of effector T cells, preventing their entry into the pancreatic islets [92, 111]. The VDR agonist BXL-219 significantly downregulates *in vitro* and *in vivo* pro-inflammatory chemokine production by islet cells, inhibiting T-cell recruitment into the pancreatic islets and T1D development [119]. The inhibition of CXCL10 is particularly relevant, consistent with the decreased recruitment of Th1 cells into sites of inflammation by treatment with an anti-CXCR3 antibody [120], and with the substantial delay of T1D development observed in CXCR3-deficient mice [121]. The inhibition of islet chemokine production by BXL-219 treatment *in vivo* is associated with upregulation of IκBα transcription, an inhibitor of nuclear factor κB (NF-κB), and with arrest of NF-κBp65 nuclear translocation [119], highlighting a novel mechanism of action exerted by VDR agonists potentially relevant for the treatment of T1D and other autoimmune diseases. These observations expand the known mechanisms of action exerted by vitamin D analogs in the treatment of T1D and other autoimmune diseases, that include arrest of DC maturation, inhibition of Th1 cell responsiveness, and enhancement of regulatory T cells [97, 98, 100]. In addition to modulating chemokine production in target tissues such as pancreatic islets, it is also possible that VDR agonists can affect the migration of CD25$^+$Ts cells by regulating their chemokine receptor expression, a hypothesis that we are currently testing.

Both human [122] and mouse (N. Giarratana et al., submitted for publication) CD4$^+$CD25$^+$ Ts cells express CCR4, and selectively migrate in response to CCR4 agonists like CCL22. An interesting confirmation to this finding is provided by the observation that human ovarian tumors produce CCL22, the cognate ligand of the CCR4 receptor, promoting the recruitment of CCR4$^+$ CD4$^+$CD25$^+$ Ts cells that act as a tumor-protective mechanism [123].

We have found that, in contrast to the high production by circulating human myeloid DCs (M-DCs), the CCR4 agonists CCL17 and CCL22 are poorly produced by plasmacytoid DCs (P-DCs) [124]. It is noteworthy that blood-borne M-DCs, in contrast to P-DCs, constitutively produce CCL17 and CCL22 *ex vivo* [124]. This selective constitutive production of CCR4 agonists by immature M-DCs could lead to the preferential attraction of CD4$^+$CD25$^+$ Ts cells, a mechanism expected to favor tolerance induction. This has been observed in ovarian carcinoma patients, in which Foxp3$^+$CCR4$^+$CD25$^+$Ts cells are selectively recruited by tumor-produced CCL22, and suppress antitumor responses leading to reduced patient survival [123]. Intriguingly, the production of CCL22 is markedly enhanced by 1,25(OH)$_2$D$_3$ in blood M-DCs but not P-DCs (Penna et al., manuscript in preparation). Besides maintaining peripheral immunological tolerance in homeostatic conditions, CD4$^+$CD25$^+$ Ts cells could turn-off and limit ongoing inflammatory responses. Inflammatory signals strongly induce maturation and influx of both M-DCs and P-DCs to secondary lymphoid tissues [8], and maturation of M-DCs and P-DCs enhances their production of several pro-inflammatory chemokines that can potentially attract different T-cell subsets. Interestingly, maturing P-DCs, similarly to activated B cells, produce large quantities of the CCR5 agonist CCL4 [124]. Thus, in analogy with the proposed role for CCL4 in CD4$^+$CD25$^+$ Ts cells attraction by activated B cells, mature P-DCs could recruit these cells to limit ongoing inflammatory responses.

27.5
Common Features of Agents Leading to Induction of Tolerogenic DCs

As summarized in Table 27.1, common features shared by biological and pharmacological agents favoring the induction of tolerogenic DCs are their capacity to inhibit differentiation, maturation, co-stimulatory molecule expression, and IL-12 production, leading to decreased allostimulatory capacity.

Co-stimulatory molecule expression is almost invariably reduced in tolerogenic DCs, with the exception of exposure to agents inducing semi-mature DCs [6]. In any case, all the tolerogenic agents tested inhibit DC maturation and reduce their capacity to stimulate alloreactive T cells in a mixed leukocyte reaction assay. Another common feature of DC-targeting drugs is the inhibition of IL-12, a cytokine critically involved in the development of Th1-dependent diseases [125]. In contrast, only 1,25(OH)$_2$D$_3$ and its analogues, among the agents tested, are able to enhance the secretion by DCs of IL-10, a cytokine favoring the induction of regulatory T cells (Table 27.1).

Several of these effects could be mediated by NF-κB, a signal transduction pathway crucially involved in the inflammatory response [126]. The NF-κB family member RelB is required for myeloid DC differentiation, and antigen-pulsed DCs in which RelB function is inhibited can induce regulatory CD4$^+$ T cells able to transfer tolerance to primed recipients in an IL-10-dependent fashion [127]. Our data showing upregulation of transcripts encoding IκBα and inhibition of RelA nuclear translocation by BXL-219 in pancreatic islet cells [119] demonstrate a novel mechanism of action in the targeting NF-κB by VDR ligands, in addition to inhibition of NF-κB1 and c-Rel [128], as well as RelB [129] expression. Interestingly, this mechanism of action has been previously demonstrated for glucorticoids, anti-inflammatory drugs that bind to a nuclear receptor in the same superfamily as the

VDR, by showing that dexamethasone upregulates the transcription of *Nfkbia*, which results in increased rate of IκBα synthesis and in reduced NF-κB translocation to the nucleus [130, 131]. The upregulation of transcripts encoding IκBα and the inhibition of RelA translocation to the nucleus by BXL-219 prevent activation of NF-κB, a transcription factor that also regulates chemokine production by pancreatic β cells [132]. The promoter of the *Nfkbia* gene encoding IκBα contains, as the *Relb* gene [129], several vitamin D responsive elements, some of which are highly conserved between human and mouse homologs, suggesting a direct transcriptional regulation of IκBα by BXL-219 [119]. The direct targeting of NF-κB components by VDR agonists could thus contribute to explain their inhibition of pro-inflammatory cytokine and chemokine production by DCs, as well as the inhibition of DC maturation, and could open new avenues in the use of VDR agonists as anti-inflammatory agents.

27.6
Conclusions

Several immunomodulatory agents, and in particular immunosuppressive and anti-inflammatory drugs, share the capacity to target DCs, rendering them tolerogenic and fostering the induction of regulatory rather than effector T cells. Multiple mechanisms contribute to induction of DC tolerogenicity, from downregulation of co-stimulatory molecules, both membrane-bound as CD40, CD80, CD86 and secreted as IL-12, to upregulation of inhibitory molecules like IDO and ILT3, to modulation of chemokine secretion, enhancing the production of chemokines able to recruit regulatory/suppressor T cells, and inhibiting chemokine production by the target organ in inflammatory conditions.

In principle, these mechanisms favoring DC tolerogenicity could be exploited in two ways. The first could rely on the *in vitro* manipulation of DCs to promote tolerogenic properties, followed by reinfusion into the patient. However, this, as any cell-based therapy, poses tremendous hurdles for clinical applicability. The standardization of effective and reproducible protocols would be very difficult, but even more problematic will be to ensure the capacity of the reinfused DCs to maintain a tolerogenic function *in vivo*, under inflammatory conditions. In contrast to therapies based on *ex-vivo* manipulation of DCs, treatments with immunosuppressive and anti-inflammatory drugs able to promote tolerogenic DCs have been in clinical use for decades to control allograft rejection and autoimmune diseases. Administration of these agents can directly target both DCs and T cells, leading to the inhibition of pathogenic effector T cells and enhancing the frequency of T cells with suppressive properties, effects that appear to be largely mediated via induction of tolerogenic DCs. Thus, direct treatment of patients with DC-tolerizing agents appears to represent a preferable therapeutic option. In addition to low molecular weight drugs, biological agents, in particular cytokines, could be directly administered *in vivo* to promote induction of tolerogenic DCs. However, issues related to immunogenicity, short and long-term adverse events, and high cost are likely to place important limitations on their clinical applicability.

Our own work has explored the immunoregulatory activities of VDR agonists, secosteroid hormones able to induce tolerogenic DCs and regulatory T cells. VDR agonists have been proven effective and safe drugs in a variety of autoimmune disease [96, 98] and graft rejection [99, 133] models, highlighting their potential applicability in chronic inflammatory conditions sustained by autoreactive or alloreactive immune responses. In addition to the topical treatment of psoriasis, a Th1-mediated autoimmune disease of the skin where VDR agonists are the most used topical drugs, these agents might eventually find a broader application in the treatment of inflammatory conditions, where their modulatory effects on DCs enhancing T cells with regulatory functions could turn out to be highly beneficial.

References

1 Banchereau, J., Briere, F., Caux, C., Davoust, J., Lebecque, S., Liu, Y. J., Pulendran, B., Palucka, K., Immunobiology of dendritic cells. *Annu. Rev. Immunol.* **2000**. 18: 767–811.

2 Steinman, R. M., Hawiger, D., Nussenzweig, M. C., Tolerogenic dendritic cells. *Annu. Rev. Immunol.* **2003**. 21: 685–711.

3 de Jong, E. C., Smits, H. H., Kapsenberg, M. L., Dendritic cell-mediated T cell polarization. *Springer Semin. Immunopathol.* **2005**. 26: 289–307.

4 Jonuleit, H., Schmitt, E., Schuler, G., Knop, J., Enk, A. H., Induction of interleukin 10-producing, nonproliferating CD4(+) T cells with regulatory properties by repetitive stimulation with allogeneic immature human dendritic cells. *J. Exp. Med.* **2000**. 192: 1213–1222.

5 Yamazaki, S., Iyoda, T., Tarbell, K., Olson, K., Velinzon, K., Inaba, K., Steinman, R. M., Direct expansion of functional CD25+ CD4+ regulatory T cells by antigen-processing dendritic cells. *J. Exp. Med.* **2003**. 198: 235–247.

6 Lutz, M. B., Schuler, G., Immature, semi-mature and fully mature dendritic cells: which signals induce tolerance or immunity? *Trends Immunol.* **2002**. 23: 445–449.

7 Reid, S. D., Penna, G., Adorini, L., The control of T cell responses by dendritic cell subsets. *Curr. Opin. Immunol.* **2000**. 12: 114–121.

8 Shortman, K., Liu, Y. J., Mouse and human dendritic cell subtypes. *Nature Rev. Immunol.* **2002**. 2: 151–161.

9 Colonna, M., Trinchieri, G., Liu, Y. J., Plasmacytoid dendritic cells in immunity. *Nat. Immunol.* **2004**. 5: 1219–1226.

10 Asselin-Paturel, C., Boonstra, A., Dalod, M., Durand, I., Yessaad, N., Dezutter-Dambuyant, C., Vicari, A., O'Garra, A., Biron, C., Briere, F., Trinchieri, G., Mouse type I IFN-producing cells are immature APCs with plasmacytoid morphology. *Nat. Immunol.* **2001**. 2: 1144–1150.

11 Shevach, E. M., CD4+ CD25+ suppressor T cells: more questions than answers. *Nat. Rev. Immunol.* **2002**. 2: 389–400.

12 Sakaguchi, S., Regulatory T cells: key controllers of immunologic self-tolerance. *Cell* **2000**. 101: 455–458.

13 Hori, S., Nomura, T., Sakaguchi, S., Control of regulatory T cell development by the transcription factor Foxp3. *Science* **2003**. 299: 1057–1061.

14 Fontenot, J. D., Gavin, M. A., Rudensky, A. Y., Foxp3 programs the development and function of CD4+CD25+ regulatory T cells. *Nat. Immunol.* **2003**. 4: 330–336.

15 Khattri, R., Cox, T., Yasayko, S. A., Ramsdell, F., An essential role for Scurfin in CD4+CD25+ T regulatory cells. *Nat. Immunol.* **2003**. 4: 337–342.

16 Read, S., Malmstrom, V., Powrie, F., Cytotoxic T lymphocyte-associated antigen 4 plays an essential role in the function of CD25(+)CD4(+) regulatory cells that control intestinal inflammation. *J. Exp. Med.* **2000**. 192: 295–302.

17 Stephens, L. A., Mason, D., CD25 is a marker for CD4+ thymocytes that

prevent autoimmune diabetes in rats, but peripheral T cells with this function are found in both CD25+ and CD25– subpopulations. *J. Immunol.* **2000**. 165: 3105–3110.

18 Salomon, B., Lenschow, D. J., Rhee, L., Ashourian, N., Singh, B., Sharpe, A., Bluestone, J. A., B7/CD28 costimulation is essential for the homeostasis of the CD4+CD25+ immunoregulatory T cells that control autoimmune diabetes. *Immunity* **2000**. 12: 431–440.

19 Wu, A. J., Hua, H., Munson, S. H., McDevitt, H. O., Tumor necrosis factor-alpha regulation of CD4+CD25+ T cell levels in NOD mice. *Proc. Natl. Acad. Sci. USA* **2002**. 99: 12287–12292.

20 Gregori, S., Giarratana, N., Smiroldo, S., Adorini, L., Dynamics of pathogenic and suppressor T cells in autoimmune diabetes development. *J. Immunol.* **2003**. 171: 4040–4047.

21 Herman, A. E., Freeman, G. J., Mathis, D., Benoist, C., CD4+CD25+ T Regulatory Cells Dependent on ICOS Promote Regulation of Effector Cells in the Prediabetic Lesion. *J. Exp. Med.* **2004**. 199: 1479–1489.

22 Kukreja, A., Cost, G., Marker, J., Zhang, C., Sun, Z., Lin-Su, K., Ten, S., Sanz, M., Exley, M., Wilson, B., Porcelli, S., Maclaren, N., Multiple immuno-regulatory defects in type-1 diabetes. *J. Clin. Invest.* **2002**. 109: 131–140.

23 Arif, S., Tree, T. I., Astill, T. P., Tremble, J. M., Bishop, A. J., Dayan, C. M., Roep, B. O., Peakman, M., Autoreactive T cell responses show proinflammatory polarization in diabetes but a regulatory phenotype in health. *J. Clin. Invest.* **2004**. 113: 451–463.

24 Goronzy, J. J., Weyand, C. M., T-cell regulation in rheumatoid arthritis. *Curr. Opin. Rheumatol.* **2004**. 16: 212–217.

25 Sospedra, M., Martin, R., Immunology of Multiple Sclerosis. *Annu. Rev. Immunol.* **2004**.

26 Hackstein, H., Thomson, A. W., Dendritic cells: emerging pharmacological targets of immunosuppressive drugs. *Nat. Rev. Immunol.* **2004**. 4: 24–34.

27 Adorini, L., Giarratana, N., Penna, G., Pharmacological induction of tolerogenic dendritic cells and regulatory T cells. *Semin. Immunol.* **2004**. 16: 127–134.

28 Terness, P., Bauer, T. M., Rose, L., Dufter, C., Watzlik, A., Simon, H., Opelz, G., Inhibition of allogeneic T cell proliferation by indoleamine 2,3-dioxy-genase-expressing dendritic cells: mediation of suppression by tryptophan metabolites. *J. Exp. Med.* **2002**. 196: 447–457.

29 Mellor, A. L., Munn, D. H., IDO expression by dendritic cells: tolerance and tryptophan catabolism. *Nat. Rev. Immunol.* **2004**. 4: 762–774.

30 Fallarino, F., Grohmann, U., Hwang, K. W., Orabona, C., Vacca, C., Bianchi, R., Belladonna, M. L., Fioretti, M. C., Alegre, M. L., Puccetti, P., Modulation of tryptophan catabolism by regulatory T cells. *Nat. Immunol.* **2003**. 4: 1206–1212.

31 Grohmann, U., Orabona, C., Fallarino, F., Vacca, C., Calcinaro, F., Falorni, A., Candeloro, P., Belladonna, M. L., Bianchi, R., Fioretti, M. C., Puccetti, P., CTLA-4-Ig regulates tryptophan catabolism in vivo. *Nat. Immunol.* **2002**. 3: 1097–1101.

32 Mellor, A. L., Chandler, P., Baban, B., Hansen, A. M., Marshall, B., Pihkala, J., Waldmann, H., Cobbold, S., Adams, E., Munn, D. H., Specific subsets of murine dendritic cells acquire potent T cell regulatory functions following CTLA4-mediated induction of indoleamine 2,3 dioxygenase. *Int. Immunol.* **2004**. 16: 1391–1401.

33 Mellor, A. L., Baban, B., Chandler, P., Marshall, B., Jhaver, K., Hansen, A., Koni, P. A., Iwashima, M., Munn, D. H., Cutting edge: induced indoleamine 2,3 dioxygenase expression in dendritic cell subsets suppresses T cell clonal expansion. *J. Immunol.* **2003**. 171: 1652–1655.

34 Fallarino, F., Asselin-Paturel, C., Vacca, C., Bianchi, R., Gizzi, S., Fioretti, M. C., Trinchieri, G., Grohmann, U., Puccetti, P., Murine plasmacytoid dendritic cells initiate the immuno-suppressive pathway of tryptophan catabolism in response to CD200 receptor engagement. *J. Immunol.* **2004**. 173: 3748–3754.

35 Hwu, P., Du, M. X., Lapointe, R., Do, M., Taylor, M. W., Young, H. A., Indole-amine 2,3-dioxygenase production by human dendritic cells results in the inhibition of T cell proliferation. *J. Immunol.* **2000**. 164: 3596–3599.

36 Munn, D. H., Sharma, M. D., Lee, J. R., Jhaver, K. G., Johnson, T. S., Keskin, D. B., Marshall, B., Chandler, P., Antonia, S. J., Burgess, R., Slingluff, C. L., Jr., Mellor, A. L., Potential regulatory function of human dendritic cells expressing indoleamine 2,3-dioxygenase. *Science* **2002**. 297: 1867–1870.

37 Yokoyama, W. M., What goes up must come down: the emerging spectrum of inhibitory receptors. *J. Exp. Med.* **1997**. 186: 1803–1808.

38 Colonna, M., Nakajima, H., Navarro, F., Lopez-Botet, M., A novel family of Ig-like receptors for HLA class I molecules that modulate function of lymphoid and myeloid cells. *J. Leukoc. Biol.* **1999**. 66: 375–381.

39 Colonna, M., Nakajima, H., Cella, M., A family of inhibitory and activating Ig-like receptors that modulate function of lymphoid and myeloid cells. *Semin. Immunol.* **2000**. 12: 121–127.

40 Ravetch, J. V., Lanier, L. L., Immune inhibitory receptors. *Science* **2000**. 290: 84–89.

41 Colonna, M., Samaridis, J., Cella, M., Angman, L., Allen, R. L., O'Callaghan, C. A., Dunbar, R., Ogg, G. S., Cerundolo, V., Rolink, A., Human myelomonocytic cells express an inhibitory receptor for classical and nonclassical MHC class I molecules. *J. Immunol.* **1998**. 160: 3096–3100.

42 Shiroishi, M., Tsumoto, K., Amano, K., Shirakihara, Y., Colonna, M., Braud, V. M., Allan, D. S., Makadzange, A., Rowland-Jones, S., Willcox, B., Jones, E. Y., van der Merwe, P. A., Kumagai, I., Maenaka, K., Human inhibitory receptors Ig-like transcript 2 (ILT2) and ILT4 compete with CD8 for MHC class I binding and bind preferentially to HLA-G. *Proc. Natl. Acad. Sci. USA* **2003**. 100: 8856–8861.

43 Cella, M., Dohring, C., Samaridis, J., Dessing, M., Brockhaus, M., Lanzavecchia, A., Colonna, M., A novel inhibitory receptor (ILT3) expressed on monocytes, macrophages, and dendritic cells involved in antigen processing. *J. Exp. Med.* **1997**. 185: 1743–1751.

44 Chang, C. C., Ciubotariu, R., Manavalan, J. S., Yuan, J., Colovai, A. I., Piazza, F., Lederman, S., Colonna, M., Cortesini, R., Dalla-Favera, R., Suciu-Foca, N., Tolerization of dendritic cells by T(S) cells: the crucial role of inhibitory receptors ILT3 and ILT4. *Nat. Immunol.* **2002**. 3: 237–243.

45 Suciu-Foca, N., Manavalan, J. S., Scotto, L., Kim-Schulze, S., Galluzzo, S., Naiyer, A. J., Fan, J., Vlad, G., Cortesini, R., Molecular characterization of allospecific T suppressor and tolerogenic dendritic cells: review. *Int. Immunopharmacol.* **2005**. 5: 7–11.

46 Manavalan, J. S., Kim-Schulze, S., Scotto, L., Naiyer, A. J., Vlad, G., Colombo, P. C., Marboe, C., Mancini, D., Cortesini, R., Suciu-Foca, N., Alloantigen specific CD8+CD28– FOXP3+ T suppressor cells induce ILT3+ ILT4+ tolerogenic endo-thelial cells, inhibiting alloreactivity. *Int. Immunol.* **2004**. 16: 1055–1068.

47 Takai, T., Ono, M., Activating and inhibitory nature of the murine paired immunoglobulin-like receptor family. *Immunol. Rev.* **2001**. 181: 215–222.

48 Liu, J., Liu, Z., Witkowski, P., Vlad, G., Manavalan, J. S., Scotto, L., Kim-Schulze, S., Cortesini, R., Hardy, M. A., Suciu-Foca, N., Rat CD8+ FOXP3+ T suppres-sor cells mediate tolerance to allogeneic heart transplants, inducing PIR-B in APC and rendering the graft invulner-able to rejection. *Transpl. Immunol.* **2004**. 13: 239–247.

49 Woltman, A. M., van Kooten, C., Func-tional modulation of dendritic cells to suppress adaptive immune responses. *J. Leukoc. Biol.* **2003**. 73: 428–441.

50 Moore, K. W., de Waal Malefyt, R., Coffman, R. L., O'Garra, A., Interleukin-10 and the interleukin-10 receptor. *Annu. Rev. Immunol.* **2001**. 19: 683–765.

51 Groux, H., O'Garra, A., Bigler, M., Rouleau, M., Antonenko, S., de Vries, J. E., Roncarolo, M. G., A CD4+ T-cell subset inhibits antigen-specific T-cell responses and prevents colitis. *Nature* **1997**. 389: 737–742.

52 Pestka, S., Krause, C. D., Sarkar, D., Walter, M. R., Shi, Y., Fisher, P. B., Interleukin-10 and related cytokines and receptors. *Annu. Rev. Immunol.* **2004**. 22: 929–979.

53 Asadullah, K., Sterry, W., Volk, H. D., Interleukin-10 therapy – review of a new approach. *Pharmacol. Rev.* **2003**. 55: 241–269.

54 Chen, W., Wahl, S. M., TGF-beta: receptors, signaling pathways and autoimmunity. *Curr. Dir. Autoimmun.* **2002**. 5: 62–91.

55 Wu, H. Y., Weiner, H. L., Oral tolerance. *Immunol. Res.* **2003**. 28: 265–284.

56 Calabresi, P. A., Fields, N. S., Maloni, H. W., Hanham, A., Carlino, J., Moore, J., Levin, M. C., Dhib-Jalbut, S., Tranquill, L. R., Austin, H., McFarland, H. F., Racke, M. K., Phase 1 trial of transforming growth factor beta 2 in chronic progressive MS. *Neurology* **1998**. 51: 289–292.

57 Geissmann, F., Revy, P., Regnault, A., Lepelletier, Y., Dy, M., Brousse, N., Amigorena, S., Hermine, O., Durandy, A., TGF-beta 1 prevents the noncognate maturation of human dendritic Langerhans cells. *J. Immunol.* **1999**. 162: 4567–4575.

58 Strobl, H., Knapp, W., TGF-beta1 regulation of dendritic cells. *Microbes Infect.* **1999**. 1: 1283–1290.

59 Lyakh, L. A., Sanford, M., Chekol, S., Young, H. A., Roberts, A. B., TGF-{beta} and Vitamin D3 Utilize Distinct Pathways to Suppress IL-12 Production and Modulate Rapid Differentiation of Human Monocytes into CD83+ Dendritic Cells. *J. Immunol.* **2005**. 174: 2061–2070.

60 Kehrl, J. H., Wakefield, L. M., Roberts, A. B., Jakowlew, S., Alvarez-Mon, M., Derynck, R., Sporn, M. B., Fauci, A. S., Production of transforming growth factor beta by human T lymphocytes and its potential role in the regulation of T cell growth. *J. Exp. Med.* **1986**. 163: 1037–1050.

61 Fantini, M. C., Becker, C., Monteleone, G., Pallone, F., Galle, P. R., Neurath, M. F., Cutting edge: TGF-beta induces a regulatory phenotype in CD4+CD25– T cells through Foxp3 induction and down-regulation of Smad7. *J. Immunol.* **2004**. 172: 5149–5153.

62 Strobl, H., Riedl, E., Bello-Fernandez, C., Knapp, W., Epidermal Langerhans cell development and differentiation. *Immunobiology* 1998. 198: 588–605.

63 Adorini, L., Immunotherapeutic approaches in multiple sclerosis. *J. Neurol. Sci.* **2004**. 223: 13–24.

64 Menges, M., Rossner, S., Voigtlander, C., Schindler, H., Kukutsch, N. A., Bogdan, C., Erb, K., Schuler, G., Lutz, M. B., Repetitive injections of dendritic cells matured with tumor necrosis factor alpha induce antigen-specific protection of mice from autoimmunity. *J. Exp. Med.* **2002**. 195: 15–21.

65 Lutz, M. B., Suri, R. M., Niimi, M., Ogilvie, A. L., Kukutsch, N. A., Rossner, S., Schuler, G., Austyn, J. M., Immature dendritic cells generated with low doses of GM-CSF in the absence of IL-4 are maturation resistant and prolong allograft survival in vivo. *Eur J. Immunol.* **2000**. 30: 1813–1822.

66 Sato, K., Yamashita, N., Baba, M., Matsuyama, T., Regulatory dendritic cells protect mice from murine acute graft-versus-host disease and leukemia relapse. *Immunity* **2003**. 18: 367–379.

67 Franzke, A., Piao, W., Lauber, J., Gatzlaff, P., Konecke, C., Hansen, W., Schmitt-Thomsen, A., Hertenstein, B., Buer, J., Ganser, A., G-CSF as immune regulator in T cells expressing the G-CSF receptor: implications for transplantation and autoimmune diseases. *Blood* **2003**. 102: 734–739.

68 Rutella, S., Lemoli, R. M., Regulatory T cells and tolerogenic dendritic cells: from basic biology to clinical applications. *Immunol. Lett.* **2004**. 94: 11–26.

69 Rutella, S., Pierelli, L., Bonanno, G., Sica, S., Ameglio, F., Capoluongo, E., Mariotti, A., Scambia, G., d'Onofrio, G., Leone, G., Role for granulocyte colony-stimulating factor in the generation of human T regulatory type 1 cells. *Blood* **2002**. 100: 2562–2571.

70 Rutella, S., Bonanno, G., Pierelli, L., Mariotti, A., Capoluongo, E., Contemi, A. M., Ameglio, F., Curti, A., De Ritis, D. G., Voso, M. T., Perillo, A., Mancuso, S., Scambia, G., Lemoli, R. M., Leone, G.,

Granulocyte colony-stimulating factor promotes the generation of regulatory DC through induction of IL-10 and IFN-alpha. *Eur J. Immunol.* **2004**. 34: 1291–1302.

71 Kared, H., Masson, A., Adle-Biassette, H., Bach, J. F., Chatenoud, L., Zavala, F., Treatment with granulocyte colony-stimulating factor prevents diabetes in NOD mice by recruiting plasmacytoid dendritic cells and functional CD4(+)CD25(+) regulatory T-cells. *Diabetes* **2005**. 54: 78–84.

72 Allison, A. C., Immunosuppressive drugs: the first 50 years and a glance forward. *Immunopharmacology* **2000**. 47: 63–83.

73 Lagaraine, C., Lebranchu, Y., Effects of immunosuppressive drugs on dendritic cells and tolerance induction. *Transplantation* **2003**. 75: 37S-42S.

74 Vieira, P. L., Kalinski, P., Wierenga, E. A., Kapsenberg, M. L., de Jong, E. C., Glucocorticoids inhibit bioactive IL-12p70 production by in vitro-generated human dendritic cells without affecting their T cell stimulatory potential. *J. Immunol.* **1998**. 161: 5245–5251.

75 Matasic, R., Dietz, A. B., Vuk-Pavlovic, S., Dexamethasone inhibits dendritic cell maturation by redirecting differentiation of a subset of cells. *J. Leukoc. Biol.* **1999**. 66: 909–914.

76 Piemonti, L., Monti, P., Allavena, P., Sironi, M., Soldini, L., Leone, B. E., Socci, C., Di Carlo, V., Glucocorticoids affect human dendritic cell differentiation and maturation. *J. Immunol.* **1999**. 162: 6473–6481.

77 Woltman, A. M., de Fijter, J. W., Kamerling, S. W., Paul, L. C., Daha, M. R., van Kooten, C., The effect of calcineurin inhibitors and corticosteroids on the differentiation of human dendritic cells. *Eur J. Immunol.* **2000**. 30: 1807–1812.

78 Mehling, A., Grabbe, S., Voskort, M., Schwarz, T., Luger, T. A., Beissert, S., Mycophenolate mofetil impairs the maturation and function of murine dendritic cells. *J. Immunol.* **2000**. 165: 2374–2381.

79 Colic, M., Stojic-Vukanic, Z., Pavlovic, B., Jandric, D., Stefanoska, I., Mycophenolate mofetil inhibits differentiation, maturation and allostimulatory function of human monocyte-derived dendritic cells. *Clin. Exp. Immunol.* **2003**. 134: 63–69.

80 Woltman, A. M., de Fijter, J. W., Kamerling, S. W., van Der Kooij, S. W., Paul, L. C., Daha, M. R., van Kooten, C., Rapamycin induces apoptosis in monocyte- and CD34-derived dendritic cells but not in monocytes and macrophages. *Blood* **2001**. 98: 174–180.

81 Hackstein, H., Taner, T., Zahorchak, A. F., Morelli, A. E., Logar, A. J., Gessner, A., Thomson, A. W., Rapamycin inhibits IL-4—induced dendritic cell maturation in vitro and dendritic cell mobilization and function in vivo. *Blood* **2003**. 101: 4457–4463.

82 Taner, T., Hackstein, H., Wang, Z., Morelli, A. E., Thomson, A. W., Rapamycin-treated, alloantigen-pulsed host dendritic cells induce ag-specific T cell regulation and prolong graft survival. *Am. J. Transplant.* **2005**. 5: 228–236.

83 Lee, J. I., Ganster, R. W., Geller, D. A., Burckart, G. J., Thomson, A. W., Lu, L., Cyclosporine A inhibits the expression of costimulatory molecules on in vitro-generated dendritic cells: association with reduced nuclear translocation of nuclear factor kappa B. *Transplantation* **1999**. 68: 1255–1263.

84 Szabo, G., Gavala, C., Mandrekar, P., Tacrolimus and cyclosporine A inhibit allostimulatory capacity and cytokine production of human myeloid dendritic cells. *J. Investig. Med.* **2001**. 49: 442–449.

85 Contreras, J. L., Wang, P. X., Eckhoff, D. E., Lobashevsky, A. L., Asiedu, C., Frenette, L., Robbin, M. L., Hubbard, W. J., Cartner, S., Nadler, S., Cook, W. J., Sharff, J., Shiloach, J., Thomas, F. T., Neville, D. M., Jr., Thomas, J. M., Peritransplant tolerance induction with anti-CD3-immunotoxin: a matter of proinflammatory cytokine control. *Transplantation* **1998**. 65: 1159–1169.

86 Thomas, J. M., Contreras, J. L., Jiang, X. L., Eckhoff, D. E., Wang, P. X., Hubbard, W. J., Lobashevsky, A. L., Wang, W., Asiedu, C., Stavrou, S., Cook, W. J., Robbin, M. L., Thomas, F. T., Neville, D. M., Jr., Peritransplant tolerance induction in macaques: early events reflecting the unique synergy between immunotoxin and deoxyspergualin. *Transplantation* 1999. 68: 1660–1673.

87 Thomas, J. M., Hubbard, W. J., Sooudi, S. K., Thomas, F. T., STEALTH matters: a novel paradigm of durable primate allograft tolerance. *Immunol. Rev.* 2001. 183: 223–233.

88 Matasic, R., Dietz, A. B., Vuk-Pavlovic, S., Cyclooxygenase-independent inhibition of dendritic cell maturation by aspirin. *Immunology* 2000. 101: 53–60.

89 Hackstein, H., Morelli, A. E., Larregina, A. T., Ganster, R. W., Papworth, G. D., Logar, A. J., Watkins, S. C., Falo, L. D., Thomson, A. W., Aspirin inhibits in vitro maturation and in vivo immunostimulatory function of murine myeloid dendritic cells. *J. Immunol.* 2001. 166: 7053–7062.

90 Millard, A. L., Mertes, P. M., Ittelet, D., Villard, F., Jeannesson, P., Bernard, J., Butyrate affects differentiation, maturation and function of human monocyte-derived dendritic cells and macrophages. *Clin. Exp. Immunol.* 2002. 130: 245–255.

91 Verhasselt, V., Vanden Berghe, W., Vanderheyde, N., Willems, F., Haegeman, G., Goldman, M., N-acetyl-L-cysteine inhibits primary human T cell responses at the dendritic cell level: association with NF-kappaB inhibition. *J. Immunol.* 1999. 162: 2569–2574.

92 Gregori, S., Casorati, M., Amuchastegui, S., Smiroldo, S., Davalli, A. M., Adorini, L., Regulatory T cells induced by 1α,25-Dihydroxyvitamin D$_3$ and mycophenolate mofetil treatment mediate transplantation tolerance. *J. Immunol.* 2001. 167: 1945–1953.

93 Izawa, A., Sayegh, M. H., Chandraker, A., The antagonism of calcineurin inhibitors and costimulatory blockers: fact or fiction? *Transplant. Proc.* 2004. 36: 570S-573S.

94 Saemann, M. D., Kelemen, P., Bohmig, G. A., Horl, W. H., Zlabinger, G. J., Hyporesponsiveness in alloreactive T-cells by NF-kappaB inhibitor-treated dendritic cells: resistance to calcineurin inhibition. *Am. J. Transplant.* 2004. 4: 1448–1458.

95 Game, D. S., Hernandez-Fuentes, M. P., Lechler, R. I., Everolimus and Basiliximab Permit Suppression by Human CD4CD25 Cells in vitro. *Am. J. Transplant.* 2005. 5: 454–464.

96 Deluca, H. F., Cantorna, M. T., Vitamin D: its role and uses in immunology. *Faseb J.* 2001. 15: 2579–2585.

97 Mathieu, C., Adorini, L., The coming of age of 1,25-dihydroxyvitamin D(3) analogs as immunomodulatory agents. *Trends Mol. Med* . 2002. 8: 174–179.

98 Adorini, L., Immunomodulatory effects of vitamin D receptor ligands in autoimmune diseases. *Int. Immunopharmacol.* 2002. 2: 1017–1028.

99 Adorini, L., 1,25-Dihydroxyvitamin D3 analogs as potential therapies in transplantation. *Curr. Opin. Investig. Drugs* 2002. 3: 1458–1463.

100 Griffin, M. D., Xing, N., Kumar, R., Vitamin D and its Analogs as Regulators of Immune Activation and Antigen Presentation. *Annu. Rev. Nutr.* 2003.

101 Carlberg, C., Polly, P., Gene regulation by vitamin D3. *Crit. Rev. Eukaryot. Gene Expr.* 1998. 8: 19–42.

102 Penna, G., Adorini, L., 1,25-dihydroxyvitamin D3 inhibits differentiation, maturation, activation and survival of dendritic cells leading to impaired alloreactive T cell activation. *J. Immunol.* 2000. 164: 2405–2411.

103 Piemonti, L., Monti, P., Sironi, M., Fraticelli, P., Leone, B. E., Dal Cin, E., Allavena, P., Di Carlo, V., Vitamin D3 affects differentiation, maturation, and function of human monocyte-derived dendritic cells. *J. Immunol.* 2000. 164: 4443–4451.

104 Griffin, M. D., Lutz, W. H., Phan, V. A., Bachman, L. A., McKean, D. J., Kumar, R., Potent inhibition of dendritic cell differentiation and maturation by vitamin D analogs. *Biochem. Biophys. Res. Commun.* 2000. 270: 701–708.

105 Berer, A., Stockl, J., Majdic, O., Wagner, T., Kollars, M., Lechner, K., Geissler, K., Oehler, L., 1,25-Dihydroxyvitamin D(3) inhibits dendritic cell differentiation and maturation in vitro. *Exp. Hematol.* 2000. 28: 575–583.

106 Canning, M. O., Grotenhuis, K., de Wit, H., Ruwhof, C., Drexhage, H. A., 1-alpha,25-Dihydroxyvitamin D3 (1,25(OH)(2)D(3)) hampers the maturation of fully active immature dendritic cells from monocytes. *Eur. J. Endocrinol.* **2001.** 145: 351–357.

107 van Halteren, A. G., van Etten, E., de Jong, E. C., Bouillon, R., Roep, B. O., Mathieu, C., Redirection of human autoreactive T-cells Upon interaction with dendritic cells modulated by TX527, an analog of 1,25 dihydroxyvitamin D(3). *Diabetes* **2002.** 51: 2119–2125.

108 Hewison, M., Freeman, L., Hughes, S. V., Evans, K. N., Bland, R., Eliopoulos, A. G., Kilby, M. D., Moss, P. A., Chakraverty, R., Differential regulation of vitamin D receptor and its ligand in human monocyte-derived dendritic cells. *J. Immunol.* **2003.** 170: 5382–5390.

109 Griffin, M. D., Lutz, W., Phan, V. A., Bachman, L. A., McKean, D. J., Kumar, R., Dendritic cell modulation by 1alpha,25 dihydroxyvitamin D3 and its analogs: A vitamin D receptor-dependent pathway that promotes a persistent state of immaturity in vitro and in vivo. *Proc. Natl. Acad. Sci. USA* **2001.** 22: 22.

110 Dhodapkar, M. V., Steinman, R. M., Krasovsky, J., Munz, C., Bhardwaj, N., Antigen-specific inhibition of effector T cell function in humans after injection of immature dendritic cells. *J. Exp. Med.* **2001.** 193: 233–238.

111 Gregori, G., Giarratana, N., Smiroldo, S., Uskokovic, M., Adorini, L., A 1α,25-Dihydroxyvitamin D₃ analog enhances regulatory T cells and arrests autoimmune diabetes in NOD mice. *Diabetes* **2002.** 51: 1367–1374.

112 Barrat, F. J., Cua, D. J., Boonstra, A., Richards, D. F., Crain, C., Savelkoul, H. F., de Waal-Malefyt, R., Coffman, R. L., Hawrylowicz, C. M., O'Garra, A., In vitro generation of interleukin 10-producing regulatory CD4(+) T cells is induced by immunosuppressive drugs and inhibited by T helper type 1 (Th1)- and Th2-inducing cytokines. *J. Exp. Med.* **2002.** 195: 603–616.

113 Cella, M., Jarrossay, D., Facchetti, F., Alebardi, O., Nakajima, H., Lanzavecchia, A., Colonna, M., Plasmacytoid monocytes migrate to inflamed lymph nodes and produce large amounts of type I interferon. *Nat. Med.* **1999.** 5: 919–923.

114 Penna, G., Sozzani, S., Adorini, L., Cutting edge: selective usage of chemokine receptors by plasmacytoid dendritic cells. *J. Immunol.* **2001.** 167: 1862–1866.

115 Penna, G., Roncari, A., Colonna, M., Adorini, L., 1,25 dihydroxyvitamin D3 upregulates the expression of the inhibitory receptor ILT3 on dendritic cells. *Minerva Biotec* **2002.** 14: 71.

116 Gilliet, M., Liu, Y. J., Generation of human CD8 T regulatory cells by CD40 ligand-activated plasmacytoid dendritic cells. *J. Exp. Med.* **2002.** 195: 695–704.

117 Manavalan, J. S., Rossi, P. C., Vlad, G., Piazza, F., Yarilina, A., Cortesini, R., Mancini, D., Suciu-Foca, N., High expression of ILT3 and ILT4 is a general feature of tolerogenic dendritic cells. *Transpl. Immunol.* **2003.** 11: 245–258.

118 Xing, N., ML, L. M., Bachman, L. A., McKean, D. J., Kumar, R., Griffin, M. D., Distinctive dendritic cell modulation by vitamin D(3) and glucocorticoid pathways. *Biochem. Biophys. Res. Commun.* **2002.** 297: 645–652.

119 Giarratana, N., Penna, G., Amuchastegui, S., Mariani, R., Daniel, K. C., Adorini, L., A vitamin D analog downregulates proinflammatory chemokine production by pancreatic islets inhibiting T cell recruitment and type 1 diabetes development. *J. Immunol.* **2004.** 173: 2280–2287.

120 Xie, J. H., Nomura, N., Lu, M., Chen, S. L., Koch, G. E., Weng, Y., Rosa, R., Di Salvo, J., Mudgett, J., Peterson, L. B., Wicker, L. S., DeMartino, J. A., Antibody-mediated blockade of the CXCR3 chemokine receptor results in diminished recruitment of T helper 1 cells into sites of inflammation. *J. Leukoc. Biol.* **2003.** 73: 771–780.

121 Frigerio, S., Junt, T., Lu, B., Gerard, C., Zumsteg, U., Hollander, G. A., Piali, L., Beta cells are responsible for CXCR3-mediated T-cell infiltration in insulitis. *Nat. Med.* **2002.** 8: 1414–1420.

122 D'Ambrosio, D., Sinigaglia, F., Adorini, L., Special attractions for suppressor T

cells. *Trends Immunol.* **2003**. 24: 122–126.

123 Curiel, T. J., Coukos, G., Zou, L., Alvarez, X., Cheng, P., Mottram, P., Evdemon-Hogan, M., Conejo-Garcia, J. R., Zhang, L., Burow, M., Zhu, Y., Wei, S., Kryczek, I., Daniel, B., Gordon, A., Myers, L., Lackner, A., Disis, M. L., Knutson, K. L., Chen, L., Zou, W., Specific recruitment of regulatory T cells in ovarian carcinoma fosters immune privilege and predicts reduced survival. *Nat. Med.* **2004**. 10: 942–949.

124 Penna, G., Vulcano, M., Roncari, A., Facchetti, F., Sozzani, S., Adorini, L., Differential chemokine production by myeloid and plasmacytoid dendritic cells. *J. Immunol.* **2002**. 169: 6673–6676.

125 Adorini, L., Interleukin-12, a key cytokine in Th1-mediated autoimmune diseases. *Cell Mol. Life Sci.* **1999**. 55: 1610–1625.

126 Li, Q., Verma, I. M., NF-kappaB regulation in the immune system. *Nat. Rev. Immunol.* **2002**. 2: 725–734.

127 Martin, E., O'Sullivan, B., Low, P., Thomas, R., Antigen-specific suppression of a primed immune response by dendritic cells mediated by regulatory T cells secreting interleukin-10. *Immunity* **2003**. 18: 155–167.

128 Yu, X. P., Bellido, T., Manolagas, S. C., Down-regulation of NF-kappa B protein levels in activated human lymphocytes by 1,25-dihydroxyvitamin D3. *Proc. Natl. Acad. Sci. USA* **1995**. 92: 10 990–10 994.

129 Dong, X., Craig, T. A., Xing, N., Bachman, L. A., Paya, C. V., Weih, F., McKean, D. J., Kumar, R., Griffin, M. D., Direct transcriptional regulation of RelB by 1alpha,25-dihydroxyvitamin D3 and its analogs: Physiologic and therapeutic implications for dendritic cell function. *J. Biol. Chem.* **2003**. 278: 49 378–49 385.

130 Scheinman, R. I., Cogswell, P. C., Lofquist, A. K., Baldwin, A. S., Jr., Role of transcriptional activation of I kappa B alpha in mediation of immunosuppression by glucocorticoids. *Science* **1995**. 270: 283–286.

131 Auphan, N., DiDonato, J. A., Rosette, C., Helmberg, A., Karin, M., Immunosuppression by glucocorticoids: inhibition of NF-kappa B activity through induction of I kappa B synthesis. *Science* **1995**. 270: 286–290.

132 Cardozo, A. K., Heimberg, H., Heremans, Y., Leeman, R., Kutlu, B., Kruhoffer, M., Orntoft, T., Eizirik, D. L., A comprehensive analysis of cytokine-induced and nuclear factor-kappa B-dependent genes in primary rat pancreatic beta-cells. *J. Biol. Chem.* **2001**. 276: 48 879–48 886.

133 Becker, B. N., Hullett, D. A., O'Herrin, J. K., Malin, G., Sollinger, H. W., DeLuca, H., Vitamin D as immunomodulatory therapy for kidney transplantation. *Transplantation* **2002**. 74: 1204–1206.

134 Vulcano, M., Struyf, S., Scapini, P., Cassatella, M., Bernasconi, S., Bonecchi, R., Calleri, A., Penna, G., Adorini, L., Luini, W., Mantovani, A., Van Damme, J., Sozzani, S., Unique regulation of CCL18 production by maturing dendritic cells. *J. Immunol.* **2003**. 170: 3843–3849.

135 Penna, G., Roncari, A., Amuchastegui, S., Daniel, K. C., Berti, E., Colonna, M., Adorini, L., Expression of the inhibitory receptor ILT3 on dendritic cells is dispensable for induction of CD4+ Foxp3+ regulatory T cells by 1,25-dihydroxyvitamin D3. *Blood* **2005**. *106:* 3490–3497.

28
Surface Molecules Involved in the Induction of Tolerance by Dendritic Cells

Laura C. Bonifaz

28.1
Introduction

To prevent autoimmunity and collateral damage during ongoing immune respons-es, T lymphocytes, as components of the adaptive immune system, need to remain tolerant to self- and innocuous environmental antigens (Ag). As a specific process, Ag are needed for the induction of tolerance. Although for central tolerance, the transcription factor autoimmune regulator protein (AIRE) [1], and possibly others, permits the expression of genes that encode tissue-specific Ag by thymic medul-lary epithelial cells, it is unlikely that all self-Ag and innocuous environmental Ag can be present in the thymus at any given time. Therefore, additional mechanisms of tolerance in peripheral lymphoid tissue are important to avoid undesirable im-mune responses.

The first step in the induction of tolerance to Ag that are not expressed by anti-gen-presenting cells is the sampling of these exogenous proteins by endocytic re-ceptors, which is followed by processing and presentation to antigen-specific T cells of peptides, bound to MHC molecules. Dendritic cells (DC), as efficient anti-gen-capturing and processing cells, are appropriate candidates to participate in the induction of the tolerance process. Several questions need to be answered in order to understand the circumstances under which DC, known for their specialized im-munizing properties, also participate in the opposing processes of tolerance.

Tolerance, by definition, is a physiological state of unresponsiveness. To ensure this state, three major mechanisms are currently accepted, which are thought to in-volve: (a) clonal deletion, (b) clonal anergy, and (c) different types of T regulatory cells such as the natural $CD4^+ CD25^+$ variety.

Clonal deletion appears to be the main mechanism for central tolerance in the thymus [2]. However, it has also been shown to operate in the periphery. Anergy has been described mainly in the periphery, and in some experimental models is still regarded as controversial [3]. $CD4^+CD25^+$ T cells are generated during develop-ment in the thymus, but they also can be expanded and possibly induced *de novo* by DC in the periphery [4].

Handbook of Dendritic Cells. Biology, Diseases, and Therapies.
Edited by M. B. Lutz, N. Romani, and A. Steinkasserer.
Copyright © 2006 WILEY-VCH Verlag GmbH & Co. KGaA, Weinheim
ISBN: 3-527-31109-2

28.2
Dendritic Cells and Central Tolerance

In addition to medullary epithelial cells, bone-marrow derived DC play an important role in the induction of central tolerance. As mentioned, the main mechanism of tolerance in the thymus is activation-induced cell death, which leads to deletion of those T cells that recognize self-peptides bound to MHC class I and class II molecules on the surface of APC. For instance, experiments carried out with the fifth component of the complement C5 as a model Ag showed that DC that are pulsed with low doses of C5 in thymic organ cultures are able to delete C5-specific transgenic T cells. Macrophages lacked this capacity but medullary epithelium was also able to delete T cells. These experiments, supported by others, suggest that DC together with epithelial medullary cells are pivotal in the induction of central tolerance [5–7].

A question remains: are all DC capable of inducing tolerance in the thymus or do thymic DC have special characteristics? The studies of Matzinger and Guerder [8] showed that allogenic DC from the spleen, when introduced into thymic organ cultures, were able to induce tolerance to alloantigens. Thus, a "professional" antigen-presenting cell from peripheral lymphoid tissue is able to activate peripheral T cells and also render thymocytes tolerant. These studies and some others [9] suggested that it is the developmental stage of the T cells which determines the negative outcome of Ag presentation rather than the nature of the DC. Additional evidence supports the notion that the threshold for tolerance induction in the thymus is lower than the threshold required by mature T cells for activation [10]. Thus, it is possible that specialized antigen-presenting cells are not required in the thymus for the induction of tolerance, except for the need of APC to efficiently capture, process and present self-Ag. Because of their efficient endocytic activity, due in part to the expression of specific receptors, and also to their ability to present exogenous Ag with MHC class I molecules (crosspresentation), it is possible that DC are crucial in the thymus for sampling the majority of self-Ag, and even to crosspresent self-Ag expressed in the thymic medullary epithelium [11].

Another important process during T-cell development is the generation of regulatory T cells, a type of suppressor cells. Growing evidence suggests an important role for these cells in the maintenance of peripheral tolerance. New evidence from Yong Jun Liu and coworkers indicates that thymic DC, when conditioned by thymic stromal lymphopoietin made by Hassall's corpuscles, play an important role in the generation of these thymic-born CD4$^+$ CD25$^+$ regulatory T cells from CD4$^+$ CD25$^-$ single positive thymocytes [12].

28.3
Dendritic Cells and Peripheral Tolerance

DC have been defined as the most potent professional antigen-presenting cells as well as "natural adjuvants" in the induction of antigen-specific immune responses

[13–16]. Surprisingly, they are also involved in the opposite outcome: the induction of peripheral tolerance [17–19]. Three questions arise:

What is the difference between "immune" DC and "tolerogenic" DC?

Are they the same cell type with distinct maturation states and phenotypes?

Conversely, are there distinct tolerogenic and immunogenic DC?

Some experimental evidence supports the idea for different maturation states between "resting" tolerogenic DC versus "activated" immunogenic DC. Others also suggest the presence of special DC for tolerance. However, an alternative point of view involves a common cell type phenotypically defined by the balance of positive versus negative signals at the time of the interaction between the DC and T cell.

Much like tolerance in the thymus, the sampling of exogenous Ag by APC in the periphery is a crucial point in the induction of specific immune tolerance. The capture of Ag by endocytic receptors on DC and their further presentation by MHC class II and class I molecules is essential; otherwise this process could be inefficient and not antigen-specific. DC express different groups of receptors: some of which are for capturing Ag; others are involved in increasing efficiency of presentation by class II molecules; and one last group is involved in Ag presentation by class I molecules. Ag presentation by both MHC pathways is crucial for tolerance induction in both CD4$^+$ and CD8$^+$ compartments. Three major groups of Ag receptors expressed by DC are involved in tolerance induction: (a) c-type lectins, which bind glycoproteins; (b) integrins, which contribute to the uptake of apoptotic cells at least *in vitro*; (c) Fc receptors that bind complexes of Ag with antibodies.

28.4
C-type Lectin Receptors

DC express a number of C-type lectin receptors (CLR) that contain single or multiple carbohydrate recognition domains. CLR bind structures on pathogens as well as self-glycoproteins, and this binding can be specific for the type of glycan and their organization patterns, thereby creating unique sets of carbohydrate recognition profiles. For instance, on DC, mannose receptors (MR) recognize single mannose moieties, whereas DC-SIGN has high affinity for more complex mannose residues in specific arrangements [20–22]. CLR have been shown to be involved in endocytosis for Ag presentation by class II molecules. This has been shown for MR (CD206), DEC-205 (CD205), DC-SIGN (CD209) and BDCA-2 [23–26]. Endocytosis by these receptors is mediated by conserved motifs in their cytoplasmic regions. Interestingly, after endocytosis, different intracellular routes are utilized for the different receptors. In bone-marrow DC, MR recycles from the early endosomes to the cell surface. By contrast, DEC-205 is able to further penetrate the endocytic pathway. This leads to a more efficient antigen presentation by class II molecules compared with MR [27].

Another important characteristic of the CLR is their ability to send intracellular signals at the time of the antigen capture. To date, cytoplasmic motifs with signaling capacities have been described in some of these receptors. DC-SIGN, the asialoglycoprotein receptor, Dectin-1, CLEC-1 and 2 contain cytoplasmic tyrosine residues that are part of the so-called immunoreceptor tyrosine-based activation motifs (ITAM) as well as the DC immunoreceptors (DCIR), which have potential inhibitory motifs (ITIM) [28, 29]. However, the role of these signaling motifs in the balance between tolerance and immunity has not been examined. In contrast, some other CLRs, including DEC-205 and BDCA-2, are devoid of signaling motifs.

Although CLRs are expressed by DC, some are also expressed by other cells. Thus, MR is expressed by macrophages but not by DC from the T-cell areas of the LN. Expression of most of these receptors is high in immature DC and decreases after maturation. An exception is DEC-205, which is expressed more by mature DC. In addition, some CLR are differentially expressed by certain subsets of DC.

28.4.1
Advantages of DEC-205 as an Endocytic Receptor for Antigen Presentation

Because of their characteristics, CLRs are interesting candidates to target Ag into DC. At this time, DEC-205 is the most widely studied receptor for targeting Ag into DC *in vivo*, while preserving their resting or steady state. The characterization of DEC-205 as a potent Ag receptor involved in antigen presentation expressed by DC but not by macrophages, and its lack of signaling motifs, are the factors backing this assertion. As the natural ligand of DEC-205 is still unknown, trials to target Ag into DC through this receptor have been carried out using mAb as a surrogate ligand. The first studies using this system involved the subcutaneous (s.c.) injection of the antibody, NLDC [30]. After injection of 10 μg of anti-DEC-205, most DC in the T-cell area of the LN were labeled. This was specific for DC as no label could be shown on T cells, B cells or macrophages [31]. This experiment was extended so that after harvesting different lymphoid tissues from mice injected s.c. with fluorescent anti-DEC-205, it was possible to see labels in ~60% DC from the draining LN. It was also possible to visualize DC from all LN including mediastinal, mesenteric and splenic nodes. This study showed the potential to load large numbers of DC systemically. After chemical conjugation of the Ag ovalbumin (OVA) with anti-DEC-205, it was also possible to detect the systemic distribution of the coupled Ag in DC [32].

The next set of experiments focused on the ability of this receptor to enhance antigen presentation *in vivo*. Thus, a MHC II restricted hen egg lysozyme (HEL) peptide expressed as a fusion protein with the heavy chain of anti-DEC-205 required only a few nanograms to induce efficient class II presentation as revealed by the proliferation of HEL antigen-specific 3A9 transgenic CD4+ T cells [31]. After chemical coupling of OVA with anti-DEC-205, it was also possible to see proliferation of OVA-specific transgenic CD4+ T cells and OVA-specific CD8+ T cells. These studies indicated the potential of this receptor to mediate presentation on both MHC class II and, interestingly, class I (crosspresentation). In the case of OVA, the increase in the efficiency of Ag presentation, compared with the uncoupled Ag , was 500 times for MHC class II and 10 000 times for MHC class I. This antigen presentation *in vivo* was mediated by DC since only CD11c$^+$ cells were capable of inducing proliferation of specific T cells *in vitro* after *in vivo* injection of the Ab OVA conjugates. The ability to present the Ag by DC was also systemic, as DC taken from distal lymph nodes and from the spleen induced Ag-specific T-cell proliferation after s.c. injection [33].

28.4.2
DEC-205: an Endocytic Receptor that Preserves the Steady State in the DC after the Capture of the Antigen

The next experiments using the DEC-205 model to target Ag to DC *in situ* were designed to determine the DC phenotype after targeting by Ab alone or conjugated with the Ag. The delivery through anti-DEC-205 did not change the phenotype of the DC, suggesting that after ligation the DEC-205 receptor is not able to send a maturation signal [31, 33]. Similar results were achieved in the presence of antigen-specific T cells. The fact that after Ag targeting, DC do not undergo any apparent change creates the opportunity to use this model to evaluate the role of resting DC in the induction of peripheral tolerance.

28.5
Induction of Peripheral Tolerance by Resting Dendritic Cells

Experimentally, tolerance is defined as antigen-specific nonresponsiveness to a challenge with Ag in the presence of potent adjuvants. Strong experimental evidence suggests the participation of DC in the induction of peripheral tolerance [34–39]. The first experiments to evaluate the consequences of Ag delivery specifically to resting DC *in situ* were made by Hawiger and coworkers. After CD45.1$^+$ CD4$^+$ transgenic HEL-specific T cells were adoptively transferred into a mouse injected with anti-DEC-205 HEL peptide, it was possible to see a vigorous proliferation of antigen-specific T cells. The proliferation was comparable to that of mice injected with the HEL peptide in CFA. HEL-specific T cells were not detected 7–14 - days after the initial injection with anti-DEC-HEL. More importantly, at day 7, these cells were unable to respond to the challenge with the HEL peptide in CFA [31]. Although these results show that resting DC are efficient antigen-presenting cells, able to induce strong T-cell proliferation, the final consequence of this initial antigen presentation was antigen-specific unresponsiveness. These studies were confirmed using OVA conjugated with anti-DEC-205. In this case the tracked cells were OVA-specific CD8$^+$ T cells. Thus, Ag taken up by DEC-205 can induce tolerance also in the CD8$^+$ compartment [33]. In both systems, the mechanism proposed for tolerance was clonal deletion, as it was not possible to detect the relevant T cells 7–14 days post-injection even in nonlymphoid tissues.

28.5.1

The Same Dendritic Cells Could Operate in the Induction of Immunity

One important point to clarify is whether the DC involved in the induction of tolerance were also able to induce immunity and reverse the tolerance upon a maturation stimulus. After coinjection of an agonistic Ab against CD40, it was possible to induce a strong CD4 response after injection of anti-DEC-205 HEL [31]. Moreover, a combined CD4 and CD8 immune response, including IFNγ release and cytolytic CD8+ T cells was induced after the injection of anti-DEC-OVA [33]. Cumulatively, these results show that resting or steady state DC are able to induce peripheral tolerance and suggest that the same DC, depending on their status or active phenotype, can also induce an immune response (Fig. 28.1).

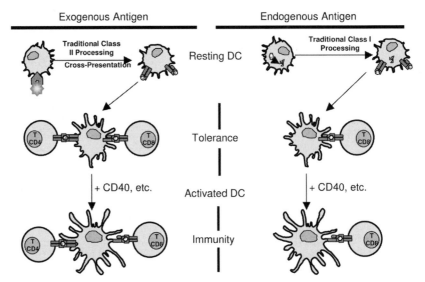

Fig. 28.1 Resting versus activated DC in the induction of tolerance or immunity. Endogenous or exogenous Ag are processed and presented, in the context of class I and class II molecules, by resting DC. The consequence of this antigen presentation is the induction of antigen-specific tolerance in the CD4 and CD8 compartments, accordingly. Activation of DC by agonistic anti-CD40 leads to the induction of immune responses.

28.5.2

The Induction of Tolerance by Steady-state Dendritic Cells Promotes Avoidance of the Induction of Autoimmunity

The next question, when using the anti-DEC-205 targeting system, is how deep is the induced tolerance? Is it sufficient to protect from autoimmune disease? Nussenzweig and coworkers showed, by using the acute EAE model, that tolerance induced after injection of just 15 _g of encephalitogenic oligodendrocyte glycoprotein (MOG) peptide coupled to anti-DEC-205 was sufficient to protect mice from autoimmune disease [40]. After examination of the spinal cord of treated and control mice an increased numbers of CD4+ T cells could be identified in spinal cord of mice that show symptoms of EAE, no such cells were found in spinal cords of mice that had been pre-treated with anti-DEC-205 MOG peptide. Thus targeting resting DC with MOG peptide prevents accumulation of effector T cells in the nervous system preventing the induction of EAE.

28.5.3
Surface Molecules are Involved in Peripheral Tolerance Induction by Resting Dendritic Cells through DEC-205

As previously mentioned, in the first two systems using anti-DEC HEL peptide or anti-DEC coupled with OVA, the proposed mechanism for tolerance was deletion because of the inability to find the adoptively transferred antigen-specific T cells 7–14 days after the induction of tolerance. Interestingly, in the system using anti-DEC-205 MOG peptide, antigen-specific T cells remain in the mice 3–7 days after injection. These cells were not fully anergic since they remain responsive to T-cell receptor stimulation *in vitro* but they failed to respond to Ag *in vivo* [40]. The *in vivo* phenotype of unresponsive T cells was similar to responsive cells in the expression of all the classical activation markers such as CD69, CD25 CD44 and CD62L. The only difference was the expression of CD5 by tolerant T cells, which is an inducible negative regulator during selection of thymocytes. Additional experiments suggested that CD5 was involved in the induction of tolerance after injection of anti-DEC MOG. Therefore, assuming that the same resting DC are involved in the presentation of all the Ag coupled to anti-DEC-205, the question is: how can the same DC induce different tolerance mechanisms? One possibility is that the affinity of the TCR at the time of the interaction dictates the final outcome.

28.5.4
Additional Evidence Supports the Role of Resting DC in the Induction of Peripheral Tolerance

Additional studies by Probst et al. support the hypothesis that the status of DC determines the induction of peripheral tolerance versus immunity. The system used was an inducible transgenic model, which permits the expression and further Ag presentation of peptides from lymphocytic choriomeningitis virus (LCMV) exclusively by resting DC. After presentation of LCMV peptides in the absence of infection, specific peripheral tolerance was observed. The tolerance was deep because it could not be overcome by a subsequent infection with LCMV [41]. These experiments were carried out without adoptively transferring T cells. In this system, the Ag is expressed in the cytosol and it is presented by the classical endogenous pathway by MHC class I molecules. Approximately 5% of DC express the Ag without restriction to a special subset, suggesting that this conclusion can be applied to practically any DC. On the other hand, antigen presentation of the LCMV peptides by activated DC with the agonist anti CD40 Ab leads to the induction of protective immunity, providing further evidence that the same cells can participate in the induction of tolerance or immunity (Fig. 28.1).

28.6
Surface Molecules Involved in the Induction of Peripheral Tolerance

Although it is clear that steady-state DC are involved in the induction of peripheral tolerance, more information is needed about the surface molecules on DC and T cells that are important for the initial proliferation and activation but also for the final disappearance or functional inactivation of the T cells. One simple explanation could be the difference in the level of co-stimulatory molecules expressed by resting versus activated DC. Considering that resting DC are immature or semi-immature with low expression of co-stimulatory molecules compared with mature or activated DC, it is possible to explain the difference in outcome after T-cell interaction by the classical second-signal dogma. However, the lack of a second signal from the traditional co-stimulatory molecules does not necessarily lead to the induction of peripheral tolerance even in the absence of co-stimulation by the B7-CD28 pathway. When stimulated solely though the TCR, CD4 T cells respond poorly but the surviving cells are not tolerant, and respond like naïve lymphocytes [42], which supports the notion that it is not the lack of co-stimulatory molecules, but the presence of some other trigger, able to interact with T lymphocytes, which sends negative signals leading to the induction of tolerance rather than immunity.

In T cells, the cytotoxic T lymphocyte associated CTLA-4 or CD152 has been shown to participate in self-tolerance since CTLA-4 deficient mice develop lymphoproliferation and lethal autoimmunity [43]. This suggests its participation in the maintenance of T-cell tolerance *in vivo* [44]. The ligands described for CTLA-4 are CD80 and CD86 expressed by DC. Recent evidence suggests that CD80 rather than CD86 acts as the main ligand for CTLA-4 [45, 46]. Another molecule involved in the negative control of T-cell responses is the protein, programmed cell death (PD-1). Mice deficient in this molecule also develop autoimmune disorders [47, 48]. The ligands for PD-1 are B7-H1 and B7-DC. Based on their characteristics, both CTLA-4 and PD-1 are appropriate candidates to participate in tolerance induction. Recently, additional experiments by Probst using the inducible transgenic model for the expression of LCMV peptides on resting DC showed that tolerance induction depends on signaling through the inhibitory receptor, PD-1 [49]. Blocking of CTLA-4 also resulted in impaired tolerance. Interestingly, PD-1 and CTLA-4 appear to act synergistically, as the effect on tolerance induction was more dramatic in the absence of both interactions. This is the first evidence of the involvement of these molecules in the induction of peripheral tolerance by resting DC. However; these results need to be expanded to other systems to be considered as a general mechanism since blocking of CTLA-4 during tolerance induction by anti-DEC-205 OVA injection did not impair the tolerance outcome (Bonifaz, L. unpublished data). The role of inhibitory receptors in the induction of tolerance supports the idea of a balance of negative versus positive signals at the time of the DC–T cell interaction for the induction of tolerance versus immunity (Fig. 28.2a).

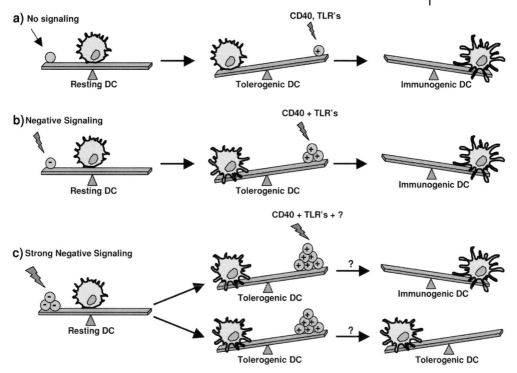

Fig. 28.2 Balance of negative versus positive signals in the induction of tolerogenic or immunogenic DC. (a) Endogenous constitutive antigen presentation or presentation after endocytosis in absence of signaling (i.e. via DEC-205) preserves the resting DC phenotype. This tolerogenic phenotype can be reversed to an immunogenic phenotype by positive signals (CD40, TLR ligands, etc). (b) The tolerogenic phenotype can be strength- ened by endocytosis via receptors with ITIM motifs. To be reversed, this phenotype may need stronger signaling or an accumulation of positive signals. (c) The presence of strong negative signals could induce expression of inhibitory molecules (ILT-3, ILT-4) that lead to a block of the NFκB pathway. This strong tolerogenic phenotype could be irreversible or reversed only by powerful (?) positive signals.

28.7
Other Receptors Involved in the Induction of Tolerance that can Preserve the Resting of DC or Induce Negative Signaling

28.7.1
Integrins

Integrins are a family of transmembrane receptors in which distinct α and β units are combined resulting in individual ligand specificity. Recently it was described a role for integrins in the uptake of cellular debris and apoptotic material by DC, which is carried out through different receptors such as thrombospondin receptor (CD36), αvb3 and αvb5 receptors, the complement receptors 3 and 4 [50, 51] and others [52]. Therefore, *in vivo*, DC constantly transport apoptotic self-Ag from tissue sites to the T-cell areas of draining LN and present peptide–MHC complexes to T cells [53, 54]. However, this presentation does not lead to the induction of an immune response. In fact, there is strong evidence supporting the induction of tolerance as a safeguard mechanism to avoid autoimmunity [55, 56].

Uptake of dead cells by complement receptors does not lead to DC activation, thus further antigen presentation would be in a tolerogenic context [57]. It has also been reported that uptake through CD36 integrin blocks DC maturation [58]. In addition, uptake of apoptotic cells by CD11b/CD18 integrin inhibits production of pro-inflammatory cytokines [59]. The role of integrins in the induction of tolerance is extensively reviewed in Chapter 29. Nevertheless, the findings mentioned herein are additional examples that the endocytosis of apoptotic material either sends negative signals to ensure the induction of peripheral tolerance or does not alter the resting state of DC, thus not inducing immunity to such material (Fig. 28.2b).

28.7.2
Fc Receptors

Fc receptors comprise a family of proteins capable of interacting with the constant region of immunoglobulins. Fc receptors are not expressed exclusively by DC and have been extensively studied in B cells. These receptors have an important role in both efficient endocytosis of antigen–antibody complexes as well as in efficient antigen presentation by class II and class I molecules [60–62]. The role of these receptors in the induction of an efficient immune response has been extensively studied. However, because of the existence of inhibitory Fc receptors containing ITM motifs, a role for FcγR in the induction of tolerance also needs to be considered. The physiological consequences of cell-bound IgG immune complexes are modulated by a balance between activating and inhibitory Fc gamma receptors [63, 64]. Studies by Kalergis and coworkers showed that targeting immune complexes to DC from mice genetically lacking inhibitory FcγRIIb leads to enhanced generation of antigen-specific CD8⁺ T-cell immunity *in vivo* [65]. Genetic deletion of FCγRIIb leads to spontaneous autoimmunity in mice [66]. Recently, an interesting study by Dhopakar and coworkers showed the role on inhibitory Fc receptors in maintaining human DC in an immature state. Selective blockade of this inhibitor receptor enables DC maturation leading to a protective immune response [67]. Therefore, these kinds of receptors, involved in both endocytosis and processing, are another clear example of a balance of inhibitory and activator signals responsible for the final outcome after T-cell proliferation and activation (Fig. 28.2b).

28.7.3
Suppressor and Regulatory T Cells

Another important tolerance mechanism to avoid immune response against self- and innocuous Ag is the existence of regulatory cell populations. At least two populations of T cells with regulatory functions have been described. The first one is generated in the thymus during T-cell development and are called suppressor cells with the CD4+ CD25+ phenotype. The second population of regulatory cells, called Tr1, can be generated and expanded in the periphery from CD25– precursors. The function of these cells is to block the function of other effector T cells. Both populations have been implicated in suppressing the immune response *in vivo* [68, 69]. The inhibition of the immune response by regulatory T cells is believed to be central to the prevention of autoimmune diseases [70, 71]. The mechanism of action of both populations of regulatory cells appears to be distinct. Thus, after TCR-triggering, that is MHC-restricted, CD4+ CD25+ regulatory cells (Tr) inhibited the immune response through an Ag and MHC restricted independent mechanism. These cells, which are apparently anergic, express the CTLA-4 molecule, which appears to play an important role in the regulatory process. On the other hand, Tr1 cells suppress the immune response in an antigen-specific manner by a mechanism dependent on cytokines such as IL-10 and TGFβ.

Immature DC have been implicated in the generation and expansion of Tr1. The studies by Jonuleit and coworkers [72] and Dhodapkar and coworkers [73, 74] showed participation of immature DC in the induction of human Tr1-like cells. Interestingly, experiments by Yamazaki and coworkers [75] showed direct expansion of functional CD25+ CD4+ regulatory T cells by DC. Such expansion could be achieved in an antigen-specific manner. CD25+ CD4+ T cells proliferated extensively in response to steady state or mature DC . The induction of these regulatory T cells by DC can suppress autoimmune diabetes [76]. These findings challenge some of the previous models of Tr function and emphasize the role of DC in the maintenance of tolerance by the induction of regulatory T cells.

28.8
Notch Ligands as Surface Molecules Involved in the Induction of Regulatory T Cells

Notch signaling is an evolutionary pathway used to direct developmental cell fate decisions in multiple organs [77]. Notch signaling is initiated through ligand–receptor interactions, leading to proteolytic cleavage of the receptor, a process that liberates the cytoplasmic domain of Notch. Notch encodes an evolutionary conserved transmenbrane receptor activated by two distinct cell surface ligands called Delta and Serrate. Mammals have four receptors (Nothc 1-4) and five ligands: Jagged1 and Jagged 2 (homologs of Serrate) and Delta 1, Delta 3 and Delta 4 (homologs of Delta). In the immune system the Notch pathway is involved at different levels. Early during T-cell ontogeny in the thymus, Nocht 1 signaling is necessary and sufficient for T-cell lineage commitment [78].

The identification of Notch receptor on peripheral T cells and its ligands on APC suggested their participation in peripheral T-cell decisions. Recent evidence suggests the participation of the Notch pathway in the differentiation of naïve T cells into regulatory T cells. Constitutive over-expression of Jagged 1 on murine dendritic cells induces differentiation of T cells into long-lived Tr1 cells, which can transfer tolerance to naïve mice [79]. In addition transgenic mice expressing Notch3 intracellular domain under control of the proximal Lck promoter have increased regulatory T-cell function, as shown by a higher degree of protection than control mice in a model of inducible autoimmune diabetes [80]. Regulatory T cells have higher Notch expression than CD4$^+$CD25$^-$ conventional T cells, raising the possibility that signaling through the Notch 3 receptor regulate the development, expansion and function of CD4$^+$CD25$^+$ regulatory T cells. Finally, priming of naïve human T cells in the presence of Jagged 1 expressing APC lead to reduced proliferation and partial inhibition of the cytotoxic effector function of CD8 T cells through the induction of regulatory T cells [81, 82].

Although DC and other APCs can express several Notch ligands [79], induction of regulatory T cells via Notch has not yet been formally demonstrated. Studies on the role of the Notch pathway in the terminal differentiation of DC showed that Notch signaling is necessary but not sufficient for their final differentiation. Interestingly, in the presence of Notch signaling, DC keep an immature phenotype [83]. Therefore one possibility for DC to induce regulatory T cells via Notch is the inability of DC to mature, hence promoting the induction of regulatory T cells as has been reported. Another possibility is that Notch ligands could be differentially induced on DC under different stimuli, as has been shown for Th1 and Th2-type T cells [84]. After NotchL expression, DC could acquire the potential to induce regulatory T cells. Even though this field is at its beginning, Notch ligands on DC and Notch receptor in T cells are good candidates to participate in the decision of tolerance versus immunity through the induction of regulatory T cells.

28.9
ILT-3 and ILT-4: Two Inhibitory Molecules Involved in Tolerance Induction

A third population of regulatory CD8 T cells has been described in humans. These cells have the noncommon phenotype CD8$^+$ CD28$-$ T cells and have been also referred to as suppressor T (Ts) cells. Ts cells can be generated *in vitro* after multiple rounds of stimulation of human peripheral blood mononuclear cells with either allogenic or xenogenic donor APCs [85, 86]. Ts CD8$^+$ cells are MHC class I restricted. They can suppress the response of CD4$^+$ T cells in an antigen-specific manner through a mechanism dependent on the presence of dendritic cells as bridge. After Ag presentation by either an immature dendritic cell or a monocyte, Ts cells induce in the APC the expression of two molecules: immunoglobulin-like transcript 3 and 4 (ILT3 and ILT4). These molecules have been described by Colona and are involved in negative signaling by the presence of ITIMs in their cytoplasmic tails [87]. Interestingly, the interaction of Ts cells with DC results in inhibition of the NFkB pathway and in downregulation of the B7 co-stimulatory molecules [88]. As a consequence of this interaction, the resulting DC is a tolerogenic DC able to induce unresponsiveness of CD4$^+$ T cells (Fig. 28.2c). After Ag presentation by these tolerogenic DC, CD4$^+$ T cells become anergic and this state could be overcome by the addition of IL-2. The relevance of this regulatory population *in vivo* has been documented through the isolation of CD8$^+$ CD28– suppressor T cells from transplant patients who did not undergo acute rejection [89]. Ts cells isolated from patients were shown *in vitro* to induce the upregulation of ILT3/ILT4 on MHC-matched APCs. In addition ILT-3 and ILT-4 have been detected on APC from transplant donors *in vivo* [90].

The generation of a DC with a special phenotype expressing specific inhibitory molecules and with impaired signaling through the NF_B pathway could support the notion of the existence of a special tolerogenic dendritic cell. The questions are: Do DC with this phenotype exist under steady state conditions? If not, under which conditions can they be generated? Are these DC also able to induce an immune response under strong inflammatory conditions? Alternatively, after their induction, is their only function the induction of tolerance? These questions still require investigation but some new evidence suggests the existence of DC with regulatory–tolerogenic functions.

28.10
Special DC for Tolerance?

The existence of specialized DC for tolerance induction has been assumed for a long time as an easy explanation for the induction of tolerance versus immunity. Initial *in vitro* studies by Shortman and coworkers [91], demonstrated that splenic CD8– DC induced a strong proliferative CD4$^+$ T-cell response, whereas CD8$^+$ DC induced a lesser response. Another report showed that CD8$^+$ DC downregulated the secretion of IL-2 by CD8$^+$ T cells [92]. These findings led to the hypothesis that CD8$^+$ DC are a subset preferentially involved in the induction of tolerance than in the induction of immunity. However, numerous reports have shown that CD8$^+$ DC are the major producers of IL-12 [93, 94]. Moreover, several reports have shown the participation of this subset in the induction of a potent immune response including IFNγ-producing T cells. To date, is clear that CD8$^+$ DC are able to induce both tolerance and immunity [95]. It is possible that this particular subset of DC has a more tolerogenic phenotype compare with the CD8$^-$. As an example, a CD11c$^+$ CD8$^+$ subset of DC both in mice and humans express the inducible enzyme indoleamine 2,3-dioxygenase (IDO) that may be involved in a novel suppression mechanism of T-cell proliferation [96, 97]. In addition, CD8$^+$ cells could have a high expression of ligands involved in the interaction with inhibitory receptors on T cells, such as the PD-1 ligands or CD80 that interact with CTLA-4. Another possibility is the role of CD8$^+$ DC in the endocytosis of apoptotic material, during which negative signaling could occur. Therefore, it is possible that strong positive signals are required to induce immune responses mediated by these cells (Fig. 28.2b).

28.11
Regulatory–tolerogenic DC

Is spite of the strong evidence supporting the notion that DC are capable of inducing tolerance or immune response depending on their activation status, the existence of DC with toleregenic phenotype has recently been reported, which have been nicknamed "regulatory DC" for their ability to differentiate naïve T cells into regulatory cells. Initial evidence for the generation of DC with regulatory–tolerogenic functions emerged from *in vitro* cultures of bone-marrow cells, and has been further supported by *in vivo* identification of similar populations. Wakkach and coworkers showed the presence of DC with an immature phenotype and with low expression of CD11c and high expression of CD45RB in cultures in the presence of IL-10 [98]. These cells developed *in vitro* in the presence of IL-10 but could also be identified *in vivo* in the spleen and lymph nodes of normal mice and are increased in the spleens of IL-10 Tg mice. These DC induce tolerance though the differentiation of Tr1 both *in vitro* and *in vivo*.

Studies by Zhang and coworkers [99] showed the potential of stromal cells to induce differentiation of DC into regulatory DC with the ability to suppress T-cell responses. Svensson and coworkers [100] found that DC could be differentiated into regulatory DC in the presence of stromal cells. These DC also had the ability to suppress T-cell responses and can also be identified *in vivo*, especially after infection with *Leishmania donovani*, where a more effective differentiation of this highly potent population of regulatory DC was observed.

Populations of regulatory DC with the ability to secrete IL-10 but not IL-12 have also been identified in the airways and in the intestinal tract [101, 102], where the balance between immune response and tolerance has to be tipped toward tolerance to prevent extensive permanent inflammation. These initial findings suggest that both in the steady state and after induction it is possible to differentiate an *in vivo* population of DC with a strong tolerogenic potential. As attractive as it seems, this model does not rule out the possibility that these findings might only reflect a functional status of a cell with otherwise immunogenic potential (Fig. 28.2c).

28.12
Concluding Remarks

Peripheral tolerance mediated by DC is still a field under extensive investigation, with many questions remaining unanswered. However, the same DC might play a crucial role both in the induction of a deep peripheral tolerance and in the induction of efficient immune responses depending on the cell's activation state. It is quite possible that the balance of negative versus positive signals sent at the time of the T cell–DC interaction is the key for the induction of tolerance versus immune responses. This balance can be achieved by the expression of certain surface molecules involved in Ag capture and presentation that either preserve the steady state of the DC or send negative signals to induce a more tolerogenic DC phenotype. The mechanism involved in the induction of tolerance versus immune response might also be dependent on the surface molecules expressed by resting DC or by surface molecules expressed as a consequence of negative signaling. The main question yet to be answered is whether tolerogenic DC are always capable of becoming immunogenic DC. In any case, the study of surface molecules involved in the induction of tolerance versus immunity is a broad field of investigation that will lead to a better understanding of the process as well as possible experimental manipulations in the future.

Acknowledgments

The author thanks Gibran Perez and Juliana Idoyaga for their help with the figures and Dr Ralph Steinman and Dr Jose Moreno for their critical review of this chapter.

References

1 Anderson, M. S., Venanzi, E. S., Klein, L., Chen, Z., Berzins, S. P., Turley, S. J., von Boehmer, H., Bronson, R., Dierich, A., Benoist, C., Mathis, D., Projection of an immunological self shadow within the thymus by the aire protein. *Science* **2002**. 298: 1395–1401.

2 Green, D. R., Droin, N., Pinkoski, M., Activation-induced cell death in T cells. *Immunol Rev* **2003**. 193: 70–81.

3 Schwartz, R. H., T cell anergy. *Annu Rev Immunol* **2003**. 21: 305–334.

4 Fehervari, Z., Sakaguchi, S., Development and function of CD25+CD4+ regulatory T cells. *Curr Opin Immunol* **2004**. 16: 203–208.

5 Brocker, T., Survival of mature CD4 T lymphocytes is dependent on major histocompatibility complex class II-expressing dendritic cells. *J Exp Med* **1997**. 186: 1223–1232.

6 Volkmann, A., Zal, T., Stockinger, B., Antigen-presenting cells in thymus that can negatively select MHC class II-restricted T cells recognizing a circulating self antigen. *J Immunol* **1997**. 158: 693–706.

7 Zal, T., Volkmann, A., Stockinger, B., Mechanisms of tolerance induction in major histocompatibility complex class II-restricted T cells specific for a blood-

borne self-antigen. *J Exp Med* **1994**. 180: 2089–2099.

8 Matzinger, P., Guerder, S., Does T-cell tolerance require a dedicated antigen-presenting cell? *Nature* **1989**. 338: 74–76.

9 Swat, W., Ignatowicz, L., vonBoehmer, H., Kisielow, P., Clonal deletion of immature CD4+8+ thymocytes in suspension culture by extrathymic antigen-presenting cells. *Nature (London)* **1991**. 351: 150–153.

10 Hogquist, K. A., Signal strength in thymic selection and lineage commitment. *Curr Opin Immunol* **2001**. 13: 225–231.

11 Gallegos, A. M., Bevan, M. J., Central tolerance to tissue-specific antigens mediated by direct and indirect antigen presentation. *J Exp Med* **2004**. 200: 1039–1049.

12 Watanabe, N., Hanabuchi, S., Soumelis, V., Yuan, W., Ho, S., de Waal Malefyt, R., Liu, Y. J., Human thymic stromal lymphopoietin promotes dendritic cell-mediated CD4+ T cell homeostatic expansion. *Nat Immunol* **2004**. 5: 426–434.

13 Banchereau, J., Steinman, R. M., Dendritic cells and the control of immunity. *Nature* **1998**. 392: 245–252.

14 Banchereau, J., Briere, F., Caux, C., Davoust, J., Lebecque, S., Liu, Y.-J.,

Pulendran, B., Palucka, K., Immuno-biology of dendritic cells. *Annu Rev Immunol* **2000**. 18: 767–811.

15 Lanzavecchia, A., Sallusto, F., The instructive role of dendritic cells on T cell responses: lineages, plasticity and kinetics. *Curr Opin Immunol* **2001**. 13: 291–298.

16 Rescigno, M., Borrow, P., The host-pathogen interaction: new themes from dendritic cell biology. *Cell* **2001**. 106: 267–270.

17 Steinman, R. M., Hawiger, D., Liu, K., Bonifaz, L., Bonnyay, D., Mahnke, K., Iyoda, T., Ravetch, J., Dhodapkar, M., Inaba, K., Nussenzweig, M., Dendritic cell function in vivo during the steady state: a role in peripheral tolerance. *Ann N Y Acad Sci* **2003**. 987: 15–25.

18 Steinman, R. M., Nussenzweig, M. C., Avoiding horror autotoxicus: the importance of dendritic cells in peripheral T cell tolerance. *Proc Natl Acad Sci USA* **2002**. 99: 351–358.

19 Steinman, R. M., Hawiger, D., Nussenzweig, M. C., Tolerogenic dendritic cells. *Annu Rev Immunol* **2003**. 21: 685–711.

20 Mitchell, D. A., Fadden, A. J., Drickamer, K., A novel mechanism of carbohydrate recognition by the C-type lectins DC-SIGN and DC-SIGNR. Subunit organization and binding to multivalent ligands. *J Biol Chem* **2001**. 276: 28 939–28 945.

21 Feinberg, H., Mitchell, D. A., Drickamer, K., Weis, W. I., Structural Basis for Selective Recognition of Oligosaccharides by DC- SIGN and DC-SIGNR. *Science* **2001**. 294: 2163–2166.

22 Leteux, C., Chai, W., Loveless, R. W., Yuen, C. T., Uhlin-Hansen, L., Combarnous, Y., Jankovic, M., Maric, S. C., Misulovin, Z., Nussenzweig, M. C., Ten, F., The cysteine-rich domain of the macrophage mannose receptor is a multispecific lectin that recognizes chondroitin sulfates A and B and sulfated oligosaccharides of blood group Lewis(a) and Lewis(x) types in addition to the sulfated N-glycans of lutropin. *J Exp Med* **2000**. 191: 1117–1126.

23 Tan, M. C., Mommaas, A. M., Drijfhout, J. W., Jordens, R., Onderwater, J. J., Verwoerd, D., Mulder, A. A., van der Heiden, A. N., Scheidegger, D., Oomen, L. C., Ottenhoff, T. H., Tulp, A., Neefjes, J. J., Koning, F., Mannose receptor-mediated uptake of antigens strongly enhances HLA class II-restricted antigen presentation by cultured dendritic cells. *Eur J Immunol* **1997**. 27: 2426–2435.

24 Jiang, W., Swiggard, W. J., Heufler, C., Peng, M., Mirza, A., Steinman, R. M., Nussenzweig, M. C., The receptor DEC-205 expressed by dendritic cells and thymic epithelial cells is involved in antigen processing. *Nature* **1995**. 375: 151–155.

25 Engering, A., Geijtenbeek, T. B., van Vliet, S. J., Wijers, M., van Liempt, E., Demaurex, N., Lanzavecchia, A., Fransen, J., Figdor, C. G., Piguet, V., van Kooyk, Y., The dendritic cell-specific adhesion receptor DC-SIGN internalizes antigen for presentation to T cells. *J Immunol* **2002**. 168: 2118–2126.

26 Dzionek, A., Sohma, Y., Nagafune, J., Cella, M., Colonna, M., Facchetti, F., Gunther, G., Okada, T., Winkels, G., Yamamoto, T., Zysk, M., Yamaguchi, Y., Schmitz, J., BDCA-2, a novel plasmacytoid dendritic cell-specific type II C-type lectin, mediates antigen-capture and is a potent inhibitor of interferon-α/β induction **2001**.

27 Mahnke, K., Guo, M., Lee, S., Sepulveda, H., Swain, S. L., Nussenzweig, M., Steinman, R. M., The dendritic cell receptor for endocytosis, DEC-205, can recycle and enhance antigen presentation via major histocompatibility complex class II-positive lysosomal compartments. *J Cell Biol* **2000**. 151: 673–684.

28 Kanazawa, N., Tashiro, K., Inaba, K., Miyachi, Y., Dendritic cell immuno-activating receptor, a novel C-type lectin immunoreceptor, acts as an activating receptor through association with Fc receptor gamma chain. *J Biol Chem* **2003**. 278: 32 645–32 652.

29 Kanazawa, N., Tashiro, K., Miyachi, Y., Signaling and immune regulatory role of the dendritic cell immunoreceptor (DCIR) family lectins: DCIR, DCAR, dectin-2 and BDCA-2. *Immunobiology* **2004**. 209: 179–190.

30 Kraal, G., Breel, M., Janse, M., Bruin, G., Langerhans' cells, veiled cells, and

interdigitating cells in the mouse recognized by a monoclonal antibody. *J Exp Med* **1986**. 163: 981–997.

31 Hawiger, D., Inaba, K., Dorsett, Y., Guo, M., Mahnke, K., Rivera, M., Ravetch, J. V., Steinman, R. M., Nussenzweig, M. C., Dendritic cells induce peripheral T cell unresponsiveness under steady state conditions in vivo. *J Exp Med* **2001**. 194: 769–779.

32 Bonifaz, L. C., Bonnyay, D. P., Charalambous, A., Darguste, D. I., Fujii, S., Soares, H., Brimnes, M. K., Moltedo, B., Moran, T. M., Steinman, R. M., In vivo targeting of antigens to maturing dendritic cells via the DEC-205 receptor improves T cell vaccination. *J Exp Med* **2004**. 199: 815–824.

33 Bonifaz, L., Bonnyay, D., Mahnke, K., Rivera, M., Nussenzweig, M. C., Steinman, R. M., Efficient targeting of protein antigen to the dendritic cell receptor DEC-205 in the steady state leads to antigen presentation on major histocompatibility complex class I products and peripheral CD8+ T cell tolerance. *J Exp Med* **2002**. 196: 1627–1638.

34 Adler, A. J., Marsh, D. W., Yochum, G. S., Guzzo, J. L., Nigam, A., Nelson, W. G., Pardoll, D. M., CD4+ T cell tolerance to parenchymal self-antigens requires presentation by bone marrow-derived antigen-presenting cells. *J Exp Med* **1998**. 187: 1555–1564.

35 Finkelman, F. D., Lees, A., Birnbaum, R., Gause, W. C., Morris, S. C., Dendritic cells can present antigen in vivo in a tolerogenic or immunogenic fashion. *J. Immunol.* **1996**. 157: 1406–1414.

36 Kurts, C., Kosaka, H., Carbone, F. R., Miller, J. F., Heath, W. R., Class I-restricted cross-presentation of exogenous self-antigens leads to deletion of autoreactive CD8(+) T cells. *J Exp Med* **1997**. 186: 239–245.

37 Morgan, D. J., Kreuwel, H. T., Sherman, L. A., Antigen concentration and precursor frequency determine the rate of CD8+ T cell tolerance to peripherally expressed antigens. *J Immunol* **1999**. 163: 723–727.

38 Sallusto, F., Lanzavecchia, A., Mobilizing dendritic cells for tolerance, priming, and chronic inflammation. *J Exp Med* **1999**. 189: 611–614.

39 Belz, G. T., Behrens, G. M., Smith, C. M., Miller, J. F., Jones, C., Lejon, K., Fathman, C. G., Mueller, S. N., Shortman, K., Carbone, F. R., Heath, W. R., The CD8alpha(+) dendritic cell is responsible for inducing peripheral self-tolerance to tissue-associated antigens. *J Exp Med* **2002**. 196: 1099–1104.

40 Hawiger, D., Masilamani, R. F., Bettelli, E., Kuchroo, V. K., Nussenzweig, M. C., Immunological unresponsiveness characterized by increased expression of CD5 on peripheral T cells induced by dendritic cells in vivo. *Immunity* **2004**. 20: 695–705.

41 Probst, H. C., Lagnel, J., Kollias, G., van den Broek, M., Inducible transgenic mice reveal resting dendritic cells as potent inducers of CD8+ T cell tolerance. *Immunity* **2003**. 18: 713–720.

42 Perez, V. L., Van Parijs, L., Biuckians, A., Zheng, X. X., Strom, T. B., Abbas, A. K., Induction of peripheral T cell tolerance in vivo requires CTLA-4 engagement. *Immunity* **1997**. 6: 411–417.

43 Ueda, H., Howson, J. M., Esposito, L., Heward, J., Snook, H., Chamberlain, G., Rainbow, D. B., Hunter, K. M., Smith, A. N., Di Genova, G., Herr, M. H., Dahlman, I., Payne, F., Smyth, D., Lowe, C., Twells, R. C., Howlett, S., Healy, B., Nutland, S., Rance, H. E., Everett, V., Smink, L. J., Lam, A. C., Cordell, H. J., Walker, N. M., Bordin, C., Hulme, J., Motzo, C., Cucca, F., Hess, J. F., Metzker, M. L., Rogers, J., Gregory, S., Allahabadia, A., Nithiyananthan, R., Tuomilehto-Wolf, E., Tuomilehto, J., Bingley, P., Gillespie, K. M., Undlien, D. E., Ronningen, K. S., Guja, C., Ionescu-Tirgoviste, C., Savage, D. A., Maxwell, A. P., Carson, D. J., Patterson, C. C., Franklyn, J. A., Clayton, D. G., Peterson, L. B., Wicker, L. S., Todd, J. A., Gough, S. C., Association of the T-cell regulatory gene CTLA4 with susceptibility to auto-immune disease. *Nature* **2003**. 423: 506–511.

44 Greenwald, R. J., Boussiotis, V. A., Lorsbach, R. B., Abbas, A. K., Sharpe, A. H., CTLA-4 regulates induction of anergy in vivo. *Immunity* **2001**. 14: 145–155.

45 Pentcheva-Hoang, T., Egen, J. G., Wojnoonski, K., Allison, J. P., B7-1 and B7-2 selectively recruit CTLA-4 and CD28 to the immunological synapse. *Immunity* **2004**. 21: 401–413.

46 Collins, A., Brodie, D., Gilbert, R., Iaboni, A., Manso-Sancho, R., Walse, B., Stuart, D., van der Merwe, P., Davis, S., The interaction properties of costimulatory molecules revisited. *Immunity* **2002**. 17: 201.

47 Nishimura, H., Okazaki, T., Tanaka, Y., Nakatani, K., Hara, M., Matsumori, A., Sasayama, S., Mizoguchi, A., Hiai, H., Minato, N., Honjo, T., Autoimmune dilated cardiomyopathy in PD-1 receptor-deficient mice. *Science* **2001**. 291: 319–322.

48 Nishimura, H., Nose, M., Hiai, H., Minato, N., Honjo, T., Development of lupus-like autoimmune diseases by disruption of the PD-1 gene encoding an ITIM motif-carrying immunoreceptor. *Immunity* **1999**. 11: 141–151.

49 Probst, H. C., McCoy, K., Okazaki, T., Honjo, T., van den Broek, M., Resting dendritic cells induce peripheral CD8+ T cell tolerance through PD-1 and CTLA-4. *Nat Immunol* **2005**. 6: 280–286.

50 Nauta, A. J., Daha, M. R., van Kooten, C., Roos, A., Recognition and clearance of apoptotic cells: a role for complement and pentraxins. *Trends Immunol* **2003**. 24: 148–154.

51 Albert, M. L., Pearce, S. F. A., Francisco, L. M., Sauter, B., Roy, P., Silverstein, R. L., Bhardwaj, N., Immature dendritic cells phagocytose apoptotic cells via $\alpha_v\beta_5$ and CD36, and cross-present antigens to cytotoxic T lymphocytes. *J Exp Med* **1998**. 188: 1359–1368.

52 Lu, Q., Lemke, G., Homeostatic regulation of the immune system by receptor tyrosine kinases of the Tyro 3 family. *Science* **2001**. 293: 306–311.

53 Huang, F.-P., Platt, N., Wykes, M., Major, J. R., Powell, T. J., Jenkins, C. D., MacPherson, G. G., A discrete subpopulation of dendritic cells transports apoptotic intestinal epithelial cells to T cell areas of mesenteric lymph nodes. *J Exp Med* **2000**. 191: 435–444.

54 Scheinecker, C., McHugh, R., Shevach, E. M., Germain, R. N., Constitutive presentation of a natural tissue autoantigen exclusively by dendritic cells in the draining lymph node. *J Exp Med* **2002**. 196: 1079–1090.

55 Legge, K. L., Gregg, R. K., Maldonado-Lopez, R., Li, L., Caprio, J. C., Moser, M., Zaghouani, H., On the role of dendritic cells in peripheral T cell tolerance and modulation of autoimmunity. *J Exp Med* **2002**. 196: 217–227.

56 Liu, K., Iyoda, T., Saternus, M., Kimura, Y., Inaba, K., Steinman, R. M., Immune tolerance after delivery of dying cells to dendritic cells in situ. *J Exp Med* **2002**. 196: 1091–1097.

57 Verbovetski, I., Bychkov, H., Trahtemberg, U., Shapira, I., Hareuveni, M., Ben-Tal, O., Kutikov, I., Gill, O., Mevorach, D., Opsonization of apoptotic cells by autologous iC3b facilitates clearance by immature dendritic cells, down-regulates DR and CD86, and up-regulates CC chemokine receptor 7. *J Exp Med* **2002**. 196: 1553–1561.

58 Urban, B. C., Willcox, N., Roberts, D. J., A role for CD36 in the regulation of dendritic cell function. *Proc Natl Acad Sci USA* **2001**. 98: 8750–8755.

59 Morelli, A. E., Larregina, A. T., Shufesky, W. J., Zahorchak, A. F., Logar, A. J., Papworth, G. D., Wang, Z., Watkins, S. C., Falo, L. D., Jr., Thomson, A. W., Internalization of circulating apoptotic cells by splenic marginal zone dendritic cells: dependence on complement receptors and effect on cytokine production. *Blood* **2003**. 101: 611–620.

60 Regnault, A., Lankar, D., Lacabanne, V., Rodriguez, A., Thery, C., Rescigno, M., Saito, T., Verbeek, S., Bonnerot, C., Ricciardi-Castagnoli, P., Amigorena, S., Fcγ receptor-mediated induction of dendritic cell maturation and major histocompatibility complex class I-restricted antigen presentation after immune complex internalization. *J Exp Med* **1999**. 189: 371–380.

61 Amigorena, S., Bonnerot, C., Fc receptors for IgG and antigen presentation on MHC class I and class II molecules. *Semin Immunol* **1999**. 11: 385–390.

62 Amigorena, S., Fc gamma receptors and cross-presentation in dendritic cells. *J Exp Med* **2002**. 195: F1–3.

63 Ravetch, J. V., A full complement of receptors in immune complex diseases. *J Clin Invest* **2002**. 110: 1759–1761.

64 Ravetch, J. V., Lanier, L. L., Immune inhibitory receptors. *Science* **2000**. 290: 84–89.

65 Kalergis, A. M., Ravetch, J. V., Inducing tumor immunity through the selective engagement of activating Fcgamma receptors on dendritic cells. *J Exp Med* **2002**. 195: 1653–1659.

66 Bolland, S., Ravetch, J. V., Spontaneous autoimmune disease in Fc(gamma)RIIB-deficient mice results from strain-specific epistasis. *Immunity* **2000**. 13: 277–285.

67 Dhodapkar, K. M., Kaufman, J. L., Ehlers, M., Banerjee, D. K., Bonvini, E., Koenig, S., Steinman, R. M., Ravetch, J. V., Dhodapkar, M. V., Selective blockade of inhibitory Fcgamma receptor enables human dendritic cell maturation with IL-12p70 production and immunity to antibody-coated tumor cells. *Proc Natl Acad Sci USA* **2005**. 102: 2910–2915.

68 Groux, H., O'Garra, A., Bigler, M., Rouleau, M., Antonenko, S., de Vries, J. E., Roncarolo, M. G., A CD4+ T-cell subset inhibits antigen-specific T-cell responses and prevents colitis. *Nature* **1997**. 389: 737–742.

69 Cottrez, F., Hurst, S. D., Coffman, R. L., Groux, H., T regulatory cells 1 inhibit a Th2-specific response In vivo. *J. Immunol.* **2000**. 165: 4848–4853.

70 Chen, Y., Kuchroo, V. K., Inobe, J.-i., Hafler, D. A., Weiner, H. L., Regulatory T cell clones induced by oral tolerance: suppression of autoimmune encephalomyelitis. *Science* **1994**. 265: 1237–1240.

71 Sakaguchi, S., Regulatory T cells: key controllers of immunologic self-tolerance. *Cell* **2000**. 101: 455–458.

72 Jonuleit, H., Schmitt, E., Schuler, G., Knop, J., Enk, A. H., Induction of human IL-10-producing, non-proliferating CD4+ T cells with regulatory properties by repetitive stimulation with allogeneic immature dendritic cells. *J Exp Med* **2000**. 192: 1213–1222.

73 Dhodapkar, M. V., Steinman, R. M., Krasovsky, J., Munz, C., Bhardwaj, N., Antigen specific inhibition of effector T cell function in humans after injection of immature dendritic cells. *J Exp Med* **2001**. 193: 233–238.

74 Dhodapkar, M. V., Steinman, R. M., Antigen-bearing, immature dendritic cells induce peptide-specific, CD8+ regulatory T cells *in vivo* in humans. *Blood* **2002**. 100: 174–177.

75 Yamazaki, S., Iyoda, T., Tarbell, K., Olson, K., Velinzon, K., Inaba, K., Steinman, R. M., Direct expansion of functional CD25+ CD4+ regulatory T cells by antigen-processing dendritic cells. *J Exp Med* **2003**. 198: 235–247.

76 Tarbell, K. V., Yamazaki, S., Olson, K., Toy, P., Steinman, R. M., CD25+ CD4+ T cells, expanded with dendritic cells presenting a single autoantigenic peptide, suppress autoimmune diabetes. *J Exp Med* **2004**. 199: 1467–1477.

77 Maillard, I., Adler, S. H., Pear, W. S., Notch and the immune system. *Immunity* **2003**. 19: 781–791.

78 Radtke, F., Wilson, A., Mancini, S. J., MacDonald, H. R., Notch regulation of lymphocyte development and function. *Nat Immunol* **2004**. 5: 247–253.

79 Hoyne, G. F., Le Roux, I., Corsin-Jimenez, M., Tan, K., Dunne, J., Forsyth, L. M., Dallman, M. J., Owen, M. J., Ish-Horowicz, D., Lamb, J. R., Serrate1-induced notch signalling regulates the decision between immunity and tolerance made by peripheral CD4(+) T cells. *Int Immunol* **2000**. 12: 177–185.

80 Anastasi, E., Campese, A. F., Bellavia, D., Bulotta, A., Balestri, A., Pascucci, M., Checquolo, S., Gradini, R., Lendahl, U., Frati, L., Gulino, A., Di Mario, U., Screpanti, I., Expression of activated Notch3 in transgenic mice enhances generation of T regulatory cells and protects against experimental autoimmune diabetes. *J Immunol* **2003**. 171: 4504–4511.

81 Yvon, E. S., Vigouroux, S., Rousseau, R. F., Biagi, E., Amrolia, P., Dotti, G., Wagner, H. J., Brenner, M. K., Over-expression of the Notch ligand, Jagged-1, induces alloantigen-specific human regulatory T cells. *Blood* **2003**. 102: 3815–3821.

82 Vigouroux, S., Yvon, E., Wagner, H. J., Biagi, E., Dotti, G., Sili, U., Lira, C., Rooney, C. M., Brenner, M. K.,

Induction of antigen-specific regulatory T cells following overexpression of a Notch ligand by human B lymphocytes. *J Virol* **2003**. 77: 10872–10880.

83 Cheng, P., Nefedova, Y., Miele, L., Osborne, B. A., Gabrilovich, D., Notch signaling is necessary but not sufficient for differentiation of dendritic cells. *Blood* **2003**. 102: 3980–3988.

84 Amsen, D., Blander, J. M., Lee, G. R., Tanigaki, K., Honjo, T., Flavell, R. A., Instruction of distinct CD4 T helper cell fates by different notch ligands on antigen-presenting cells. *Cell* **2004**. 117: 515–526.

85 Liu, Z., Tugulea, S., Cortesini, R., Suciu-Foca, N., Specific suppression of T helper alloreactivity by allo-MHC class I-restricted CD8+CD28– T cells. *Int Immunol* **1998**. 10: 775–783.

86 Damle, N. K., Mohagheghpour, N., Hansen, J. A., Engleman, E. G., Alloantigen-specific cytotoxic and suppressor T lymphocytes are derived from phenotypically distinct precursors. *J Immunol* **1983**. 131: 2296–2300.

87 Colonna, M., Navarro, F., Lopez-Botet, M., A novel family of inhibitory receptors for HLA class I molecules that modulate function of lymphoid and myeloid cells. *Curr Top Microbiol Immunol* **1999**. 244: 115–122.

88 Li, J., Liu, Z., Jiang, S., Cortesini, R., Lederman, S., Suciu-Foca, N., T suppressor lymphocytes inhibit NF-kappa B-mediated transcription of CD86 gene in APC. *J Immunol* **1999**. 163: 6386–6392.

89 Colovai, A. I., Mirza, M., Vlad, G., Wang, S., Ho, E., Cortesini, R., Suciu-Foca, N., Regulatory CD8+CD28– T cells in heart transplant recipients. *Hum Immunol* **2003**. 64: 31–37.

90 Manavalan, J. S., Rossi, P. C., Vlad, G., Piazza, F., Yarilina, A., Cortesini, R., Mancini, D., Suciu-Foca, N., High expression of ILT3 and ILT4 is a general feature of tolerogenic dendritic cells. *Transpl Immunol* **2003**. 11: 245–258.

91 Suss, G., Shortman, K., A subclass of dendritic cells kills CD4 T cells via Fas/Fas-ligand induced apoptosis. *J Exp Med* **1996**. 183: 1789–1796.

92 Kronin, V., Winkel, K., Suss, G., Kelso, A., Heath, W., Kirberg, J., von Boehmer, H., Shortman, K., A subclass of dendritic cells regulates the response of naive CD8 T cells by limiting their IL-2 production. *J Immunol* **1996**. 157: 3819–3827.

93 Iwasaki, A., Kelsall, B. L., Unique functions of cd11b(+), cd8alpha(+), and double-negative Peyer's patch dendritic cells. *J Immunol* **2001**. 166: 4884–4890.

94 Maldonado-Lopez, R., De Smedt, T., Michel, P., Godfroid, J., Pajak, B., Heirman, C., Thielemans, K., Leo, O., Urbain, J., Moser, M., CD8α+ and CD8α– subclasses of dendritic cells direct the development of distinct T helper cells in vivo. *J Exp Med* **1999**. 189: 587–592.

95 Martin, P., Del Hoyo, G. M., Anjuere, F., Arias, C. F., Vargas, H. H., Fernandez, L. A., Parrillas, V., Ardavin, C., Characterization of a new subpopulation of mouse CD8alpha+ B220+ dendritic cells endowed with type 1 interferon production capacity and tolerogenic potential. *Blood* **2002**. 100: 383–390.

96 Fallarino, F., Vacca, C., Orabona, C., Belladonna, M. L., Bianchi, R., Marshall, B., Keskin, D. B., Mellor, A. L., Fioretti, M. C., Grohmann, U., Puccetti, P., Functional expression of indoleamine 2,3-dioxygenase by murine CD8 alpha(+) dendritic cells. *Int Immunol* **2002**. 14: 65–68.

97 Munn, D. H., Sharma, M. D., Lee, J. R., Jhaver, K. G., Johnson, T. S., Keskin, D. B., Marshall, B., Chandler, P., Antonia, S. J., Burgess, R., Slingluff, C. L., Jr., Mellor, A. L., Potential regulatory function of human dendritic cells expressing indoleamine 2,3-dioxygenase. *Science* **2002**. 297: 1867–1870.

98 Wakkach, A., Fournier, N., Brun, V., Breittmayer, J. P., Cottrez, F., Groux, H., Characterization of dendritic cells that induce tolerance and T regulatory 1 cell differentiation in vivo. *Immunity* **2003**. 18: 605–617.

99 Zhang, M., Tang, H., Guo, Z., An, H., Zhu, X., Song, W., Guo, J., Huang, X., Chen, T., Wang, J., Cao, X., Splenic stroma drives mature dendritic cells to differentiate into regulatory dendritic cells. *Nat Immunol* **2004**. 5: 1124–1133.

100 Svensson, M., Maroof, A., Ato, M., Kaye, P. M., Stromal cells direct local

differentiation of regulatory dendritic cells. *Immunity* **2004**. 21: 805–816.

101 Akbari, O., DeKruyff, R. H., Umetsu, D. T., Pulmonary dendritic cells producing IL-10 mediate tolerance induced by respiratory exposure to antigen. *Nat Immunol* **2001**. 2: 725–731.

102 Williamson, E., Bilsborough, J. M., Viney, J. L., Regulation of mucosal dendritic cell function by receptor activator of NF-kappa B (RANK)/RANK ligand interactions: impact on tolerance induction. *J Immunol* **2002**. 169: 3606–3612.

29
Interaction Between Dendritic Cells and Apoptotic Cells

Adriana T. Larregina and Adrian E. Morelli

29.1
Introduction

Apoptosis, or programmed cell death, constitutes a series of genetically-controlled, energy-dependent cellular events by which cells actively orchestrate their own demise. During this process, cells undergo biochemical and morphological changes, including redistribution of lipids of the plasma membrane, loss of mitochondrial membrane potential, inter-nucleosomal DNA cleavage, fragmentation of the nucleus and cytoplasm into membrane-enclosed apoptotic bodies, and shedding of apoptotic blebs from the cell surface [1]. The changes in the molecular composition of the surface membrane of cells undergoing apoptosis initiates the recognition and removal of apoptotic cells by professional phagocytes including macrophages and immature dendritic cells (DC) and by neighboring parenchymal or stromal cells (e.g. fibroblasts, epithelial cells, endothelial cells, glomerular mesangial cells, etc) able to function as semi-professional phagocytes.

Clearance of apoptotic cells is an essential process during embryogenesis, morphogenesis and maintenance of tissue homeostasis. In physiological conditions, apoptotic cells are phagocytosed rapidly, while their plasma membrane retains its integrity, before leakage of toxic mediators like proteases, cationic proteins and oxidizing molecules into the extracellular space. Thus, the rapid clearance of cells in early apoptosis prevents the potential tissue damage resulting from cells that disintegrate by late apoptosis or secondary necrosis. Moreover, release of proteolytic fragments of self-proteins and oligonucleosomal DNA to the extracellular space may trigger an autoimmune response. In this regard, it is known that apoptotic cells externalize and concentrate nuclear autoantigens on the surface of apoptotic blebs that are shed to the extracellular milieu [2].

In physiological conditions, the clearance of cells in early apoptosis is so efficient that phagocytes with internalized apoptotic cells are only detected histologically in tissues undergoing a high rate of apoptosis, such as thymus, bone marrow (BM) and germinal centers of B cell follicles. The speed and efficiency of removal of apoptotic cells in the steady state is probably one of the reasons by which this type of

Handbook of Dendritic Cells. Biology, Diseases, and Therapies.
Edited by M. B. Lutz, N. Romani, and A. Steinkasserer.
Copyright © 2006 WILEY-VCH Verlag GmbH & Co. KGaA, Weinheim
ISBN: 3-527-31109-2

cell death was not identified until recently [3, 4]. More recent data indicate that apoptotic cells may actively suppress inflammation and participate in induction/maintenance of T and B cell peripheral tolerance. This phenomenon is mediated by a highly specialized interaction between apoptotic cells and phagocytes of the innate and adaptive immune system. However, depending on the characteristics of the apoptotic cells and the mediators present in the extracellular milieu, apoptotic cells may eventually trigger inflammation and promote immunity [5].

29.2
Dendritic Cells Phagocytose and Process Apoptotic Cells

In 1986, the studies of Fossum and Rolstad [6] on allogeneic lymphocyte cytotoxicity demonstrated for the first time that interdigitating DC of lymph nodes and spleen were able to internalize fragments derived from allogeneic lymphocytes killed by host natural killer cells. However, the conclusive evidence that DC phagocytose apoptotic cells was obtained later from studies on Langerhans cells (LC) in the vaginal epithelium of rodents. By means of electron microscopy, Parr and colleagues [7] demonstrated that LC of the vaginal epithelium internalized apoptotic epithelial cells during late metestrus and early diestrus in mice. However, the physiological importance of this observation remained unclear until 1998 when Albert et al. [8] demonstrated *in vitro* that apoptotic cell-derived antigens (Ag) are acquired by human BM-derived DC and presented in MHC class-I molecules to CD8+ cytotoxic T lymphocytes (CTL). Presentation of peptides derived from extracellular Ag via MHC class-I molecules to CD8+ T cells is known as "cross-priming". This pathway of Ag processing allows CD8+ T cells to recognize peptides that do not have access to the cytosol of Ag presenting cells (APC) and therefore, that are unable to follow the endogenous pathway of antigenic processing. Albert et al. [8] also demonstrated that, although macrophages were more efficient at phagocytosis of apoptotic cells than DC, only DC efficiently crosspresented apoptotic cell-derived Ag to CD8+ T cells. A similar observation was made in a different experimental model where DC, but not macrophages, crosspresented Ag from *Salmonella*-infected apoptotic cells via MHC class-I molecules to specific CD8+ T cells [9]. Furthermore, *in vivo* studies have shown that unlike macrophages, DC pulsed with apoptotic cells induce the generation of antitumor CTL [10]. Addition of macrophages to DC cultures impaired the ability of DC to crosspresent apoptotic cell-derived peptides [8]. It is likely that, due to their high phagocytic ability, macrophages sequester (and degrade completely) apoptotic cells from DC [8]. Unlike monocyte/macrophages, the ability of DC to crosspresent internalized Ag is in part due to a unique membrane transport pathway. This mechanism allows internalized Ag to gain access to the cytosolic Ag-processing machinery and to the conventional MHC class-I molecule presentation pathway [11].

Alternatively, DC can process apoptotic cell fragments on MHC class-II molecules with high efficiency for presentation to CD4+ T cells by using the endocytic pathway of antigenic processing. Inaba et al. [12] have shown that murine BM-derived DC present Ag from internalized apoptotic cells to CD4$^+$ T cells 1–10000 times better that preprocessed peptide.

Immature DC phagocytose apoptotic cell fragments more efficiently than mature DC [13]. This phenomenon requires the presence of extracellular calcium and rearrangement of the DC cytoskeleton, as it is inhibited by incubation with EDTA or cytochalasin D [13, 14]. The $\alpha_v\beta_5$ integrin not only mediates the binding of apoptotic cells to DC but also their internalization. Engagement of the apoptotic cell to the $\alpha_v\beta_5$ heterodimer on the phagocyte surface results in the recruitment of the CrkII-Dock180 molecular complex, which triggers Rac1 activation and phagosome formation [15–17].

Once internalized by DC, apoptotic cells are processed within MHC class-II-rich compartments (MHC-II$^+$ LAMP$^+$ H-2M$^+$ vesicles) [12]. Incubation of DC with ammonium chloride, which inhibits acidification and proteolysis within endocytic vacuoles, blocked presentation of apoptotic cell-derived peptides through MHC class-I and II [8, 12]. The intracellular route by which apoptotic cell-derived peptides generated within endocytic vacuoles of DC gain access to the lumen of the endoplasmic reticulum to be loaded into MHC class-I molecules for crosspresentation remains unclear. The fact that lactacystin, a 26S proteasome inhibitor, partially blocked crosspresentation of apoptotic cell-derived peptides by DC indicates that both classical and nonclassical MHC class-I pathways participate in this process [8].

29.3
The Phagocytic Synapse

A critical feature of apoptotic cells is their ability to be recognized specifically and removed by professional phagocytes (e.g. macrophages, immature DC) or by neighboring parenchymal or stromal cells. This mechanism of recognition and uptake of apoptotic cells is very complex and has not been fully characterized. Although there is extensive literature regarding the receptors used by macrophages for phagocytosis of apoptotic cells, little is known about the mechanisms employed by DC.

The surface of apoptotic cells displays a series of Apoptotic-Cell-Associated Molecular Patterns (ACAMP) that function as "eat me" signals that interact with Pattern-Recognition Receptors (PRR) expressed on the surface of phagocytes [18]. In general, these eat me signals do not require de novo protein synthesis by dying cells, instead they are generated by modification or translocation of preexisting plasma membrane components or by deposition of soluble extracellular molecules (opsonins) on the surface of apoptotic cells. Fadok et al. [19] have coined the term "phagocytic synapse" for this highly sophisticated molecular interaction between phagocytes and apoptotic cells.

During the last decade, numerous ACAMP and their corresponding PRR have been identified. Some of them are ligand–receptor pairs composed of receptors expressed by the phagocytes that directly bind ligands present on the apoptotic cell surface (Fig. 29.1A). Other receptors require the recruitment of "bridging molecules" from the extracellular fluid that function as opsonins, linking ligands on the surface of apoptotic cells with PRR expressed by the phagocytes (Fig. 29.1B). The following are some of the most extensively studied ligand–receptor interactions that participate in phagocytosis of apoptotic cells by phagocytes of the immune system.

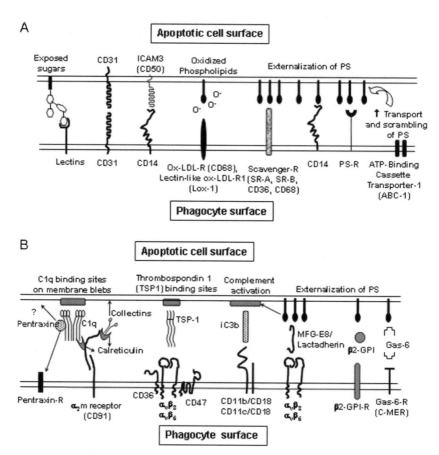

Fig. 29.1 Ligands and receptors involved in clearance of apoptotic cells by phagocytes. (A) Pattern-recognition receptors expressed by phagocytes and their ligands on the apoptotic cell surface. (B) Bridging molecules that participate in recognition, attachment and internalization of apoptotic cells by phago-cytes including DC. PS: phosphatidylserine; Ox-LDL: Oxidized low density lipoprotein; -R: receptor; MFG: milk-fat globule protein; β2-GPI: β2 Glycoprotein I; Gas-6: growth arrest specific gene-6 product; C-MER: MER tyrosine kinase receptor; α2m: α2 macro-globulin.

29.3.1
Externalized Phosphatidylserine (PS) and Receptors for PS

One of the earliest changes in apoptotic cells is the loss of phospholipid asymmetry of the plasma membrane caused by the translocation of PS, phosphatidylethanolamine (PE) and phosphatidylcholine to the outer leaflet of the plasma membrane [20]. Externalization of PS occurs hours before the characteristic morphological changes of apoptosis. In viable cells, an ATP- and magnesium-dependent aminophospholipid translocase flips PS, and to a lesser extent PE, from the outer to the inner leaflet of the plasma membrane maintaining the phospholipid symmetry of the membrane [19]. On the other hand, a phospholipid scramblase moves phospholipids bidirectionally across the plasma membrane [19]. The entry of calcium that occurs during the initial stages of apoptosis inhibits the lipid translocase and activates a lipid scramblase resulting in the accumulation of PS on outer leaflet of the plasma membrane. Phospholipid scrambling seems to be facilitated by the ABC-1 transporter, a member of the ATP-binding cassette superfamily of membrane transporters. There is evidence that the facilitating effect of ABC-1 on clearance of apoptotic cells depends on the effects of ABC-1 transporter on the phospholipid asymmetry in both the phagocyte and target apoptotic cell [21, 22].

Binding of externalized PS with fluorochrome-conjugated annexin V has been a very useful tool to detect cells in early and late apoptosis by microscopy and flow cytometry [20]. The presence of PS on the surface of apoptotic cells is recognized by phagocytes via different receptors, including a stereospecific PS receptor (PS-R) [20, 23]. Recent evidence indicates that apoptotic cells also externalize annexin I (also known as lipocortin), a molecule that probably facilitates the interaction of the extremely small polar group of PS with the PS-R [24]. Mice lacking PS-R exhibited a severe deficiency in the clearance of apoptotic cells [25]. Several scavenger receptors (SR) (e.g. SR-A, SR-B1, CD36 and CD68), CD14 and soluble opsonins may also bind externalized PS and trigger internalization of apoptotic cells by phagocytes (Fig. 29.1). Most of the receptors for PS have been studied in macrophages, and little is known regarding their function during internalization of apoptotic cells by DC. Murine splenic DC transcribe mRNA for PS-R, SR-AII, SR-BI, CD14 and CD68 [26]. Whether DC translate all these mRNA transcripts into functional receptors able to recognize PS on apoptotic cells is still unknown, although there is evidence that DC express the stereospecific PS-R [19]. The presence of receptors for PS on the surface of DC is supported by the fact that human monocyte-derived DC change their phenotype and function following incubation with liposomes containing PS [27].

The serum β_2-glycoprotein I (β_2-GPI) is an aminophospholipid ligand that recognizes PS on the surface of apoptotic cells [28]. Opsonization of dying cells with β_2-GPI facilitates DC uptake of apoptotic cells and enhances presentation of MHC class-II-restricted apoptotic cell-derived peptides by DC to $CD4^+$ T cells [29].

Cells undergoing apoptosis are under considerable oxidative stress and generate reactive oxygen species that oxidize aminophospholipids (including PS) exposed on the cell surface. These oxidized lipids are structurally analogous to moieties detected on the oxidized low-density lipoprotein particle (Ox-LDL), which are recognized by SR-A, SR-B1, CD36, CD68 and lectin-like O-LDL-R1 (LOX-1 or SR-E). It is believed that these phagocyte receptors may also bind oxidized aminophospholipids on the surface of apoptotic cells (Fig. 29.1). In fact, apoptotic cell uptake by macrophages can be inhibited *in vitro* by blocking the oxidized phospholipids present on the surface of apoptotic by means of specific monoclonal antibodies (mAb) [30]. The role of receptors for oxidized lipids in apoptotic cell recognition by DC is still unknown.

29.3.2
Thrombospondin-1 (TSP-1), CD36 and the Integrins $\alpha_v\beta_3$ and $\alpha_v\beta_5$

TSP-1 is a platelet-derived multifunctional trimeric glycoprotein that interacts with cell surfaces and the extracellular matrix. The surface of apoptotic cells exposes anionic sites that allow deposition of TSP-1. By this mechanism, TSP-1 functions as a molecular bridge linking apoptotic cells with phagocytes. TSP-1 is recognized by the integrins $\alpha_v\beta_3$ (vitronectin receptor) and $\alpha_v\beta_5$ and the SR CD36 on the phagocyte surface. The $\alpha_v\beta_3$ integrin and CD36 recognize different regions of the TSP-1 trimer and only the $\alpha_v\beta_3$-TSP-1 interaction is blocked by Arg-Gly-Glu (RGD)-bearing peptides [31]. CD36 plays an amplifying role in the $\alpha_v\beta_3$-TSP-1 recognition, increasing the ability of phagocytes to internalize apoptotic cells [32, 33]. Human immature DC express $\alpha_v\beta_3$ [34] and $\alpha_v\beta_5$ [35] that cooperate with CD36 during recognition of TSP-1 on the surface of apoptotic cells. Human monocyte-derived DC downregulate surface expression of $\alpha_v\beta_5$ and CD36 following maturation *in vitro*, a fact that correlates with the reduced capability of mature DC for phagocytosis of apoptotic cells [35].

In mice, blockade of $\alpha_v\beta_3$ by means of mAb decreased the uptake of apoptotic cells by splenic DC *in vitro* [14] and only splenic CD8α^+ DC express high levels of CD36 on their surface [36, 37]. Interestingly, the subpopulation of CD8α^+ DC are the primary APC involved in crosspresentation of extracellular Ag via MHC class-I molecules to CD8$^+$ T lymphocytes [38]. In humans, it has been suggested that crosspresentation of apoptotic cell-associated Ag depends on the uptake of apoptotic cell fragments by DC via CD36 and $\alpha_v\beta_3/\beta_5$ integrins [34, 35]. However, splenic CD8α^+ DC from mice lacking CD36 or β_3/β_5 integrins crosspresent extracellular self- or foreign-Ag (associated with apoptotic or viable cells) as efficiently as their wild type counterparts [36, 37]. These results indicate that CD8α^+ DC must employ other receptors for internalization of extracellular Ag to be processed for crosspresentation. Alternatively, receptors required for apoptotic cell-clearance may operate at a high level of redundancy and other DC surface molecules may compensate for the lack of CD36 or β_3/β_5 integrins.

29.3.3
Complement Factors and Complement Receptors (CR)

The presence of heat-labile serum factors increases the uptake of apoptotic cells by phagocytes [39]. These factors have been identified as components of the complement system [40]. There is evidence that the surface of apoptotic cells can activate the complement cascade via the classical and the alternative pathways [40]. C1q, the first component of the classical pathway, binds to the surface of apoptotic cells by different mechanisms. Lysophosphatidylcholine exposed on the surface of apoptotic cells can be recognized by natural IgM antibodies [41, 42]. Binding of C1q to IgM attached to apoptotic cells initiates complement activation via the classical pathway, generating C3-derived opsonins that are recognized by phagocytes through CR.

The globular head of C1q may bind directly (without intermediate IgM) to poorly-characterized ligands that cluster on the surface membrane of apoptotic cell blebs [43, 44], a phenomenon that may trigger complement activation [45]. Alternatively, the collagen-like tail of C1q can bind calreticulin (also known as cC1qR) attached to the endocytic receptor CD91 (α-2-macroglobulin receptor) on the phagocyte surface [46] (Fig. 29.1B). By means of these mechanisms, C1q increases uptake of apoptotic cells by immature DC [47]. Interestingly, immature DC synthesize higher amounts of C1q than macrophages [48].

Other molecules in the plasma structurally related to C1q are the collectins, which include mannose-binding lectin (MBL) and lung surfactant protein A (SP-A) and D (SP-D). Collectins are characterized by their amino-terminal collagen-like region followed by a globular head with C-type lectin binding properties [49]. MBL and SP-A participate in uptake of apoptotic cells by attaching their globular domains directly to the surface of apoptotic cells. On the other end of the molecule, their collagen-like tails bind to the calreticulin/CD91 receptor on the surface of the phagocyte [50–53]. A recent report indicates that MBL enhances uptake of apoptotic cell by immature DC [47].

Mevorach et al. [40] have shown that externalization of PS by cells in early apoptosis was responsible in part for complement activation and iC3b deposition on the surface of apoptotic cells. In support of a direct role for PS in complement activation, PS-bearing micelles and cardiolipin (an anionic phospholipid similar to PS) were shown to activate complement [53]. Human monocyte-derived DC and mouse splenic DC employ the iC3b receptors CR3 (CD11b/CD18) and CR4 (CD11c/CD18) to uptake iC3b-opsonized apoptotic cells *in vitro* [14, 54]. *In vivo*, splenic DC of mice made hypocomplementemic after injection of cobra venom factor (CVF) significantly decreased their ability to internalize circulating apoptotic cells [14]. CVF is a C3-like polypeptide that functions as an inactivation-resistance C3 convertase that, in mice, results in a severe inactivation of C3 (and secondary C5) without affecting factors like C1q that precedes C3 in the complement cascade. Therefore, complement factors derived from C3 or later in the cascade are of key importance for uptake of apoptotic cells by DC *in vivo* [14].

In physiological conditions, complement activation on the surface of apoptotic cells does not cause cell lysis, leakage of toxic mediators into the extracellular milieu or inflammation. One possible explanation for this paradoxical effect is that apoptotic cells control the generation of complement factors beyond iC3b by means of complement regulatory molecules. These regulatory factors prevent assembly of the C5b-C9 membrane attack complex (MAC) on the plasma membrane of the apoptotic cell. One of these molecules is Factor H, a complement regulatory protein that accelerates the decay of the C3 and C5 convertases. Interestingly, the pentraxin C-reactive protein (CRP) binds to apoptotic cells and recruits Factor H on the surface of apoptotic cells, accelerating the dissociation of the C3 and C5 convertases and therefore decreasing generation of MAC [55]. A rapid clearance of apoptotic cells may also contribute to prevent generation of MAC on apoptotic cells [53]. The lack of an inflammatory reaction is due to the fact that binding of apoptotic cells via C3bi downregulates production of pro-inflammatory mediators by DC and macrophages [14].

29.3.4
Pentraxins

Pentraxins are acute-phase proteins secreted in response to inflammation. The name of this family of molecules is derived from the characteristic cyclic pentameric structure. It has been demonstrated that the pentraxins C reactive protein (CRP), serum amyloid P (SAP) and pentraxin 3 (PTX3) bind apoptotic cells *in vitro* [55–58]. Besides their binding to anionic phospholipids, CRP recognizes small nuclear ribonucleoproteins and SAP binds to chromatin/nucleolar components [57]. These intracellular molecules are mobilized to the cell surface during apoptosis and become accessible to pentraxins. The ligand(s) recognized by PTX3 on the apoptotic cell membrane is still unknown.

In the absence of complement, opsonization with SAP or CRP increases uptake of apoptotic cells by macrophages, indicating the existence of specific receptors for pentraxins [59]. It has been shown that pentraxins may also bind Fcγ receptors [60]. Unlike the pro-phagocytic effect of SAP and CRP, it has been shown that binding of PTX3 to apoptotic cells masks eat me signals required for uptake of apoptotic cells by immature DC [57].

29.3.5
Milk-fat Globule Protein Epidermal Growth Factor 8 (MFG-E8)/lactadherin

MFG-E8/lactadherin was originally identified on the surface of epithelial mammary cells and it is believed to participate in the apoptosis associated with mammary gland involution. MFG-E8 is secreted in two molecular variants, each containing two factor VIII-homologous regions and two epidermal growth factor (EGF) domains. MFG-E8 binds selectively to anionic phosholipids (PS and PE) by means of the VIII-homologous region [61]. The EGF domain includes an RDG motif that is recognized by the $\alpha_v\beta_3$ integrin [61]. Thus, MFG-E8 functions as a bridge between externalized PS on the surface of apoptotic cells and the $\alpha_v\beta_3$ integrin expressed by the phagocytes. Mouse immature DC and LC synthesize MFG-E8, and DC from MFG-E8-deficient mice are decreased drastically in their capability to internalize apoptotic cells [62]. In agreement with the fact that mature DC exhibit a limited ability for clearance of apoptotic cells, expression of MFG-E8 decreases in DC induced to mature by LPS [62]. Molecules like MFG-E8 and C1q, which are synthesized by immature DC, may be critical for the uptake of apoptotic cells in those tissues where serum-derived factors may have limited access [63].

29.3.6
Other Apoptotic Cell Recognition Signals

A plethora of other, not so well characterized, ligand–receptor interactions has been observed to function in recognition and uptake of apoptotic cells. Most of these receptors have been studied in macrophages and information regarding their possible role in uptake of apoptotic cells by DC is limited or still unknown.

Apoptotic cells change the composition of sugars (N-acetylglucosamine, N-acetylgalactosamine) on the cell membrane. These modified carbohydrates can be recognized specifically by lectins expressed by phagocytes [64]. An asialoglycoprotein receptor identified on macrophages [65] and the lectin DEC-205 (CD205) expressed by CD8α^+ DC, LC and thymic epithelial cells are good candidate receptors for apoptotic cell uptake via lectin–sugar recognition. DEC-205 is a receptor for endocytosis and delivery of Ag via DEC-205 to DC facilitates the recycling of the Ag through endosomes/lysosomes and enhances Ag presentation in the context of MHC class-II molecules [66]. Based on the fact that blockade of DEC-205 reduces uptake of apoptotic thymocytes by thymic epithelial cells *in vitro* [67], it is tempting to postulate that DEC-205 may participate in engulfment of apoptotic cells by CD8α^+ DC. However, Iyoda and colleagues [26] have demonstrated that DC from DEC-205 deficient mice are still able to uptake apoptotic splenocytes. A possible explanation for this result is that DC from DEC-205 KO mice express other receptors for clearance of apoptotic cells that may compensate the lack of DEC-205.

The intracytoplasmic domain of the receptor Mer, a member of the Axl/Mer/Tyro3 receptor tyrosine kinase family, has been shown to be critical for uptake of apoptotic cells in macrophages [68]. On the phagocyte surface Mer binds the soluble ligand Gas6, a product of the growth arrest-specific gene 6 that recognizes PS on the surface of apoptotic cells [69, 70]. By mediating the binding between Mer and PS, Gas6 functions as a molecular bridge between apoptotic cells and macrophages. Although murine BM-derived and splenic DC express Mer on their cell surface, DC from Mer-deficient mice engulf apoptotic cell as efficiently as DC of wild type controls [71].

The macrophage marker CD14 binds a structurally modified form of ICAM-3 that is present on the surface of apoptotic cells [72, 73]. The authors suggest that during apoptosis, ICAM-3 exposes/generates a new epitope that binds to CD14 and promotes apoptotic cell engulfment by macrophages [72, 73]. Apoptotic cell-clearance by DC is believed to be CD14-independent since human and murine DC do not express (or exhibit very low levels) of CD14.

Phagocytes can discriminate between apoptotic and viable cells by delivering detachment signals to living cells via homophilic interactions of CD31, a marker expressed by DC [74, 75]. By a poorly understood mechanism, apoptotic cells can switch the function of surface CD31 from repulsive to adhesive promoting tethering of dying cells to phagocytes [74].

29.4
Redundant Receptors and Backup Mechanisms for Apoptotic Cell Clearance

The numerous receptor–ligand interactions that participate in recognition and internalization of apoptotic cells by professional or semi-professional phagocytes may explain why, in normal conditions, the uptake of apoptotic cells is so efficient. The redundancy of receptors and backup mechanisms for apoptotic cell clearance is also an indirect indicator that removal of apoptotic cells must be extremely important to preserve the steady-state condition that characterizes healthy organisms [31]. In some cases, different receptors (i.e. CD36 and the $\alpha_v\beta_3$ integrin) cooperate mutually to increase the efficiency of apoptotic cell uptake [32, 33]. Receptors like Mer may be critical for some phagocytes (e.g. macrophages) [68] and play a redundant role or no function for other cell types (e.g. immature DC) [71]. Different ligand–receptor interactions may be required to recognize molecules expressed on the surface of apoptotic cells at different stages of apoptosis. Receptors that bind to PS may be important at early stages of apoptosis, when anionic phospholipids are exposed on the surface of apoptotic cells. At later stages of apoptosis, when intracellular molecules are exposed de novo on the surface of apoptotic cells, the binding of soluble pentraxins may play an important role in apoptotic cell recognition and clearance by phagocytes.

29.5
Regulatory Effects of Early Apoptotic Cells on Dendritic Cells

The normal turnover of billions of tissue cells that die by apoptosis every day in our body is not accompanied by inflammation. Likewise, conditions causing apoptotic cell death (e.g. moderate UV-B irradiation, certain viral infection, tumor growth, etc) are characterized by lack of inflammation and suppression of T cell-mediated immunity. It was originally thought that the high efficiency of the clearance of apoptotic cells was the simple reason by which apoptotic cells failed to provide the danger signals required to initiate inflammation. Voll and colleagues [76] were the first to demonstrate that apoptotic cells suppress the inflammatory response through an active mechanism that includes delivery of inhibitory signals to the phagocytes of the immune system. More recently, several laboratories have demonstrated that internalization of cells in early apoptosis by immature DC is not accompanied by DC activation in humans and rodents [77, 78]. Internalization of apoptotic cells by immature DC is not an immunologically null event. Immature DC that have internalized cells in early apoptosis fail to upregulate the expression of MHC-II, CD80, CD86, CD40 and CD83 (in humans) even after subsequent stimulation with different DC-maturation factors including LPS, CD40 ligation, TNF-α and monocyte-conditioned medium [54, 78–80]. This inhibitory effect is not merely the consequence of phagocytosis of particulate Ag, since immature DC do not exhibit defects in phenotype or maturation driven by LPS after ingestion of control latex beads of similar size than apoptotic cell fragments [78]. Following uptake of apoptotic cells, DC decrease their capacity to stimulate Ag-specific TCRtg T cells and allogeneic T cells [78, 79]. DC exposed to apoptotic cells exhibit a limited capacity for presentation of antigenic peptides to T-cell clones that have less stringent requirements for co-stimulation than primary T lymphocytes [80]. Collectively, these results suggest that the defect of DC exposed to apoptotic cells is probably at the level of expression of MHC-peptide complexes and co-stimulatory molecules on the cell surface, rather than a deficiency in the Ag-processing capacity of DC [80]. By contrast, immature DC exposed to necrotic cells or their supernatants, increase expression of MHC and co-stimulatory molecules and exhibit an augmented capacity to stimulate CD4$^+$ or CD8$^+$ T cells [13].

As described initially in macrophages [76, 81, 82], DC that have interacted with and/or internalized cells in early apoptosis secrete significantly lower levels of pro-inflammatory mediators, including IL-1α, IL-1β, IL-6, IL-12p70 and TNF-α [14, 80].

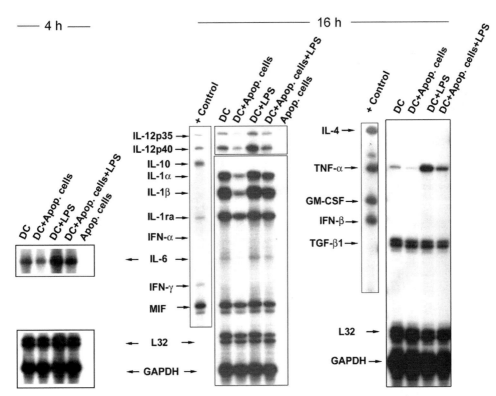

Fig. 29.2 Effect of apoptotic cells on DC cytokine gene mRNA transcription or stabilization. Comparative RNAse protection assay analysis of cytokine mRNA transcribed by DC following phagocytosis of apoptotic splenocytes. mRNA was isolated from immunomagnetic bead-sorted BM-derived DC after co-incubation with apoptotic cells for 4 or 16 h, in the absence or presence of 200 μg ml^{-1} LPS. For most cytokines, the changes in mRNA levels were evident after 16 h with the exception of IL-6. Reproduced with permission from *Blood* 2003. 101: 611–620 (A.E. Morelli et al.).

Other groups have reported that DC exposed to apoptotic cells may increase secretion of IL-6, IL-12p40 and TNF-α [78, 79]. These contrasting results may be due to different experimental conditions or exposure of DC to cells in later stages of apoptosis. The inhibitory effect of apoptotic cells on secretion of pro-inflammatory cytokines by DC was maintained even in the presence of LPS [14]. Interestingly, internalization of apoptotic cells did not interfere with secretion of immunosuppressive cytokines like TGF-β1 by murine DC [14] and increased release of IL-10 by human DC [80]. At least part of the inhibitory effect of apoptotic cells on cytokine production by DC is at the level of cytokine mRNA transcription or stabilization (Fig. 29.2) [14].

It has been shown that macrophages with internalized apoptotic cells inhibit release of pro-inflammatory mediators and augment secretion of TGF-β1, prostaglandin E2 (PGE2) and platelet-activating factor [82]. Release of soluble TGF-β1, PGE2 and platelet-activating factor inhibited by means of a paracrine mechanism secretion of pro-inflammatory cytokines by bystander macrophages that were not exposed to apoptotic cells [82]. By contrast, modulation of the DC function by apoptotic cells appears to be exclusively a cell contact-dependent process. Stuart and colleagues [78] have demonstrated that the inhibitory effects of apoptotic cells were restricted to those DC that have engulfed apoptotic cells, while bystander DC remained unaffected. Unlike macrophages, the inhibitory effects of apoptotic cells on DC were not mediated by soluble cytokines like TGF-β1 or IL-10 [78, 80].

29.6
Molecular Mechanisms of the Interaction between Dendritic Cells and Apoptotic Cells

As were originally demonstrated in macrophages [76], apoptotic cells signal DC through some of the surface receptors that participate in recognition/uptake of apoptotic cells. Urban et al. [80] have demonstrated that human immature DC exposed to mAb against CD36 or the α$_v$-integrin chain (CD51) evoked a minimal allogeneic T-cell response and failed to mature in response to LPS, CD40 ligation, TNF-α and monocyte-conditioned medium. Interestingly, exposure of DC to anti-CD36 mAb also increased secretion of IL-10 and inhibited release of IL-12p70 in response to LPS [80].

Apoptotic cells can interact with DC via CR. In humans, DC exposed to iC3b-opsonized apoptotic cells remained immature, did not secrete IL-12 and failed to upregulate surface HLA-DR and CD86 after stimulation with anti-CD40 mAb or LPS [54]. In mice, co-incubation of BM-derived DC with erythrocytes bearing IgM and mouse iC3b on their surface inhibited DC secretion of IL-1α, IL-1β, IL-6, IL-12p70 and TNF-α even in the presence of LPS, but it did not affect release of TGF-β1 [14]. The effect of iC3b on DC cytokines was at least in part mediated via CD11b/CD18 (CR3), since a similar modulation of cytokines can be obtained following exposure of DC to polystyrene beads coated with mAb that binds to CD11b or CD18 [14]. These results agree with previous reports that showed that binding of iC3b to CR3 inhibits IL-12p70 secretion by macrophages [83, 84] and that local deposition of iC3b is critical for UV-B-mediated skin immunosuppression, an effect that requires interaction of iC3b with a subset of dermal macrophages [85, 86]. The internalization of C1q-opsonized apoptotic cells, but not nonopsonized apoptotic cells, by DC stimulated production of IL-6, IL-10 and TNF-α without increasing IL-12p70 secretion [87].

CD47 (also known as integrin-associated protein) is a molecule present on the plasma membrane that is physically and functionally coupled to the $\alpha_v\beta_3$ integrin (Fig. 29.1). This multispan transmembrane protein is expressed by immature DC and recognizes TSP as its ligand. Although blockade of CD47 does not affect the ability of human DC to engulf apoptotic cells, there is evidence that apoptotic cells might modulate the function of DC through CD47. Engagement of CD47 on the surface of human monocyte-derived immature DC, by means of mAb or a TSP-derived synthetic peptide, inhibited DC maturation and secretion of IL-6, IL-12p70 and TNF-α in response to bacterial stimulation [88].

Fadok and colleagues have demonstrated that, in macrophage cell lines, binding of the stereospecific receptor for PS decreases secretion of TFN-α and augments release TGF-β1 in response to LPS [23].

29.7
Dendritic Cells, Apoptotic Cells and Peripheral Tolerance

DC play a dual role in the immune system, they initiate T-cell immunity and participate in central and peripheral T-cell tolerance. In the thymus, DC delete those thymocytes with high affinity for self MHC-peptide complexes. However, central deletion of self-reactive thymocytes is incomplete and not all self-Ag have permanent access to thymic APC because they are expressed later in life (e.g. puberty), are synthesized transiently (e.g. pregnancy) or remain sequestered in peripheral tissues (e.g. myelin basic protein). Self-reactive T cells that escape thymic deletion have access to peripheral tissues and may eventually trigger autoimmunity. An efficient mechanism of peripheral T-cell tolerance must constantly monitor and inactivate/delete any possible self-reactive T cells and/or induce/amplify T cells with regulatory function [regulatory T (T_{reg}) cells].

The current DC-based models of peripheral T-cell tolerance propose that immature/semi-mature DC migrate constitutively from peripheral tissues to secondary lymphoid organs as "Ag transporting cells" [89–91]. These migratory DC, expressing self-peptides-MHC complexes and with weak T-cell stimulatory function, would induce anergy and/or apoptosis of autoreactive T cells or generation of self-Ag-specific T_{reg} cells [89, 90]. Experiments in mice expressing transgenic model Ag controlled by tissue-specific promoters have demonstrated that, in the steady state, constitutively migratory DC transport and process tissue-specific Ag from periphery to lymph nodes and spleen. These DC silence, instead of stimulate, autoreactive CD4$^+$ or CD8$^+$ T cells [92, 93]. In humans, repetitive stimulation of naïve CD4$^+$ T cells by allogeneic immature DC reportedly generates T_{reg} cells [95]. Administration of immature DC pulsed with an MHC class-I restricted influenza-derived peptide leads to specific inhibition of Ag-specific CD8+ T cells [96, 97].

A critical point of the current model of peripheral T-cell tolerance is that migratory DC transport self-Ag from the periphery to secondary lymphoid organs. Apoptotic cells that result from the normal cell turnover of peripheral tissues are a rich source of self-Ag and are internalized efficiently by immature DC of peripheral tissues. Interestingly, Ip and colleagues [98] have reported that following ingestion of apoptotic cells, DC decreased expression of the chemokine receptor CCR5 and increased the levels of CCR7. This result indicates that DC with internalized apoptotic cells, although remaining immature, acquire the ability to home to draining lymph nodes. Huang et al. [99] reported that intestinal DC with intracellular apoptotic cell fragments (derived from intestinal epithelial cells) traffic to mesenteric lymph nodes. This observation suggests that internalization of apoptotic cells by DC in peripheral tissues followed by transport and presentation of self-peptides to naïve T cells in secondary lymphoid organs plays a role in induction/maintenance of peripheral T-cell tolerance [99]. Experiments in gnotobiotic rats under germ-free conditions indicate that migration of intestinal DC with phagocytosed apoptotic cells towards the mesenteric lymph nodes is a constitutive phenomenon, independent of the presence of DC-maturation stimuli derived from the intestinal bacterial flora [99]. The presence of DC with engulfed apoptotic bodies or with intracellular inclusions derived from neighboring parenchymal cells has been reported in other peripheral tissues. LC with internalized apoptotic keratinocytes derived from epithelial cells have been detected in the vaginal epithelium of mice [7]. In the skin, epidermal LC internalize melanosomes released by surrounding melanocytes and skin-draining lymph nodes contain DC with phagocytosed melanosomes [100]. In mice, Scheinecker et al. [101] have detected DC with intracytoplasmic fragments containing proton pump H+/K+-ATPase (a self-Ag expressed by parietal cells) close to the gastric epithelium and in T-cell areas of stomach-draining lymph nodes. These DC efficiently presented an H+/K+-ATPase-derived peptide to specific TCR tg T cells [101]. Blood-borne apoptotic/dying cells are also captured efficiently from circulation by splenic DC (Fig. 29.3) [14, 26]. Iyoda et al. [26] have shown that circulating leukocytes subjected to apoptosis by osmotic shock are internalized *in vivo* exclusively by splenic CD8α+ DC within the T cell-areas. By contrast, our group [14] has shown that blood-borne (UV-B-induced) allogeneic apoptotic leukocytes are captured initially by splenic CD8α− DC of the marginal zone. These marginal zone DC might later acquire CD8α and mobilize to T-cell areas, or alternatively, transfer the apoptotic cell fragments to CD8α+ DC, as suggested by others [102]. However, it is likely that those dying blood-borne cells still able to traffic to T cells areas of the spleen before becoming apoptotic are phagocytosed directly by CD8α+ DC and those circulating apoptotic cells unable to migrate to T-cell areas are captured easily by marginal zone CD8α− DC.

Internalization of apoptotic cells by splenic DC is associated with the presentation of apoptotic cell-derived peptides in MHC class-I and -II molecules to CD8+ and CD4+ T cells, respectively [26]. Liu et al. [103] have shown that, in the steady state, presentation of apoptotic cell-derived peptides by immature/semi-mature

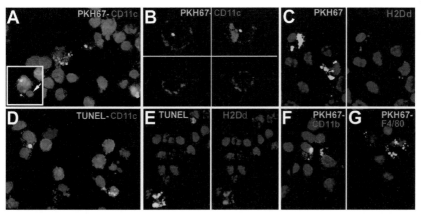

Fig. 29.3 Entrapment of apoptotic cells by splenic DC *in vivo*. Internalization of apoptotic cells by splenic DC was analyzed in cytospins of immunobead-sorted DC 1 h after injection of PKH67-labeled (green) apoptotic (BALB/c) splenocytes in (B10) mie. (A) CD11c⁺ DC with apoptotic cell fragments (green) and with DAPI⁺ intracytoplasmic inclusions, -likely DNA from ingested apoptotic cells (in blue indicated by arrow in inset). (B) Serial sections analyzed by confocal microscopy confirmed the intracellular localization of PKH67-labeled fragments in splenic CD11c⁺ DC. (C) The donor origin (BALB/c) of the intracytoplasmic inclusions in (B10) DC was confirmed by H2D^d expression (in red) in PKH67-labeled (green) fragments. (D & E) FITC-TUNEL staining in combination with Cy3-anti-CD11c or Cy3-anti-H2D^d confirmed the presence of donor (BALB/c)-derived apoptotic cells within (B10) DC. (F & G) One h after i.v. injection of apoptotic cells, DC that internalized apoptotic cells expressed CD11b^hi and F4/80^lo/−. Nuclei were counterstained with DAPI (1000x). Reproduced with permission from *Blood* **2003**. 101: 611–620 (A.E. Morelli et al.).

DC in secondary lymphoid organs leads to peripheral CD8⁺ T-cell tolerance. Following i.v. injection of apoptotic/dying leukocytes loaded with the model Ag OVA, splenic DC internalized the circulating apoptotic cells and presented an OVA-derived peptide to TCR tg CD8⁺ cells [103]. Since internalization of apoptotic cells did not lead to APC maturation, DC expressing inadequate levels of co-stimulatory molecules induced abortive proliferation of OVA-specific CD8⁺ T cells followed by deletion and CD8⁺ T-cell tolerance to antigenic challenge [103].

29.8
The Potential Therapeutical Use of Apoptotic Cells for Peripheral Tolerance

Based on the current model of peripheral T-cell tolerance, specific targeting of Ag to immature/semi-mature DC in secondary lymphoid organs *in situ* (in the steady state) is a novel approach to induce T-cell tolerance for treatment of transplant rejection or autoimmune disorders [91]. To adopt this principle to achieve T-cell tolerance, the method employed to deliver the Ag to the host DC *in situ* should avoid the activation of the acceptor DC, while maintaining their intrinsic capacity to present the Ag in a tolerogenic fashion [91, 104]. Importantly, this new approach would prevent the spontaneous DC activation induced by manipulation of DC generated *in vitro*, which has been a drawback in the current DC-based therapies for tolerance induction [91]. Interestingly, it has been demonstrated in mice that specific targeting of model Ag to immature/semi-mature DC in lymphoid tissues by delivering the HEL_{46-61} peptide attached to an anti-DEC-205 mAb (DEC-205 is expressed by DC in secondary lymphoid organs) induced a sustained reduction of the T-cell response to HEL_{46-61} challenge [105, 106]. Ferguson et al. [107] have demonstrated that i.v. administration of trinitrophenyl (TNP)-coupled splenocytes that become apoptotic (via a Fas-FasL dependent pathway) soon after injection induce hapten-specific T-cell unresponsiveness in a model of contact hypersensitivity in mice. The phenomenon was mediated by host splenic $CD8\alpha^+$ DC and the inhibitory effect on contact hypersensitivity could be adoptively transferred from unresponsive animals to untreated hosts, suggesting the generation of T_{reg} cells [107].

Based on the current model of peripheral T-cell tolerance, administration of donor cells (expressing MHC class-I and –II) in early apoptosis to deliver alloAg to host semi-mature DC in secondary lymphoid tissues (without disrupting the steady state) may be an effective way to induce donor-specific tolerance for transplantation. Early apoptotic cells have the following advantages for specific targeting of alloAg to DC *in vivo* compared to other systems: (i) apoptotic cells are a rich source of alloAg; (ii) unlike soluble MHC, MHC peptides, chimeric MHC molecules and Ag tagged to mAb, early apoptotic cells are easy to prepared in the laboratory; (iii) i.v. administration of allogeneic apoptotic cells is relatively safe; (iv) once injected i.v., blood-borne apoptotic cells are captured efficiently by splenic DC [14, 26]; (v) early apoptotic cells deliver a potent immunosuppressive signal to DC [14, 54, 78]; and (vi) DC present apoptotic cell-derived allopeptides [108].

In this regard, we have demonstrated that i.v. injection of donor apoptotic splenocytes (in early apoptosis) 7 days before transplant surgery prolongs survival of cardiac allografts in the absence of immunosuppression in a murine experimental model [109]. The effect was donor-specific and depended on injection of donor cells in early apoptosis, since administration of donor necrotic cells did not affect graft survival. In this model, splenic DC rapidly ingested the i.v. injected apoptotic cells, processed apoptotic cell-derived peptides into MHC molecules and mobilized to T-cell areas of the splenic follicle. Administration of donor apoptotic leukocytes before transplant surgery drastically reduces the systemic and local anti-donor response. The effect requires the interaction of circulating apoptotic donor cells with recipient splenic $CD8\alpha^+$ DC [109]. In agreement with our results, it has been shown that human immature DC loaded with allogeneic apoptotic cells induce cross-tolerance to allospecific T cells *in vitro* [108]. Using a more indirect approach, Xu et al. [110] have recently shown in mice that intra-portal administration of recipient immature BM-derived DC loaded *in vitro* with donor apoptotic splenocytes prolongs survival of cardiac allografts in the absence of pharmacological immmunosuppression.

In a different transplantation model, de Carvalho Bittencourt and colleagues [111] have shown that i.v. injection of donor apoptotic splenocytes enhances BM engraftment in mice. Interestingly in this study, the therapeutic effect of i.v. administration of apoptotic leukocytes was nonspecific and independent of the origin of the apoptotic bodies (donor, recipient, third party or xenogeneic) [111]. Of note, in this study apoptotic cells were administered simultaneously with the BM inoculum [111]. Therefore, it is possible that the interaction of host DC with a dose of self, third party or xenogeneic apoptotic cells may have exerted a bystander inhibitory effect on DC while presenting donor BM-derived alloAg to anti-donor T cells.

In humans, there is evidence that photopheresis exerts beneficial prophylactic and therapeutic effects on survival of cardiac allografts [112]. During this process, extracorporeal UV-irradiation of mononuclear cells triggers their apoptosis after re-infusion into the patient. Mechanisms underlying the inhibitory effects of apoptotic cells in photopheresis-treated patients have yet to be clarified, but modulation of the recipient DC function by circulating apoptotic cells may probably be involved in the therapeutic effect.

29.9
Pathogens and Apoptotic Cell-like Mimicry

Clearance of apoptotic cells by phagocytes is a system highly conserved through evolution, extremely efficient and designed to prevent tissue damage and development of immunity. Therefore it is not strange that certain pathogens have evolved to target this mechanism to suppress immunity and to establish infection. The malaria parasite *Plasmodium falciparum* employs some of the mechanisms used for clearance of apoptotic cells to evade the host immune response. DC exposed to malaria-infected red blood cells failed to mature in response to LPS and to induce primary and secondary Ag-specific T-cell responses [113]. The inhibitory effect of parasite-infected erythrocytes on DC was mediated by direct binding of the erythrocytes to CD36 and indirectly, through attachment of the bridging molecule TSP-1, to CD36 and $\alpha_v\beta_3/\beta_5$ on the DC surface [80]. By this mechanism malaria-infected erythrocytes prevented DC maturation and shifted secretion of cytokines from IL-12p70 to IL-10 [80]. Thus, parasite-infected erythrocytes downregulated the function of DC as occurs during interaction with apoptotic cells, a phenomenon that may explain the impairment of T-cell immunity in patients with malaria. *Trypanosoma cruzi* is another example of a parasite that takes advantage of the mechanism of apoptotic cell clearance for its survival. In a model of Chagas disease, phagocytosis of apoptotic cells by macrophages via the $\alpha_v\beta_3$ integrin induces secretion of TGF-β1 and prostaglandin E_2 that promote growth of *T. cruzi* [114].

P. Vanlandschoot and G. Leroux-Roels [115] have postulated the hypothesis that the hepatitis B virus (HBV) may affect the function of DC and alter the Th1/Th2 balance in detriment of secretion of IL-12 and IFN-γ by mimicking mechanisms employed by apoptotic cells to maintain self-tolerance. HBV-infected hepatocytes release noninfectious subviral particles into circulation, known as hepatitis surface Ag (HBsAg). It is unclear, how the massive number of HBsAg particles produced during hepatitis B participates in establishment and maintenance of HBV infection. Interestingly, blood-borne HBsAg display phospholipids and oxidized phospholipids on their surface, similar to those present on the surface of apoptotic cells [115]. These HBsAg phospholipids bind soluble β2-GPI and CD14 (and probably other receptors for phospholipids) as demonstrated in apoptotic cells. Alternatively, circulating complexes of HBsAg and IgM/IgG may induce deposition of complement factors that opsonize HBsAg particles facilitating their interaction with DC [115]. Yeast-expressed HBsAg behave as apoptotic cell-like particles inhibiting LPS-induced secretion of pro-inflammatory cytokines and augmenting IL-10 produced by monocytes [116]. The possible apoptotic cell-like interaction of HBsAg particles with DC and its impact on DC function are still unknown.

29.10
Dead Cells and the Delicate Balance between Immunity and Tolerance

Dead cells exert different effects on DC. Several groups have demonstrated that exposure of DC to necrotic cells, but not apoptotic cells, induce APC maturation and generation of Ag-specific CD4[+] and CD8[+] T cells [13, 77]. Necrotic cell death, unlike apoptosis, has been traditionally associated with inflammation and initiation of adaptive immunity. Since the initial observation by Sauter et al. [13] demonstrating that filtered supernatants from necrotic cell lines induced DC maturation, the nature of soluble mediators released by necrotic cells and responsible for DC activation are beginning to be elucidated. Cells undergoing necrosis passively release uric acid, heat shock proteins (HSP) and high mobility group box 1 (HMGB1), all molecules with potent pro-inflammatory effects [117–119]. HSP are released from necrotic or stressed cells but are confined within apoptotic cells, unless the latter are disrupted or undergo secondary necrosis [117]. Once in the extracellular milieu, HSP bind to receptors on the surface of DC (e.g. CD91) and induce their activation [117]. HMGB1 is a chromatin-binding protein that binds the receptor for glycosylation end products (RAGE), activates the nuclear factor-κB (NF-κB) signaling pathways and induces DC maturation [120]. Interestingly, apoptotic cells do not release HMGB1, a phenomenon due to active de-acetylation of histones that result in tight binding of HMGB1 to the chromatin of apoptotic bodies.

The idea that DC distinguish between the two types of cell death and decide the outcome of adaptive immunity accordingly, with necrotic cells providing information for initiation of immunity and apoptotic cells for induction/maintenance of tolerance, is a generalization of a far more complex phenomenon. If DC-activating signals came exclusively from necrotic cells, all pathogens that induce apoptosis would escape immunity. Several studies have shown that DC loaded with apoptotic cells activate CD4[+] [9, 10] and cross-prime CD8[+] T cells efficiently [8–10]. Although, it may be argued that some of these studies may have employed a mixed population of apoptotic and necrotic cells or that the apoptotic cells have been stressed or infected with pathogens, there is evidence that, under certain conditions, apoptotic cells may initiate inflammation and/or adaptive immunity. In this regard, the type of opsonin deposited on the surface of apoptotic cells may exert different effects on the function of the acceptor phagocytes. Endocytosis of IgG-opsonized apoptotic cells by macrophages via Fc receptors did not prevent LPS-induced secretion of pro-inflammatory cytokines, as occurs after internalization of control apoptotic cells [82]. In certain autoimmune disorders, the phagocyte receptors employed to internalize apoptotic cells may activate different downstream signaling pathways and perturb the anti-inflammatory effect of apoptotic cells. Patients with systemic lupus erythematosus (SLE) exhibit circulating anti-phospholipid auto-antibodies that recognize externalized phospholipids on the surface of apoptotic cells [121]. Internalization of these Ig-opsonized apoptotic cells can stimulate secretion, instead of suppression, of TNF-α by macrophages and other phagocytes [121].

Impairment or saturation of the mechanism of clearance of cells in early apoptosis may lead to accumulation of cells in late apoptosis (or secondary necrosis) with the consequent release of toxic mediators that may eventually induce DC activation. This phenomenon may be triggered by the simple increase of the ratio between apoptotic cells and DC. Incubation of DC with an excess of apoptotic cells was associated with DC maturation and efficient presentation to both MHC class I- and class-II restricted T cells [10]. The effect was apoptotic cell dose-dependent and accompanied by an autocrine/paracrine secretion of IL-1β and TNF-α [10]. Moreover, immunization with high numbers of apoptotic cells or with DC exposed to high numbers of apoptotic cells primed tumor-specific CTL and conferred protection against tumor challenge in mice [122]. Alternatively, accumulation of cells in late apoptosis may be caused by a deficiency in the mechanism of recognition/docking/internalization of apoptotic cells. In humans and mice, defective clearance of apoptotic cells is associated with development of an SLE-like autoimmune disorder. Apoptotic cells are thought to be a major source of auto-Ag in SLE in humans. Mice deficient in molecules involved in the uptake of apoptotic cells such as C1q, C4, IgM, SAP, and c-Mer develop systemic autoimmunity [123–128]. SLE in humans is associated with genetic deficiencies in the complement factors C1q, C1r, C1s, C4 and C2 [129], and phagocytes from patients with SLE exhibit impaired uptake of apoptotic cells [130, 131]. There is a hierarchical association between SLE susceptibility and severity of the disease according to the position of the complement factor in the activation pathway, with the highest incidence of the disease in patients with C1q deficiency [132].

Exposure of DC to APC-activating signals while ingesting apoptotic cells may also affect the outcome of the T-cell response. Several viruses induce apoptosis of their host cells and different types of virus-infected cells secrete IFN-α. DC that have internalized apoptotic fibroblasts and that otherwise would remain immature, become activated following incubation with IFN-α [77]. Therefore, pathogen-induced secretion of DC-activating cytokines (e.g. IL-1β, TNF-α, IFN-α) may help to explain those studies where DC primed specific T cells efficiently following uptake of apoptotic cells derived from monocytes infected by influenza A virus [8] or from macrophages infected by *Salmonella typhimurium* [9]. Alternatively, DC also could receive maturation-signals via CD4[+] T cell-dependent CD40 stimulation, viral double stranded RNA, CpG DNA motifs, LPS from gram-negative bacteria, HSP from stressed apoptotic cells or HSP and HMGB1 from necrotic cells.

The intrinsic characteristics of the apoptotic cells may also influence the capability of the acceptor APC to process and present apoptotic cell-derived Ag and to shift the balance between immunity and tolerance [133]. Gao et al. [134] and Chen and colleagues [135] have shown that apoptotic leukocytes can release IL-10 and bio-active TGF-β1, two cytokines that exert immunoregulatory effects on DC. By contrast, heat stress can upregulate expression of molecules that induce DC activation, such as HSP72 and HSP60 on the surface of apoptotic cells [136]. Vaccination with DC pulsed with heat-stressed apoptotic tumor cells results in development of tumor-specific immunity in mice [136]. A study by Lauber and colleagues [137] showed that apoptotic cells release lysophosphatidylcholine, a lipid chemotactic for monocytes and macrophages. Lysophosphatidylcholine is produced by hydrolysis of membrane phosphatidylcholine by phospholipase A2 (PLA2) and calcium-independent PLA2 is activated by cleavage by caspase 3 during apoptosis. By binding the immunoregulatory receptor G2A, lysophosphatidylcholine can activate pro-inflammatory transcription factors like NF-κB and signaling through mitogen-activated protein kinase (MAPKs) pathways. Therefore, local release of lysophosphatidylcholine by apoptotic cells may eventually create a pro-inflammatory environment for macrophages and DC.

29.11
Concluding Remarks

During the last ten years, the better understanding of the mechanisms of interaction of professional APC with apoptotic cells has opened new possibilities to deliver exogenous Ag to DC for immunization or negative vaccination. Many variables including the intrinsic properties of apoptotic cells and their interactions with receptors on the surface of DC will have to be optimized in view of possible therapeutic uses of apoptotic cells in the future. These approaches may include administration of DC loaded with apoptotic cells for treatment of infectious diseases or malignant neoplasms, or alternatively, for induction of Ag-specific tolerance for prevention/treatment of autoimmune disorders or allograft rejection. Still from the grave, a dead cell has stories to tell.

Acknowledgements

Supported by grants from the National Institutes of Health: R01 HL077545, R01 HL075512, R21 HL69725 and R21 AI55027 (A.E.M) and R01 CA100893 and R21 AI57958 (A.T.L.). We thank B.L. Colvin and W.J. Shufesky for their comments during the preparation of the manuscript.

References

1 Wyllie, A.H., Duvall, E., Blow. J.J., Intracellular mechanisms in cell death in normal and pathological tissues. In Davis, A.I., Sigel, D.C. (Eds.) *Cell aging and cell death*. Cambridge Press, Cambridge **1980**.

2 Casciola-Rosen, L.A., Anhalt, G., Rosen, A., Autoantigens targeted in systemic lupus erythematosus are clustered in two populations of surface structures on apoptotic keratinocytes. *J. Exp. Med.* **1994**. 179: 1317–1330.

3 Kerr, J.F., Wyllie, A.H., Currie, A.R., Apoptosis: a basic biological phenomenon with wide-ranging implications in tissue kinetics. *Br. J. Cancer* **1972**. 26: 239–257.

4 Wyllie, A.H., Kerr, J.F.R., Currie, A.R., Cell death: the significance of apoptosis. *Int. Re.v Cytol.* **1980**. 68: 251–306.

5 Restifo, N.P., Building better vaccines: how apoptotic cell death can induce inflammation and activate innate and adaptive immunity. *Curr. Opin. Immunol.* **2000**. 12: 597–603.

6 Fossum, S., Rolstad, B., The roles of interdigitating cells and natural killer cells in the rapid rejection of allogeneic lymphocytes. *Eur. J. Immunol.* **1986**. 16: 440–450.

7 Parr, M.B., Kepple, L., Parr, E.L., Langerhans cells phagocytose vaginal epithelial cells undergoing apoptosis during the murine estrous cycle. *Biol. Reprod.* **1991**. 45: 252–260.

8 Albert, M.L., Sauter, B., Bhardwaj, N., Dendritic cells acquire antigen from apoptotic cells and induce class I-restricted CTLs. *Nature* **1998**. 392: 86–89.

9 Yrlid, U., Wick, M.J., Salmonella-induced apoptosis of infected macrophages results in presentation of a bacteria-encoded antigen after uptake by bystander dendritic cells. *J. Exp. Med.* **2000**. 191: 613–623.

10 Rovere, P., Vallinoto, C., Bondanza, A., Crosti, M.C., Rescigno, M., Ricciardi-Castagnoli, P., Rugarti, C., Manfredi, A.A., Bystander apoptotic cells trigger dendritic cell maturation and antigen-presenting function. *J. Immunol.* **1998**. 161: 4467–4471.

11 Rodriguez, A., Regnault, A., Kleijmeer, M., Ricciardi-Castagnoli, P., Amingorena, S., Selective transport of internalized antigens to the cytosol for MHC class I presentation in dendritic cells. *Nat. Cell Biol.* **1999**. 1: 362–368.

12 Inaba, K., Turlev. S., Yamaide, F., Iyoda, T., Mahnke, K., Inaba, M., Pack, M., Subklewe, M., Sauter, B., Sheff, D., et al., Efficient presentation of phagocytosed cellular fragments on the major histocompatibility complex class II products of dendritic cells. *J. Exp. Med.* **1998**. 11: 2163–2173.

13 Sauter, B., Albert, M., Francisco, L., Larsson, M., Somersan, S., Bhardwaj, N., Consequences of cell death: exposure to necrotic tumor cells, but not primary tissue cells or apoptotic cells, induces the maturation of immunostimulatory dendritic cells. *J. Exp. Med.* **2000**. 191: 423–433.

14 Morelli, A.E., Larregina, A.T., Shufesky, W.J., Zahorchack, A., Logar, A., Papworth, G.D., Wang, Z., Watkins, S.C., Falo Jr., L.D., Thomson, A.W., Internalization of circulating apoptotic cells by splenic marginal zone dendritic cells: dependence on complement receptors and effect on cytokine production. *Blood* **2003**. 101: 611–620.

15 Albert, M.L., Kim, J.-I., Bridges, R.B., αvβ5 integrin recruits the CrkII-Dock180-Rac1 complex for phagocytosis of apoptotic cells. *Nat. Cell Biol.* **2000**. 2: 899–905.

16 Akakura, S., Singh, S., Spataro, M., Akakura, R., Kin, J.I, Albert, M.L., Birge, R.B., The opsonin MFG-E8 is a ligand for the alphavbeta5 integrin and triggers DOCK180-dependent Rac-1 activation for the phagocytosis of apoptotic cells. *Exp. Cell Res.* **2004**. 292: 403–416.

17 Savill, J., Fadok, V., Corpse clearance defines the meaning of cell death. *Nature* **2000**. 407: 784–788.

18 Gregory, C.D., CD14-dependent clearance of apoptotic cells: relevance to the immune system. *Curr. Opin. Immunol.* **2001**. 12: 27–34.

19 Fadok, V.A., Bratton, D.L., Henson P.M., Phagocyte receptors for apoptotic cells:

recognition, uptake, and consequences. *J. Clin. Invest.* **2001.** 108: 957–962.

20 Fadok, V.A., Voelker, D.R., Campbell, P.A., Cohen, J.J., Bratton, D.L., Henson P.M., Exposure of phosphatidylserine on the surface of apoptotic lymphocytes triggers specific recognition and removal by macrophages. *J. Immunol.* **1992.** 148: 2207–2216.

21 Hamon, Y., Broccardo., C., Chambenoit, O., Luciani, M.F., Toti, F., Chaslin, S., Freyssinet, J.M., Devaux, P.F., McNeish, J., Marguet, D., Chimini, G. ABC1 promotes engulfment of apoptotic cells and transbilayer redistribution of phosphatidylserine. *Nat. Cell. Biol.* **2000.** 2: 399–406.

22 Marguet, D., Luciani, M.F., Moynaault, A., Williamson, P., Chimini, G. Engulfment of apoptotic cells involves the redistribution of phosphatidylserine on phagocyte and prey. *Nat. Cell Biol.* **1999.** 1: 454–456.

23 Fadok V.A., Bratton D.L., Rose D.M., Pearson, A., Ezekewitz, A.B., Henson, P.A., A receptor for phosphatidylserine-specific clearance of apoptotic cells. *Nature* **2000.** 405: 85–90.

24 Arur, S., Uche, U.E., Rezauk, K., Fong, M., Scranton, V., Cowan, A.E., Mohler, W., Han, D.K., Annexin I is an endogenous ligand that mediates apoptotic cell engulfment. *Dev. Cell* **2003.** 4: 587–598.

25 Li, M.O., Sarkisian, M.R., Mehal, W.Z., Rakic, P., Flavell, R.A., Phosphatyidylserine receptor is required for clearance of apoptotic cells. *Science* **2003.** 302: 1560–1563.

26 Iyoda, T., Shimoyama, S., Liu, K., Omatsu, Y., Akiyama, Y., Maeda, Y., Takahara, K., Steinman, R.M., Inaba, K., The CD8+ dendritic cell subset selectively endocytoses dying cells in culture and in vivo. *J. Exp. Med.* **2002.** 195: 1289–1302.

27 Chen, X., Doffek, K., Sugg, S.L., Shilyansky J., Phosphatidylserine regulates the maturation of human dendritic cells. *J. Immunol.* **2004.** 173: 2985–2994.

28 Balasubramanian, K., Schroit, A.J. Characterization of phosphatidylserine-dependent β2-glycoprotein I macrophage interactions. Implications for apoptotic

cell clearance by phagocytes. *J. Biol. Chem.* **1998.** 273: 29272–29277.

29 Rovere, P., Grazia Sabbadini, M., Vallinoto, C., Fascio, U., Rescigno, M., Crosti, M., Ricciardi-Casstagnoli, P., Balestrieri, G., Tincani, A., Manfredi, A.A, Dendritic cell presentation of antigens from apoptotic cells in a proinflammatory context. *Arth. Rheum.* **1999.** 42: 1412–1420.

30 Chang, M.K., Bergmark, C., Laurila, A., Horkko, S., Han, K.H., Friedman, P., Dennis, E.A., Witztum, J.L., Monoclonal antibodies against oxidized low-density lipoprotein bind to apoptotic cells and inhibit their phagocytosis by elicited macrophages: evidence that oxidation-specific epitopes mediate macrophage recognition. *Proc. Nat. Acad. Sci. USA* **1999.** 96: 6353–6358.

31 Savill, J., Recognition and phagocytosis of cells undergoing apoptosis. *Br. Med. Bull.* **1997.** 53: 491–508.

32 Ren, Y., Silverstein, R.L., Allen, J., Savill, J., CD36 gene transfer confers capacity for phagocytosis of cells undergoing apoptosis. *J. Exp. Med.* **1995.** 181: 1857–1862.

33 Savill, J., Hogg, N., Ren, Y., Haslett, C., Thrombospondin cooperates with CD36 and the vitronectin receptor in macrophage recognition of neutrophils undergoing apoptosis. *J. Clin. Invest.* **1992.** 90: 1513–1522.

34 Rubartelli, A., Poggi, A., Zocchi, M.R., The selective engulfment of apoptotic cells by dendritic cells is mediated by the αvβ3 integrin and requires intracellular and extracellular calcium. *Eur. J. Immunol.* **1997.** 27: 1893–1900.

35 Albert, M.L., Pearce, S.F.A., Francisco, L.M., Sauter, B., Roy, P., Silverstein, R.L., Bhardwaj, N., Immature dendritic cells phagocytose apoptotic cells via αvβ5 and CD36, and cross-present antigens to cytotoxic T lymphocytes. *J. Exp. Med.* **1998.** 188: 1359–1368.

36 Belz, G.T., Vremec, D., Febbraio, M., Corcoran, L., Shortman, K., Carbone, F.R., Heath, W.R., CD36 is differentially expressed by CD8+ splenic dendritic cells but is not required for cross-presentation in vivo. *J. Immunol.* **2002.** 168: 6066–6070.

37 Schulz, O., Pennington, D.J., Hodivala-Dilke, K., Febbraio, M., Reis, E., Sousa, C., CD36 or alphavbeta3 and alphavbeta5 integrins are not essential for MHC class I cross-presentation of cell-associated antigen by CD8 alpha+ murine dendritic cells. *J. Immunol.* **2002**. 168: 6057–6065.

38 den Haan, J.M., Lehar, S.M., Bevan, M.J., CD8(+) but not CD8(-) dendritic cells cross-prime cytotoxic T cells in vivo. *J. Exp. Med.* **2000**. 192: 1685–1696.

39 Takizawa, F., Tsuji, S., Nagasawa, S., Enhancement of macrophage phago-cytosis upon iC3b deposition on apoptotic cells. *FEBS Let.* **1996**. 397: 269–272.

40 Mevorach, D., Mascarenhas, J.O., Gershov, D., Elkon, K.B., Complement-dependent clearance of apoptotic cells by human macrophages. *J. Exp. Med.* **1998**. 188: 2313–2320.

41 Kim, S.J., Gershov, D., Ma, X., Brot, N., Elkon, K.B., I-PLA2 activation during apoptosis promotes exposure of membrane lysophosphatidylcholine leading to binding by natural immunoglobulin M antibodies and complement activation. *J. Exp. Med.* **2002**. 196: 655–665.

42 Zwart, B., Ciurana, C., Rensink, I., Manoe, R., Hack, C.E., Aarden, L.A., Complement activation by apoptotic cells occurs predominantely via IgM and is limited to late apoptotic (secondary necrotic) cells. *Autoimmunity* 2004. 37: 95–102.

43 Korb, L.C., Ahearn, J.M., C1q binds directly and specifically to surface blebs of apoptotic human keratinocytes: complement deficiency and systemic lupus erythematous revisited. *J. Immunol* **1997**. 158: 4525–4528.

44 Navratil, J.S., Watkins, S.C., Wisnieski, J.J., Ahearn J.M., The globular heads of C1q specifically recognize surface blebs of apoptotic vascular endothelial cells. *J. Immunol.* **2001**. 166: 3231–3239.

45 Nauta, A.J., Trouw, L.A., Daha, M.R., Tijsma, O., Mieuwland, R., Schwaeble, W.J., Gingras, A.R., Mantovani, A., Hack, E.C., Roos, A., Direct binding of C1q to apoptotic cells an cell blebs induces complement activation. *Eur. J. Immunol.* **2002**. 32: 1726–1736.

46 Ogden, C.A., deCathelineau, A., Hoffmanm, P.R., Bratton, D., Ghebrehiwet, B., Fadok, V.A., Henson, P.M., C1q and mannose binding lectin engagement of cell surface calreticulin and CD91 initiates macropinocytosis and uptake of apoptotic cells. *J. Exp. Med.* **2001**. 194: 781–795.

47 Roos, A., Castellano, G., Nauta, A.J., Garred, P., Daha, M.R., vanKooten, C., A pivotal role for innate immunity in the clearance of apoptotic cells. *Eur. J. Immunol.* **2004**. 34: 921–929.

48 Castellano, G., Woltman, A.M. Nauta, A.J., Roos, A., Trouw, L.A., Seelen, M.A., Schena, F.P., Daha, M.R., Van Kooten, C., Maturation of denditic cells abrogates C1q production in vivo and in vitro. *Blood* 2004. 103: 3813–3820.

49 Nicholson-Weller, A., Klickstein, L.B., C1q-binding proteins and C1q receptors. *Curr. Opin. Immunol.* **1999**. 11: 42–46.

50 Nauta, A.J., Raascou-Jensen, N., Roos, A., Daha, M.R., Madsen, H.O., Borrias-Esser, M.C., Ryder, L.P., Koch, C., Garred, P., Mannose-binding lectin engagement with late apoptotic cells and necrotic cells. *Eur. J. Immunol.* **2003**. 33: 2853–2863.

51 Schagat, T.L., Wofford, J.A., Wright, J.R., Surfactant protein A enhances alveolar macrophage phagocytosis of apoptotic neutrophils. *J. Immunol.* **2001**. 166: 2727–2733.

52 Vandivier, R.W., Ogden, C.A., Fadok, V.A., Hoffman, P.R., Brown, K.K., Botto, M., Walport, M., Fisher, J.H., Henson, P.M., Greene, K.E., Role of surfactant proteins A, D, and C1q in the clearance of apoptotic cells in vivo and in vitro: calreticulin and CD91 as a common collectin receptor complex. *J. Immunol.* **2002**. 169: 3978–3986.

53 Fishelson, Z., Attali, G., Mevorach, D., Complement and apoptosis. *Mol. Immunol.* **2001**. 38: 207–219.

54 Verbovetski, I., Bychkov, H., Trahtemberg, U., Shapira, I., Hareuveni, M., Ben-Tal, O., Kutikov, I., Gill, O., Mevorach, D., Opsonization of apoptotic cells by autologous iC3b facilitates clearance by immature dendritic cells, down-regulates DR and CD86, and up-regulates CC chemokine receptor 7. *J. Exp. Med.* **2002**. 196: 1553–1561.

55 Gershov, D., Kim, S., Brot, N., Elkon, K.B., C-Reactive protein binds to apoptotic cells, protects the cells from assembly of the terminal complement components, and sustains an anti-inflammatory innate immune response: implications for systemic autoimmunity. *J. Exp. Med.* **2000**. 192: 1353–1363.

56 Familian, A., Zwart, B., Huisman, H.G., Rensink, I., Roem, D., Hordijk, P.L., Aarden, L.A., Hack, C.E., Chromatin-independent binding of serum amyloid P component (SAP) to apoptotic cells. *J. Immunol.* **2001**. 167: 647–654.

57 Rovere, P., Peri, G., Fazzini, F., Bottazzi, B., Doni, A., Bondanza, A., Zimmermann, V.S., Garlanda, C., Fascio, U., Grazia Sabbadini, M., et al, The long pentraxin PTX3 binds to apoptotic cells and regulates their clearance by antigen-presenting dendritic cells. *Blood* **2000**. 96: 4300–4306.

58 Nauta, A.J., Daha, M.H., van Kooten, C., Roos, A., Recognition and clearance of apoptotic cells: a role for complement and pentraxins. *Trends Immunol.* **2003**. 24: 148–154.

59 Mold, C., Baca, R., Du Clos, T.W., Serum amyloid P component and C-reactive protein opsonize apoptotic cells for phagocytosis through Fcγ receptors. *J. Autoimmunity* **2002**. 19: 147–154.

60 Bharadwaj, D., Mold, C., Markham, E., Du Clos, T.W., Serum amyloid P component binds to Fcγ receptors and opsonizes particles for phagocytosis. *J. Immunol.* **2001**. 166: 6735–6741.

61 Hanayama, R., Tanaka, M., Miwa, K., Shinohara, A., Iwamatsu, A., Nagata, S., Identification of a factor that links apoptotic cells to phagocytes. *Nature* **2002**. 417: 182–187.

62 Miyasaka, K., Hanayama, R., Tanaka, M., Nagata, S., Expression of milk fat globule epidermal groth factor 8 in immature dendritic cells for engulfment of apoptotic cells. *Eur. J. Immunol.* **2004**. 34: 1414–1422.

63 Ezekowitz, R.A., Local opsonization for apoptosis? *Nat. Immunol.* **2002**. 3: 510–512.

64 Duvall, E., Wyllie, A.H., Morris, R.G., Macrophage recognition of cells undergoing programmed cell death (apoptosis). *Immunology* **1985**. 56: 351–358.

65 Ii, M., Kurata, H., Itoh, N., Yamashima I., Kawasaki, T. Molecular cloning and sequence analysis of cDNA encoding the macrophage lectin specific for galactose and N-acetylgalactosamine. *J. Biol. Chem.* **1990**. 265: 11 295–11 298.

66 Mahnke, K., Guo, M., Lee, S., Sepulveda, H., Swain, S.L., Nussenzweig, M., Steinman, R.M., The dendritic cell receptor for endocytosis, DEC-205, can recycle and enhance antigen presentation via major histocompatibility complex class II-positive lysosomal compartments. *J. Cell Biol.* **2000**. 151: 673–684.

67 Small, M., Kraal, G., In vitro evidence from participation of DEC-205 expressed by thymic cortical epithelial cells in clearance of apoptotic thymocytes. *Int. Immunol.* **2003**. 15: 197–203.

68 Scott, R.S., McMahon, E.J., Pop, S.M., Reap, E.A., Caricchio, R., Cohen, P.L., Earp, H.S., and Matsushima, G.K., Phagocytosis and clearance of apoptotic cells is mediated by MER. *Nature* **2001**. 411: 207–211.

69 Nagata, K., Ohashi, K., Nakano, T., Arita, H., Zong, C., Hanafusa, H., Mizuno, K., Identification of the product of the growth arrest-specific gene 6 as a common ligand for Axl, Sky, and Mer receptor tyrosine kinases. *J. Biol. Chem.* **1996**. 271: 30 022–30 027.

70 Nakano, T., Ishimoto, Y., Kishimo, J., Umeda, M., Inoue, K., Nagata, K., Ohashi, K., Mizumo, K., Arita, H., Cell adhesion to phosphatidylserine mediated by a product of growth arrest-specific gene 6. *J. Biol. Chem.* **1997**. 272: 29 411–29 414.

71 Behrens, E.M., Gadue, P., Gong, S., Garret, S., Stein, G., Cohen, P.L., The mer receptor tyrosine kinase: expression and function suggest a role in innate immunity. *Eur. J. Immunol.* **2003**. 33: 2160–2167.

72 Devitt, A., Moffatt, O.D., Raykundalia, C., Capra, J.D., Simmons, D.L., Gregory C.D., Human CD14 mediates recognition and phagocytosis of apoptotic cells. *Nature* **1998**. 392: 505–509.

73 Moffatt, O.D., Devitt, A., Bell, E.D., Simmons, D.L., Gregory C.D., Macrophage recognition of ICAM-3 on apoptotic leukocytes. *J. Immunol.* **1999**. 162: 6800–6810.

74 Brown, S., Heinisch, I., Ross, E., Shaw, K., Buckley, C.D., Savill, J., Apoptosis disables CD31-mediated cell detachment from phagocytes promoting binding and engulfment. *Nature* **2002**. 418: 139–141.

75 Colvin, B.L., Lau, A.H., Schell, A.M., Thomson, A.W., Disparate ability of murine CD8alpha- and CD8alpha+ dendritic cell subsets to traverse endothelium is not determined by differential CD11b expression. *Immunology* **2004**. 113: 328–337.

76 Voll, R.E., Herrmann, M., Roth, E.A., Stach, J., Kalden, R., Immunosuppressive effects of apoptotic cells. *Nature* **1997**. 390: 350–351.

77 Gallucci, S., Lolkema, M., Matzinger, P., Natural adjuvants: endogenous activators of dendritic cells. *Nat. Med.* **1999**. 5: 1249–1255.

78 Stuart, L.M., Lucas, M., Simpson, C., Lamb, J., Savill, J., Lacy-Hulbert, A., Inhibitory effects of apoptotic cell ingestion upon endotoxin-driven myeloid dendritic cell maturation. *J. Immunol.* **2002**. 168: 1627–1635.

79 Takahashi, M., Kobayashi, Y., Cytokine production in association with phagocytosis of apoptotic cells by immature dendritic cells. *Cell. Immunol.* **2003**. 226: 105–115.

80 Urban, B.C., Willcox, N., Roberts, D.J., A role for CD36 in the regulation of dendritic cell function. *Proc. Nat. Acad. Sci. USA* **2001**. 98: 8750–8755.

81 Fadok, V.A., McDonald, P.P., Bratton, D.L., Henson, P.M., Regulation of macrophage cytokine production by phagocytosis of apoptotic and post-apoptotic cells. *Bioch. Soc. Trans.* **1998**. 26: 653–656.

82 Fadok, V.A., Bratton, D.L., Konowal, A., Freed, P.W., Westcott, J.Y., Henson, P.M. Macrophages that have ingested apoptotic cells in vitro inhibit proinflammatory cytokine production through autocrine/paracrine mechanisms involving TGF-beta, PGE2, and PAF. *J. Clin. Invest.* **1998**. 101: 890–898.

83 Sutterwala, F.S., Noel, G.J., Clynes, R., Mosser, D.M., Selective suppression of interleukin-12 induction after macrophage receptor ligation. *J. Exp. Med.* **1997**. 185: 1977–1985.

84 Marth, T., Kelsall, B.L. Regulation of interleukin-12 by complement receptor 3 signaling. *J. Exp. Med.* **1997**. 185: 1987–1995.

85 Hammerberg, C., Katiyan, S.K., Carrol, M.C., Cooper, K., Activated complement component 3 (C3) is required for ultraviolet induction of immunosuppression and antigenic tolerance. *J. Exp. Med.* **1998**. 187: 1133–1138.

86 Yoshida, Y., Kang, K., Berger, M., Chen, G., Gilliam, A.C., Moser, A., Wu. L.. Hammerberg, C., Cooper, K.D., Monocyte induction of IL-10 and down-regulation of IL-12 by iC3b deposited in ultraviolet-exposed human skin. *J. Immunol.* **1998**. 161: 5873–5879.

87 Nauta, A.J., Castellano, G., Xu, W., Woltman, A.M., Borrias, M.C., Daha, M.R., van Kooten, C., Roos, A., Opsonization with C1q and mannose-binding lectin targets apoptotic cells to dendritic cells. *J. Immunol.* **2004**. 173: 3044–3050.

88 Demeure, C.E., Tanaka, H., Mateo, V., Rubio, M., Delespese, G., Sarfati, M., CD47 engagement inhibits cytokine production and maturation of human dendritic cells. *J. Immunol.* **2000**. 164: 2193–2139.

89 Steinman, R.M., Nussenzweig, M.C., Avoiding horror autotoxicus: the importance of dendritic cells in peripheral T cell tolerance. *Proc. Nat. Acad. Sci. USA* **2002**. 99: 351–358.

90 Steinman, R.M., Hawiger, D., Nussenzweig, M.C., Tolerogenic dendritic cells. *Annu. Rev. Immunol.* **2003**. 21: 685–671.

91 Morelli, A.E., Thomson, A.W., Dendritic cells: regulators of alloimmunity ad opportunities for tolerance induction. *Immunol. Rev.* **2003**. 196: 125–146.

92 Adler, A.J., Marsh, D.W., Yochum, G.S., Guzzo, J.L., Nigam, A., Nelson W.G., Pardoll D.M., CD4+ T cell tolerance to parenchymal self-antigens requires presentation by bone marrow-derived antigen-presenting cells. *J. Exp. Med.* **1998**. 187: 1555–1564.

93 Kurts, C., Kosaka, H., Carbone, F.R., Miller, J.F., Heath W.R., Class I-restricted cross-presentation of exogenous self-antigens leads to deletion of autoreactive CD8(+) T cells. *J. Exp. Med.* **1997**. 186: 239–245.

94 Jonuliet, H., Schmitt, E., Schuler, G., Knop J., Enk, A.H., Induction of interleukin 10-producing, nonproliferating CD4+ T cells with regulatory properties by repetitive stimulation with allogeneic immature human dendritic cells. *J. Exp. Med.* **2000**. 192: 1213–1222.

95 Jonuleit, H.E., Schmitt, E., Schuler, G., Knop, J., Enk, A.H., Induction of interleukin 10-producing, non-proliferating CD4+ T cells with regulatory properties by repetitive stimulation with allogeneic immature human dendritic cells. *J. Exp. Med.* **2000**. 192: 1213–1222.

96 Dhodapkar, M.V., Steinman, R.M., Krasovsky, J., Munz, C., Bhardwaj, N., Antigen-specific inhibition of effector T cell function in humans after injection of immature dendritic cells. *J. Exp. Med.* **2001**. 193: 233–238.

97 Dhodapkar, M.V., Steinman, R.M., Antigen-bearing immature dendritic cells induce peptide-specific CD8(+) regulatory T cells in vivo in humans. *Blood* **2002**. 100: 174–177.

98 Ip, W.K., Lau, Y.-L., Distinct maturation of, but not migration between, human monocyte-derived dendritic cells upon ingestion of apoptotic cells of early or late phases. *J. Immunol.* **2004**. 173: 189–196.

99 Huang, F.-P., Platt, N., Wykes, M., Major, J.R., Powell, T.J., Jenkins, C.D., MacPherson, G.G., A discrete subpopulation of dendritic cells transports apoptotic intestinal epithelial cells to T cell areas of mesenteric lymph nodes. *J. Exp. Med.* **2000**. 191: 435–443.

100 Mishima, Y., Melanosomes in phagocytic vacuoles in Langerhans cells. Electron microscopy of keratin-stripped human epidermis. *J. Cell Biol.* **1966**. 30: 417–423.

101 Scheinecker, C., McHugh, R., Shevach, E.M., Germain, R.N., Constitutive presentation of a natural tissue auto-antigen exclusively by dendritic cells in the draining lymph node. *J. Exp. Med.* **2002**. 196: 1079–1090.

102 Steinman, R.M., Turley, S., Mellman, I., Inaba, K., The induction of tolerance by dendritic cells that have captured apoptotic cells. *J. Exp. Med.* **2000**. 191: 411–416.

103 Liu, K., Iyoda, T., Saternus, M., Kimura, Y., Inaba, K., Steinman, R.M., Immune tolerance after delivery of dying cells to dendritic cells in situ. *J. Exp. Med.* **2002**. 196: 1091–1097.

104 Lutz, M.B., Schuler, G., Immature, semi-mature and fully mature dendritic cells: which signals induce tolerance or immunity? *Trends Immunol.* **2002**. 23: 445–449.

105 Hawiger, D., Inaba,K., Dorsett, Y., Guo, M., Mahnke, K., Rivera, M., Ravetch, J.V., Steinman, R.M., and Nussenzweig, M.C., Dendritic cells induce peripheral T cell unresponsiveness under steady state conditions in vivo. *J. Exp. Med.* **2001**. 194: 769–779.

106 Bonifaz, L., Bonnyay, D., Mahnke, K., Rivera, M., Nussenzweig, M.C., Steinman, R.M., Efficient targeting of protein antigen to the dendritic cell receptor DEC-205 in the steady state leads to antigen presentation on major histocompatibility complex calss I products and peripheral CD8+ T cell tolerance. *J. Exp. Med.* **2002**. 196: 1627–1638.

107 Ferguson, T.A., Herndon, J., Elzey, B., Griffith, T.S., Schoenberger, S., Green, D.R., Uptake of apoptotic antigen-coupled cells by lymphoid dendritic cells and cross-priming of CD8(+) T cells produce active immune unresponsiveness. *J. Immunol.* **2002**. 168: 5589–5595.

108 Nouri-Shirazi, M., Guinet, E., Direct and indirect cross-tolerance of alloreactive T cells by dendritic cells retained in the immature stage. *Transplantation* **2002**. 74: 1035–1044.

109 Morelli, A.E., Wang, Z., Shufesky, W.J., Larregina, A.T., Zahorchak, A.F., Logar, A.J., Papworth, G.D., Thomson, A.W. Use of donor apoptotic cells is a safe and effective means to prolong graft survival through interaction with recipient dendritic cells. *Am. J. Transplantation* **2003**. 3: 195 Suppl. 5 (Abst).

110 Xu, D.-L., Liu, Y., Tan, J.-M., Li, B., Zhong, C.-P., Zhang, X.H., Wu, C.-Q.,

Tang, X.-D., Marked prolongation of murine cardiac allograft survival using recipient immature dendritic cells loaded with donor-derived apoptotic cells. *Scand. J. Immunol.* **2004**. 59: 536–544.

111 de Carvalho Bittencourt, M., Perruche, S., Contassot, E., Fresnay, S., Baron, M.H., Angonin, R., Aubin, F., Herve, P., Tiberghien, P., Saas, P., Intravenous injection of apoptotic leukocytes enhances bone marrow engraftment across major histocompatibility barriers. *Blood* **2001**. 98: 224–230.

112 Barr, M.L., Meiser, B.M., Eisen, H.J., Roberts, R.F., Livi, U., Dall'Amico, R., Dorent, R., Rogers, J.G., Radovancevic, B., Taylor, D.O., Jeevanandam, V., Marboe C.C., Photopheresis for the prevention of rejection in cardiac transplantation. *New Engl. J. Med.* **1998**. 339: 1744–1751.

113 Urban, B.C., Ferguson, D.J.P., Pain, A., Willcox, N., Plebanski, M., Austyn, J.M., Roberts, D.J., Plasmodium falciparum-infected erythrocytes modulate the maturation of dendritic cells. *Nature* **1999**. 400: 73–77.

114 Freire-de-Lima, C.G., Nascimento, D.O., Soares, M.B.P., Bozza, P.T., Castro-Faria-Neto, H.C., de Mello, F.G., DosReis, G.A., Lopes, M.F., Uptake of apoptotic cells drives the growth of a pathogenic trypanosome in macrophages. *Nature* **2000**. 403: 199–203.

115 Vanlandschoot, P., Leroux-Roels, G., Viral apoptotic mimicry: an immune evasion strategy developed by the hepatitis B virus? *Trends Immunol.* **2003**. 24: 144–147.

116 Vanlandschoot, P., Roobrouck, A., Van Houtte, F., Leroux-Roels, G. Recombinant HBsAg, an apoptotic-like lipoprotein, interferes with LPS-induced phosphorylation of the ERK-1/2 and JNK-1/2 in monocytes. *Biochem. Biophys. Res Commun.* **2002**. 297: 486–491.

117 Basu, S., Binder, R.J., Suto, R., Anderson, K.M., Srivastava, P.K., Necrotic but not apoptotic cell death releases heat shock proteins, which deliver a partial maturation signal to dendritic cells and activate the NF-kappa B pathway. *Int. Immunol.* **2000**. 12: 1539–1546.

118 Shi, Y., Evans, J.E., Rock, K.L., Molecular identification of a danger signal that alerts the immune system to dying cells. *Nature* **2003**. 425: 516–521.

119 Scaffidi, P., Misteli, T., Bianchi, M.E., Release of chromatin protein HMGB1 by necrotic cells triggers inflammation. *Nature* **2002**. 418: 191–195.

120 Messmer, D., Yang, H., Telusma, G., Knoll, F., Li, J., Messmer, B., Tracey, K.J., Chiorazzi, N., High mobility group box protein 1: an endogenous signal for dendritic cell maturation and Th1 polarization. *J. Immunol.* **2004**. 173: 307–313.

121 Manfredi, A.A., Rovere, P., Galati, G., Heltai, S., Bozzolo, E., Soldini, L., Davoust, J., Balestrieri, G., Tincani, A., Sabbadini, M.G., Apoptotic cell clearance in systemic lupus erythematosus. I. Opsonization by antiphospholipid antibodies. *Arth. Rheum.* **1998**. 41: 205–214.

122 Ronchetti, A., Rovere, P., Iezzi, G., Galati, G., Heltai, S., Protti, M.P., Garanchi, M.P., Manfredi, A.A., Rugarli, C., Bellone, M., Immunogenicity of apoptotic cells in vivo: role of antigen load, antigen-presenting cells, and cytokines. *J. Immunol.* **1999**. 163: 130–136.

123 Botto, M., Dell'Agnola, C., Bygrave, A.E., Thompson, E.M>, Cook, H.T., Petry, F., Loss, M., Pandolfi, P.P., Walport, M.J., Homozygous C1q deficiency causes glomerulonephritis associated with multiple apoptotic bodies. *Nat. Genetics* **1998**. 19: 56–59.

124 Chen, Z., Koralov, S.B., Kelsoe, G., Complement C4 inhibits systemic autoimmunity through a mechanism independent of complement receptors CR1 and CR2. *J. Exp. Med.* **2000**. 192: 1339–1352.

125 Bickerstaff, M.C., Botto, M., Hutchinson, W.L., Herbert, J., Tennent, G.A., Bybee, A., Mitchell, D.A., Cook, H.T., Butler, P.J., Walport, M.J., Pepys, M.B., Serum amyloid O component controls chromatin degradation and prevents antinuclear autoimmunity. *Nat. Med.* **1999**. 5: 694–697.

126 Ehrenstein, M.R., Cook, H.T., Neuberger, M.S., Deficiency in serum

immunoglobulin (Ig)M predisposes to development of IgG autoantibodies. *J. Exp. Med.* **2000**. 191: 1253–1258.

127 Boes, M., Schmidt, T., Linkemann, K., Beaudette, B.C., Marshak-Rothstein, A., Chen, J., Accelerated development of IgG autoantibodies and autoimmune disease in the absence of secreted IgM. *Proc. Natl. Acad. Sci USA* **2000**. 97: 1184–1189.

128 Cohen, P.L., Caricchio, R., Abraham, V., Camenisch, T.D., Jennette, J.C., Roubey, R.A., Earp, H.S., Matsushima, G., Reap, E.A., delayed apoptotic cell clearance and lupus-like autoimmunity in mice lacking the c-mer membrane tyrosine kinase. *J. Exp. Med.* **2002**. 196: 135–140.

129 Pickering, M.C., Walport, M.J. Links between complement abnormalities and systemic lupus erythematosus. *Rheumatology* **2000**. 39: 133–141.

130 Baumann, I., Kolowos, W., Voll, R.E., Manger, B., Gaipl, U., Neuhuber, W.L., Kirchner, T. Kalden, J.R., Herrmann, M., Impaired uptake of apoptotic cells into tingible body macrophages in germinal centers of patients with systemic lupus erythematosus. *Arth. Rheum.* **2002**. 46: 191–201.

131 Shoshan, Y., Shapira, I., Toubi, E., Frolkis, I., Yaron, M., and Mevorach, D. Accelerated Fas-mediated apoptosis of monocytes and maturing macrophages from patients with systemic lupus erythematosus: relevance to in vitro impairment of interaction with iC3b-opsonized apoptotic cells. *J. Immunol.* **2001**. 167: 5963–5969.

132 Taylor, P. R., Carugati, A., Fadok, V., Cook, H.T., Andrews, M., Carroll, M.C., Savill, J.S., Henson, P.M., Botto, M., Walport, M.J., A hierarchical role for classical pathway complement proteins in the clearance of apoptotic cells in vivo. *J. Exp. Med.* **2000**. 192: 359–366.

133 Albert, M., Death-defying immunity: do apoptotic cells influence antigen processing and presentation? *Nat. Rev. Immunol.* **2004**. 4: 223–231.

134 Gao, Y., Herndon, J.M., Zhang, H., Griffith, T.S., Ferguson, T.A. Antiinflammatory effects of CD95 ligand (FasL)-induced apoptosis. *J. Exp. Med.* **1998**. 188: 887–896.

135 Chen, W., Frank, M.E., Jin, W., Wahl, S.M., TGF-beta released by apoptotic T cells contributes to an immunosuppressive milieu. *Immunity* **2001**. 14: 715–725.

136 Feng, H., Zeng, Y., Whitesell, L., Katsanis, E., Stressed apoptotic tumor cells express heat shock proteins and elicited tumor-specific immunity. *Blood* **2001**. 97: 3505–3512.

137 Lauber, K., Bohn, E., Krober, S.M., Xiao, Y.J., Blumenthal, S.G., Lindemann, R.K., Marini, P., Wiedig, C., Zobywalski, A., Baksh, S., et al., Apoptotic cells induce migration of phagocytes via caspase-3-mediated release of a lipid attraction signal. *Cell.* **2003**. 113: 717–730.

30
Pharmacologically Modified Dendritic Cells:
A Route to Tolerance-associated Genes

Kathleen F. Nolan, Stephen F. Yates, Alison M. Paterson, Paul J. Fairchild and Herman Waldmann

30.1
Dendritic Cells, Maturation and Tolerance

Just as the immune system has evolved to ensure wide-ranging efficacy in combating harmful agents, so it has also incorporated multiple layers of negative control to minimize self-harm, localizing and limiting the extent of productive responses, as well as preventing the initiation of inappropriate immunity. For example, although autoreactive T cells are deleted in the thymus during the establishment of central tolerance, this process is incomplete and peripheral tolerance mechanisms are required to control escaping self-reactive T cells [1], in addition to tempering T cells potentially reactive against harmless ubiquitous agents such as inhaled antigens and gut flora.

It has been demonstrated that dendritic cells (DCs) play a role in the induction and maintenance of peripheral T-cell tolerance by a variety of incompletely understood mechanisms, including deletion, induction of anergy or hyporesponsiveness and the generation of regulatory T cells, Tregs [2, 3]. $CD4^+CD25^+Foxp3^+$ T cells, "natural" regulators, although typically considered to arise in the thymus, do also appear to be generated extrathymically [4, 5], and while it is known that they require antigen stimulation and IL-2 to trigger their suppressive effects, the ensuing suppression of pathogenic T cells is mediated, at least in *in vitro* readouts, by an as yet elusive, antigen-nonspecific, cell-contact-dependent mechanism [6]. Tr1-like cells, "adaptive" regulators on the other hand, are differentiated extrathymically from $CD4^+CD25^-$ cells, typically in the presence of IL-10 [7], and are considered to represent a specialized activation state of conventional $CD4^+$ cells. Indeed T helper (Th) 1/Th2 polarized Tr1-like cells have been differentially generated following respiratory challenge [8, 9]. Tr1-like cells are thought to regulate T-cell responses in an antigen-specific, contact-dependent manner through the action of IL-10 and/or TGFβ [10]. The exact molecular signals conveyed by DCs leading to the generation of Tregs are currently unknown.

Handbook of Dendritic Cells. Biology, Diseases, and Therapies.
Edited by M. B. Lutz, N. Romani, and A. Steinkasserer.
Copyright © 2006 WILEY-VCH Verlag GmbH & Co. KGaA, Weinheim
ISBN: 3-527-31109-2

DCs distribute throughout the periphery where they act as "sentinels" for both immunity and tolerance [2, 11]. While immature DCs exhibit immense capacity to continuously sample their antigenic micro-environment, these cells exhibit low surface MHC class II and co-stimulatory molecule expression and the sampled antigen is not displayed at the cell surface. In response to appropriate stimuli DCs mature, rapidly losing their capacity for antigen uptake and acquiring increased surface expression of MHC and co-stimulatory molecules. This maturation is accompanied by alterations in chemokine and chemokine receptor expression that facilitate exit from the inflammatory site and homing to the T-cell areas of the lymph node where they can now potently stimulate naïve T cells (Fig. 30.1) [11–15]. IL-6 generated by DCs in response to microbial stimuli causes CD4+CD25− effector T cells to become refractory to suppression by CD4+CD25+ Tregs promoting T-cell immunity [16]. Surface MHC:peptide and co-stimulatory molecules provide what have been termed signals 1 and 2 that are necessary for effective T-cell priming, while a third signal is provided in the form of polarizing Th1 or Th2 cytokines and is determined by the nature of the activation stimulus [17, 18]. The capacity of DCs to activate T cells appears closely related to their level of maturation and this in turn is influenced by a variety of factors, including the nature of the stimulus, the relative responsiveness of the DC lineage or subtype and the cytokine micro-environment in which the stimulus is encountered (Fig. 30.1). In the absence of relevant pathogen derived stimuli, DCs require a maturation boost from Th cells to achieve priming of naïve CD8+ cells and the development of effector cytotoxic T lymphocytes (CTLs), a process that has been termed "conditioning" or "licensing" of the DC [19–21]. Cross-presentation of self-antigen in the absence of licensing results in deletion of self-reactive CTLs and "cross-tolerance" [22–24]. In a similar manner, activation of DCs by inflammatory mediators in the absence of microbial stimulation is sufficient to generate DCs that are phenotypically mature with respect to upregulation of MHC II and co-stimulatory molecules, and which can support CD4+ T-cell clonal expansion, but which fail to direct Th differentiation [25].

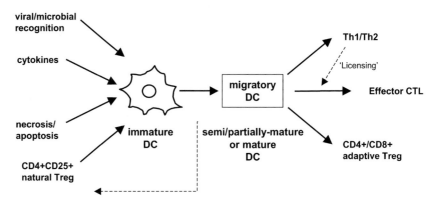

Fig. 30.1 The functional phenotype of dendritic cells is related to their maturation state, the level of which represents a balance of environmental influences and cellular responsiveness.

Regulating the state of DC maturity appears to be a key pivot point in the maintenance of an appropriate balance between immune activation and tolerance. The tolerogenic nature of antigen presentation by immature DCs has been illustrated by repetitive *in vitro* stimulation of naïve CD4⁺ T cells, which caused the differentiation of Tr1-like cells [26, 27], while a single injection of influenza matrix peptide-pulsed immature DCs into human volunteers caused transient, antigen-specific inhibition of CTL responses and the induction of IL10-producing, contact-dependent, regulatory CD8⁺ T cells [28, 29]. Using an inducible transgenic mouse model to avoid perturbations that could be associated with *ex vivo* manipulation and adoptive transfer of resting DCs, presentation of an LCMV-derived CTL epitope similarly resulted in antigen-specific CD8⁺ tolerance, which could not be broken by subsequent infection with LCMV [30]. Conjugation of antigen or antigenic peptide to an antibody targeting the DC-specific lectin DEC-205 has been used to deliver antigen to immature DC *in vivo via* endocytosis, a process that in itself does not induce DC maturation. While the simultaneous delivery of a maturation stimulus generated immunity, delivery of the antigen-anti-DEC-205 species alone initiated a burst of T-cell proliferation followed by deletion and the induction of antigen-specific unresponsiveness [31, 32], and the induction of CD25⁺ Treg cells [33].

It has been proposed that apoptotic cells can act as a reservoir of self-antigen for the maintenance of peripheral tolerance and that while circulating apoptotic particles can be captured and presented directly by lymph-node or spleen resident DCs in the steady state [34, 35], under non-inflammatory conditions self-antigens can also be captured by DCs in the periphery and transported to lymph nodes *via* the afferent lymph [36]. Uptake of apoptotic cells, in contrast to the uptake of necrotic cells, does not mediate maturation of DCs, but does result in a decreased production of the pro-inflammatory cytokines IL-1α, IL-1β, IL-6, IL-12p70 and TNFα [34, 37, 38]. Rather than becoming resident in the T-cell areas of the nodes themselves, it has been suggested that these migratory DCs may actually die rapidly and be processed by lymph-node resident DCs that then tolerize autoreactive CD4⁺ and cross-tolerize autoreactive CD8⁺ T cells [22, 35, 39]. For steady-state migration to occur at least some level of DC maturation is required. Following intranasal application of antigen, pulmonary DCs migrating with an apparently mature phenotype mediate tolerance by the induction of IL-10 producing Tr1-like cells [8, 9, 40]. Repeated injections of *in vitro* generated DCs matured in the presence of TNFα prevented the induction of experimental autoimmune encephalomyelitis (EAE) in a murine model, with the induction of antigen specific, IL-10 producing regulatory CD4⁺ T cells [41]. The term "semi-mature" has been used to describe these apparently mature, tolerogenic, migratory cells [42]. They are mature with respect to upregulation of surface MHC, co-stimulatory and adhesion molecules, but characteristically exhibit low or absent production of pro-inflammatory cytokines.

The maintenance of incomplete signals between T cells and APCs appears to be a key factor in contributing to the induction of tolerance. This phenomena has been indicated even from early reports of co-stimulation blockade [43] and reiterated in more recent work involving altered peptide ligands [44]. Administration of co-stimulatory molecule-deficient or antisense-targeted DCs mediated prolonged cardiac allograft survival [45, 46], while CD40-deficient DCs suppressed a primed delayed type hypersensitivity response [47]. Inhibitors of the transcription factor NF-κB mediate prolonged allograft survival [48, 49], and NF-κB has been implicated as an important regulator in determining DC tolerance or immunity, at least in part by modulating surface levels of CD40 [47, 50]. It has been reported that CD4$^+$CD25$^+$ Tregs impact on DC function by down-modulating the co-stimulatory capacity of DCs [51]. A more complete picture of factors contributing to a DC's ability to induce tolerance probably involves not simply a reduction in co-stimulation, but also an alteration in the balance between stimulatory and inhibitory molecules that work in concert to establish activation thresholds and tune cellular responses. Such molecules implicated in DC biology include members of the related leukocyte immunoglobulin receptor (LIR) or immunoglobin-like transcript (ILT) family and paired immunoglobulin-like receptor (PIR) molecules and also activatory and inhibitory FcγRs [52–58]. Production of the tryptophan-catabolizing enzyme, indoleamine-2,3-dioxygenase (IDO), has also been suggested to play a role in DC mediated suppression of T-cell proliferation [59, 60].

The plasticity of DC maturation means that the balance between immunity and tolerance will also be influenced by the overall context of antigen exposure, and it may be that a DC capable of mediating tolerance on maturation in one context might still be able to prime when matured in another. This makes it difficult to generalize about DCs as being universally tolerogenic or immunogenic. However, within any one model or context it should be possible to make comparisons between two sets of DCs with immunogenic and tolerogenic outcomes and thereby draw conclusions on mechanisms that distinguish them.

30.2
Gene Profiling

Comparison of global gene expression patterns provides an essentially unbiased approach to investigating biological phenomena. Unrestricted by current dogma, this type of approach has the potential to reveal previously unanticipated areas for further study and has already proved informative in elucidating aspects of DC biology.

In response to various microbial and inflammatory stimuli a coordinated program of gene expression changes has been revealed that corresponds with the temporal and spatial segregation of different functional aspects of DC maturation [13, 61]. These changes include a transient increase in pro-inflammatory genes, in addition to waves of gene changes required to downregulate phagocytic activity, facilitate migration to draining lymph nodes and to promote T-cell priming, with a previously unappreciated wave of IL-2 expression compatible, with priming of both CD4$^+$ and CD8$^+$ T-cell responses, revealed [61]. Moreover, while it was revealed that a core set of genes change in response to all stimuli, pathogen-specific changes were demonstrated that presumably act to tailor the resulting immune response [13].

DCs matured using the inflammatory mediator TNFα alone, although capable of maturing DCs with respect to upregulation of surface MHC class II and co-stimulatory molecules, cannot drive maturation to a stage necessary for initiation of an optimal immune response. Gene profiling has indicated that TNFα stimulation is insufficient to induce expression of genes encoding pro-inflammatory cytokines, such as IL-6, IL-12p40 and IL-1β, and this is consistent with the designation of these cells as "semi-mature" [42, 62].

Transcript profiling of DCs modulated by a CD8$^+$CD28$^-$ T-cell population with suppressive function revealed a downregulation of co-stimulatory molecules, including CD40, CD80, CD86, OX-40L and an upregulation of transcripts regulating NF-κB activity, anti-apoptotic genes and inhibitory molecules, including ILT3 and ILT4 [63]. Transduction of ILT3 and ILT4 into a DC line resulted in cells that, like ILT3^{high}ILT4^{high} APC, induced anergy in CD4$^+$ helper T cells and were able to induce the generation of antigen specific CD4$^+$CD25$^+$ regulatory T cells, as well as further CD8$^+$CD28$^-$ suppressor T cells, in a feedback type manner [56, 63, 64]. Upregulation of ILT3 and ILT4, in addition to other inhibitory molecules bearing immunoreceptor tyrosine based inhibitory motifs (ITIMs), has also been demonstrated on DC populations modulated for tolerance using IL-10, interferon-α or the biologically active metabolite of vitamin D$_3$, 1α,25-dihydroxyvitaminD3, (VD3) [64, 65], with a particularly potent tolerogenic subset of IL-10-modulated DCs identified as ILT^{high}BDCA3$^+$ [65].

The above profiling studies were all facilitated by the availability of *in vitro* culture systems to provide ready access to populations of DCs in ample numbers, at defined stages of activation and at sufficient homogeneity for reproducible analyses. The *in vitro* systems used have either been based on well-defined long-term DC lines, such as D1 and KG-1 [66, 67], or on short-term culture systems developed to generate immature DCs from human monocytes or murine bone marrow [68, 69]. Although profiling of more abundant primary DC populations is possible [70], the time and manipulation involved in isolating these cells is not ideal and the homogeneity of the resulting populations is limited by available sorting and isolation criteria [38]. Using *in vitro* culture systems a number of agents have been reported to modulate the immunostimulatory potential of DCs [50], apparently forcing them to become functionally tolerogenic. These agents include cytokines such as IL-10 [71, 72], TGFβ [73–75] and vascular endothelial growth factor (VEGF) [76], in addition to more traditional pharmacological agents, such as aspirin [77], nicotine [78], LF15-0195 [79], dexamethasone [80, 81] and VD3 [82, 83]. While the exact physiological subset or relative maturation state of DCs responsible for peripheral tolerance *in vivo* remains controversial the generation of tolerogenic DCs using such intervention *in vitro* provides a readily accessible source of potentially regulatory cells for the investigation of tolerance mechanisms using gene expression profiling.

30.2.1
Gene Profiling Technologies

Ideally expression profiling would provide comprehensive information on protein expression patterns, including information on post-translational modifications and other features that reflect functional activity, but as yet this level of analysis remains beyond the capabilities of proteomic technologies. The alternative is transcript profiling which reveals context-dependent patterns of gene expression, variations in which provide informative predictors of changes in cell function. Several large-scale genomics platforms are currently available for the assessment of gene expression patterns and selection tends to depend both on experimental considerations, such as the requirement for depth of genome penetration and/or the identification of novel genes *versus* analysis of multiple samples, and practical considerations, such as the availability of technology and cost. The most widely used applications are based either on hybridization to known sequences, as in oligonucleotide and cDNA microarray analyses, or on the sampling of transcript information from total mRNA pools. While the approach of expressed sequence tag (EST) enumeration samples partial transcripts, approaches such as serial analysis of gene expression (SAGE) and massively parallel signature sequencing (MPSS™) more efficiently sample short representative transcript tags. Although the depth of transcriptome penetration and procedural automation achieved by MPSS™ make it a very attractive approach (it readily samples millions of short transcript tag sequences in a single machine run) [84, 85], the costs associated with this proprietary technology have so far precluded its more general use. Hybridization-based approaches offer advantages relating to convenience, and the extent of their genome coverage has improved in line with increasing availability of genome data, however this technology is still limited to analysis of known or available transcripts and has proved difficult to standardize across multiple datasets.

30.2.2
Serial Analysis of Gene Expression (SAGE)

SAGE represents a particularly versatile tool for elucidating gene expression patterns. Both quantitative and qualitative transcript information are generated and stored in the form of electronic databases or "libraries" that accumulate as a permanent, accessible resource. Any SAGE library can be readily compared with any other library, or combination of libraries, including those generated in different laboratories, provided the libraries have all been generated using the same anchor enzyme (AE) and are derived from the same species [86]. "Virtual northern" analyses based on SAGE data are now found as a standard component of public transcript/unigene databases. This compatibility across wide data sets is a powerful feature of the technology.

Comparisons of libraries from related but functionally distinct cell populations can be used to focus on transcript changes underlying the functional distinction, while comparison of libraries from cells with related functions can be used to focus on genes responsible for functional similarities. Phenotypically unrelated libraries are often included in comparisons to remove "noise" generated by abundant, widely-expressed "house-keeping" type genes. An expanding resource of libraries (currently including 246 human and 81 mouse *Nla* III libraries) is available at the gene expression omnibus (GEO) data repository at the national center for biotechnology information (NCBI) and can be downloaded *via* the internet for inclusion in comparative analyses.

Unlike hybridization-based technologies, SAGE samples from the entire mRNA pool with no requirement for prerequisite sequence information and indeed SAGE tags can be used to clone previously uncharacterized transcripts [87]. Depending on the relative positioning, SAGE tags can also distinguish alternatively spliced forms. It has been suggested that 40–60% of multi-exon genes are alternatively spliced, and that of these at least 49% undergo alternative splicing of a terminal exon [88]. While the 3' bias of the original procedure makes it amenable to the cloning and characterization of 3' ends, SAGE has more recently been adapted to generate comprehensive analysis of 5' ends, or alternatively the very extremes of transcripts, a task that has so far thwarted automated genome annotation procedures, providing information on transcription start sites and the alternative use of promoters as well as polyadenylation sites [89–91].

30.2.2.1 SAGE Methodology

The original SAGE approach was based on three main principles [92]: Firstly, that a short sequence tag of 9–10 bp is sufficient to uniquely identify a transcript, provided the tag is isolated from a defined position; secondly, that efficient sequencing of these short tags can be achieved by concatenation, provided that there is a means to register the boundaries of each tag; and thirdly, that the number of times a tag is sampled reflects the relative frequency of the associated transcript in the starting mRNA pool. The availability of commercial SAGE kits has now brought this technology within the capabilities of most laboratories, particularly if commercial outsourcing is considered for the large-scale sequencing component.

Typical libraries are generated from 2–5 µg of total RNA, although pre-amplification protocols have enabled libraries to be generated from as little as 40–50 ng [95–98]. RNA integrity is essential for authentic profiling and the quality of each starting RNA population should be routinely established. This can be achieved using as little as 200 pg of material using instrumentation such as the Agilent Technologies 2100 bioanalyzer. A schematic representation of the SAGE procedure is outlined in Fig. 30.2. Modifications from the original protocol [92], reflect streamlining to increase efficiency. Poly A+ mRNA is magnetically isolated using (poly-T)-coated beads with double-stranded cDNA generated directly on the beads. The cDNA is digested using a frequent cutting restriction enzyme (most commonly *Nla* III); this enzyme defines the position within the transcript from which the tag will be isolated and is referred to as the "anchor enzyme" (AE). The most extreme 3' AE restriction fragments are isolated magnetically, divided in two and unique linkers (1 and 2) ligated onto either half. The samples are then recombined and digested using a type IIS restriction enzyme that recognizes a non-palindromic sequence located towards the 3' end of each linker and cleaves a defined distance downstream to release a "linker+tag" unit; this enzyme defines the length of the SAGE tag and is referred to as the "tagging enzyme" (TE). The remaining 3' ends, still bound to the (poly-T)-coated beads, are removed magnetically and the linker-

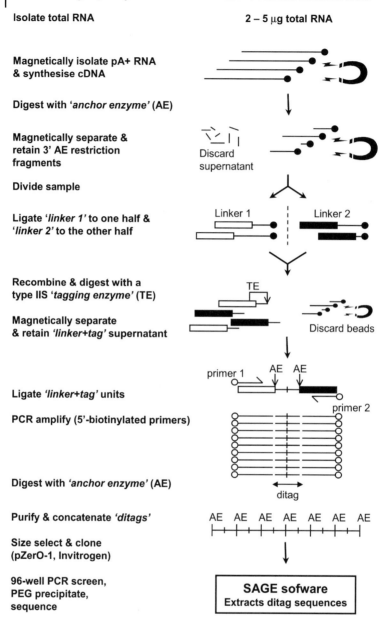

Isolate total RNA

2 – 5 μg total RNA

**Magnetically isolate pA+ RNA
& synthesise cDNA**

Digest with 'anchor enzyme' (AE)

**Magnetically separate &
retain 3' AE restriction
fragments**

Discard
supernatant

Divide sample

**Ligate 'linker 1' to one half &
'linker 2' to the other half**

Linker 1   Linker 2

**Recombine & digest with a
type IIS 'tagging enzyme' (TE)**

TE

**Magnetically separate
& retain 'linker+tag' supernatant**

Discard beads

primer 1   AE   AE

Ligate 'linker+tag' units

primer 2

PCR amplify (5'-biotinylated primers)

Digest with 'anchor enzyme' (AE)

ditag

Purify & concatenate 'ditags'

AE   AE   AE   AE   AE   AE   AE

**Size select & clone
(pZerO-1, Invitrogen)**

**96-well PCR screen,
PEG precipitate,
sequence**

SAGE sofware
Extracts ditag sequences

Fig. 30.2 Schematic representation of the SAGE experimental procedure.

tag units purified from the supernatant by precipitation. These units are ligated to generate PCR templates that are amplified using 5'-biotinylated primers located within the linker sequences. "Ditags" are released from the amplified products by again digesting with the AE. Following gel purification of the "ditags", any remaining traces of linker are removed using streptavidin-coated magnetic beads to facilitate efficient concatenation [94]. A commercial PCR clean-up column is used to remove small concatenation fragments prior to size selection and cloning [93]. To facilitate high throughput analysis, cloned inserts are amplified from individual colonies using PCR in a 96-well plate format, the amplified products are precipitated using polyethylene glycol (PEG) and directly sequenced using a single primer. Sequence data is transferred directly to SAGE software (provided by K.W. Kinzler, Johns Hopkins Oncology Center, Baltimore, MD, USA, http://www.sagenet.org), which locates the punctuating AE sites in the concatemers and extracts the intervening "ditag" sequence information. The generation of ditags prior to PCR amplification provides an important safeguard against PCR-generated data distortions and/or inadvertent multiple analysis of the same clone [92]. In theory a 4 bp restriction site should occur on average every 256 bp (4^4), and as such the anchor enzyme (AE) should cleave at least once in every transcript [92]. Although it is appreciated that this assumption is inaccurate it would appear that the number of transcripts missed due to the absence of an AE site is small, <1% [102].

The original tagging enzyme (TE) used for SAGE was *Bsm* F1 which generates tags of ~10 bp (+4 bp contributed from the AE site). More recently TEs that extend further away from the recognition site have been incorporated resulting in "long" and "super" SAGE [99, 100]. *Mme* I cuts 7 bp further away than *Bsm* FI to yield ~(17+4) bp, "long" tags, while *EcoP* 15I generates "super" tags of ~(22+4) bp. So long as the AE is retained in common, these "long" and "super" SAGE libraries remain directly comparable with each other and with the original "short" SAGE libraries over the common AE+10 bp sequence. The "long" tags confer distinct benefits with regard to tag-to-gene annotation, including the ability to map tags directly to the genome, avoiding inconsistencies encountered from the use of incomplete cDNA/Unigene database entries [99, 101]. Justification however for the additional cost and labor involved in sequencing "super" libraries remains to be demonstrated.

30.2.2.2 Handling Raw SAGE Data

Data entered into the SAGE software is single-pass and unedited. Low sequence quality will either obscure the serial pattern of AE punctuation or lead to the introduction of ambiguous bases, both of which cause the sequence to be discarded and not entered into the final dataset. Although tags generated as a result of random sequence errors fail to accumulate statistical counts and are ignored, non-random sequence errors arising from highly abundant tags can significantly influence the final data [86, 103]. For "short" SAGE a numerical manipulation excluding tags matching 9 out of 10 bases of any other tag occurring at a >10-fold frequency, has been incorporated to account for accumulated non-random errors [86]. In practice it has been established that once a depth equivalent to 0.1–1% of the most abundant transcript has been achieved (in practice between 10 000 and 30 000 tags) then artifacts accumulate more rapidly than novel gene tags, making additional statistically relevant tag accumulation both inefficient and expensive [86].

The advent of "long" SAGE has essentially resolved issues associated with sequencing errors. The longer tag length has facilitated the incorporation of a cluster algorithm for the development of the corrective software, SAGEScreen, [104]. "Long" SAGE ditag information is extracted from the raw sequence data using SAGE300 software and is passed through the SAGEScreen software to generate an error-corrected tag list.

In an attempt to account for inconsistencies, such as partial entries in cDNA databases, the corrected SAGE tag lists are subjected to an automated hierarchical tag-to-gene annotation process based on Unigene full mapping files and a search algorithm giving priority to a list of hand-annotated entries. For each hand-annotated entry the existence of a plausible polyadenylation sequence has been confirmed, as has the tag location, immediately downstream of the most 3' AE (Nla III) site. The next annotation priority is given to cDNAs with a polyadenylation signal in the correct orientation, then EST clusters with a probable polyadenylation signal, cDNAs with no polyadenylation signal and finally other matching ESTs. The annotated tag list is imported into a custom-written software package, SAGEClus, that removes specified artifact tags, for example those potentially arising from linkers or primers, and facilitates statistical library comparisons (http://www.molbiol.ox.ac.uk/pathology/tig/software/softlist.html) [86].

30.2.3
Accumulation of a Comparative SAGE Resource for Identifying Tolerance-associated Genes

The capacity of SAGE to make comparisons across wide data sets offers particular potential to correlate gene expression with functional phenotype. By comparing immune and non-immune cell populations, individual genes and "signatures" of genes associated with pro-tolerogenic rather then immunogenic phenotypes can be elucidated to provide much needed novel markers of immune regulation and mechanistic insight into the body's natural tolerance machinery.

For informative library comparisons the authenticity and homogeneity of the starting cell populations must be well established. A culture system adapted from that previously described by Inaba et al. [69] has been used to generate a series of functionally distinct murine bone marrow derived DC populations. Following 7 days culture in GM-CSF cells exhibit surface and functional phenotypes characteristic of immature DCs, while the inclusion of lipopolysaccharide (LPS) for the final 18–20 h causes them to acquire characteristics typical of mature DCs; that is increased surface expression of MHC class II and co-stimulatory molecules (CD40, CD80 and CD86), diminished capacity to take up and present new antigen and enhanced capacity to stimulate naive T cells. DC populations grown under these conditions are >90% homogeneous with respect to expression of CD11c and SAGE libraries have been generated from both the immature and mature states [87].

Consistent with previous findings, bmDCs counter-modulated by each of the three pharmacological agents IL-10, TGFβ and VD3, (as outlined in Fig. 30.3), are reminiscent of immature bmDC, with moderate to low MHC class II and low lev-

Fig. 30.3 Generation of bmDC populations for SAGE analysis. Cells were prepared using an adaptation of the culture system previously described by Inaba *et al* [69, 87]. Immature cells were harvested on day 7. IL-10 was added (20 ng ml⁻¹) from day 6 and the cells harvested on day 9. Alternatively, 1α,25-dihydroxyvitamin D3, VD3 (10⁻⁷ M) was added from day 3, or TGFβ (2 ng ml⁻¹) added from day 0, with the cells harvested on day 7. When added, LPS (1 µg/ml) was included for the final 18–20 h of culture.

els of surface co-stimulatory molecules. However, unlike immature bmDCs the levels of these markers do not increase in response to LPS and the modulated cells are impaired in their ability to generate IL-12p70 and to support naïve T-cell proliferation [72, 87, 105–109]. The modulated cells were also tested for their ability to mediate tolerance in a male antigen specific, murine TCR transgenic skin transplant model [110]. While the administration of LPS-matured male bmDC prior to grafting did not affect rejection of a male graft, administration of modulated DCs, cultured with or without the addition of LPS, mediated graft survival (unpublished data). SAGE libraries have been generated from bmDCs modulated with IL-10, TGFβ and VD3, and from IL-10-modulated cells treated with LPS.

An accumulated resource comprising almost fifty murine SAGE libraries has now been generated [86, 87, 111]. In addition to libraries generated from the untreated and pharmacologically-modulated bmDC populations, with or without the addition of LPS, described above, the resource also includes libraries derived from embryonic stem (ES) cells and DCs differentiated from these cells, again with and without maturation in response to LPS [112, 113], purified T-cell populations and polarized T-cell clones, including defined "adaptive" Tr1-like and "natural" CD4⁺CD25⁺ regulatory populations [86, 114–116], in addition to a B cell population and a number of non-immune related populations, such as a fibroblast line and organs such as the heart and brain. Both "long" and "short" SAGE libraries are represented, and new libraries are continually added to the resource as they become available. For the purpose of SAGE, house-keeping genes are defined as genes corresponding to tags with significant expression in all the libraries of the resource and for clarity are removed from many of the analyses.

An overview of processes involved in using the SAGE resource to identify and follow up on tolerance-associated genes is shown in Fig. 30.4. The hand-annotated gene list used for the hierarchical tag-to-gene annotation process was initially established to ensure accurate assessment of immunological genes of interest, however as unknown tags are assigned or automated assignments confirmed they are also added to this list. Statistical, pairwise and global library comparisons are facilitated by SAGEClus software [86]. Pairwise scatter plots, associated with automated annotation links, readily identify genes differentially expressed between two

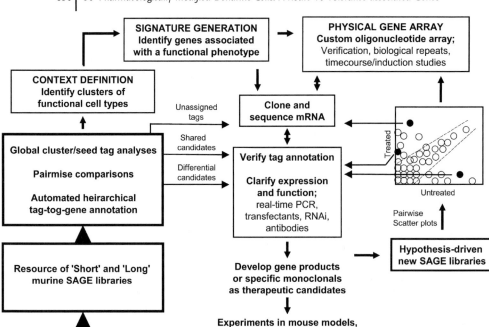

Fig. 30.4 A generalized strategy for the identification and analysis of candidate genes associated with tolerance. This strategy relies on comparative analysis of an accumulated resource of murine SAGE libraries (anchor enzyme *Nla* III).

populations, such as tolerogenic and non-tolerogenic. Global cluster analyses provide a measure of the similarity/relatedness of different cell populations with respect to their gene expression patterns, and can identify clusters of genes associated with related populations. Alternatively "seed" tags exhibiting idealized hypothetical expression profiles that mirror a particular phenotype, e.g. tolerogenic or non-tolerogenic, can be introduced and tags with the closest matching profiles identified. Combinations of these methods can be used to highlight groups of tolerance associated candidate genes, or "signatures". Signature genes can be brought together on a custom oligonucleotide microarray to facilitate rapid simultaneous analysis in various different samples, including biological replicates of the populations used for the SAGE, kinetic time-course or induction studies and to investigate the representation of the candidate genes in samples from in vivo investigations.

The annotation of each candidate tag is verified by hand. When no unique transcript can be identified or the short tag assignment is inconclusive then the "long" SAGE tag can be used to query the mouse genome directly. If the assignment is ambiguous, then the corresponding transcript can be revealed using rapid amplification of cDNA ends (RACE) procedures [117]. Differential expression patterns are clarified by real-time quantitative PCR methods.

Where the candidate gene is known, reagents such as antibodies may be available and biological information accessible from the literature. For less well characterized/unknown sequences computational analyses can reveal potential open reading frames and gene/protein homologies, as well as information regarding structural and functional features of the putative protein that may provide insight into cellular location and function. Biological tools such as antibodies and chimeric constructs, the expression of mutant or dominant negative forms, or the use of gene silencing and RNAi knock-down technologies, can be further utilized to unravel biological relevance. Custom SAGE libraries can be generated to reveal molecular mechanism by which specific genes affect regulation. As therapeutic targets are revealed human homologues can be identified and investigated.

30.2.3.1 Relationship of Modulated DC Populations based on Gene Expression Patterns

The relationship of different cell populations based on their relative gene expression patterns can be established using clustering programs within the SAGEClus software package [86]. Using this analysis, DC libraries represented within the current SAGE resource cluster together, away from a small cluster of T cell-related libraries and a further small cluster of embryonic cell libraries (Fig. 30.5). As expected the two independently generated "long" and "short" LPS matured bmDC libraries cluster together.

Within the DC cluster, the non-immunogenic libraries, generated from populations that were unable to stimulate proliferation of naïve T cells and which mediated tolerance when transferred to the TCR transgenic skin transplant model, clus-

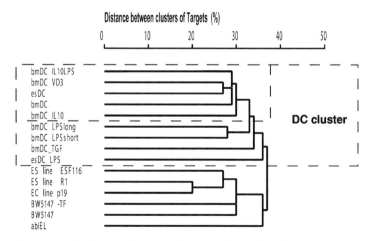

Fig. 30.5 A dendrogram illustrating the relative relationship of immature, LPS-matured and modulated DC populations based on their relative gene expression patterns. Clustering was performed using SAGEClus software [86]. (Libraries downloaded from public sites; R1-ES cell line (GSM580), EC line p19 (GSM1682), αβ intra-epithelial lymphocytes (IEL) [118].)

ter together; that is immature bmDC and ES-derived DC, IL-10- and VD3-modulated bmDC and bmDC modulated with IL-10 and subsequently exposed to LPS. The exception is the TGFβ-modulated library which clusters independently (Fig. 30.5). This would imply that the mechanisms by which TGFβ invokes immune-modulation are at least to some degree distinct from that of IL-10 and VD3.

30.2.3.2 Elucidation of "Signatures" of Genes Associated with Tolerance

Introducing hypothetical "seed" profiles into the SAGEClus analysis can be used to identify clusters of genes whose patterns of expression correlate with a functional phenotype, such as tolerance inducing capability. Libraries are selected as "containing", "not containing", or "no preference", and tags matching the selected profile are listed in order of compliance with the selected profile (Fig. 30.6).

Tags associated with pharmacologically-modulated DC populations, but not with immunogenic LPS-matured bmDC or esDC populations have been identified using a hypothetical test pattern and the most closely matching tags are shown in Fig. 30.6. The alternative use of polyadenylation sites generates two tags for the chemokine CCL6, both of which are represented in this cluster. Consistent with the SAGE data, it has previously been reported that CCL6 expression is not induced in response to LPS, but is increased in response to the cytokines IL-3, IL-4 and GMCSF [119]. CCL6 is particularly chemotactic for monocytes and macrophages and significantly enhances macrophage phagocytic activity [120, 121]. It belongs to a subgroup of chemokines that each contain a unique second exon and there is evidence that, like other members of this subgroup, CCL6 mediates at least some of its functional activity through CCR1 (the tag for which is also represented in this modulated DC gene cluster) [122]. While CCR1 is known to mediate recruitment and maintenance of immature DC at inflamed sites, exposure to IL-10 causes this receptor to become functionally uncoupled and to take on a ligand scavenging role [123]. It is unclear what the functional state of this receptor is in other modulated populations although the relative expression of CCL6 in the selected DC populations has been confirmed by real-time quantitative PCR and western blot analyses (unpublished data).

TGFβ-induced 68-KDa protein, also known as beta Ig-H3, is an extracellular matrix (ECM) protein that mediates cell adhesion and migration through interactions with integrins and may play a role in regulating angiogenesis [124]. Embigin is also implicated in cell binding *via* integrins [125], while ring finger protein-130 is a zinc-finger protein, cloned from a myeloid precursor line following removal of IL-3 to induce apoptotic death [126]. Transglutaminase-2, Tgm2, belongs to a family of molecules that catalyze post-translational modifications that result in polymerized, cross-linked proteins. It functions both intracellularly and extracellularly and has been implicated in stabilization of the ECM [127]. Tgm2 would appear to play an important role in regulating the bioavailability of TGFβ, as only after Tgm2-catalysed linkage to the matrix does latent transforming growth factor binding protein-1 release active TGFβ [128]. It has also been implicated in cytoskeletal polymerization during the final steps of apoptosis, preventing the release of cell com-

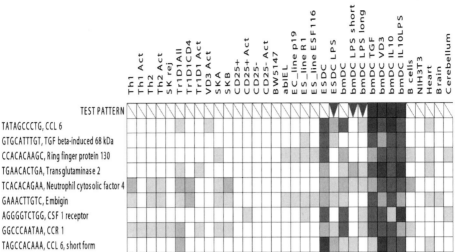

TEST PATTERN

TATAGCCCTG, CCL 6
GTGCATTTGT, TGF beta-induced 68 kDa
CCACACAAGC, Ring finger protein 130
TGAACACTGA, Transglutaminase 2
TCACACAGAA, Neutrophil cytosolic factor 4
GAAACTTGTC, Embigin
AGGGGTCTGG, CSF 1 receptor
GGCCCAATAA, CCR 1
TAGCCACAAA, CCL 6, short form
GTCTGCAAGG, Ly86
TTACTTTAAT, Fc gamma receptor 2b

Key to Gene Expression Chart:

Abundantly Expressed Gene (>1%)
Abundantly Expressed Gene (>0.3%)
Moderately Expressed Gene (>0.1%)
Significant Expression (>=7 tags)
Positive Expression (>=3 tags)
Positive (not significant <3 tags)
Negative (undetected or no tags)

Fig. 30.6 SAGE tags associated with modulated DCs. Using SAGEClus software, tags associated with the pharmacologically-modulated DC populations have been selected based on the relatedness of their expression profiles to the idealized test pattern indicated in the top row of the clustergram. A blue square in the test pattern indicates moderate tag representation in that library, as indicated in the expression key, while an inverted grey triangle indicates no tag representation and a diagonal line no preference. Details of the comparator libraries indicated can be found in [86].

ponents that may give rise to inflammatory or autoimmune responses, and in the process of receptor-mediated endocytosis [129]. Neutrophil cytosolic factor-4 is otherwise known as p40phox and is a regulatory component of the NADPH oxidase complex that is activated during phagocytosis to mediate the generation of mirobicidal antioxidants [130]. Signaling through the colony stimulating factor-1 receptor, CSF1 receptor, regulates the survival, proliferation and chemotaxis of macrophages and supports their activation [131]. MD-1 (Ly86) has been implicated in mediating responses to LPS. As MD-2 is required for the surface expression of TLR4, so the related protein MD-1 associates with the protein RP105, and mice lacking either MD-1 or RP105 have been shown to exhibit a decreased immune response against LPS [132, 133]. Despite this, structural features of MD-2 required for binding of LPS are absent from MD-1 and it has recently been proposed that signals transduced through RP105/MD-1 by an unknown ligand may somehow act collaboratively on the LPS-activation pathway through TLR4/MD-2 [134]. Finally shown in this limited cluster is the tag corresponding to the inhibitory Fcγ receptor. Selective blockade of this receptor leads to spontaneous and full maturation of DCs, requiring the presence of endogenous plasma IgG and it is suggested that the activation status of DCs in normal human serum depends on the balance between activating and inhibitory FcγRs [58].

While the functional relevance of these genes requires experimental clarification, such signatures can also serve as a useful diagnostic to facilitate the identification of functionally related populations or to monitor the appearance of tolerogenic DCs *in vivo* using custom microarray analysis as described below (Section 30.3.1).

30.2.3.3 Identification of Novel Genes

By sampling tags from the entire transcriptome SAGE provides for the possibility of identifying and cloning previously undescribed transcripts. The pro-inflammatory chemokine DCIP-1 was first revealed as a 10 bp SAGE tag, represented in LPS-matured, but not immature bmDC libraries, for which no annotation could be made from the existing databases [87]. The corresponding transcript was subsequently cloned, initially using the tag itself to prime amplification of the 3′ end. The novel transcript revealed an ELR$^+$ CXC chemokine, that by homology is a novel member of a subgroup of pro-inflammatory chemokines, comprising murine CXCL1 (MIP-2) and CXCL2 (KC), the rat CINC and the human GRO proteins. It was subsequently demonstrated that this chemokine is a ligand for CXCR2, and that it mediates selective recruitment of neutrophils *in vivo*. The kinetics of induction of the DCIP-1 transcript are consistent with expression during the innate pro-inflammatory phase of DC maturation.

30.2.3.4 SAGE Library Comparisons Provide Insights to Biological Mechanism

Although bmDC modulated with IL-10 do not acquire the ability to stimulate naïve T cells in response to LPS, and similarly do not increase their levels of surface MHC class II or co-stimulatory molecules, the global changes in gene expression patterns observed by multiple pairwise scatter-plot comparisons are immediately supportive of a mechanism more subtle than simple blocking of maturation. On addition of IL-10 to bmDCs, ~23% of the tag changes were also observed in response to LPS, suggesting that at least some features of IL-10 modulation are actually in common with the LPS maturation process. After LPS treatment of the IL-10-modulated DCs, ~37% of the tag changes were in common with those mediated by addition of LPS to unmodulated bmDC, and this represented ~56% of the unmodulated bmDC response to LPS [87]. While classic maturation of IL-10-modulated bmDCs does not occur, aspects of the normal response do appear to be retained.

Closer inspection of individual tag modulations revealed that although CCR7 SAGE tag levels were markedly increased in response to LPS, this increase was not seen in bmDCs previously modulated by exposure to IL-10. A similar block was also observed in the appearance of SAGE tags for the chemokines CCL17, CCL21 and CCL22 that promote various DC:T cell interactions, suggesting that the ability of DCs to recruit adaptive immunity in response to a microbial stimulus is hindered as a consequence of IL-10 modulation [87]. Indeed it has been demonstrated that IL-10 does inhibit maturation-induced migration of DCs to lymph nodes by blocking the switch from inflammatory chemokine receptors, such as CCR1 and CCR5, to the lymphoid homing receptor CCR7, [123, 135]. In mice infected with *Leshmania donovani*, IL-10 mediated inhibition of CCR7 expression results in a spatial segregation of DCs and T cells that contributes to immune-suppression and promotes pathogenesis [136]. Tag levels for the chemokine receptors CCR1 and CCR5 were not reduced in the IL-10-modulated DCs in response to LPS, consistent with reports that these receptors are not downregulated following IL-10 modulation, but functionally uncoupled, acquiring a scavenger role and possibly contributing to subsequent dampening of potentially destructive inflammation [123].

Multiple pairwise SAGE comparisons combined with quantitative real-time PCR data, indicated that in contrast to their impaired ability to recruit adaptive immunity following exposure to microbes, IL-10 modulated DCs are not only not impaired, but indeed appear to be enhanced in their ability to mount innate inflammatory responses [87]. Transcription of the genetically-linked inflammatory chemokines DCIP-1, CXCL2, CXCL4 and CXCL5, involved in neutrophil recruitment and activation, and the pro-inflammatory cytokine IL-1, are all increased in IL-10-modulated DC in response to LPS, while a decrease in tag numbers for the decoy-receptor IL-1R2 in response to IL-10 alone is suggestive of priming for IL-1 responsiveness. An increase in tags for both membrane-bound and soluble innate pattern recognition molecules, such as MARCO, CD14, TLR2 and galectin-3, in combination with a general increase in tags corresponding to degrading lysosomal enzymes, is suggestive of an enhanced potential for phagocytic bacterial clearance in LPS-treated, IL-10-modulated DCs, consistent with observations of increased antigen uptake by IL-10-treated DCs in the presence of bacteria [87, 137]. These observations are in contrast to reports of IL-10 as an anti-inflammatory cytokine and probably reflect the importance of context in determining the outcome of cytokine exposure.

The increase in inflammatory potential resulting from exposure to IL-10 appears to be tempered by a co-ordinate increase in the production of anti-inflammatory agents [87]. Tags corresponding to hemoxygenase-1, reportedly responsible for IL-10-mediated protection of mice from LPS-induced septic shock [138], and arginase-1, which competes with iNOS for the substrate arginine, downregulating the production of NO [139], are both increased in the IL-10-modulated, LPS-exposed library. Carbon monoxide generated by hemoxygenase-1 breakdown of heme also inhibits T-cell proliferation [140, 141]. Tags for the chemokine CXCL7 were uniquely acquired to abundant levels in the IL-10-modulated DC library and this was reversed following the addition of LPS [87]. Differential rates of processing of various precursor forms of this chemokine result in pro- and anti-inflammatory scenarios [142]. Inflammatory responses reflect a complex network of chemokine/protease interactions and while the substantial increase in cathepsins D, S, L, B and C may be related to increased antigen processing in the IL-10-modulated DCs, such proteases could also mediate as yet unrealized aspects of the IL-10-modulated DC phenotype, including roles in inflammation, cell migration and transcriptional control of the cell cycle [143–145].

The outcome of antigen presentation to T cells is at least to some extent determined by a balance of inhibitory and stimulatory receptor ligand interactions. An important contribution to the IL-10-mediated decrease in DC stimulatory capacity is provided by the interaction of PD-L1 and PD-L2 with their cognate inhibitory receptor PD-1, expressed on T cells [146]. This was reflected in the SAGE data by increased levels of the PD-L1 tag in the IL-10-modulated SAGE library and in particular in the IL-10-modulated library exposed to LPS [87]. PD-1 expression is reported to be more strongly induced on T cells that have received a weak antigenic signal implying that interactions with PD-1 will preferentially inhibit low avidity antigen receptors [146, 147].

30.3
Downstream Assessment of Tolerance Associated Candidate Genes

The immediate action on identifying a candidate tag is to verify the automated gene assignment and to clarify the authenticity of the transcript candidacy. While verification of the assignment can frequently be achieved by critical reference to existing databases, it can also on occasions require experimental clarification using 3′ RACE, to ensure the assignment is indeed derived from a full-length transcript. The transcript expression pattern is verified using quantitative real-time PCR analysis, which can also provide information on the relative distribution of alternatively spliced versions of a candidate indistinguishable by SAGE.

30.3.1
Simultaneous Assessment of Multiple Candidate Gene Expression Levels using a Custom "Immunochip"

Limited commercial microarrays have been used to screen biological replicates to validate a number of the SAGE libraries included in the SAGE resource. A further application of microarray technology is being adopted through the generation of an "in-house" custom "immunochip". SAGE has implicated multiple candidate tags, corresponding to genes of known and unknown function, as having a potential role in mediating tolerance. Oligonucleotides representing these candidates, as well as candidates identified from parallel SAGE studies investigating regulatory T cells and also candidates reported in the literature, are being combined onto a custom array. It is envisaged that this "immunochip" will be used to facilitate rapid, simultaneous analysis of genes of interest across a range of different samples and will thus allow assessment of the kinetics of induction or repression of candidates by time-course studies and the relative expression of candidates in samples obtained from *in vivo* tolerance models. The convenience of arrays, combined with the inclusion of SAGE-derived diagnostic gene "signatures", should facilitate a more speculative use of the "immunochip", such as during the preliminary stages of an experimental investigations to consider the effects of variations of culture conditions or treatment regimes on the induction of tolerance.

30.3.2
Assessing the Functional Relevance of Tolerance Candidates by Genetic Manipulation of DCs

Expression profiling and gene discovery form only the first steps in elucidating molecular mechanisms of tolerance. The ultimate challenge is to establish informative functional assays in order to investigate downstream biological relevance. This process has proven particularly difficult in the DC field, largely due to the inherent resistance of primary DCs to the introduction of heterologous genes, their susceptibility to mature upon manipulation and their short life span following terminal maturation. While efficient transfer of DNA constructs to primary DCs has been reported using a number of viral vectors, the transduced cells are often perturbed in their function and/or limit investigations to certain stages of DC development [148, 149].

Although tools such as antibodies and chimeric constructs can be used to investigate the contribution of candidates to tolerance, genetic manipulations, such as over-expression or expression of mutant/dominant negative forms and the use of RNAi knock-down approaches, are essentially precluded in primary DCs. An alternative has been presented by the demonstration that DCs can be differentiated in culture from ES cells [113]. These embryonic stem cell-derived DC (esDC) undergo significant expansion *in vitro*, are phenotypically stable over time and retain the capacity to mature in response to LPS [111–113]. Genetic manipulations can be achieved at the level of the ES cell using standard methodologies, avoiding the inherent difficulties associated with manipulating primary DCs, or the investment of generating manipulated mice as a source of bone-marrow from which to generate genetically modified DCs. The feasibility of generating stable lines of genetically-modified esDC has been verified using EGFP [112]. The stable EGFP-esDC lines maintained their capacity for maturation in response to LPS with no associated loss of transgene expression and retained their migration patterns *in vivo*. The capacity of different candidate transgenes to modulate or skew this process of maturation *in vitro*, in combination with investigation of the effects of these modified cells in an *in vivo* transplant tolerance model (see below) will be informative as to their functional role in the balance of immunity *versus* tolerance.

Comparative SAGE analysis of the parent ES cell line and differentiated esDC, has provided further supportive data for the integrity of this esDC system [111]. These SAGE libraries are available for subtractive comparisons and can be used to identify genes responsive to a transfected candidate transcript. The feasibility of long-term stable expression of knock-down hairpin RNAs in this system is currently under investigation.

30.3.3
Assessing the Functional Impact of Candidates in an in vivo Tolerance Model

A murine TCR transgenic skin transplant model has been established to assess the tolerogenic potential of manipulated DC populations and of genetically-modified esDC. Female CBA/Ca.A1.RAG1$^{-/-}$ mice possess T cells that are specific for a male peptide, HY, presented in the context of H-2E^k, and provide a model of transplant rejection mediated by CD4$^+$ T cells [110]. The RAG1$^{-/-}$ background ensures the monospecificity of this model and allows for a reductionist approach to be taken to the investigation of conditions influencing the establishment of T-cell tolerance. In addition to lacking CD8$^+$ T cells and B cells, these RAG1$^{-/-}$ mice also notably lack naturally occurring CD4$^+$CD25$^+$Foxp3$^+$ regulatory T cells. While control female CBA/Ca.A1.RAG1$^{-/-}$ mice receiving PBS alone all promptly reject male skin grafts with a mean survival time of 14 days, tolerance has been demonstrated in this system through prior treatment with a blocking antibody directed to CD4 [5], using an altered-peptide ligand [44] and more recently, by the injection of immature and pharmacologically-modulated male DC populations (unpublished data). The tolerance induced is antigen-specific, indefinite and dominant and involves an accumulation of FoxP3$^+$ regulatory T cells in the accepted graft. By administering genetically-modified male CBA/Ca-derived esDCs prior to skin grafting the impact of individual candidate molecules on graft survival and the induction of tolerance can be assessed.

30.4
Downstream Clinical Relevance

An incentive driving investigations into tolerance mechanisms is the potential to facilitate improved clinical outcome in situations of dysregulated immunity, such as allergy and autoimmunity, and in the more contrived situation of transplant rejection, where current drug therapies are both relatively non-specific and require long-term, repeated administration, whilst conferring a multitude of deleterious side effects. The ultimate goal is to develop short-term therapies that harness natural tolerance mechanisms to mediate long-term, antigen-specific re-education of the immune system. For application in the clinic this re-education will have to be achievable in mature immune systems and in the case of allergy and autoimmunity, in systems already primed to respond to the deleterious antigen.

In pharmacological terms, the use of drugs that block pathways in DCs critical to the activation of effector T cells, whilst sparing elements that vaccinate Tregs is a clear target direction for the future and could yet prove decisive in the much-heralded era of stem cell-derived organ replacement therapies. Although acute rejection can be avoided by the generation of organs devoid of DCs, a more robust solution, also encompassing issues of chronic rejection, may be to pre-tolerize by inoculation with immature esDCs, or esDCs manipulated for tolerance, that are derived from the same parent ES cell line as that used to derive the replacement organ itself [150].

References

1 Bouneaud, C., Kourilsky, P., Bousso, P., Impact of negative selection on the T cell repertoire reactive to a self-peptide: a large fraction of T cell clones escapes clonal deletion. *Immunity* **2000**. 13: 829–840.

2 Steinman, R. M., Hawiger, D., Nussenzweig, M. C., Tolerogenic dendritic cells. *Annu Rev Immunol* **2003**. 21: 685–711.

3 Rutella, S., Lemoli, R. M., Regulatory T cells and tolerogenic dendritic cells: from basic biology to clinical applications. *Immunol Lett* **2004**. 94: 11–26.

4 Apostolou, I., von Boehmer, H., *In vivo* instruction of suppressor commitment in naive T cells. *J Exp Med* **2004**. 199: 1401–1408.

5 Cobbold, S. P., Castejon, R., Adams, E., Zelenika, D., Graca, L., Humm, S., Waldmann, H., Induction of foxP3+ regulatory T cells in the periphery of T cell receptor transgenic mice tolerized to transplants. *J Immunol* **2004**. 172: 6003–6010.

6 Fehervari, Z., Sakaguchi, S., CD4+ Tregs and immune control. *J Clin Invest* **2004**. 114: 1209–1217.

7 Groux, H., O'Garra, A., Bigler, M., Rouleau, M., Antonenko, S., de Vries, J. E., Roncarolo, M. G., A CD4+ T-cell subset inhibits antigen-specific T-cell responses and prevents colitis. *Nature* **1997**. 389: 737–742.

8 Stock, P., Akbari, O., Berry, G., Freeman, G. J., Dekruyff, R. H., Umetsu, D. T., Induction of T helper type 1-like regulatory cells that express Foxp3 and protect against airway hyper-reactivity. *Nat Immunol* **2004**. 5: 1149–1156.

9 Akbari, O., DeKruyff, R. H., Umetsu, D. T., Pulmonary dendritic cells producing IL-10 mediate tolerance induced by respiratory exposure to antigen. *Nat Immunol* **2001**. 2: 725–731.

10 Roncarolo, M. G., Bacchetta, R., Bordignon, C., Narula, S., Levings, M. K., Type 1 T regulatory cells. *Immunol Rev* **2001**. 182: 68–79.

11 Banchereau, J., Steinman, R. M., Dendritic cells and the control of immunity. *Nature* **1998**. 392: 245–252.

12 Sozzani, S., Allavena, P., D'Amico, G., Luini, W., Bianchi, G., Kataura, M., Imai, T., Yoshie, O., Bonecchi, R., Mantovani, A., Differential regulation of chemokine receptors during dendritic cell maturation: a model for their trafficking properties. *J Immunol* **1998**. 161: 1083–1086.

13 Huang, Q., Liu, D., Majewski, P., Schulte, L. C., Korn, J. M., Young, R. A., Lander, E. S., Hacohen, N., The plasticity of dendritic cell responses to pathogens and their components. *Science* **2001**. 294: 870–875.

14 Pierre, P., Turley, S. J., Gatti, E., Hull, M., Meltzer, J., Mirza, A., Inaba, K., Steinman, R. M., Mellman, I., Developmental regulation of MHC class II transport in mouse dendritic cells. *Nature* **1997**. 388: 787–792.

15 Cella, M., Engering, A., Pinet, V., Pieters, J., Lanzavecchia, A., Inflammatory stimuli induce accumulation of MHC class II complexes on dendritic cells. *Nature* **1997**. 388: 782–787.

16 Pasare, C., Medzhitov, R., Toll pathway-dependent blockade of CD4+CD25+ T cell-mediated suppression by dendritic cells. *Science* **2003**. 299: 1033–1036.

17 Reis e Sousa, C., Dendritic cells as sensors of infection. *Immunity* **2001**. 14: 495–498.

18 Reis e Sousa, C., Toll-like receptors and dendritic cells: for whom the bug tolls. *Semin Immunol* **2004**. 16: 27–34.

19 Bennett, S. R., Carbone, F. R., Karamalis, F., Flavell, R. A., Miller, J. F., Heath, W. R., Help for cytotoxic-T-cell responses is mediated by CD40 signalling. *Nature* **1998**. 393: 478–480.

20 Ridge, J. P., Di Rosa, F., Matzinger, P., A conditioned dendritic cell can be a temporal bridge between a CD4+ T-helper and a T-killer cell. *Nature* **1998**. 393: 474–478.

21 Schoenberger, S. P., Toes, R. E., van der Voort, E. I., Offringa, R., Melief, C. J., T-cell help for cytotoxic T lymphocytes is mediated by CD40-CD40L interactions. *Nature* **1998**. 393: 480–483.

22 Albert, M. L., Jegathesan, M., Darnell, R. B., Dendritic cell maturation is required for the cross-tolerization of CD8+ T cells. *Nat Immunol* **2001**. 2: 1010–1017.

23 Kurts, C., Kosaka, H., Carbone, F. R., Miller, J. F., Heath, W. R., Class I-restricted cross-presentation of exogenous self-antigens leads to deletion of autoreactive CD8(+) T cells. *J Exp Med* **1997**. 186: 239–245.

24 Kurts, C., Carbone, F. R., Barnden, M., Blanas, E., Allison, J., Heath, W. R., Miller, J. F., CD4+ T cell help impairs CD8+ T cell deletion induced by cross-presentation of self-antigens and favors autoimmunity. *J Exp Med* **1997**. 186: 2057–2062.

25 Sporri, R., Reis e Sousa, C., Inflammatory mediators are insufficient for full dendritic cell activation and promote expansion of CD4+ T cell populations lacking helper function. *Nat Immunol* **2005**. 6: 163–170.

26 Jonuleit, H., Schmitt, E., Schuler, G., Knop, J., Enk, A. H., Induction of interleukin 10-producing, nonproliferating CD4(+) T cells with regulatory properties by repetitive stimulation with allogeneic immature human dendritic cells. *J Exp Med* **2000**. 192: 1213–1222.

27 Levings, M. K., Gregori, S., Tresoldi, E., Cazzaniga, S., Bonini, C., Roncarolo, M. G., Differentiation of Tr1 cells by immature dendritic cells requires IL-10 but not CD25+CD4+ Tr cells. *Blood* **2005**. 105: 1162–1169.

28 Dhodapkar, M. V., Steinman, R. M., Antigen-bearing immature dendritic cells induce peptide-specific CD8(+) regulatory T cells in vivo in humans. *Blood* **2002**. 100: 174–177.

29 Dhodapkar, M. V., Steinman, R. M., Krasovsky, J., Munz, C., Bhardwaj, N., Antigen-specific inhibition of effector T cell function in humans after injection of immature dendritic cells. *J Exp Med* **2001**. 193: 233–238.

30 Probst, H. C., Lagnel, J., Kollias, G., van den Broek, M., Inducible transgenic mice reveal resting dendritic cells as potent inducers of CD8+ T cell tolerance. *Immunity* **2003**. 18: 713–720.

31 Bonifaz, L., Bonnyay, D., Mahnke, K., Rivera, M., Nussenzweig, M. C., Steinman, R. M., Efficient targeting of protein antigen to the dendritic cell receptor DEC-205 in the steady state leads to antigen presentation on major histocompatibility complex class I products and peripheral CD8+ T cell tolerance. *J Exp Med* **2002**. 196: 1627–1638.

32 Hawiger, D., Inaba, K., Dorsett, Y., Guo, M., Mahnke, K., Rivera, M., Ravetch, J. V., Steinman, R. M., Nussenzweig, M. C., Dendritic cells induce peripheral T cell unresponsiveness under steady state conditions in vivo. *J Exp Med* **2001**. 194: 769–779.

33 Mahnke, K., Qian, Y., Knop, J., Enk, A. H., Induction of CD4+/CD25+ regulatory T cells by targeting of antigens to immature dendritic cells. *Blood* **2003**. 101: 4862–4869.

34 Morelli, A. E., Larregina, A. T., Shufesky, W. J., Zahorchak, A. F., Logar, A. J., Papworth, G. D., Wang, Z., Watkins, S. C., Falo, L. D., Jr., Thomson, A. W., Internalization of circulating apoptotic cells by splenic marginal zone dendritic cells: dependence on complement receptors and effect on cytokine production. *Blood* **2003**. 101: 611–620.

35 Inaba, K., Turley, S., Yamaide, F., Iyoda, T., Mahnke, K., Inaba, M., Pack, M., Subklewe, M., Sauter, B., Sheff, D., Albert, M., Bhardwaj, N., Mellman, I., Steinman, R. M., Efficient presentation of phagocytosed cellular fragments on the major histocompatibility complex class II products of dendritic cells. *J Exp Med* **1998**. 188: 2163–2173.

36 Huang, F. P., Platt, N., Wykes, M., Major, J. R., Powell, T. J., Jenkins, C. D., MacPherson, G. G., A discrete subpopulation of dendritic cells transports apoptotic intestinal epithelial cells to T cell areas of mesenteric lymph nodes. *J Exp Med* **2000**. 191: 435–444.

37 Sauter, B., Albert, M. L., Francisco, L., Larsson, M., Somersan, S., Bhardwaj, N., Consequences of cell death: exposure to necrotic tumor cells, but not primary tissue cells or apoptotic cells, induces the maturation of immunostimulatory dendritic cells. *J Exp Med* **2000**. 191: 423–434.

38 Gallucci, S., Lolkema, M., Matzinger, P., Natural adjuvants: endogenous activators of dendritic cells. *Nat Med* **1999**. 5: 1249–1255.

39 Steinman, R. M., Turley, S., Mellman, I., Inaba, K., The induction of tolerance by

dendritic cells that have captured apoptotic cells. *J Exp Med* **2000**. 191: 411–416.

40 Akbari, O., Freeman, G. J., Meyer, E. H., Greenfield, E. A., Chang, T. T., Sharpe, A. H., Berry, G., DeKruyff, R. H., Umetsu, D. T., Antigen-specific regulatory T cells develop via the ICOS-ICOS-ligand pathway and inhibit allergen-induced airway hyperreactivity. *Nat Med* **2002**. 8: 1024–1032.

41 Menges, M., Rossner, S., Voigtlander, C., Schindler, H., Kukutsch, N. A., Bogdan, C., Erb, K., Schuler, G., Lutz, M. B., Repetitive injections of dendritic cells matured with tumor necrosis factor alpha induce antigen-specific protection of mice from autoimmunity. *J Exp Med* **2002**. 195: 15–21.

42 Lutz, M. B., Schuler, G., Immature, semi-mature and fully mature dendritic cells: which signals induce tolerance or immunity? *Trends Immunol* **2002**. 23: 445–449.

43 Larsen, C. P., Elwood, E. T., Alexander, D. Z., Ritchie, S. C., Hendrix, R., Tucker-Burden, C., Cho, H. R., Aruffo, A., Hollenbaugh, D., Linsley, P. S., Winn, K. J., Pearson, T. C., Long-term acceptance of skin and cardiac allografts after blocking CD40 and CD28 pathways. *Nature* **1996**. 381: 434–438.

44 Chen, T. C., Waldmann, H., Fairchild, P. J., Induction of dominant transplantation tolerance by an altered peptide ligand of the male antigen Dby. *J Clin Invest* **2004**. 113: 1754–1762.

45 Liang, X., Lu, L., Chen, Z., Vickers, T., Zhang, H., Fung, J. J., Qian, S., Administration of dendritic cells transduced with antisense oligodeoxyribonucleotides targeting CD80 or CD86 prolongs allograft survival. *Transplantation* **2003**. 76: 721–729.

46 Fu, F., Li, Y., Qian, S., Lu, L., Chambers, F., Starzl, T. E., Fung, J. J., Thomson, A. W., Costimulatory molecule-deficient dendritic cell progenitors (MHC class II+, CD80dim, CD86–) prolong cardiac allograft survival in nonimmunosuppressed recipients. *Transplantation* **1996**. 62: 659–665.

47 Martin, E., O'Sullivan, B., Low, P., Thomas, R., Antigen-specific

suppression of a primed immune response by dendritic cells mediated by regulatory T cells secreting interleukin-10. *Immunity* **2003**. 18: 155–167.

48 Giannoukakis, N., Bonham, C. A., Qian, S., Chen, Z., Peng, L., Harnaha, J., Li, W., Thomson, A. W., Fung, J. J., Robbins, P. D., Lu, L., Prolongation of cardiac allograft survival using dendritic cells treated with NF-κB decoy oligodeoxyribonucleotides. *Mol Ther* **2000**. 1: 430–437.

49 Tiao, M. M., Lu, L., Tao, R., Wang, L., Fung, J. J., Qian, S., Prolongation of cardiac allograft survival by systemic administration of immature recipient dendritic cells deficient in NF-kappaB activity. *Ann Surg* **2005**. 241: 497–505.

50 Hackstein, H., Thomson, A. W., Dendritic cells: emerging pharmacological targets of immunosuppressive drugs. *Nat Rev Immunol* **2004**. 4: 24–34.

51 Cederbom, L., Hall, H., Ivars, F., CD4+CD25+ regulatory T cells downregulate co-stimulatory molecules on antigen-presenting cells. *Eur J Immunol* **2000**. 30: 1538–1543.

52 Nakamura, A., Kobayashi, E., Takai, T., Exacerbated graft-*versus*-host disease in Pirb-/- mice. *Nat Immunol* **2004**. 5: 623–629.

53 Brown, D., Trowsdale, J., Allen, R., The LILR family: modulators of innate and adaptive immune pathways in health and disease. *Tissue Antigens* **2004**. 64: 215–225.

54 Borges, L., Cosman, D., LIRs/ILTs/MIRs, inhibitory and stimulatory Ig-superfamily receptors expressed in myeloid and lymphoid cells. *Cytokine Growth Factor Rev* **2000**. 11: 209–217.

55 Ujike, A., Takeda, K., Nakamura, A., Ebihara, S., Akiyama, K., Takai, T., Impaired dendritic cell maturation and increased T(H)2 responses in PIR-B(–/–) mice. *Nat Immunol* **2002**. 3: 542–548.

56 Chang, C. C., Ciubotariu, R., Manavalan, J. S., Yuan, J., Colovai, A. I., Piazza, F., Lederman, S., Colonna, M., Cortesini, R., Dalla-Favera, R., Suciu-Foca, N., Tolerization of dendritic cells by T(S) cells: the crucial role of inhibitory receptors ILT3 and ILT4. *Nat Immunol* **2002**. 3: 237–243.

57 Beinhauer, B. G., McBride, J. M., Graf, P., Pursch, E., Bongers, M., Rogy, M., Korthauer, U., de Vries, J. E., Aversa, G., Jung, T., Interleukin 10 regulates cell surface and soluble LIR-2 (CD85d) expression on dendritic cells resulting in T cell hyporesponsiveness in vitro. *Eur J Immunol* **2004**. 34: 74–80.

58 Dhodapkar, K. M., Kaufman, J. L., Ehlers, M., Banerjee, D. K., Bonvini, E., Koenig, S., Steinman, R. M., Ravetch, J. V., Dhodapkar, M. V., Selective blockade of inhibitory Fcgamma receptor enables human dendritic cell maturation with IL-12p70 production and immunity to antibody-coated tumor cells. *Proc Natl Acad Sci USA* **2005**. 102: 2910–2915.

59 Mellor, A. L., Baban, B., Chandler, P., Marshall, B., Jhaver, K., Hansen, A., Koni, P. A., Iwashima, M., Munn, D. H., Cutting edge: induced indoleamine 2,3 dioxygenase expression in dendritic cell subsets suppresses T cell clonal expansion. *J Immunol* **2003**. 171: 1652–1655.

60 Mellor, A. L., Chandler, P., Baban, B., Hansen, A. M., Marshall, B., Pihkala, J., Waldmann, H., Cobbold, S., Adams, E., Munn, D. H., Specific subsets of murine dendritic cells acquire potent T cell regulatory functions following CTLA4-mediated induction of indoleamine 2,3 dioxygenase. *Int Immunol* **2004**. 16: 1391–1401.

61 Granucci, F., Vizzardelli, C., Pavelka, N., Feau, S., Persico, M., Virzi, E., Rescigno, M., Moro, G., Ricciardi-Castagnoli, P., Inducible IL-2 production by dendritic cells revealed by global gene expression analysis. *Nat Immunol* **2001**. 2: 882–888.

62 Granucci, F., Vizzardelli, C., Virzi, E., Rescigno, M., Ricciardi-Castagnoli, P., Transcriptional reprogramming of dendritic cells by differentiation stimuli. *Eur J Immunol* **2001**. 31: 2539–2546.

63 Suciu-Foca, N., Manavalan, J. S., Scotto, L., Kim-Schulze, S., Galluzzo, S., Naiyer, A. J., Fan, J., Vlad, G., Cortesini, R., Molecular characterization of allospecific T suppressor and tolerogenic dendritic cells: review. *Int Immunopharmacol* **2005**. 5: 7–11.

64 Manavalan, J. S., Rossi, P. C., Vlad, G., Piazza, F., Yarilina, A., Cortesini, R., Mancini, D., Suciu-Foca, N., High expression of ILT3 and ILT4 is a general feature of tolerogenic dendritic cells. *Transpl Immunol* **2003**. 11: 245–258.

65 Velten, F. W., Duperrier, K., Bohlender, J., Metharom, P., Goerdt, S., A gene signature of inhibitory MHC receptors identifies a BDCA3(+) subset of IL-10-induced dendritic cells with reduced allostimulatory capacity in vitro. *Eur J Immunol* **2004**. 34: 2800–2811.

66 St Louis, D. C., Woodcock, J. B., Franzoso, G., Blair, P. J., Carlson, L. M., Murillo, M., Wells, M. R., Williams, A. J., Smoot, D. S., Kaushal, S., Grimes, J. L., Harlan, D. M., Chute, J. P., June, C. H., Siebenlist, U., Lee, K. P., Evidence for distinct intracellular signaling pathways in CD34+ progenitor to dendritic cell differentiation from a human cell line model. *J Immunol* **1999**. 162: 3237–3248.

67 Winzler, C., Rovere, P., Rescigno, M., Granucci, F., Penna, G., Adorini, L., Zimmermann, V. S., Davoust, J., Ricciardi-Castagnoli, P., Maturation stages of mouse dendritic cells in growth factor-dependent long-term cultures. *J Exp Med* **1997**. 185: 317–328.

68 Romani, N., Gruner, S., Brang, D., Kampgen, E., Lenz, A., Trockenbacher, B., Konwalinka, G., Fritsch, P. O., Steinman, R. M., Schuler, G., Proliferating dendritic cell progenitors in human blood. *J Exp Med* **1994**. 180: 83–93.

69 Inaba, K., Inaba, M., Romani, N., Aya, H., Deguchi, M., Ikehara, S., Muramatsu, S., Steinman, R. M., Generation of large numbers of dendritic cells from mouse bone marrow cultures supplemented with granulocyte/macrophage colony-stimulating factor. *J Exp Med* **1992**. 176: 1693–1702.

70 Edwards, A. D., Chaussabel, D., Tomlinson, S., Schulz, O., Sher, A., Reis e Sousa, C., Relationships among murine CD11c(high) dendritic cell subsets as revealed by baseline gene expression patterns. *J Immunol* **2003**. 171: 47–60.

71 Muller, G., Muller, A., Tuting, T., Steinbrink, K., Saloga, J., Szalma, C., Knop, J., Enk, A. H., Interleukin-10-treated dendritic cells modulate immune responses of naive and sensitized T cells in vivo. *J Invest Dermatol* **2002**. 119: 836–841.

72 Steinbrink, K., Wolfl, M., Jonuleit, H., Knop, J., Enk, A. H., Induction of tolerance by IL-10-treated dendritic cells. *J Immunol* **1997**. 159: 4772–4780.

73 Geissmann, F., Revy, P., Regnault, A., Lepelletier, Y., Dy, M., Brousse, N., Amigorena, S., Hermine, O., Durandy, A., TGF-beta 1 prevents the noncognate maturation of human dendritic Langerhans cells. *J Immunol* **1999**. 162: 4567–4575.

74 Alard, P., Clark, S. L., Kosiewicz, M. M., Mechanisms of tolerance induced by TGF beta-treated APC: CD4 regulatory T cells prevent the induction of the immune response possibly through a mechanism involving TGF beta. *Eur J Immunol* **2004**. 34: 1021–1030.

75 Kosiewicz, M. M., Alard, P., Tolerogenic antigen-presenting cells: regulation of the immune response by TGF-beta-treated antigen-presenting cells. *Immunol Res* **2004**. 30: 155–170.

76 Oyama, T., Ran, S., Ishida, T., Nadaf, S., Kerr, L., Carbone, D. P., Gabrilovich, D. I., Vascular endothelial growth factor affects dendritic cell maturation through the inhibition of nuclear factor-kappa B activation in hemopoietic progenitor cells. *J Immunol* **1998**. 160: 1224–1232.

77 Hackstein, H., Morelli, A. E., Larregina, A. T., Ganster, R. W., Papworth, G. D., Logar, A. J., Watkins, S. C., Falo, L. D., Thomson, A. W., Aspirin inhibits *in vitro* maturation and *in vivo* immuno-stimulatory function of murine myeloid dendritic cells. *J Immunol* **2001**. 166: 7053–7062.

78 Hogg, N., Nicotine has suppressive effects on dendritic cell function. *Immunology* **2003**. 109: 329–330.

79 Yang, J., Bernier, S. M., Ichim, T. E., Li, M., Xia, X., Zhou, D., Huang, X., Strejan, G. H., White, D. J., Zhong, R., Min, W. P., LF15-0195 generates tolerogenic dendritic cells by suppression of NF-kappaB signaling through inhibition of IKK activity. *J Leukoc Biol* **2003**. 74: 438–447.

80 Matyszak, M. K., Citterio, S., Rescigno, M., Ricciardi-Castagnoli, P., Differential effects of corticosteroids during different stages of dendritic cell maturation. *Eur J Immunol* **2000**. 30: 1233–1242.

81 Piemonti, L., Monti, P., Allavena, P., Sironi, M., Soldini, L., Leone, B. E., Socci, C., Di Carlo, V., Glucocorticoids affect human dendritic cell differentiation and maturation. *J Immunol* **1999**. 162: 6473–6481.

82 Penna, G., Adorini, L., 1 Alpha,25-dihydroxyvitamin D3 inhibits differentiation, maturation, activation, and survival of dendritic cells leading to impaired alloreactive T cell activation. *J Immunol* **2000**. 164: 2405–2411.

83 Griffin, M. D., Lutz, W., Phan, V. A., Bachman, L. A., McKean, D. J., Kumar, R., Dendritic cell modulation by 1alpha,25 dihydroxyvitamin D3 and its analogs: a vitamin D receptor-dependent pathway that promotes a persistent state of immaturity *in vitro* and *in vivo*. *Proc Natl Acad Sci USA* **2001**. 98: 6800–6805.

84 Brandenberger, R., Khrebtukova, I., Thies, R. S., Miura, T., Jingli, C., Puri, R., Vasicek, T., Lebkowski, J., Rao, M., MPSS profiling of human embryonic stem cells. *BMC Dev Biol* **2004**. 4: 10.

85 Jongeneel, C. V., Iseli, C., Stevenson, B. J., Riggins, G. J., Lal, A., Mackay, A., Harris, R. A., O'Hare, M. J., Neville, A. M., Simpson, A. J., Strausberg, R. L., Comprehensive sampling of gene expression in human cell lines with massively parallel signature sequencing. *Proc Natl Acad Sci USA* **2003**. 100: 4702–4705.

86 Cobbold, S. P., Nolan, K. F., Graca, L., Castejon, R., Le Moine, A., Frewin, M., Humm, S., Adams, E., Thompson, S., Zelenika, D., Paterson, A., Yates, S., Fairchild, P. J., Waldmann, H., Regulatory T cells and dendritic cells in transplantation tolerance: molecular markers and mechanisms. *Immunol Rev* **2003**. 196: 109–124.

87 Nolan, K. F., Strong, V., Soler, D., Fairchild, P. J., Cobbold, S. P., Croxton, R., Gonzalo, J. A., Rubio, A., Wells, M., Waldmann, H., IL-10-conditioned dendritic cells, decommissioned for recruitment of adaptive immunity, elicit innate inflammatory gene products in response to danger signals. *J Immunol* **2004**. 172: 2201–2209.

88 Zavolan, M., Kondo, S., Schonbach, C., Adachi, J., Hume, D. A., Hayashizaki, Y.,

Gaasterland, T., Impact of alternative initiation, splicing, and termination on the diversity of the mRNA transcripts encoded by the mouse transcriptome. *Genome Res* **2003**. 13: 1290–1300.

89 Hashimoto, S., Suzuki, Y., Kasai, Y., Morohoshi, K., Yamada, T., Sese, J., Morishita, S., Sugano, S., Matsushima, K., 5'-end SAGE for the analysis of transcriptional start sites. *Nat Biotechnol* **2004**. 22: 1146–1149.

90 Ng, P., Wei, C. L., Sung, W. K., Chiu, K. P., Lipovich, L., Ang, C. C., Gupta, S., Shahab, A., Ridwan, A., Wong, C. H., Liu, E. T., Ruan, Y., Gene identification signature (GIS) analysis for transcriptome characterization and genome annotation. *Nat Methods* **2005**. 2: 105–111.

91 Wei, C. L., Ng, P., Chiu, K. P., Wong, C. H., Ang, C. C., Lipovich, L., Liu, E. T., Ruan, Y., 5' Long serial analysis of gene expression (LongSAGE) and 3' LongSAGE for transcriptome characterization and genome annotation. *Proc Natl Acad Sci USA* **2004**. 101: 11701–11706.

92 Velculescu, V. E., Zhang, L., Vogelstein, B., Kinzler, K. W., Serial analysis of gene expression. *Science* **1995**. 270: 484–487.

93 Du, Z., Scott, A. D., May, G. D., Amplification of high-quantity serial analysis of gene expression ditags and improvement of concatemer cloning efficiency. *Biotechniques* **2003**. 35: 66–67, 70–62.

94 Powell, J., Enhanced concatemer cloning-a modification to the SAGE (Serial Analysis of Gene Expression) technique. *Nucleic Acids Res* **1998**. 26: 3445–3446.

95 Heidenblut, A. M., Luttges, J., Buchholz, M., Heinitz, C., Emmersen, J., Nielsen, K. L., Schreiter, P., Souquet, M., Nowacki, S., Herbrand, U., Kloppel, G., Schmiegel, W., Gress, T., Hahn, S. A., aRNA-longSAGE: a new approach to generate SAGE libraries from microdissected cells. *Nucleic Acids Res* **2004**. 32: e131.

96 Neilson, L., Andalibi, A., Kang, D., Coutifaris, C., Strauss, J. F., 3rd, Stanton, J. A., Green, D. P., Molecular phenotype of the human oocyte by PCR-SAGE. *Genomics* **2000**. 63: 13–24.

97 Peters, D. G., Kassam, A. B., Yonas, H., O'Hare, E. H., Ferrell, R. E., Brufsky, A. M., Comprehensive transcript analysis in small quantities of mRNA by SAGE-lite. *Nucleic Acids Res* **1999**. 27: e39.

98 Vilain, C., Libert, F., Venet, D., Costagliola, S., Vassart, G., Small amplified RNA-SAGE: an alternative approach to study transcriptome from limiting amount of mRNA. *Nucleic Acids Res* **2003**. 31: e24.

99 Saha, S., Sparks, A. B., Rago, C., Akmaev, V., Wang, C. J., Vogelstein, B., Kinzler, K. W., Velculescu, V. E., Using the transcriptome to annotate the genome. *Nat Biotechnol* **2002**. 20: 508–512.

100 Matsumura, H., Reich, S., Ito, A., Saitoh, H., Kamoun, S., Winter, P., Kahl, G., Reuter, M., Kruger, D. H., Terauchi, R., Gene expression analysis of plant host-pathogen interactions by SuperSAGE. *Proc Natl Acad Sci USA* **2003**. 100: 15718–15723.

101 Lu, J., Lal, A., Merriman, B., Nelson, S., Riggins, G., A comparison of gene expression profiles produced by SAGE, long SAGE, and oligonucleotide chips. *Genomics* **2004**. 84: 631–636.

102 Boon, K., Osorio, E. C., Greenhut, S. F., Schaefer, C. F., Shoemaker, J., Polyak, K., Morin, P. J., Buetow, K. H., Strausberg, R. L., De Souza, S. J., Riggins, G. J., An anatomy of normal and malignant gene expression. *Proc Natl Acad Sci USA* **2002**. 99: 11287–11292.

103 Velculescu, V. E., Zhang, L., Zhou, W., Vogelstein, J., Basrai, M. A., Bassett, D. E., Jr., Hieter, P., Vogelstein, B., Kinzler, K. W., Characterization of the yeast transcriptome. *Cell* **1997**. 88: 243–251.

104 Akmaev, V. R., Wang, C. J., Correction of sequence-based artifacts in serial analysis of gene expression. *Bioinformatics* **2004**. 20: 1254–1263.

105 Piemonti, L., Monti, P., Sironi, M., Fraticelli, P., Leone, B. E., Dal Cin, E., Allavena, P., Di Carlo, V., Vitamin D3 affects differentiation, maturation, and function of human monocyte-derived dendritic cells. *J Immunol* **2000**. 164: 4443–4451.

106 Bonham, C. A., Lu, L., Banas, R. A., Fontes, P., Rao, A. S., Starzl, T. E., Zeevi,

A., Thomson, A. W., TGF-beta 1 pretreatment impairs the allostimulatory function of human bone marrow-derived antigen-presenting cells for both naive and primed T cells. *Transpl Immunol* **1996**. 4: 186–191.

107 Yamaguchi, Y., Tsumura, H., Miwa, M., Inaba, K., Contrasting effects of TGF-beta 1 and TNF-alpha on the development of dendritic cells from progenitors in mouse bone marrow. *Stem Cells* **1997**. 15: 144–153.

108 Steinbrink, K., Graulich, E., Kubsch, S., Knop, J., Enk, A. H., CD4(+) and CD8(+) anergic T cells induced by interleukin-10-treated human dendritic cells display antigen-specific suppressor activity. *Blood* **2002**. 99: 2468–2476.

109 Koch, F., Stanzl, U., Jennewein, P., Janke, K., Heufler, C., Kampgen, E., Romani, N., Schuler, G., High level IL-12 production by murine dendritic cells: upregulation via MHC class II and CD40 molecules and downregulation by IL-4 and IL-10. *J Exp Med* **1996**. 184: 741–746.

110 Zelenika, D., Adams, E., Mellor, A., Simpson, E., Chandler, P., Stockinger, B., Waldmann, H., Cobbold, S. P., Rejection of H-Y disparate skin grafts by monospecific CD4+ Th1 and Th2 cells: no requirement for CD8+ T cells or B cells. *J Immunol* **1998**. 161: 1868–1874.

111 Fairchild, P. J., Nolan, K. F., Waldmann, H., Probing dendritic cell function by guiding the differentiation of embryonic stem cells. *Methods Enzymol* **2003**. 365: 169–186.

112 Fairchild, P. J., Nolan, K. F., Cartland, S., Graca, L., Waldmann, H., Stable lines of genetically modified dendritic cells from mouse embryonic stem cells. *Transplantation* **2003**. 76: 606–608.

113 Fairchild, P. J., Brook, F. A., Gardner, R. L., Graca, L., Strong, V., Tone, Y., Tone, M., Nolan, K. F., Waldmann, H., Directed differentiation of dendritic cells from mouse embryonic stem cells. *Curr Biol* **2000**. 10: 1515–1518.

114 Zelenika, D., Adams, E., Humm, S., Graca, L., Thompson, S., Cobbold, S. P., Waldmann, H., Regulatory T cells overexpress a subset of Th2 gene transcripts. *J Immunol* **2002**. 168: 1069–1079.

115 Zelenika, D., Adams, E., Humm, S., Lin, C. Y., Waldmann, H., Cobbold, S. P., The role of CD4+ T-cell subsets in determining transplantation rejection or tolerance. *Immunol Rev* **2001**. 182: 164–179.

116 Graca, L., Thompson, S., Lin, C. Y., Adams, E., Cobbold, S. P., Waldmann, H., Both CD4(+)CD25(+) and CD4(+)CD25(−) regulatory cells mediate dominant transplantation tolerance. *J Immunol* **2002**. 168: 5558–5565.

117 Frohman, M. A., Dush, M. K., Martin, G. R., Rapid production of full-length cDNAs from rare transcripts: amplification using a single gene-specific oligonucleotide primer. *Proc Natl Acad Sci USA* **1988**. 85: 8998–9002.

118 Shires, J., Theodoridis, E., Hayday, A. C., Biological insights into TCRgammadelta+ and TCRalphabeta+ intraepithelial lymphocytes provided by serial analysis of gene expression (SAGE). *Immunity* **2001**. 15: 419–434.

119 Orlofsky, A., Lin, E. Y., Prystowsky, M. B., Selective induction of the beta chemokine C10 by IL-4 in mouse macrophages. *J Immunol* **1994**. 152: 5084–5091.

120 LaFleur, A. M., Lukacs, N. W., Kunkel, S. L., Matsukawa, A., Role of CC chemokine CCL6/C10 as a monocyte chemoattractant in a murine acute peritonitis. *Mediators Inflamm* **2004**. 13: 349–355.

121 Steinhauser, M. L., Hogaboam, C. M., Matsukawa, A., Lukacs, N. W., Strieter, R. M., Kunkel, S. L., Chemokine C10 promotes disease resolution and survival in an experimental model of bacterial sepsis. *Infect Immun* **2000**. 68: 6108–6114.

122 Ma, B., Zhu, Z., Homer, R. J., Gerard, C., Strieter, R., Elias, J. A., The C10/CCL6 chemokine and CCR1 play critical roles in the pathogenesis of IL-13-induced inflammation and remodeling. *J Immunol* **2004**. 172: 1872–1881.

123 D'Amico, G., Frascaroli, G., Bianchi, G., Transidico, P., Doni, A., Vecchi, A., Sozzani, S., Allavena, P., Mantovani, A., Uncoupling of inflammatory chemokine receptors by IL-10: generation of functional decoys. *Nat Immunol* **2000**. 1: 387–391.

124 Nam, J. O., Kim, J. E., Jeong, H. W., Lee, S. J., Lee, B. H., Choi, J. Y., Park, R. W., Park, J. Y., Kim, I. S., Identification of the alphavbeta3 integrin-interacting motif of betaIg-h3 and its anti-angiogenic effect. *J Biol Chem* **2003**. 278: 25 902–25 909.

125 Huang, R. P., Ozawa, M., Kadomatsu, K., Muramatsu, T., Embigin, a member of the immunoglobulin superfamily expressed in embryonic cells, enhances cell-substratum adhesion. *Dev Biol* **1993**. 155: 307–314.

126 Baker, S. J., Reddy, E. P., Cloning of murine G1RP, a novel gene related to Drosophila melanogaster g1. *Gene* **2000**. 248: 33–40.

127 Aeschlimann, D., Thomazy, V., Protein crosslinking in assembly and remodelling of extracellular matrices: the role of transglutaminases. *Connect Tissue Res* **2000**. 41: 1–27.

128 Nunes, I., Gleizes, P. E., Metz, C. N., Rifkin, D. B., Latent transforming growth factor-beta binding protein domains involved in activation and transglutaminase-dependent cross-linking of latent transforming growth factor-beta. *J Cell Biol* **1997**. 136: 1151–1163.

129 Esposito, C., Caputo, I., Mammalian transglutaminases. Identification of substrates as a key to physiological function and physiopathological relevance. *Febs J* **2005**. 272: 615–631.

130 Vignais, P. V., The superoxide-generating NADPH oxidase: structural aspects and activation mechanism. *Cell Mol Life Sci* **2002**. 59: 1428–1459.

131 Pixley, F. J., Stanley, E. R., CSF-1 regulation of the wandering macrophage: complexity in action. *Trends Cell Biol* **2004**. 14: 628–638.

132 Nagai, Y., Shimazu, R., Ogata, H., Akashi, S., Sudo, K., Yamasaki, H., Hayashi, S., Iwakura, Y., Kimoto, M., Miyake, K., Requirement for MD-1 in cell surface expression of RP105/CD180 and B-cell responsiveness to lipopoly-saccharide. *Blood* **2002**. 99: 1699–1705.

133 Ogata, H., Su, I., Miyake, K., Nagai, Y., Akashi, S., Mecklenbrauker, I., Rajewsky, K., Kimoto, M., Tarakhovsky, A., The toll-like receptor protein RP105 regulates lipopolysaccharide signaling in B cells. *J Exp Med* **2000**. 192: 23–29.

134 Tsuneyoshi, N., Fukudome, K., Kohara, J., Tomimasu, R., Gauchat, J. F., Nakatake, H., Kimoto, M., The functional and structural properties of MD-2 required for lipopolysaccharide binding are absent in MD-1. *J Immunol* **2005**. 174: 340–344.

135 Takayama, T., Morelli, A. E., Onai, N., Hirao, M., Matsushima, K., Tahara, H., Thomson, A. W., Mammalian and viral IL-10 enhance C-C chemokine receptor 5 but down-regulate C-C chemokine receptor 7 expression by myeloid dendritic cells: impact on chemotactic responses and in vivo homing ability. *J Immunol* **2001**. 166: 7136–7143.

136 Ato, M., Stager, S., Engwerda, C. R., Kaye, P. M., Defective CCR7 expression on dendritic cells contributes to the development of visceral leishmaniasis. *Nat Immunol* **2002**. 3: 1185–1191.

137 Faulkner, L., Buchan, G., Baird, M., Interleukin-10 does not affect phagocytosis of particulate antigen by bone marrow-derived dendritic cells but does impair antigen presentation. *Immunology* **2000**. 99: 523–531.

138 Lee, T. S., Chau, L. Y., Heme oxygenase-1 mediates the anti-inflammatory effect of interleukin-10 in mice. *Nat Med* **2002**. 8: 240–246.

139 Sonoki, T., Nagasaki, A., Gotoh, T., Takiguchi, M., Takeya, M., Matsuzaki, H., Mori, M., Coinduction of nitric-oxide synthase and arginase I in cultured rat peritoneal macrophages and rat tissues in vivo by lipopolysaccharide. *J Biol Chem* **1997**. 272: 3689–3693.

140 Pae, H. O., Oh, G. S., Choi, B. M., Chae, S. C., Kim, Y. M., Chung, K. R., Chung, H. T., Carbon monoxide produced by heme oxygenase-1 suppresses T cell proliferation via inhibition of IL-2 production. *J Immunol* **2004**. 172: 4744–4751.

141 Song, R., Mahidhara, R. S., Zhou, Z., Hoffman, R. A., Seol, D. W., Flavell, R. A., Billiar, T. R., Otterbein, L. E., Choi, A. M., Carbon monoxide inhibits T lymphocyte proliferation *via* caspase-dependent pathway. *J Immunol* **2004**. 172: 1220–1226.

142 Ehlert, J. E., Ludwig, A., Grimm, T. A., Lindner, B., Flad, H. D., Brandt, E., Down-regulation of neutrophil functions by the ELR(+) CXC chemokine platelet basic protein. *Blood* **2000**. 96: 2965–2972.

143 Chapman, H. A., Cathepsins as transcriptional activators? *Dev Cell* **2004**. 6: 610–611.

144 Nomura, T., Katunuma, N., Involvement of cathepsins in the invasion, metastasis and proliferation of cancer cells. *J Med Invest* **2005**. 52: 1–9.

145 Wolf, M., Clark-Lewis, I., Buri, C., Langen, H., Lis, M., Mazzucchelli, L., Cathepsin D specifically cleaves the chemokines macrophage inflammatory protein-1 alpha, macrophage inflammatory protein-1 beta, and SLC that are expressed in human breast cancer. *Am J Pathol* **2003**. 162: 1183–1190.

146 Brown, J. A., Dorfman, D. M., Ma, F. R., Sullivan, E. L., Munoz, O., Wood, C. R., Greenfield, E. A., Freeman, G. J., Blockade of programmed death-1 ligands on dendritic cells enhances T cell activation and cytokine production. *J Immunol* **2003**. 170: 1257–1266.

147 Latchman, Y., Wood, C. R., Chernova, T., Chaudhary, D., Borde, M., Chernova, I., Iwai, Y., Long, A. J., Brown, J. A., Nunes, R., Greenfield, E. A., Bourque, K., Boussiotis, V. A., Carter, L. L., Carreno, B. M., Malenkovich, N., Nishimura, H., Okazaki, T., Honjo, T., Sharpe, A. H., Freeman, G. J., PD-L2 is a second ligand for PD-1 and inhibits T cell activation. *Nat Immunol* **2001**. 2: 261–268.

148 Hirschowitz, E. A., Weaver, J. D., Hidalgo, G. E., Doherty, D. E., Murine dendritic cells infected with adenovirus vectors show signs of activation. *Gene Ther* **2000**. 7: 1112–1120.

149 Jenne, L., Schuler, G., Steinkasserer, A., Viral vectors for dendritic cell-based immunotherapy. *Trends Immunol* **2001**. 22: 102–107.

150 Fairchild, P. J., Cartland, S., Nolan, K. F., Waldmann, H., Embryonic stem cells and the challenge of transplantation tolerance. *Trends Immunol* **2004**. 25: 465–470.

Part B
Dendritic Cells in Disease

Handbook of Dendritic Cells. Biology, Diseases, and Therapies.
Edited by M. B. Lutz, N. Romani, and A. Steinkasserer.
Copyright © 2006 WILEY-VCH Verlag GmbH & Co. KGaA, Weinheim
ISBN: 3-527-31109-2

XI
Parasites

31
Malaria

Britta C. Urban and Francis M. Ndungu

31.1
Introduction to Malaria

Malaria remains the world's worst public health problem after HIV/AIDS. Global-ly 2.4 billion people (40% of the world's population in 1999) from more than 90 countries of the world are at risk of being infected and up to 500 million suffer from the disease with varying degree of severity. It is estimated that malaria kills 1 million people annually, mainly children below the age of 5 years and pregnant women [1, 2]. Almost all of these deaths are caused by *Plasmodium falciparum*, one of the four species of malaria parasites that infect humans. Other species infecting humans are *P. vivax*, *P. malariae* and *P. ovale*. This high burden of mortality is not evenly distributed but falls heavily on sub-Saharan Africa, where over 90% of ma-larial deaths occur and it is estimated that over 5% of children die before their fifth birthday [3] . Malaria is also a primary cause of poverty as a consequence of factors that include effects of the disease on fertility, population growth, saving and invest-ment, worker productivity, absenteeism, premature death and medical costs [4]. Malaria-specific mortality has more than doubled in the last two decades and the situation continues to worsen due to wide spread resistance of plasmodium para-sites to preventative and therapeutic drugs. It is estimated that malaria lowers the economic growth of affected countries by 1.3% of their gross domestic product [5] thus contributing to the poverty of these already impoverished communities.

   P. falciparum has a complex life cycle with many different developmental stages in the mosquito vector and human host. The bite of an infected *Anopheline* mos-quito injects infective sporozoites into the human host, where they are transported rapidly via the blood stream to hepatocytes in the liver. Within the hepatocyte, the parasite matures, differentiates, and undergoes one round of asexual multiplica-tion forming approximately 20000–40000 haploid merozoites that are released

Handbook of Dendritic Cells. Biology, Diseases, and Therapies.
Edited by M. B. Lutz, N. Romani, and A. Steinkasserer.
Copyright © 2006 WILEY-VCH Verlag GmbH & Co. KGaA, Weinheim
ISBN: 3-527-31109-2

into the blood-stream. This stage takes about 7 days and does not give rise to clinical symptoms. The released merozoites immediately invade red blood cells (RBC) and undergo a process of growth and asexual multiplication to produce between 8 and 32 daughter merozoites per every infected erythrocyte over a period of 48 h. When the daughter merozoites are fully matured (the schizont stage), the infected red cell bursts, releasing the merozoites to invade other erythrocytes. This period of exponential growth is responsible for all the clinical symptoms of malaria and continues until the parasite multiplication is controlled by drug treatment, the immune response, or death in some cases. A small proportion of the invading merozoites undergo an alternative pathway of differentiation and develop into either male or female gametocyte, which are subsequently taken up in a mosquito blood meal. In the mosquito mid-gut, the male and female gametes fuse to form a zygote, which then undergoes a series of complicated differentiation, and growth stages that results in the production of infective sporozoites in the salivary glands of the mosquito.

31.2
Antigenic Variation

Infected RBC of *P. falciparum* (Pf-iRBC), containing the late trophozoite and schizont stages (24–48 h post-invasion) sequester on endothelial cells in almost all tissues. Deep tissue sequestration of Pf-iRBC is generally believed to favor the survival of parasites by preventing Pf-iRBC passage through the spleen where they would otherwise be recognized as abnormal and removed. Sequestration in deep tissue microvasculature may also promote rapid asexual multiplication by placing the Pf-iRBC in a parasite-favoring micro-aerophilic environment.

Sequestration is mediated by adhesion of Pf-iRBC to a variety of host receptors expressed on endothelial cells, RBC and leukocytes (reviewed in [6]). Almost all field and laboratory isolates bind to CD36 and some of them can bind in addition to CD31, CD35 and CD54. Pf. iRBC insert parasite-derived proteins into the RBC membrane, the *P. falciparum* erythrocyte membrane protein 1 (PfEMP-1). PfEMP-1 is encoded by a multigene family of approximately 60 *var* genes per haploid genome [7]. *Var* genes are highly diverse but share a similar organization. They contain two to five copies of a motif denoted "duffy binding-like ligand" (DBL). A second motif common to all PfEMP-1 is the cysteine-rich interdomain region (CIDR) of which there are one or two copies termed α and β. The first CIDR, CIDR-1α is found immediately after DBL1α in most *var* genes and together they form a semiconserved head structure [8]. Adhesion to different host receptors has been located to distinct domains expressed by different var genes. For instance, CIDR1α binds to CD36 although not all CIDR1α harbor this property [9]. Apart from PfEMP-1, three additional highly variable families of *P. falciparum* proteins have been described: riffins, STEVOR and clag-9 [7]. Their functional properties are not well described yet.

31.3
Animal Models for Malaria

Several Plasmodium strains infecting the free living rodent species *Grammomys surdaster* and *Thamnomys rutilans* have been adapted to infect laboratory mice. None of these models can replicate all the aspects of disease as observed in human malaria but they are used to investigate specific aspects of the immune response to malaria parasites. Although iRBC of rodent Plasmodium species can sequester to a certain extent, particularly in the liver and the spleen, sequestration is not as pronounced as in Pf-iRBC and the parasite molecules that mediate adhesion have not been identified. However, like *P. falciparum*, rodent Plasmodium species encode variant surface proteins, which may be important for the evasion of humoral immune responses [10]. Immunity to blood-stage infection seems to vary with particular host–parasite combinations. For example, *P. yoelii* and *P. berghei* infections are not controlled in B-cell-depleted mice [11, 12] while *P.chabaudi chabaudi, P.chabaudi adami* and *P. vinckei* are partially controlled [13–16] suggesting that, although antibodies are critical for clearing infections, different parasite species vary in their ability to mount cellular-mediated mechanisms. While animal models provide valuable insights into immune response mechanisms during the course of one infection, they lack the extensive chronicity and exposure-related acquisition of immunity typical for human infections with *P. falciparum* . Therefore, studies in animal models have to be complemented by studies in human *P. falciparum* infections involving both *in vitro* and *ex vivo* systems coupled with immuno-epidemiological studies in endemic areas.

31.4
Acquired Immunity to Malaria

Despite decades of research, naturally acquired immunity to malaria is still poorly understood. *P. falciparum* infection rarely if ever induces sterile immunity but repeated infection eventually results into clinical immunity with moderate to no clinical symptoms and reduced mortality. In areas of intense, perennial transmission, the main load of malarial disease is experienced by children. Here, a certain degree of clinical immunity to severe, life-threatening malaria is apparent after a few disease episodes and is essentially complete by the age of five years [17]. However, the vast majority of adults living in endemic areas may have low levels of circulating parasites and yet few will have mild, if any, clinical symptoms. Many targets of both T-cell and B-cell responses are either polymorphic or variant antigens. Therefore, infection with different strains or expression of a different variant during chronic infection requires that a new immune response to the polymorphic or variant antigen is mounted. This will occur on the background of already existing immune responses and therefore take a different course from a primary infection. Protective immune responses in clinical immune individuals are the summary of responses to different polymorphic or variant antigens rather than a response to one particular antigen.

31.4.1
Immune Response to Liver Stages

It is not known whether sterile immunity to the liver stage of infection occurs under natural conditions although irradiated sporozoites can protect both mouse and man from challenge with homologous parasite strain and induce some cross protection against heterologous parasite strains. Evidence from mouse models suggests that liver-stage responses are primarily mediated by CD8 T cells. *In vivo* depletion of CD8 T cells completely abrogates protection, while adoptive transfer of CD8 T cells to naïve mice confers protection [18, 19]. Furthermore, β-2-microglobulin$^{-/-}$ mice, which lack MHC class I are not protected by either active immunization or passive transfer of wild type, splenic T cells from immune mice [20]. It is now well established that protection in mice induced by immunization with irradiated *P. berghei* [21, 22] or *P. yoelii* [23] sporozoites is absolutely dependent on CD8 T cells, IFN-γ and nitric oxide (NO). More recently, IL-12 and NK cells involvement in this mechanism has been demonstrated [24].

The liver stage of the *P. falciparum* infection is now an important target in experimental vaccines [19]. The efficacy of subunit vaccines is now being tested in challenge studies under experimental settings in the UK and the USA, as well as under natural transmission in endemic areas [25]. Some vaccines tested in nonimmune volunteers or in endemic areas showed reduction in infected hepatocytes, infection rates or disease severity at least for a short period of time [26, 27].

31.4.2
Cellular Immunity to the Erythrocytic Stage

During the erythrocytic stage, the malaria parasite spends most of its time in RBCs. RBC do not express appreciable levels of either MHC class I or class II and it is therefore unlikely that parasite antigens would be presented by iRBC to induce T-cell activation. However, antigen presentation by dendritic cells (DC) and macrophages will result in T-cell activation which will either have a direct or indirect effect on the control of parasitemia. Most of the available evidence for a role of T cells in blood-stage malaria comes from animal models and *in vitro* T-cell stimulations using peripheral blood mononuclear cells (PBMC) from people living in malaria endemic areas.

It is clear from murine malaria models that CD4 T cells provide help to B cells to make protective antibodies. B-cell-deficient mice failed to control *P. yoelii* parasitemia [11], and protection from challenge in naïve-irradiated mice could be transferred by a mixture of immune CD4 T and B cells [28]. In the *P. c. chabaudi* and *P. c. adami* models, infection can be controlled in mice depleted of B cells [14, 16, 29, 30]. During the acute phase, the responding T cells are predominantly Th1 which are thought to act by inducing cell-mediated parasiticidal mechanisms followed by a shift to a Th2 phenotype reflecting the importance of CD4 T-cell help to B cells in order to make antibodies to eliminate parasitemia [31, 32].

There is evidence that T cells from malaria-exposed donors proliferate or produce cytokines in response to malaria antigens (reviewed in [33]). A limited number of studies have reported *in vitro* data using T/B-cell cooperation assays suggesting that CD4 T cells give help to B cells resulting in antibody production [34, 35]. There is also limited evidence that CD4 T cells may have an effector role beyond giving help to B cells in malaria by inhibiting parasite growth *in vitro* [36].

Further evidence for the involvement of CD4 T cells in naturally-acquired immunity to malaria comes from epidemiological studies on the interaction between malaria and HIV-1. HIV-1 has been associated with increased frequency of clinical malaria and parasitemia in an adult cohort in Uganda [37, 38], and pregnant women in Malawi [39–41] and Kenya [42, 43]. In each of these studies, this association was more pronounced with increasing depletion of CD4 T cells. Collectively, these data underline a critical role for CD4 T cells in mechanisms that mediate antimalarial immunity during the asexual blood-stage cycle in people who are naturally exposed to endemic malaria.

31.4.3
Humoral Immunity to the Erythrocytic Stage

Perhaps the strongest evidence yet that antibody has an important anti-parasitic as well as an antidisease effect comes from the adoptive transfer of immune serum into malaria naïve individuals in the 1960s. Cohen and colleagues [44] purified γ-globulin from adult Gambians and transferred these preparations into Gambian children with high *P. falciparum* parasitemia. This treatment resolved fevers and reduced parasitemia in these children while nonimmune γ-globulin had no effect. Similar results were obtained in Nigeria in 1962 [45]. Both East African and Thai children could be treated with purified γ-globulin from immune West African adults, suggesting that either there are few regional differences in the distribution of parasite variants [46, 47] or that protective epitopes recognized by antibody are not strain specific.

Data from *in vitro* studies suggested that the protection seen in the passive transfer model was mediated by cytophilic antibodies that interact with monocytes. In an antibody-dependent cellular inhibition assay (ADCI), the protective IgG did not inhibit parasite growth and invasion *in vitro* when added alone to cultures, but did so when added in the presence of mononuclear cells from malaria naïve donors [48]. The protective and ADCI active sera were found to have high levels of the cytophilic antibodies IgG1 and IgG3 and relatively lower levels of IgG2 and IgM [49, 50]. IgG3 was associated with reduced frequency of malaria attacks in Senegal further strengthening the hypothesis that cytophilic IgG participates in protective mechanisms [51]. It has been suggested that the mechanism underlying ADCI involves the interaction of merozoites or Pf-iRBC with antibody and monocytes leading to the release of soluble mediator(s) responsible for parasite killing [52]. However, ADCI is not the only mechanism that could explain the massive reduction in parasitemia observed in the passive antibody transfer experiments described above. In contrast to the observation by Bouharoun-Tayoun (1990), other studies have demonstrated that immune IgG can inhibit parasite growth *in vitro* in the absence of adherent cells [53, 54].

Studies in children living in malaria endemic areas have demonstrated the presence of antibodies to variant surface antigens (VSA) of various *P. falciparum* isolates in their sera [55, 56]. Both the antigenic and functional properties of VSA can be largely attributed to PfEMP-1 (reviewed in [57]). In an early longitudinal study in the Gambia [56], the titer of VSA antibodies was shown to be the only one of a series of immune assays that was associated with subsequent protection against disease. In a large longitudinal study of surface antigens of *P. falciparum*-infected erythrocytes from Kenyan children, malaria tended to be caused by parasite isolates expressing VSA variants corresponding to gaps in the repertoire of antibodies carried by the children before they became ill [58], an observation which has been confirmed in other studies [59]. Thus, immunity is associated with piecemeal acquisition of a repertoire of variant specific antibodies [56, 58], which may contribute to the slow acquisition of protective immunity to falciparum malaria. It now emerges that antibody responses to a variety of antigens are often short-lived and can only be detected in the presence of asymptomatic infection. Importantly, children harboring parasites and antibodies are protected from disease during the next malaria season whereas children with antibodies but no parasites tend to be more susceptible [60–62]. Because children will move between these groups – asymptomatically infected or free of parasites – within a given dry season or between dry seasons, these results indicate that asymptomatic infection induces protective, but short-lived, immune responses. The mechanisms underlying these protective immune responses are not clear.

31.5
Immune Recognition of iRBC

During asexual blood-stage malaria, the host is exposed to a considerable amount of foreign antigen in the circulation. After rupture of iRBC, when merozoites leave the RBC, additional debris is released into the bloodstream, which often coincides with the induction of fever. Both, the iRBC itself and debris after schizont rupture are recognized by pattern recognition receptors on monocyte/macrophages and DCs.

31.5.1
Toll-like Receptors

In the last year, TLR-mediated recognition of iRBC has been extensively investigated. The earliest report showed that mice lacking the adaptor molecule MyD88 failed to induce IL-12 and were protected from T-cell-mediated liver injury during the intra-erythocytic stages of the parasite [63]. Now evidence is accumulating that

P. falciparum GPI bind to TLR2 and TLR4 and hemozoin binds to TLR9, respectively. *P. falciparum* schizont lysate contained a ligand for TLR9, which induced IFN-α production in peripheral blood plasmacytoid DCs [64]. It has recently been demonstrated, that the unknown TLR9 ligand is hemozoin [65]. Hemozoin is the crystallized form of heme, a by-product resulting from digestion of hemoglobin by the parasite in RBC. When mature schizonts rupture, hemozoin, together with other debris, is released and rapidly taken up by neutrophils, monocyte/macrophages and DCs. *P. falciparum* hemozoin or synthetic β-hematin activated both CD11c⁺B220⁻ myeloid and CD11c⁺B220⁺ plasmacytoid DC derived from bone marrow from wildtype, TLR2, TLR4 and TLR7 knock-out mice but not from TLR9 knock-out mice. Crosslinking of TLR9 by natural or synthetic hemozoin resulted in the secretion of pro-inflammatory cytokines such as TNF-α and IL-6.

Likewise, Plasmodium GPI has long been suspected to induce inflammatory signals. It has recently been shown that GPI bound to TLR2 and to a lesser extent to TLR4 in both mouse and human macrophages and induced TNF-α secretion [66, 67]. Mouse macrophages also produced IL-12, IL-6 and NO in response to GPI when they were first primed with IFN-γ. It is perceivable that GPI will also activate human myeloid DC via TLR2 and TLR4. It was noted that free GPI is very quickly inactivated *in vivo* by phospholipases in serum and on cell surfaces. This may explain why activation of myeloid cells by Plasmodium GPI has long been suspected but very difficult to prove.

31.5.2
CD36

At least in human, binding of *P. falciparum* iRBC to CD36 expressed on monocyte/macrophages and dendritic cells plays an important role in non-opsonic clearance of iRBC. Non-opsonic phagocytosis of iRBC is directly correlated with the expression levels of CD36 [68]. In monocyte/macrophages, this process did not result in the production of TNF-α. In falciparum malaria, ligation of CD36 is a direct consequence of PfEMP-1 mediated adhesion of iRBC. Although mature iRBC are sequestered in post-capillary venules, a proportion of mature iRBC can be found in the spleen where iRBC might be removed by macrophages and DC in the perifollicular zone and in the marginal zone. Indeed, parasitemia in the spleen tends to be higher than in the peripheral circulation [69]. Thus, adhesion to CD36 expressed on endothelial cells favors multiplication of the parasite while adhesion to CD36 on monocyte/macrophages favors the removal of iRBC without the induction of inflammatory responses. Adhesion to CD36 is functionally conserved, in that most iRBC isolated from individuals with falciparum malaria bind to CD36. In addition, iRBC from children with mild malarial disease show a higher avidity for CD36 than isolates from children with severe malaria [70, 71]. Together, these observations suggested that adhesion of iRBC to CD36 has a role beyond sequestration in the regulation of immune responses to the parasite.

The role of CD36 adhesion in rodent malaria is less clear. *P. falciparum* iRBC bind to CD36 expressed on mouse macrophages [72]. However, although one report showed that *P. chabaudi chabaudi* iRBC can bind to CD36, the evidence was indirect and awaits further investigation [73]. In mouse CD36 is expressed on monocyte/macrophages and on CD8⁺ DC in the T-cell areas of lymph nodes and spleen but not on DC in the marginal zone.

31.5.3
Other Scavenger Receptors

In malaria, the role of scavenger receptors has not been extensively studied. SR-AI/II knock-out mice showed a similar course of parasitemia when infected with *P. chabaudi* iRBC to wildtype mice. However, blocking of non-SR-AI/II macrophage receptors with poly (I) resulted in earlier peak parasitemia *in vivo* and reduced phagocytosis of iRBC *in vitro*, suggesting that CD36, MARCO or macrosilain are involved in non-opsonic clearance of iRBC. In addition, blocking of the mannose receptor *in vitro* resulted in reduced phagocytosis [74].

31.5.4
Complement and Fc Receptors

Complement receptors and Fc receptors mediate opsonin-dependent phagocytosis of iRBC by monocyte/macrophages and DCs. In this respect, Fc receptors but not complement receptors are critical for the control of parasitemia in Plasmodium infected mice [75, 76]. Their role for DC maturation, conventional antigen presentation or cross-presentation has not been studied in Plasmodium infection in mice or man.

31.6
Dendritic Cells in Malaria

31.6.1
DCs in Human Malaria

Using monocyte-derived DCs, we have shown that intact *P. falciparum*-infected erythrocytes modulate DC maturation and function [77, 78]. Parasite-modulated DCs failed to upregulate the expression of MHC, co-stimulatory and adhesion molecules in response to stimulation with LPS, CD40-ligand, TNF-α or monocyte-conditioned medium ([77] and unpublished observations). Subsequently, both naïve and memory T-cells co-cultured with parasite-modulated DCs were functionally unresponsive with respect to proliferation and secretion of IL-2. Parasite-modulated DCs secreted IL-10 rather than IL-12, a cytokine which could inhibit T-cell activation. Both control DC and parasite-modulated DC produced TNF-α. These effects were in principle dependent on the ability of Pf-iRBC to adhere to CD36 and seemed to be mediated by PfEMP-1. Lysate of iRBC allowed normal dendritic cell maturation but affected antigen-processing, possibly due to the ingestion of hemozoin [77]. In subsequent studies, similar modulation of DC had been shown with antibodies against CD36 [78] and MC-CIDR, a CD36 binding domain of one PfEMP-1 variant [Urban, unpublished observation].

A recent study by Skorokhod and colleagues showed that monocyte differentiation into DC is impaired in the presence of hemozoin [79]. In addition, DC co-cultured with hemozoin show an altered response to maturation signals. The impairment was accompanied by increased expression of the peroxisome proliferator-activated receptor-γ, upregulation of which is known to interfere with DC maturation [80, 81]. The effect on differentiation of monocytes appeared to be mediated by the same biochemical processes as hemozoin-induced changes in monocyte function. Hemozoin is not biochemically inert but reacts with membrane phospholipids and is transformed into hydroxy-polyunsaturated fatty acids, which cause membrane peroxidation [82, 83]. In addition, hemozoin catalysis induces the formation of prostaglandin PGE_2 and PGF_{2a}. While hydroxy-polyunsaturated fatty acids inhibit monocyte function such as phagocytosis, activation by inflammatory cytokines and generation of an oxidative burst, the release of PGE_2 and PGF_{2a} either by trophozoites or by monocytes, which have ingested pigment and/or trophozoites, alters T- and B-cell functions.

Plasmacytoid DC recently received attention, when a study by Pichyangkul *et al.* demonstrated that mature schizonts or lysate induced the expression of CD86 on and the secretion of IFN-α by plasmacytoid DCs *in vitro* [64]. These responses appear to be due to a soluble ligand of TLR9 in schizont lysate, most probably hemozoin. The same authors reported that the frequency of plasmacytoid DCs was reduced while the plasma levels of IFN-α were increased in Thai adults with both complicated and uncomplicated malaria.

Phenotypic analysis of DCs from patients who suffer from acute falciparum malaria can give some indication of DC function *in vivo*. In paraffin-fixed spleen sections from Vietnamese patients who died with falciparum malaria, we observed that myeloid DCs accumulate in the red pulp and in the marginal zone but not in the white pulp and T-cell areas. Furthermore, we noted a remarkable downregulation of HLA DR molecules on myeloid cells, including cordal macrophages and DCs, whereas HLA DR expression on sinusoidal lining cells was increased compared to controls [69]. This is in agreement with the observation that HLA DR expression is reduced on peripheral blood DCs in Kenyan children suffering from acute malaria [84].

31.6.2
DCs in Rodent Malaria

In contrast to human monocyte-derived DC, mouse bone-marrow derived DC incubated with the rodent parasite species *P. chabaudi chabaudi* readily increased the expression of co-stimulatory molecules and MHC molecules and produced inflammatory cytokines such as IL-12p70, TNF-α and IL-6. DC maturation in response to intact *P. chabaudi chabaudi* iRBC was further increased by the addition of LPS or TNF-α [85]. These observations were in agreement with a later study by Perry and colleagues, who purified CD11c+ DC from spleen of mice infected with *P. yoelii* on day 6 post-infection [86]. Splenic CD11c$^+$ DC showed increased surface expression of CD40 and CD80 and induced IL-2 production in T cells. Leisewitz et al. reported that CD11c$^+$ DC in the spleen of *P. chabaudi chabaudi* infected mice migrated from the marginal zone into the T-cell areas of the spleen as early as day 5 post-infection. In addition the surface expression of CD40 steadily increased from day 3 post-infection followed by CD54 and CD86 from day 7 onwards [87]. Together these results indicated that myeloid DCs are readily activated, migrate and secrete inflammatory cytokines when exposed to iRBC of rodent *Plasmodium* species either *in vitro* or *in vivo*.

However, other authors showed that bone-marrow derived DC failed to mature and to activate T-cells when co-cultured with *P. chabaudi chabaudi* or *P. yoelii* iRBC. In one study, these DC retained immunostimulatory capacity because mice injected with iRBC-loaded DC were protected from death when challenged with an otherwise lethal dose of *P. yoelii*. Nevertheless, these mice still experienced high levels of parasitemia [88]. It seems possible that in these experiments, antigen was transferred from injected, iRBC-loaded DCs to resident DCs [89]. Such a scenario might explain why apparently modulated DC allowed protection from death while affecting parasitemia only marginally: resident DC were not exposed to intact iRBC but to partially broken down antigen.

Ocana-Morgner and colleagues also observed that *P. yoelii* infected iRBC inhibited DC maturation *in vitro* in response to LPS and *in vivo*. In addition, the ratio of IL12 and IL10 secreted by DC was reversed. Subsequently, Ocana-Morgner showed, that blood-stage malaria inhibited the CD8$^+$ T-cell response to the liver-stage of the parasite. The mechanisms are not completely understood but most probably due to soluble factors secreted by DC [90].

31.7
Synopsis

Despite apparent heterogenous reports on DC function in rodent malaria, common patterns can be established (Table 31.1). *P. chabaudi* iRBC seem to be less prone to modulate DC function than *P. yoelii* iRBC. These differences could be due to parasite-encoded surface proteins involved in the interaction of iRBC with DC. Variant surface antigens of rodent Plasmodium species are now under active investigation, although a link between cytoadhesion and antigenic variation has not been established. In addition, the genetic background of mouse strains used in these experiments directly influences the kinetics and cellular composition of the immune response to a given parasite strain. This restriction may well have an effect on the responsiveness of DC to activation and determine the cytokine environment in which it occurs. In addition, DC modulation seems to be dependent on parasite inoculation rates or DC:iRBC ratio used in different experiments. Low ratios of DC to iRBC *in vitro* or low parasitemia *in vivo* seem to favor DC maturation, whereas high ratios *in vitro* or parasitemia *in vivo* appears to result in the inhibition of DC function. Whether this is due to active inhibition of DC function or a conse-

Tab. 31.1 DC function in rodent malaria.

Parasite	Assay	DC	Dose[a]	Para[b]	DC:iRBC[c]	Outcome	Ref.
P. chabaudi	*in vitro*	BM-DC			1:30	• mature iRBC induce surface expression of co-stimulatory molecules and secretion of cytokines	85
P. chabaudi	*in vivo*	CD11c[+]	10^5	>5%		• migration of CD11c[+] DC into the T-cell zones on day 5	87
				> 5%		• upregulation of CD40 on day 5	
				13%		• upregulation of CD54, CD86 on day 7	
P. chabaudi / P. yoelii	*in vitro*	BM-DC			1:10	• no upregulation of co-stimulatory molecules in response to LPS	88
	in vitro/ in vivo	BM-DC	10^6		1:10	• transfer of iRBC loaded BM-DC into mice protected from death but not parasitemia after subsequent challenge with a lethal dose of iRBC	
P. yoelii	*ex vivo*	CD11c[+]		13%		• DC upregulated surface expression of co-stimulatory cytokines • DC induced activation and cytokine secretion in T cells	86
P. yoelii	*in vitro/ in vivo*	BM-DC CD11c[+]	4 x 10^6		1:100	• iRBC inhibited increase in surface expression of co-stimulatory molecules and reversed ratio of IL12:IL10 secretion	90
	in vivo		4 x 10^6			• Inhibition of liver-stage specific CD8[+] T cells due to soluble factor secreted by DC	

[a] number of iRBC used for infection of mice where known
[b] parasitemia at the time of assay where known
[c] ration of DC to iRBC used in *in vitro* assays

quence of the resolution of inflammation will have to be established in more detailed studies.

In principle a similar effect of parasitemia on DC function should apply to falciparum malaria in humans. It may even be more pronounced, because CD36-mediated modulation of myeloid DC is a contact-dependent process. It suggests that early on during infection, plasmacytoid DC could be induced to secrete IFNα through engagement of TLR9 by hemozoin. In parallel, myeloid DC might be activated via ligation of TLR2 and 4 or cytokines such as IFN-α and IFN-γ secreted by plasmacytoid DCs, NK cells, NKT cells, secrete IL12 and other cytokines resulting in the activation of CD8⁺ and CD4⁺ T-cell responses. With increasing parasitemia, more and more myeloid DC in the spleen might be modulated either by direct interaction with iRBC or through ingestion of increasing amounts of hemozoin (Fig. 31.1). Modulation of myeloid DCs at best might prevent the induction of additional T-cell responses or at worst alter already existing T-cell responses or induce regulatory T cells. Given that severe malarial disease, at least in part, is mediated by inflammatory immune responses, modulation of DC might prove to be beneficial to the host rather than exuberating disease. Clearly, detailed analysis not only of DC function during the course of Plasmodium infection but also the consequences for T- and B-cell responses are required both in human falciparum malaria as well as in rodent models of infection.

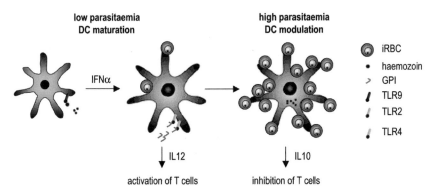

Fig. 31.1 Simplified diagram on the effect of parasitemia on DC function. Early on during infection, engagement of TLR9 by hemozoin results induces plasmacytoid DC to secrete IFNα and engagement of TLR2 and TLR4 by GPI induces myeloid DC to secrete IL12. With increasing parasitemia, more and more myeloid DC in the spleen might be modulated either directly through interaction with iRBC or through ingestion of increasing amounts of hemozoin and secrete IL10.

Acknowledgments

BCU holds a Wellcome Trust Career Development Fellowship in Basic Biomedical Science. FMN is supported by the European Union Framework 6 Network of Excellence, BioMalPar.

References

1 Snow, R. W., Craig, M., Deichmann, U., Marsh, K., Estimating mortality, morbidity and disability due to malaria among Africa's non-pregnant population. *Bull World Health Organ* **1999**. 77: 624–640.

2 Gupta, S., Snow, R. W., Donnelly, C., Newbold, C., Acquired immunity and postnatal clinical protection in childhood cerebral malaria. *Proc R Soc Lond B Biol Sci* **1999**. 266: 33–38.

3 Phillips, R. S., Current status of malaria and potential for control. *Clin Microbiol Rev* **2001**. 14: 208–226.

4 Sachs, J., Malaney, P., The economic and social burden of malaria. *Nature* **2002**. 415: 680–685.

5 Lycett, G. J., Kafatos, F. C., Anti-malarial mosquitoes? *Nature* **2002**. 417: 387–388.

6 Kyes, S., Horrocks, P., Newbold, C., Antigenic variation at the infected red cell surface in malaria. *Annu Rev Microbiol* **2001**. 55: 673–707.

7 Gardner, M. J., Hall, N., Fung, E., White, O., Berriman, M., Hyman, R. W., Carlton, J. M., Pain, A., Nelson, K. E., Bowman, S., Paulsen, I. T., James, K., Eisen, J. A., Rutherford, K., Salzberg, S. L., Craig, A., Kyes, S., Chan, M. S., Nene, V., Shallom, S. J., Suh, B., Peterson, J., Angiuoli, S., Pertea, M., Allen, J., Selengut, J., Haft, D., Mather, M. W., Vaidya, A. B., Martin, D. M., Fairlamb, A. H., Fraunholz, M. J., Roos, D. S., Ralph, S. A., McFadden, G. I., Cummings, L. M., Subramanian, G. M., Mungall, C., Venter, J. C., Carucci, D. J., Hoffman, S. L., Newbold, C., Davis, R. W., Fraser, C. M., Barrell, B., Genome sequence of the human malaria parasite *Plasmodium falciparum*. *Nature* **2002**. 419: 498–511.

8 Chen, Q., Heddini, A., Barragan, A., Fernandez, V., Pearce, S. F., Wahlgren, M., The semiconserved head structure of *Plasmodium falciparum* erythrocyte membrane protein 1 mediates binding to multiple independent host receptors. *J Exp Med* **2000**. 192: 1–10.

9 Kraemer, S. M., Smith, J. D., Evidence for the importance of genetic structuring to the structural and functional specialization of the *Plasmodium falciparum* var gene family. *Mol Microbiol* **2003**. 50: 1527–1538.

10 Janssen, C. S., Barrett, M. P., Turner, C. M., Phillips, R. S., A large gene family for putative variant antigens shared by human and rodent malaria parasites. *Proc R Soc Lond B Biol Sci* **2002**. 269: 431–436.

11 Weinbaum, F. I., Weintraub, J., Nkrumah, F. K., Evans, C. B., Tigelaar, R. E., Rosenberg, Y. J., Immunity to *Plasmodium berghei yoelii* in mice. II. Specific and nonspecific cellular and humoral responses during the course of infection. *J Immunol* **1978**. 121: 629–636.

12 Roberts, D. W., Weidanz, W. P., T-cell immunity to malaria in the B-cell deficient mouse. *Am J Trop Med Hyg* **1979**. 28: 1–3.

13 Cavacini, L. A., Parke, L. A., Weidanz, W. P., Resolution of acute malarial infections by T cell-dependent non-antibody-mediated mechanisms of immunity. *Infect Immun* **1990**. 58: 2946–2950.

14 Langhorne, J., Cross, C., Seixas, E., Li, C., von der Weid, T., A role for B cells in the development of T cell helper function in a malaria infection in mice. *Proc Natl Acad Sci USA* **1998**. 95: 1730–1734.

15 van der Heyde, H. C., Elloso, M. M., Chang, W. L., Pepper, B. J., Batchelder, J., Weidanz, W. P., Expansion of the gammadelta T cell subset in vivo during bloodstage malaria in B cell-deficient mice. *J Leukoc Biol* **1996**. 60: 221–229.

16 von der Weid, T., Langhorne, J., Altered response of CD4+ T cell subsets to *Plasmodium chabaudi chabaudi* in B cell-deficient mice. *Int Immunol* **1993**. 5: 1343–1348.

17 Gupta, S., Snow, R. W., Donnelly, C. A., Marsh, K., Newbold, C., Immunity to non-cerebral severe malaria is acquired after one or two infections. *Nat Med* **1999**. 5: 340–343.

18 Good, M. F., Berzofsky, J. A., Miller, L. H., The T cell response to the malaria circumsporozoite protein: an immunological approach to vaccine development. *Annu Rev Immunol* **1988**. 6: 663–688.

19 Hoffman, S., Franke, E. D., Hollingdale, M., Druilhe, P. (Eds.), *Attacking the infected erythrocyte*. American Society for MIcrobiology Press, Washington DC **1996**.

20 White, K. L., Snyder, H. L., Krzych, U., MHC class I-dependent presentation of exoerythrocytic antigens to CD8+ T lymphocytes is required for protective immunity against *Plasmodium berghei*. *J Immunol* **1996**. 156: 3374–3381.

21 Seguin, M. C., Klotz, F. W., Schneider, I., Weir, J. P., Goodbary, M., Slayter, M., Raney, J. J., Aniagolu, J. U., Green, S. J., Induction of nitric oxide synthase protects against malaria in mice exposed to irradiated *Plasmodium berghei* infected mosquitoes: involvement of interferon gamma and CD8+ T cells. *J Exp Med* **1994**. 180: 353–358.

22 Schofield, L., Villaquiran, J., Ferreira, A., Schellekens, H., Nussenzweig, R., Nussenzweig, V., Gamma interferon, CD8+ T cells and antibodies required for immunity to malaria sporozoites. *Nature* **1987**. 330: 664–666.

23 Doolan, D. L., Hoffman, S. L., Pre-erythrocytic-stage immune effector mechanisms in Plasmodium spp. infections. *Philos Trans R Soc Lond B Biol Sci* **1997**. 352: 1361–1367.

24 Hoffman, S. L., Crutcher, J. M., Puri, S. K., Ansari, A. A., Villinger, F., Franke, E. D., Singh, P. P., Finkelman, F., Gately, M. K., Dutta, G. P., Sedegah, M., Sterile protection of monkeys against malaria after administration of interleukin-12. *Nat Med* **1997**. 3: 80–83.

25 Ballou, W. R., Arevalo-Herrera, M., Carucci, D., Richie, T. L., Corradin, G., Diggs, C., Druihle, P., Giersing, B. K., Saul, A., Heppner, D. G., Kester, K. E., Lanar, D. E., Lyon, J., Hill, A. V. S., Pan, W., Cohen, J. D., Update on the clinical development of candidate malaria vaccines. *Am J Trop Med Hyg* **2004**. 71: 239–247.

26 Bejon, P., Andrews, L., Andersen, R. F., Dunachie, S., Webster, D., Walther, M., Gilbert, S. C., Peto, T., Hill, A. V., Calculation of Liver-to-Blood Inocula, Parasite Growth Rates, and Preerythrocytic Vaccine Efficacy, from Serial Quantitative Polymerase Chain Reaction Studies of Volunteers Challenged with Malaria Sporozoites. *J Infect Dis* **2005**. 191: 619–626.

27 Alonso, P. L., Sacarlal, J., Aponte, J. J., Leach, A., Macete, E., Milman, J., Mandomando, I., Spiessens, B., Guinovart, C., Espasa, M., Bassat, Q., Aide, P., Ofori-Anyinam, O., Navia, M. M., Corachan, S., Ceuppens, M., Dubois, M. C., Demoitie, M. A., Dubovsky, F., Menendez, C., Tornieporth, N., Ballou, W. R., Thompson, R., Cohen, J., Efficacy of the RTS,S/AS02A vaccine against *Plasmodium falciparum* infection and disease in young African children: randomised controlled trial. *Lancet* **2004**. 364: 1411–1420.

28 Jayawardena, A. N., Murphy, D. B., Janeway, C. A., Gershon, R. K., T cell-mediated immunity in malaria. I. The Ly phenotype of T cells mediating resistance to *Plasmodium yoelii*. *J Immunol* **1982**. 129: 377–381.

29 Grun, J. L., Weidanz, W. P., Immunity to *Plasmodium chabaudi adami* in the B-cell-deficient mouse. *Nature* **1981**. 290: 143–145.

30 von der Weid, T., Honarvar, N., Langhorne, J., Gene-targeted mice lacking B cells are unable to eliminate a blood stage malaria infection. *J Immunol* **1996**. 156: 2510–2516.

31 Stevenson, M. M., Tam, M. F., Differential induction of helper T cell subsets during blood-stage *Plasmodium chabaudi* AS infection in resistant and susceptible mice. *Clin Exp Immunol* **1993**. 92: 77–83.

32 Langhorne, J., Gillard, S., Simon, B., Slade, S., Eichmann, K., Frequencies of CD4+ T cells reactive with *Plasmodium chabaudi chabaudi*: distinct response kinetics for cells with Th1 and Th2 characteristics during infection. *Int Immunol* **1989**. 1: 416–424.

33 Troye-Blomberg, M., Human T-cell responses to blood stage antigens in *Plasmodium falciparum* malaria. *Immunol Lett* **1994**. 41: 103–107.

34 Kabilan, L., Troye-Blomberg, M., Patarroyo, M. E., Bjorkman, A., Perlmann, P., Regulation of the immune response in *Plasmodium falciparum* malaria: IV. T cell dependent production of immunoglobulin and anti-P. falciparum antibodies in vitro. *Clin Exp Immunol* **1987**. 68: 288–297.

35 Chougnet, C., Troye-Blomberg, M., Deloron, P., Kabilan, L., Lepers, J. P., Savel, J., Perlmann, P., Human immune response in *Plasmodium falciparum* malaria. Synthetic peptides corresponding to known epitopes of the Pf155/RESA antigen induce production of parasite-specific antibodies in vitro. *J Immunol* **1991**. 147: 2295–2301.

36 Fell, A. H., Currier, J., Good, M. F., Inhibition of *Plasmodium falciparum* growth in vitro by CD4+ and CD8+ T cells from non-exposed donors. *Parasite Immunol* **1994**. 16: 579–586.

37 Whitworth, J., Morgan, D., Quigley, M., Smith, A., Mayanja, B., Eotu, H., Omoding, N., Okongo, M., Malamba, S., Ojwiya, A., Effect of HIV-1 and increasing immunosuppression on malaria parasitaemia and clinical episodes in adults in rural Uganda: a cohort study. *Lancet* **2000**. 356: 1051–1056.

38 French, N., Nakiyingi, J., Lugada, E., Watera, C., Whitworth, J. A., Gilks, C. F., Increasing rates of malarial fever with deteriorating immune status in HIV-1-infected Ugandan adults. *Aids* **2001**. 15: 899–906.

39 Steketee, R. W., Wirima, J. J., Bloland, P. B., Chilima, B., Mermin, J. H., Chitsulo, L., Breman, J. G., Impairment of a pregnant woman's acquired ability to limit *Plasmodium falciparum* by infection with human immunodeficiency virus type-1. *Am J Trop Med Hyg* **1996**. 55: 42–49.

40 Mount, A. M., Mwapasa, V., Elliott, S. R., Beeson, J. G., Tadesse, E., Lema, V. M., Molyneux, M. E., Meshnick, S. R., Rogerson, S. J., Impairment of humoral immunity to *Plasmodium falciparum* malaria in pregnancy by HIV infection. *Lancet* **2004**. 363: 1860–1867.

41 Mwapasa, V., Rogerson, S. J., Molyneux, M. E., Abrams, E. T., Kamwendo, D. D., Lema, V. M., Tadesse, E., Chaluluka, E., Wilson, P. E., Meshnick, S. R., The effect of *Plasmodium falciparum* malaria on peripheral and placental HIV-1 RNA concentrations in pregnant Malawian women. *Aids* **2004**. 18: 1051–1059.

42 Moore, J. M., Ayisi, J., Nahlen, B. L., Misore, A., Lal, A. A., Udhayakumar, V., Immunity to placental malaria. II. Placental antigen-specific cytokine responses are impaired in human immunodeficiency virus-infected women. *J Infect Dis* **2000**. 182: 960–964.

43 Chaisavaneeyakorn, S., Moore, J. M., Otieno, J., Chaiyaroj, S. C., Perkins, D. J., Shi, Y. P., Nahlen, B. L., Lal, A. A., Udhayakumar, V., Immunity to placental malaria. III. Impairment of interleukin(IL)-12, not IL-18, and interferon-inducible protein-10 responses in the placental intervillous blood of human immunodeficiency virus/malaria-coinfected women. *J Infect Dis* **2002**. 185: 127–131.

44 Cohen, S., McGregor, I. A., Carrington, S., Gamma-globulin and acquired immunity to human malaria. *Nature* **1961**. 192: 733–737.

45 Edozien, J. C., Gilles, H. M., Udeozo, I. O. K., Adult and cord blood gamma-globulin and immunity to malaria in Nigerians. *Lancet* **1962**. 2: 951–955.

46 Jongwutiwes, S., Tanabe, K., Hughes, M. K., Kanbara, H., Hughes, A. L., Allelic variation in the circumsporozoite protein of *Plasmodium falciparum* from Thai field isolates. *Am J Trop Med Hyg* **1994**. 51: 659–668.

47 Conway, D. J., Greenwood, B. M., McBride, J. S., Longitudinal study of *Plasmodium falciparum* polymorphic antigens in a malaria-endemic population. *Infect Immun* **1992**. 60: 1122–1127.

48 Bouharoun-Tayoun, H., Attanath, P., Sabcharoen, A., Chongsuphajaisiddhi, T., Druilhe, P., Antibodies that protect humans against *Plasmodium falciparum* blood stages do not on their own inhibit parasite growth and invasion in vitro, but act in cooperation with monocytes. *J Exp Med* **1990**. 172: 1633–1641.

49 Bouharoun-Tayoun, H., Druilhe, P., *Plasmodium falciparum* malaria: evidence for an isotype imbalance which may be responsible for delayed acquisition of protective immunity. *Infect Immun* **1992**. 60: 1473–1481.

50 Bouharoun-Tayoun, H., Druilhe, P., Antibodies in falciparum malaria: what matters most, quantity or quality? *Mem Inst Oswaldo Cruz* **1992**. 87 Suppl 3: 229–234.

51 Aribot, G., Rogier, C., Sarthou, J. L., Trape, J. F., Balde, A. T., Druilhe, P., Roussilhon, C., Pattern of immunoglobulin isotype response to *Plasmodium falciparum* blood-stage antigens in

individuals living in a holoendemic area of Senegal (Dielmo, west Africa). *Am J Trop Med Hyg* **1996**. 54: 449–457.

52 Bouharoun-Tayoun, H., Oeuvray, C., Lunel, F., Druilhe, P., Mechanisms underlying the monocyte-mediated antibody-dependent killing of *Plasmodium falciparum* asexual blood stages. *J Exp Med* **1995**. 182: 409–418.

53 Brown, G. V., Anders, R. F., Mitchell, G. F., Heywood, P. F., Target antigens of purified human immunoglobulins which inhibit growth of Plasmodium falciparum in vitro. *Nature* **1982**. 297: 591–593.

54 Cohen, S., Butcher, G. A., Crandall, R. B., Action of malarial antibody in vitro. *Nature* **1969**. 223: 368–371.

55 Marsh, K., Howard, R. J., Antigens induced on erythrocytes by *P. falciparum*: expression of diverse and conserved determinants. *Science* **1986**. 231: 150–153.

56 Marsh, K., Otoo, L., Hayes, R. J., Carson, D. C., Greenwood, B. M., Antibodies to blood stage antigens of *Plasmodium falciparum* in rural Gambians and their relation to protection against infection. *Trans R Soc Trop Med Hyg* **1989**. 83: 293–303.

57 Bull, P. C., Marsh, K., The role of antibodies to *Plasmodium falciparum*-infected-erythrocyte surface antigens in naturally acquired immunity to malaria. *Trends Microbiol* **2002**. 10: 55–58.

58 Bull, P. C., Lowe, B. S., Kortok, M., Molyneux, C. S., Newbold, C. I., Marsh, K., Parasite antigens on the infected red cell surface are targets for naturally acquired immunity to malaria. *Nat Med* **1998**. 4: 358–360.

59 Dodoo, D., Staalsoe, T., Giha, H., Kurtzhals, J. A., Akanmori, B. D., Koram, K., Dunyo, S., Nkrumah, F. K., Hviid, L., Theander, T. G., Antibodies to variant antigens on the surfaces of infected erythrocytes are associated with protection from malaria in Ghanaian children. *Infect Immun* **2001**. 69: 3713–3718.

60 Kinyanjui, S. M., Mwangi, T., Bull, P. C., Newbold, C. I., Marsh, K., Protection against clinical malaria by heterologous immunoglobulin G antibodies against malaria-infected erythrocyte variant surface antigens requires interaction with asymptomatic infections. *J Infect Dis* **2004**. 190: 1527–1533.

61 Bull, P. C., Lowe, B. S., Kaleli, N., Njuga, F., Kortok, M., Ross, A., Ndungu, F., Snow, R. W., Marsh, K., Plasmodium falciparum infections are associated with agglutinating antibodies to parasite-infected erythrocyte surface antigens among healthy Kenyan children. *J Infect Dis* **2002**. 185: 1688–1691.

62 Polley, S. D., Mwangi, T., Kocken, C. H. M., Thomas, A. W., Dutta, S., Lanar, D. E., Remarque, E., Ross, A., Williams, T. N., Mwambingu, G., Human antibodies to recombinant protein constructs of *Plasmodium falciparum* Apical Membrane Antigen 1 (AMA1) and their associations with protection from malaria. *Vaccine* **2004**. 23: 718–728.

63 Adachi, K., Tsutsui, H., Kashiwamura, S.-I., Seki, E., Nakano, H., Takeuchi, O., Takeda, K., Okumura, K., Van Kaer, L., Okamura, H., Akira, S., Nakanishi, K., *Plasmodium berghei* Infection in Mice Induces Liver Injury by an IL-12- and Toll-Like Receptor/Myeloid Differentiation Factor 88-Dependent Mechanism. *J Immunol* **2001**. 167: 5928–5934.

64 Pichyangkul, S., Yongvanitchit, K., Kum-arb, U., Hemmi, H., Akira, S., Krieg, A. M., Heppner, D. G., Stewart, V. A., Hasegawa, H., Looareesuwan, S., Shanks, G. D., Miller, R. S., Malaria Blood Stage Parasites Activate Human Plasmacytoid Dendritic Cells and Murine Dendritic Cells through a Toll-Like Receptor 9-Dependent Pathway. *J Immunol* **2004**. 172: 4926–4933.

65 Coban, C., Ishii, K. J., Kawai, T., Hemmi, H., Sato, S., Uematsu, S., Yamamoto, M., Takeuchi, O., Itagaki, S., Kumar, N., Horii, T., Akira, S., Toll-like receptor 9 mediates innate immune activation by the malaria pigment hemozoin. *J Exp Med* **2005**. 201: 19–25.

66 Zhu, J., Krishnegowda, G., Gowda, D. C., Induction of proinflammatory responses in macrophages by the glycosylphosphatidylinositols (GPIs) of *Plasmodium falciparum*: The requirement of ERK, p38, JNK and NF-kappa B pathways for the expression of proinflammatory cytokines and nitric oxide. *J Biol Chem* **2004**: M413539200.

67 Krishnegowda, G., Hajjar, A. M., Zhu, J., Douglass, E. J., Uematsu, S., Akira, S.,

Woods, A. S., Gowda, D. C., Induction of proinflammatory responses in macrophages by the glycosylphosphatidylinositols (GPIs) of *Plasmodium falciparum*: Cell signaling receptors, GPI structural requirement, and regulation of GPI activity. *J Biol Chem* **2004**: M413541200.

68 McGilvray, I. D., Serghides, L., Kapus, A., Rotstein, O. D., Kain, K. C., Nonopsonic monocyte/macrophage phagocytosis of *Plasmodium falciparum*-parasitized erythrocytes: a role for CD36 in malarial clearance. *Blood* **2000**. 96: 3231–3240.

69 Urban, B. C., Hien, T. T., Day, N. P., Phu, N. H., Roberts, R., Pongponratn, E., Jones, M., Mai, N. T., Bethell, D., Turner, G. D. H., Ferguson, D. J., White, N. J., Roberts, D. J., Fatal *Plasmodium falciparum* malaria causes specific patterns of splenic architectural disorganisation. *Infect Immun* **2005**. 73: 1986–1994.

70 Newbold, C., Warn, P., Black, G., Berendt, A., Craig, A., Snow, B., Msobo, M., Peshu, N., Marsh, K., Receptor-specific adhesion and clinical disease in *Plasmodium falciparum*. *Am J Trop Med Hyg* **1997**. 57: 389–398.

71 Rogerson, S. J., Tembenu, R., Dobano, C., Plitt, S., Taylor, T. E., Molyneux, M. E., Cytoadherence characteristics of *Plasmodium falciparum*-infected erythrocytes from Malawian children with severe and uncomplicated malaria. *Am J Trop Med Hyg* **1999**. 61: 467–472.

72 Patel, S. N., Serghides, L., Smith, T. G., Febbraio, M., Silverstein, R. L., Kurtz, T. W., Pravenec, M., Kain, K. C., CD36 mediates the phagocytosis of *Plasmodium falciparum*-infected erythrocytes by rodent macrophages. *J Infect Dis* **2004**. 189: 204–213.

73 Mota, M. M., Jarra, W., Hirst, E., Patnaik, P. K., Holder, A. A., *Plasmodium chabaudi*-infected erythrocytes adhere to CD36 and bind to microvascular endothelial cells in an organ-specific way. *Infect Immun* **2000**. 68: 4135–4144.

74 Su, Z., Fortin, A., Gros, P., Stevenson, M. M., Opsonin-independent phagocytosis: an effector mechanism against acute blood-stage *Plasmodium chabaudi* AS infection. *J Infect Dis* **2002**. 186: 1321–1329.

75 Taylor, P. R., Seixas, E., Walport, M. J., Langhorne, J., Botto, M., Complement Contributes to Protective Immunity against Reinfection by *Plasmodium chabaudi chabaudi* Parasites. *Infect Immun* **2001**. 69: 3853–3859.

76 Yoneto, T., Waki, S., Takai, T., Tagawa, Y.-i., Iwakura, Y., Mizuguchi, J., Nariuchi, H., Yoshimoto, T., A Critical Role of Fc Receptor-Mediated Antibody-Dependent Phagocytosis in the Host Resistance to Blood-Stage *Plasmodium berghei* XAT Infection. *J Immunol* **2001**. 166: 6236–6241.

77 Urban, B. C., Ferguson, D. J., Pain, A., Willcox, N., Plebanski, M., Austyn, J. M., Roberts, D. J., *Plasmodium falciparum*-infected erythrocytes modulate the maturation of dendritic cells. *Nature* **1999**. 400: 73–77.

78 Urban, B. C., Willcox, N., Roberts, D. J., A role for CD36 in the regulation of dendritic cell function. *Proc Natl Acad Sci USA* **2001**. 98: 8750–8755.

79 Skorokhod, O. A., Alessio, M., Mordmuller, B., Arese, P., Schwarzer, E., Hemozoin (malarial pigment) inhibits differentiation and maturation of human monocyte-derived dendritic cells: a peroxisome proliferator-activated receptor-gamma-mediated effect. *J Immunol* **2004**. 173: 4066–4074.

80 Angeli, V., Hammad, H., Staels, B., Capron, M., Lambrecht, B. N., Trottein, F., Peroxisome proliferator-activated receptor gamma inhibits the migration of dendritic cells: consequences for the immune response. *J Immunol* **2003**. 170: 5295–5301.

81 Nencioni, A., Grunebach, F., Zobywlaski, A., Denzlinger, C., Brugger, W., Brossart, P., Dendritic cell immunogenicity is regulated by peroxisome proliferator-activated receptor gamma. *J Immunol* **2002**. 169: 1228–1235.

82 Schwarzer, E., Turrini, F., Ulliers, D., Giribaldi, G., Ginsburg, H., Arese, P., Impairment of macrophage functions after ingestion of *Plasmodium falciparum*-infected erythrocytes or isolated malarial pigment. *J Exp Med* **1992**. 176: 1033–1041.

83 Schwarzer, E., Kuhn, H., Valente, E., Arese, P., Malaria-parasitized erythrocytes and hemozoin nonenzymatically generate

large amounts of hydroxy fatty acids that inhibit monocyte functions. *Blood* **2003**. 101: 722–728.

84 Urban, B. C., Mwangi, T., Ross, A., Kinyanjui, S., Mosobo, M., Kai, O., Lowe, B., Marsh, K., Roberts, D. J., Peripheral blood dendritic cells in children with acute *Plasmodium falciparum* malaria. *Blood* **2001**. 98: 2859–2861.

85 Seixas, E., Cross, C., Quin, S., Langhorne, J., Direct activation of dendritic cells by the malaria parasite, Plasmodium chabaudi chabaudi. *Eur J Immunol* **2001**. 31: 2970–2978.

86 Perry, J. A., Rush, A., Wilson, R. J., Olver, C. S., Avery, A. C., Dendritic cells from malaria-infected mice are fully functional APC. *J Immunol* **2004**. 172: 475–482.

87 Leisewitz, A. L., Rockett, K. A., Gumede, B., Jones, M., Urban, B., Kwiatkowski, D.

P., Response of the splenic dendritic cell population to malaria infection. *Infect Immun* **2004**. 72: 4233–4239.

88 Pouniotis, D. S., Proudfoot, O., Bogdanoska, V., Apostolopoulos, V., Fifis, T., Plebanski, M., Dendritic cells induce immunity and long-lasting protection against blood-stage malaria despite an in vitro parasite-induced maturation defect. *Infect Immun* **2004**. 72: 5331–5339.

89 Carbone, F. R., Belz, G. T., Heath, W. R., Transfer of antigen between migrating and lymph node-resident DCs in peripheral T-cell tolerance and immunity. *Trends in Immunology* **2004**. 25: 655–658.

90 Ocana-Morgner, C., Mota, M. M., Rodriguez, A., Malaria blood stage suppression of liver stage immunity by dendritic cells. *J Exp Med* **2003**. 197: 143–151.

32

Dendritic Cells in Leishmaniasis: Regulators of Immunity and Tools for New Immune Intervention Strategies

Heidrun Moll

32.1
Introduction

Leishmaniasis is considered a tropical affliction that is included in the World Health Organization/Tropical Disease Research list of the six most important diseases. Disability-adjusted life years (DALYs) lost due to leishmaniasis are close to 2.4 million, the world-wide incidence is estimated to be 12–15 million cases, and a population of 350 million is at risk [1]. There has been a sharp increase in the number of recorded cases in recent years. For example, the current epidemic in Afghanistan affects hundreds of thousands of people. Leishmaniasis is endemic in Central and South America, India and the Middle East, but also in southern European countries, such as Portugal, Spain, Greece and Italy. There is an increased interest in leishmaniasis in industrialized countries, due to the importance of travel medicine and the rising incidence of human immunodeficiency virus (HIV) and *Leishmania* co-infections. Especially in south-western Europe, the mutual reinforcement associated with HIV and *Leishmania* co-infection is considered a real threat: leishmaniasis accelerates the onset of AIDS and shortens the life expectancy of HIV-infected people, and infection with HIV can increase the risk of leishmaniasis by 100–1000 times in endemic areas [1, 2].

In addition to these clinical aspects, the interest in leishmaniasis is based on its importance as a model to define the factors controlling the development of polarized T helper (Th)1 and Th2 cell responses. In fact, the relevance of the Th1/Th2 cell balance *in vivo* to the outcome of a disease was first documented in the model of murine leishmaniasis with *Leishmania major* [3, 4]. This experimental system has provided a wealth of information on the immunological mechanisms leading to the restriction or facilitation of pathogen growth, with implications not only for infectious diseases but also for general aspects of immunoregulation.

Leishmaniasis is caused by protozoa of the genus *Leishmania* which are transmitted by sand flies. The parasites alternate between the flagellated promastigote form in the insect vector and the obligatory intracellular amastigote form in the mammalian host. They induce a group of diseases (Table 32.1) that vary in severity

Tab. 32.1   Spectrum of diseases induced by *Leishmania* parasites.

Disease	CL	DCL	MCL	VL
Major parasite species	*L. major* *L. tropica* *L. mexicana* *L. aethiopica*	*L. mexicana* *L. aethiopica*	*L. amazonensis* *L. brasiliensis*	*L. donovani* *L. infantum* *L. chagasi*
Major clinical symptoms	localized skin lesions	disseminating skin lesions	destruction of oro-nasal and pharyngeal mucosal tissue	fever, weight loss, malaise, hepato-splenomegaly, anemia
Disease outcome	self-healing	chronic, progressive	chronic, progressive	progressive, may be fatal if untreated

CL, cutaneous leishmaniasis; DCL, diffuse cutaneous leishmaniasis; MCL, mucocutaneous leishmaniasis; VL, visceral leishmaniasis.

from the self-healing cutaneous leishmaniasis (oriental sore), characterized by a localized skin lesion at the site of the sand fly's bite, to potentially fatal visceral disease (kala-azar), in which the parasites disseminate and invade lymph nodes, spleen, liver and bone marrow [5]. The clinical manifestation depends not only on the parasite species, but also involves the genetic basis of the host's ability to mount an effective cell-mediated immune response, thus resembling the situation in leprosy. In contrast to viral and bacterial infections, no vaccines are available to protect humans from parasitic diseases including leishmaniasis and, therefore, control measures rely exclusively on chemotherapy. The current treatments for leishmaniasis are unsatisfactory due to their toxic side effects, expense and the increasing problems with drug resistance. Thus, there is an urgent need to develop novel strategies for the prevention and treatment of leishmaniasis and other parasite infections. A thorough understanding of the complex immune mechanisms resulting in resistance or pathology is a prerequisite for the elaboration of new approaches to be used for vaccination and immunotherapy.

32.2
Mechanisms Mediating Resistance or Susceptibility to Leishmaniasis

Leishmania parasites that are pathogenic to man also infect mice, and the spectrum of diseases seen in humans can be mimicked by infection of different strains of inbred mice with *L. major*, a cause of human cutaneous leishmaniasis. Murine infection with *L. major* parasites is perhaps the disease model that has been used most widely to study the cell populations and cytokines involved in host resistance or susceptibility to a microbial pathogen. A plethora of data demonstrated that the course of disease depends on the type of the host's immune response (Fig. 32.1).

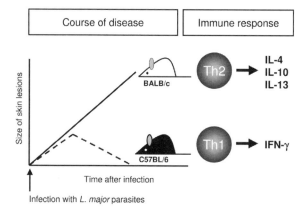

Fig. 32.1 Th-cell polarization correlates with resistance or susceptibility to experimental leishmaniasis. After infection with *L. major*, a cause of human cutaneous leishmaniasis, mice of the inbred strain C57BL/6 develop skin lesions that resolve spontaneously and lead to the generation of life-long immunity to re-infection, whereas BALB/c mice develop progressive lesions with systemic disease and fatal outcome. The ability of C57BL/6 mice to control *L. major* infection is associated with the development of Th1 cells producing IFN-γ. In contrast, the development of Th2 cells secreting IL-4, IL-10 and IL-13 predominates in susceptible BALB/c mice.

Protective immunity is associated with the development of a Th1 response, while a Th2 cytokine pattern prevails in mice that succumb to infection [6]. Interleukin (IL) 12 is considered to have a central regulatory function in directing the development of protective Th1 cells. These findings were paradigmatic for many infectious diseases.

32.2.1
The Role of T Helper Cell Subsets

Experimental infection of mice with *L. major* is an excellent model to explore the development and functions of Th-cell subsets *in vivo*. Most strains of inbred mice, such as C57BL/6 and C3H, are able to restrain the disease, with skin lesions resolving spontaneously. In contrast, mice of some other strains, such as BALB/c, are highly susceptible to infection, with parasites disseminating to the viscera and fatal outcome. The healing of lesions in resistant mice is linked with the induction of Th1 cells secreting interferon-γ (IFN-γ). This cytokine stimulates the parasiticidal activity of macrophages via expression of inducible nitric oxide synthase (iNOS) and formation of reactive nitrogen metabolites, resulting in the killing of the intracellular *Leishmania*. The inability of BALB/c mice to control *L. major* infection is associated with early IL-4 production by CD4+ T cells and the development of a sustained Th2 response. This Th2 dominance suppresses the development of Th1 cells in susceptible mice. However, it is important to note that an early, although transient, production of IL-4 can also be detected in *L. major*-infected mice of resistant strains [7–9]. Thus, a critical factor determining resistance is the ability to redirect the early IL-4 response following infection to a protective Th1 pathway.

The development of a Th1 response and resistance to leishmaniasis is driven by IL-12. Depletion of this cytokine by genetic means or antibody neutralization abrogates the ability of resistant mice to clear the infection, while treatment of susceptible mice with IL-12 induces protection [10–12]. These data show that IL-12 is necessary to redirect the early Th2 response. The essential role of IL-12 in resistance to leishmaniasis is not confined to its involvement in the initiation of a protective Th1-mediated response, but IL-12 is also required for the maintenance of immunity to *Leishmania* [13, 14]. When IL-12-deficient mice were treated with recombinant IL-12 only transiently during the initial stage of infection, they eventually developed progressive disease that was associated with a loss of the Th1 profile and the development of a Th2 cytokine pattern [13]. It has been suggested that the importance of IL-12 in maintaining an established Th1 response may be based on its ability to (1) induce optimal proliferation and IFN-γ production by Th1 cells, (2) ensure Th1 cell survival by preventing apoptotic cell death and (3) prevent Th2 cell development by replenishing the pool of Th1 cells from uncommitted Th cells or central memory T cells [15].

Another susceptibility factor for *L. major* infection is IL-13 [16]. Mice deficient for both IL-4 and IL-13 are more resistant than either single knock-out strain, suggesting that IL-13 effectively cooperates with IL-4 to promote the development of Th2 cells in *L. major*-susceptible mice. In addition, IL-10 is involved in disease exacerbation and the instruction of Th2 cells. Two sources of IL-10 may be relevant in leishmaniasis, macrophages and CD4$^+$ T cells. IL-10 can be a component of the Th2 cytokine profile, but it is also produced by other CD4$^+$ subsets, in particular CD4$^+$ CD25$^+$ regulatory T cells (Treg cells).

32.2.2
The Role of Regulatory T Cells

Treg cells are specialized subsets of CD4$^+$ T cells that negatively regulate various cell-mediated immune functions. They are characterized by the expression of high levels of IL-10 and transforming growth factor-β, cytokines involved in the down-regulation of both macrophage effector functions and Th1 differentiation. The importance of Treg cells in leishmaniasis was demonstrated by the finding that these cells represent a large proportion of CD4$^+$ T cells in chronic and in healed lesions [17, 18]. They are activated during infection to suppress *Leishmania*-specific Th1 immunity in resistant mice, thereby preventing sterile cure, and may be important to limit the tissue damage associated with sustained inflammatory immune responses. This effect of Treg cells was shown to depend on their ability to produce IL-10 [17, 19]. In conclusion, although multiple pathways control resistance or susceptibility to leishmaniasis, it is clear that the Th1/Th2 balance regulates the disease outcome *in vivo*.

32.3
Dendritic Cell Interaction with *Leishmania* Parasites

Leishmania parasites are transmitted by the bite of infected sand flies depositing the infectious load in the skin of the mammalian host. At this site, a number of distinct potential phagocytes and antigen-presenting cells (APC) are present, including epidermal Langerhans cells, dermal dendritic cells (DC) and dermal macrophages. The early interaction of the pathogen with host cells is likely to be critical for the course of infection. Therefore, it is important to understand the mechanisms facilitating or counteracting the establishment of parasite infection.

32.3.1
Parasite Uptake by Dendritic Cells

Leishmania promastigotes attach to specialized surface receptors on mononuclear phagocytes and are rapidly taken up by phagocytosis. The bulk of parasites is harbored by macrophages, but Langerhans cells and dermal DC, which are known to monitor peripheral tissues for invading pathogens [20, 21], also ingest the microorganisms. DC express a variety of receptors for pathogen-associated molecular patterns, and at least two of them are involved in the uptake of *Leishmania* parasites. The phagocytosis of *L. major* by *ex vivo*-derived murine Langerhans cells was demonstrated to be mediated by CR3, the receptor for complement component C3bi [22]. More recently, using immature human DC, it was shown that a C-type lectin, the DC-specific intercellular adhesion molecule (ICAM)-3grabbing nonintegrin (DC-SIGN, CD209), is also involved in the uptake of *Leishmania* parasites by DC [23]. DC-SIGN appears to discriminate among *Leishmania* species because it bound *L. pifanoi*, a cause of New World cutaneous leishmaniasis, as well as *L. infantum* and *L. donovani*, causing visceral leishmaniasis, with high avidity, while it interacted only poorly with *L. major*, a species responsible for cutaneous pathology in the Old World [24]. Interestingly, opsonized *L. infantum* parasites were found to exhibit a much lower capacity to bind to DC-SIGN [24], raising questions about its role in the interaction of *Leishmania* and DC *in vivo* because in physiological conditions the majority of parasites are likely to be opsonized by serum components, such as antibodies and complement factors. It has been shown that DC are able to phagocytose *Leishmania* regardless of whether they had previously been opsonized [25].

Leishmania glycosylinositolphospholipids, high mannose-containing molecules abundantly expressed on the surface of both promastigotes and amastigotes, appear not to be involved in parasite binding to DC via DC-SIGN [23]. In contrast, *L. mexicana* lipophosphoglycan (LPG), the mannose-capped major surface glycoconjugate of promastigotes, was proposed as a *Leishmania* ligand for DC-SIGN [26]. However, the binding of *L. donovani*, *L. infantum* and *L. pifanoi* to DC-SIGN was shown to be independent of LPG [24] and, because LPG-defective parasite strains bound DC-SIGN with even higher avidity than their wild-type counterparts, it was suggested that LPG may in fact mask other DC-SIGN ligands on the *Leishmania* surface membrane [24]. Thus, it remains to be clarified which parasite molecules are involved in the attachment to DC receptors.

The ability to ingest *Leishmania* parasites of different life cycle stages, amastigotes versus promastigotes, seems to depend on the origin of DC. While Langerhans cells and Langerhans cell-like fetal skin-derived DC have been reported to take up amastigotes, but not promastigotes [22, 27], DC isolated from spleen and DC generated from bone marrow or peripheral blood precursor cells by *in vitro* culture in the presence of IL-4 and/or granulocyte-macrophage colony-stimulating factor were shown to internalize promastigotes and amastigotes [25, 28–30]. However, regardless of the type of DC and the infective stage of *Leishmania*, it has generally been found that the uptake of parasites by DC is not as efficient as that seen with macrophages. The percentage of parasitized DC is significantly lower and each infected DC usually harbors only 1 to 2 microorganisms, whereas macrophages may contain a tenfold number of parasites. This is a notable finding strongly suggesting that macrophages and DC play distinct roles in leishmaniasis. In contrast to macrophages, the uptake of parasites by DC is not aimed at avid scavenging and clearance of the pathogen but at antigen processing and presentation to T cells (see Section 32.4)

32.3.2
Subcellular Location of *Leishmania* Parasites in Dendritic Cells

Inside host cells, the parasites initially reside in phagosomes which subsequently fuse with lysosomes to form phagolysosomes. While this process of endosome maturation is well characterized in parasitized macrophages [31], there is little information about the fate of phagosomes in DC harboring *Leishmania* or other intracellular pathogens. A distinct feature of DC is the variation of their properties with the different stages of their life span. DC have highly specialized endocytic structures, which are regulated upon maturation. DC in nonlymphoid tissues, such as the skin and the gut, are immature cells which can phagocytose and process particles, but are only weak stimulators of T-cell immune responses. The differentiation of DC is triggered by exposure to pathogens or inflammatory cytokines and is accompanied by the loss of endocytic activity and a marked upregulation of the expression of molecules involved in antigen presentation.

Analysis of the *Leishmania*-containing parasitophorous vacuole (PV) in DC by confocal immunofluorescence microscopy revealed that the parasites reside in an acidic compartment containing major histocompatibility complex (MHC) class II and H-2M molecules, characteristics of the MIIC [32], macrosialin (CD68) and the lysosome-associated membrane proteins (Lamp) 1 and 2, which are markers of late endosomes and lysosomes (Fig. 32.2) [25, 29, 33]. While these features are similar to those of infected macrophages, the finding that the PV in DC, in contrast to macrophages, express low levels of rab7p, a molecule involved in the fusion of late endosomes with lysosomes, and cysteine and aspartyl proteases, like cathepsin B, H, L and D, support the idea that PV biogenesis is different in DC and macrophag-

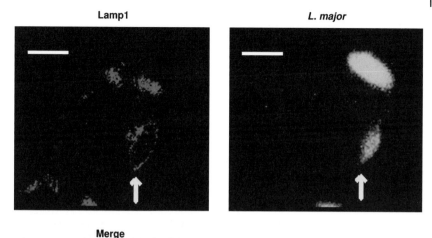

Lamp1 L. major Merge

Fig. 32.2 Endosomal compartments in DC harbor *Leishmania* parasites. DC infected for one hour with *L. major* that had been pre-stained with 5-chloromethylfluorescein-diacetat (CMFDA, green) were subjected to intracellular staining for Lamp1, a marker of late endosomes/lysosomes, using a phyco-erythrin-labeled antibody. The localization of Lamp1 (red, top left) and *L. major* parasites (green, top right) was analyzed by confocal microscopy. A parasite residing in a para-sitophorous vacuole that also contains Lamp1 is indicated by arrows (merge, bottom left). Bar: 5 μm.

es [25]. This may have consequences for the parasite antigen presentation functions of DC.

32.3.3
Dendritic Cell Subsets Involved in the Uptake of *Leishmania*

DC can be separated into subsets which differ in phenotype, function and localization [34]. Three distinct DC subpopulations have been identified in mouse spleen, as determined by the expression of CD4 and CD8α molecules on their surfaces, and the existence of five DC subpopulations was documented for skin-draining lymph nodes, with the two additional subsets originating from the skin (Langerhans cells and dermal DC). Each of these subsets was shown to exhibit a different pattern of cytokine secretion [35]. Interestingly, the three subpopulations of murine splenic DC can be ranked on the basis of their ability to internalize *L. major* parasites [29]. CD4$^+$ CD8$^-$ DC have the highest infection rate, followed by CD4$^-$ CD8$^-$ DC. The CD4$^-$ CD8$^+$ subpopulation of DC showed the lowest level of parasite uptake. This hierarchy in the phagocytosis activity of the different DC populations correlated with the levels of their CR3 expression [29], supporting the previous suggestion that this complement receptor is critical for the uptake of *Leishmania* by DC [22].

32.3.4
Dendritic Cells in *Leishmania*-infected Tissues

As summarized above, a number of studies demonstrated that DC can phagocytose *Leishmania in vitro*. Moreover, parasite antigen-containing DC can also be detected *in vivo* in tissues of *Leishmania*-infected mice and humans [22, 36–38]. Inadequate epidermal homing of Langerhans cells has been described to lead to tissue damage in patients with chronic cutaneous leishmaniasis [39, 40].

In mouse ear skin infected with *L. major*, immunohistologic staining revealed distinct clusters of CD205$^+$ Langerhans cells in the parasite-containing dermal infiltrate. These cells constituted less than 1% of the mononuclear cells in the infiltrate and some of them expressed parasite antigen, as demonstrated by double labeling [22]. Langerhans cells in the epidermis were found not to be parasitized. In contrast to the primary lesion, in which macrophages and granulocytes appear as the mainly infected cells, CD11c$^+$ DC were shown to be the most frequently infected cell population in the draining lymph nodes during the peak of *L. major* infection in both resistant C57BL/6 and susceptible BALB/c mice [37]. In lymph nodes of infected BALB/c mice, DC presented high parasite loads, suggesting that *in vivo* infection might affect the physiology of DC more strongly, possibly due to the duration of infection and/or the presence of environmental factors. In addition, CD11c$^+$ multinucleated giant cells harboring a dramatic accumulation of parasites were frequently observed [37].

The characterization of DC subpopulations purified from the draining lymph nodes of *L. major*-infected mice revealed a larger number of plasmacytoid DC in susceptible BALB/c mice than in resistant C57BL/6 mice [41]. Interestingly, when the parasite load in the different DC subpopulations of the lymph nodes draining the lesion was examined in the course of infection, the frequency of cells carrying live parasites was similar for all the subpopulations, but the plasmacytoid DC in resistant mice continued to harbor parasites for the longest period. The possible consequences of these differences remain to be elucidated but it is conceivable that the high levels of type I IFN (IFNα/β) expression known to be associated with plasmacytoid DC [42] may influence the quality of the immune response. In contrast to classical DC subsets, plasmacytoid DC lack the ability to stimulate naïve T cells, but can effectively promote the differentiation of antigen-experienced unpolarized T cells into Th1 cells [43].

32.4
Dendritic Cell Migration and Induction of a *Leishmania*-specific Immune Response

The movement of DC precursors into sites of microbial infection in peripheral tissues is regulated by the sequential involvement of chemokines, such as monocyte chemotactic protein (MCP) and macrophage inflammatory protein (MIP)-3α/CCL20, binding to the chemokine receptors CCR2 and CCR6, respectively, on the DC surface. Immature DC also respond to various inducible chemokines that are released by macrophages upon activation by microbial products, such as CCL3 (MIP-1α), CCL5 (RANTES) and CXCL8 (IL-8), the ligands of the receptors CCR1, CCR5 and CXCR1 expressed by these DC, and thus accumulate at the site of infection [44]. In the skin, Langerhans cells represent immature DC with a high endocytic activity that are able to take up microbes and/or microbial antigens for degradation and peptide loading onto MHC molecules [45]. The recognition of microbial signals induces the maturation of DC, a process that is accompanied by their efflux from the skin and migration to the T-cell areas of draining lymph nodes [46]. During maturation, the endocytic capacity of DC is downregulated, while their expression of MHC class II as well as co-stimulatory and adhesion molecules, such as CD80, CD86 and CD40, is strongly upregulated [20]. Thus, they acquire the distinct ability to trigger a primary T-cell response. The expression of the inflammatory chemokine receptors CCR1, CCR2, CCR5 and CCR6 decreases in the course of DC maturation and, therefore, the responsiveness to the corresponding ligands is reduced. At the same time, DC gain surface expression of the chemokine receptor CCR7 and responsiveness to its ligands CCL19 (MIP-3β) and CCL21 (secondary lymphoid tissue chemokine) that are produced in the T-cell areas of lymph nodes. CCR7 is a key receptor for the convergence of antigen-loaded mature DC and responder T cells, as it also mediates the homing of activated T cells to the lymph nodes [47].

32.4.1
The Role of Chemokines and Chemokine Receptors Expressed by Dendritic Cells

Leishmania parasites have been shown to selectively modulate the expression of chemokines and chemokine receptors by DC in a species-dependent manner. DC from mice infected with *L. donovani* were reported to have reduced expression of CCR7 and a decreased responsiveness to CCR7 ligands [48]. This is associated with a failure of DC from chronically infected mice to migrate from the marginal zone to the periarteriolar region of the spleen. Treatment with CCR7-expressing DC provided protection against parasite growth, demonstrating that defective CCR7-mediated DC migration plays a major role in the pathogenesis of visceral leishmaniasis. In contrast, interaction of DC with *L. major* was shown to enhance the level of CCR7 expression and the DC response to its ligand CCL21 [49]. Expression of the chemokine receptors CCR2 and CCR5, however, and the responsiveness to the respective ligands CCL2 (MCP-1) and CCL3 (MIP-1α) were downregulated by DC exposure to *L. major* [49]. These alterations in chemokine receptor expression induced by *L. major* were observed with DC from resistant and susceptible mice, indicating that the differential ability of these mice to control cutaneous leishmaniasis is not due to host-dependent effects of the parasite on chemokine receptors directing DC migration. Notably, *L. major* elicited expression of the chemokine CXCL10 (IFN-inducible protein, IP-10) only in DC from resistant mice. This observation extended earlier studies [50, 51], which showed that *L. major* infection up-regulates CXCL10 expression in the draining lymph nodes of resistant, but not susceptible mice, by demonstrating that CXCL10 derived from *L. major*-stimulated DC may account for the higher level of this chemokine in resistant mice. CXCL10 recruits and activates natural killer (NK) cells, a component of the innate immune system, that have been implicated in the development of resistance to leishmaniasis [52].

The importance of chemokine receptors for DC functions and the course of *L. major* infection was also tested with mice lacking CCR2 or CCR5 [53]. This study showed that deficiency of CCR2, a receptor for CCL2 (MCP-1), but not CCR5, or its ligand CCL3 (MIP-1α), leads to distinct defects in DC biology. While the density of Langerhans cells in the epidermis of CCR2-deficient mice and their ability to move to the dermis was normal, their migration to the draining lymph nodes was strongly impaired. CCR2 knock-out mice had decreased numbers of DC in the spleen and a block in the *L. major* infection-induced relocalization of DC from the marginal zone to the T-cell areas of the spleen. CCR2-deficient mice with a *L. major*-resistant genetic background were susceptible to disease and developed a Th2-dominated cytokine profile. These findings demonstrated that CCR2 plays an important role in DC migration and localization, with critical impact on the development of a protective immune response to *Leishmania*.

32.4.2
Transport and Presentation of *L. major* Antigen by Dendritic Cells

After intradermal infection of mice with *L. major*, a significant change in the distribution of Langerhans cells was observed [22]. A considerable loss of CD205$^+$ Langerhans cells in the segment of the epidermis overlying the parasite-containing infiltrate was concomitant with the appearance of those cells in the dermal layer of the lesion, some of which contained *L. major* antigen. These observations indicated that Langerhans cells migrate from the epidermis to the site of infection in the dermis for uptake of *L. major* parasites and/or parasite antigen. Moreover, the cells were demonstrated to migrate to the draining lymph node within 24 to 48 h [36], using double labeling immunohistochemistry as well as *in vivo* tracking of labeled Langerhans cells that had been infected with *L. major in vitro* and were reinjected into the skin. Such a translocation was not observed with infected macrophages under equivalent conditions. The migratory DC in the lymph nodes expressed *L. major* antigen and were able to stimulate resting T cells to mount a *L. major*-specific immune response *in vivo*. These findings suggested that Langerhans cells transport *L. major* parasites and/or parasite antigen from the skin to the regional lymph nodes for initiation of the specific T-cell immune response [54, 55].

The role of DC subpopulations in parasite dissemination from the infected skin to the draining lymph node was recently examined in detail [41]. Using ear skin explants from *L. major*-infected resistant or susceptible mice, it was shown that the majority of DC emigrating from the skin were Langerhans cells (20% on average), whereas only a small population of migratory dermal DC was observed (less than 5%). However, no parasites could be detected in DC emigrating from infected skin explants, while macrophages harbored significant numbers. Moreover, viable parasites were not detectable in DC subpopulations purified by flow cytometry from draining lymph nodes until 3 weeks after infection. Antigen presentation functions were not examined in this study. The authors concluded that macrophages but not DC are the vehicle that ferries parasites from skin to lymph nodes and that DC take up parasites from infected macrophages in the lymph nodes. This seems to be at odds with the earlier findings described above [22, 36]. However, the following points need to be considered. (1) Fundamentally different experimental approaches were used in these studies. The migratory behavior of DC in skin explants is likely to differ from that of DC *in vivo* because chemokines direct them from skin to lymph nodes. Furthermore, the methods used for detection of *L. major*-loaded DC in the lymph nodes (immunocytochemistry of tissue sections versus DC isolation by flow cytometry) may differ in sensitivity; *in vivo* tracking of labeled DC revealed that only a very small proportion of the cells (0.1–0.5%) migrate to the draining lymph node [36]. (2) Different time points were used for the analysis of DC in the draining lymph nodes. While no DC containing live parasites were detected at day 8, the earliest time point analyzed in the one study [41], small numbers of DC expressing *L. major* antigen could be demonstrated in the early phase of infection (1–4 days) in the other study [36]. (3) Finally, it is important to note that DC do not need to carry live parasites to fulfill their prime task, i.e. the initiation of a specific T cell-mediated immune response. It has been shown that DC, but not macrophages, pulsed with parasites or cell-free *L. major* culture filtrates are able to stimulate a primary T-cell response [28]. This idea is supported by the finding that two waves of *L. major* antigen-containing CD205$^+$ DC appear in the draining lymph nodes. The DC of the first wave become detectable at 8 h after infection but do not harbor intact parasites, while DC carrying live parasites were observed in the second wave at 24–48 h after infection [56]. The DC of the first wave were shown to prime parasite-specific CD4$^+$ T cells.

Using a mouse model in which MHC class II expression is restricted to CD11b$^+$ and CD8α$^+$ DC, it has recently been demonstrated that antigen presentation by these DC subsets is sufficient for T-cell priming, induction of Th1 differentiation and control of subcutaneous *L. major* infection [57]. Antigen presentation by macrophages was not required. In these mice, plasmacytoid DC and Langerhans cells also lack MHC class II and, thus, were not involved in the activation of parasite-specific Th cells. It will be important to assess the requirement for MHC class II expression by Langerhans cells in intradermal infection models. In another recent report, also using subcutaneous infection with *L. major*, it has been suggested that dermal CD8α$^-$ Langerin-negative DC transport *L. major* antigen to the lymph nodes and induce a secondary proliferative response of CD4$^+$ T cells [58]. However, their ability to stimulate a primary T-cell response to *L. major* and Th1-mediated control of cutaneous disease was not examined in this study.

32.4.3
Parasite Persistence in Immune Hosts

After clinical cure of cutaneous leishmaniasis in resistant mice, life-long immunity is maintained but small numbers of parasites persist, preferentially in the lymph nodes draining the prior skin lesion, even in the face of a competent immune response [59]. *L. major* parasites were found to be sequestered in macrophages, fibroblasts and DC [60–62]. The sustained expression of iNOS, IL-12 and IFN-γ is crucial for the control of latent infection, because impairment of either of these responses in cured mice has been shown to cause clinical recrudescence of the disease [14, 19, 62]. Sterile cure is achieved in IL-10-deficient mice after healing and by treatment with anti-IL-10 receptor antibodies [19], but the observation that these mice are no longer immune to re-infection [17] suggests that antigen persistence is required for the maintenance of protective immunity. Notably, the examination of latently infected host cells isolated from draining lymph nodes of cured mice demonstrated that only DC, but not macrophages, are able to present endogenous *L. major* antigen to specific T cells *in vitro* [61]. Thus, DC may contribute to the sustained stimulation of effector memory T cells and long-term control of the parasite. This extraordinary efficiency in antigen presentation by DC may be explained by the highly increased stability of MHC class II molecules loaded with immunogenic parasite peptides [33].

32.5
Regulation of the *Leishmania*-specific Immune Response by Dendritic Cells

Microbial structures are recognized by Toll-like receptors (TLR) expressed on the surface of DC. TLR signaling induces the maturation of DC and stimulates the release of inflammatory cytokines, thus enabling DC to prime T cells and shape the developing cell-mediated immune response. In this way, TLR are involved in the link between innate and adaptive immunity [63, 64]. The different TLR possess conserved cytoplasmic domains and use common intracellular signaling pathways that involve recruitment of the cytoplasmic adaptor molecule MyD88, activation of serine/threonine kinases of the IRAK family, and finally degradation of IκB and translocation of NF-κB to the nucleus. A MyD88-independent pathway has been shown to be responsible for the induction of IFN-β and IFN-inducible genes.

Systemic administration of *Leishmania* parasites (*L. brasiliensis*, *L. donovani*, *L. major* or *L. mexicana*) was shown to induce full DC maturation *in vivo*, as hallmarked by DC migration from marginal zones to T-cell areas of the spleen and upregulation of MHC class II and the co-stimulatory molecules CD40, CD80 and CD86 [65]. Comparable maturation processes were observed for DC from genetically susceptible and resistant mice, indicating that the difference in their ability to control leishmaniasis is not caused by alterations in parasite sensing by DC. The induction of DC maturation by *Leishmania* was independent of the presence of T and B cells or granulocytes, as it was also observed in RAG$^{-/-}$ and GR1-depleted mice, but was partially abolished in mice lacking MyD88 [65]. Thus, the recently reported requirement of a functional MyD88 transduction pathway for the development of a protective Th1 immune response to *L. major* [66–68] may involve MyD88-dependent DC maturation. The recognition of *Leishmania* via TLR may be mediated by the promastigote surface glycoconjugate LPG which has been demonstrated to bind to TLR2 [68].

A central aspect of the ability of DC to tailor immune responses to pathogens is their potential to release different cytokines depending on the type of microbial stimulus that is recognized. The level of cytokine expression is enhanced by T-cell feedback signals mediated by CD40 ligation. Various reports document that the choice of cytokine production by DC has a profound influence on the host's ability to control infection with *Leishmania* parasites.

32.5.1
The Role of IL-12 Production by Dendritic Cells

The synthesis of IL-12 at the early stage of infection is crucial for the determination of both innate immunity, as it activates NK cells to produce IFN-γ, and the adaptive host response, via selective induction of Th1 cell differentiation. *Leishmania* parasites have been shown to actively inhibit IL-12 production by macrophages [69] and *in situ* analysis first documented that DC are the source of early IL-12 production following *Leishmania* infection [70]. A subsequent study demonstrated that the secretion of high IL-12 levels by human myeloid DC harboring *L. major* was dependent on the interaction of CD40 expressed by DC with CD40 ligand [71]. Furthermore, DC generated from patients with cutaneous leishmaniasis caused by *L. major*, upon loading with parasites, were able to induce a CD40 ligand-dependent IFN-γ response. In contrast to *L. major*, however, the *Leishmania* species responsible for visceral disease (*L. donovani*), and the species associated with persistent cutaneous lesions (*L. tropica*) did not induce CD40 ligand-mediated production of bioactive heterodimeric IL-12p70 by DC, but primed the cells for expression of only the IL-12p40 subunit [72]. The intrinsic differences in the ability of *Leishmania* species to trigger critical DC functions may contribute to the evolution of different clinical forms of leishmaniasis. Species restriction in the interaction of *Leishmania* and DC has also been observed with DC derived from the bone marrow of mice. DC infection with *L. amazonensis*, a cause of cutaneous leishmaniasis in the New World, resulted in upregulated expression of MHC class II and co-stimulatory molecules including CD40. However, in contrast to what has been found for *L. major*, *L. amazonensis* failed to induce CD40-dependent IL-12 production by DC, but rather enhanced their secretion of IL-4 and priming of a parasite-specific Th2 response [73]. Interestingly, exposure to *L. amazonensis* increased IL-4 production and Th2 priming by DC from susceptible mice but not by those from resistant mice.

Analogous to the concept that the disease outcome, that is healing cutaneous versus nonhealing systemic leishmaniasis, may correlate with the *Leishmania* species-dependent induction of IL-12 production by DC, it is an appealing idea that differences in the IL-12 levels expressed by DC may also account for the host-dependent resistance or susceptibility of inbred mice to infection with *L. major*. However, there is presently no evidence for such a difference. Langerhans cell-like fetal skin-derived DC from susceptible BALB/c and resistant C57BL/6 mice were demonstrated to be phenotypically and functionally equivalent [74]. *L. major* infection of DC from both strains was accompanied by upregulation of MHC and co-stimulatory molecules and induction of IL-12 release, indicating that genetic susceptibility does probably not reflect intrinsic defects of BALB/c DC to respond to the parasites. In contrast, although the CD11b$^+$ CD8$^-$ subset of lymph node DC was shown to be responsible for the priming of CD4$^+$ T cells in both susceptible BALB/c and resistant B10.D2 mice, DC from susceptible mice had a more pronounced ability to polarize naïve CD4$^+$ T cells into Th2 effector cells than DC from resistant mice [75]. This difference did not correlate with different levels of IL-12 expression by the DC, but CD11b$^+$ DC from resistant B10.D2 mice were found to express significantly more IL-1β mRNA than those from BALB/c mice.

DC subpopulations were shown to secrete different levels of IL-12 in response to infection with *L. major*. The CD4$^-$ CD8$^+$ subset of mouse spleen DC was least permissive to infection but produced the highest amount of IL-12 [29]. In the lymph nodes draining the lesions, the DC subsets of susceptible BALB/c mice overall produced higher levels of IL-12p70 than those of resistant C57BL/6 mice [41]. The cells producing the least IL-12p70 were the plasmacytoid DC, although there was a marked increase at 3 weeks of infection for C57BL/6 mice. Together, the observations described above point to a complex regulation of *Leishmania*-induced IL-12 expression and Th-cell polarization by DC that depends on the parasite species, the type of DC and the stage of infection.

32.5.2
Other Parameters that may Govern the Polarization of T Helper Cells

Additional factors may be involved in the induction of Th1 differentiation by DC. For example, the co-stimulatory molecules CD80 and CD86 are required for the development of an early immune response to *L. major* [76], and expression of CD80 was reported to be downregulated on Langerhans cells from susceptible mice but not on those from resistant mice [77].

With regard to cytokines other than IL-12, members of the IL-1 cytokine system have been identified to be produced at higher levels by DC from resistant mice compared with those from susceptible mice [75, 78]. In one study, using CD11b$^+$ DC isolated from draining lymph nodes of *L. major*-infected mice, it was found that the DC from resistant B10.D2 mice expressed much higher mRNA levels of IL-1β, but not IL-1α, than the DC from susceptible BALB/c mice [75]. In the other study, using Langerhans cell-like fetal skin-derived DC stimulated with *L. major in vitro*, DC from resistant C57BL/6 mice were shown to express significantly higher amounts of IL-1α mRNA than DC from BALB/c mice [78]. Furthermore, it was demonstrated that treatment of susceptible BALB/c mice with recombinant IL-β [75] or IL-1α [78] increased the ability of these mice to control infection and shifted the Th1/Th2 balance towards a protective immune response. The efficiency of IL-1α administration was found to be strictly dependent on the presence of IL-12, indicating that both cytokines act in conjunction [78]. IL-1α and IL-1β bind to the same receptor, IL-1 receptor type I, and mice deficient for this receptor were shown to develop enhanced Th2-like cytokine responses following infection with *L. major* [79]. DC-derived IL-1 has been demonstrated to be important for IL-12-mediated Th1 differentiation [80, 81] and, thus, appears to be involved in the generation of a protective immune response during leishmaniasis.

32.6
Parasite Evasion of Dendritic Cell Function

A distinct feature of parasites is the production of long-lasting chronic infections. Their successful colonization of the host is dependent on their ability to interfere with the development of an effective immune response. As DC play a key role in connecting innate and adaptive immunity, various strategies to modify DC functions provide means to facilitate parasite survival in the host. *Leishmania* have been shown to impair the maturation, cytokine secretion, migration and antigen presentation function of DC (Table 32.2).

In a model mimicking natural infection with *L. major*, the inoculation of small numbers of promastigotes (100–1000, the amount transmitted by infected sand flies) into a dermal site resulted in a silent phase of 4–5 weeks during which no T-

Tab. 32.2 *Leishmania* parasite evasion of DC functions.

Effect	Mechanism	Parasite species	Ref.
Impairment of DC activation	Uptake of live parasites	*L. mexicana*	83
Impairment of DC activation	Uptake of parasites not opsonized with Ab	*L. amazonensis*	25
Induction of a Th2 response	Inhibition of IL-12 production, induction of IL-4 production by DC	*L. amazonensis*	73
Inhibition of IL-12 production	Limitation of IL-12p35 expression	*L. donovani, L. tropica*	72
Impairment of APC functions	Inhibition of IL-12 production, inhibition of MHC class II expression	*L. donovani*	84
Inhibition of DC migration	Secreted parasite products	*L. major*	85
Inhibition of DC migration	*Leishmania* LPG	*L. major*	86
Inhibition of DC migration	Defective CCR7 expression by DC	*L. donovani*	48
Inhibition of CD1-restricted T-cell activation	Reduction of CD1 expression by DC	*L. donovani*	89

Ab, antibodies

cell responses were observed [82]. This may indicate an impairment of DC activation during *in vivo* infection. In contrast to what has been found for *in vitro* infection of DC with *L. major* [27, 28], cultured DC infected with *L. mexicana* did not upregulate MHC class II and CD86 expression and failed to secrete IL-12, whereas their activation in response to other stimuli was not affected [83]. Uptake of *L. amazonensis* induced enhanced expression of MHC class II, CD40, CD80 and CD86 by DC, but infected DC from susceptible mice synthesized IL-4 rather than IL-12 and induced a disease-promoting Th2 response [73]. *Leishmania* species-dependent effects on DC cytokine production were also found with human DC. *In vitro* infection with *L. major*, but not *L. tropica* or *L. donovani*, triggered IL-12 secretion by blood-derived DC from healthy donors [72].

The *in vivo* modulation of DC functions has been analyzed using the model of murine visceral leishmaniasis [84]. Upon infection of mice with *L. donovani*, an impairment in the ability of splenic DC to induce allogeneic mixed lymphocyte reaction and present *L. donovani* antigen to specific T cells was observed at 2 months of infection. This defect in the antigen presentation functions of DC correlated with reduced MHC class II surface expression, lack of IL-12 production and their ability to suppress IFN-γ release by *Leishmania* antigen-primed T cells. At the onset of control over parasite replication in the spleen at 4 months post infection, however, the DC functions were found to be restored.

Impairment of the migratory activities of DC is another effective way to manipulate the host's immune response. The motility of splenic DC was shown to be inhibited by secreted products of *L. major* promastigotes [85], and the emigration of Langerhans cells from skin was significantly reduced by the phosphoglycan moiety of *L. major* LPG [86], suggesting that parasite products interfere with the ability of DC to transport antigen to or within lymphoid organs.

With regard to antigen presentation, an exquisite property of DC is their ability to present non-protein antigens to T cells via CD1 molecules which are constitutively expressed by DC. The CD1 family of non-polymorphic molecules consists of group I CD1 molecules, comprising human CD1a, CD1b and CD1c, and the group II molecule CD1d found on both human and murine DC. The best-characterized microbial antigens for CD1-mediated presentation are mycobacterial lipids and glycolipids [87]. These antigens can be recognized by conventional T cells, a subset of γδ T cells and NK T cells. *Leishmania* species also contain abundant glycolipid molecules and it was recently reported that LPG, as well as related glycoinositol phospholipids, bind to CD1d and stimulate CD1d-dependent IFN-γ production by a subset of hepatic NK T cells [88]. However, infection of human DC with *L. donovani* has been shown to downregulate CD1 expression and CD1-restricted T-cell activation by DC [89].

Finally, sand fly components may also contribute to an impairment of the host immune response. Salivary gland homogenate of *Lutzomyia longipalpis*, the New World vector of *Leishmania*, has been shown to decrease the CD40 ligand-induced expression of MHC class II and co-stimulatory molecules by human DC [90] and may thus interfere with their ability to stimulate an adaptive immune response to parasite antigens.

32.7
Dendritic Cells as Tools for Novel Immune Intervention Strategies Against Leishmaniasis

The knowledge of the crucial role of DC in the tuning of immune responses and the availability of techniques for the generation and culture of DC *ex vivo* has led to their use for specific manipulations of the immune system. Most of these pioneering studies have been performed in tumor models, and some strategies are currently being tested in clinical trials, but DC-based immune intervention approaches were also shown to mediate protection against a wide spectrum of infectious diseases caused by viral, bacterial, parasitic and fungal pathogens [91]. Experimental infection of mice with *Leishmania* is a prototype model to explore the factors driving Th1 differentiation *in vivo* and, thus, the knowledge obtained in this system may be of great value for the development of general concepts to treat disorders that are associated with misdirected Th-cell responses. DC are crucial determinants of the Th-cell effector choice and, thus, may be used as tools to dictate Th1 cell development. The overall aim is the targeting of DC *in vivo* with a prophylactic or therapeutic vaccine containing the optimal *Leishmania* antigen preparation, DC activation molecules and DC-specific ligands or promoters for targeting the desired DC subset. For rational design of such immune intervention strategies against human leishmaniasis, however, the complexity of DC immunobiology brings about the necessity for detailed understanding of the cellular and molecular mechanisms of DC-mediated induction of anti-*Leishmania* immunity. Therefore, *ex vivo* approaches in model systems need to be employed to define the parameters for DC targeting in human tissues.

32.7.1
Dendritic Cell-based Vaccination and Immunotherapy

The first studies documenting that DC can serve as a natural adjuvant to induce protective immunity against leishmaniasis were performed in the model of murine infection with *L. major*. A single treatment with Langerhans cells that had been pulsed with parasite lysate *ex vivo* was shown to induce long-lasting protection of otherwise susceptible BALB/c mice against subsequent challenges with virulent parasites [92]. The solid immunity induced by DC-based vaccination was paralleled by a pronounced shift of the cytokine expression towards a Th1-like pattern with high levels of IFN-γ and very low levels of IL-4 and IL-10. In a murine model of visceral leishmaniasis, it was subsequently shown that DC engineered to over-express IL-12 and pulsed *ex vivo* with soluble *L. donovani* antigens mediated significantly enhanced protection associated with an increased parasite-specific IFN-γ response [93].

The choice of antigen is of substantial importance for the efficacy of vaccination. The antigen preparation should be molecularly defined and it should be possible to manufacture it in a safe and reproducible manner. DC pulsed with a mixture of the recombinant *Leishmania* antigens LACK, KMP-11, gp63 and PSA, or with the single leishmanial peptide LeIF, mediated significant levels of protection against leishmaniasis, demonstrating that the development of a DC-based subunit vaccine against leishmaniasis is feasible [94]. In this study, the protective effect was found to depend on DC-derived IL-12 because antigen-pulsed Langerhans cells from IL-12-deficient mice failed to confer resistance. The importance of antigenic peptide selection for DC-based vaccination was confirmed by the finding that DC pulsed with peptide L1 (154–169aa) of *L. major* gp63 induced protection against experimental cutaneous leishmaniasis that was associated with a Th1 response, whereas DC pulsing with peptide L2 (467–482aa) of gp63 resulted in disease exacerbation and a Th2 profile [95].

DC loaded with *Leishmania* antigen can also be employed for immunotherapy of mice with established infections. This has been documented with IL-12-engineered and *L. donovani* antigen-pulsed DC [93] as well as with IL-12-producing and *L. chagasi* antigen-pulsed DC [96], both enhancing the cure of experimental visceral leishmaniasis. In contrast, the intralesional administration of *L. amazonensis* antigen-pulsed DC plus IL-12 did not promote healing of cutaneous lesions although it induced Th1 cell development [97]. Interestingly, combined treatment with *L. donovani* antigen-pulsed DC and the conventional antileishmanial compound sodium antimony gluconate was shown to result in complete clearance of parasites from the liver and spleen and cure of established murine visceral leishmaniasis [98].

32.7.2
Parameters Determining the Efficacy of Dendritic Cell-based Immune Intervention Strategies

The appropriate instruction of DC is critical for their maturation into APC that direct the development of naive T cells toward a Th1 phenotype. This notion is supported by the finding that a mere expansion of the number of mature DC, which can be achieved by treatment of mice with Flt3 ligand, is not sufficient to mediate complete protection against cutaneous leishmaniasis [99]. DC need to be educated in a specific manner to acquire the ability to drive an effective immune response. The criteria usually applied to evaluate the immunostimulatory potential of DC is their induction of cytokines associated with Th1 or Th2 cells *in vitro*. However, it has recently been shown that the immunological characteristics of DC *in vitro* are not necessarily predictive of their *in vivo* immunizing properties [100]. *L. major* antigen-pulsed DC activated by tumor necrosis factor-α, lipopolysaccharide or CD40 ligation, three prototype DC stimuli that are known to trigger enhanced expression of MHC class II and co-stimulatory molecules by DC and evoke DC cytokine production, failed to induce protection against leishmaniasis in susceptible mice. In contrast, mice vaccinated with a single dose of antigen-loaded DC stimulated by exposure to CpG motifs were completely protected and developed an antigen-specific Th1 response. These findings demonstrated that the type of stimulatory signal is critical for activating the potential of DC to induce a Th1 response *in vivo* that confers complete resistance against an intracellular pathogen. In the same study, it was also revealed that the role of IL-12 depends on the type of DC used for vaccination against cutaneous leishmaniasis. Whereas earlier reports, using Langerhans cells or Langerhans cell-like DC, emphasized the key role of IL-12 produced by the parasite-antigen presenting DC [74, 94], it was shown that the protection mediated by CpG-activated and *Leishmania* antigen-pulsed DC derived from the bone marrow was independent of their ability to release IL-12 [100]. Taken together, critical parameters determining the efficacy of DC-based vaccination against microbial pathogens include the origin and type of DC, the choice of antigen to be used for DC loading and the state of DC maturation and activation (Figure 32.3).

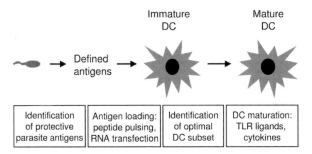

Fig. 32.3 Parameters determining the efficacy of DC-based prophylactic and therapeutic vaccination against leishmaniasis. Single *Leishmania* antigens inducing a protective Th1 response or mixtures of such antigens can be loaded into selected DC subpopulations by peptide pulsing or by transfection with mRNA encoding these antigens. Co-transfection with immuno-modulatory cytokines, such as IL-12, may enhance the potential of DC to induce Th1-mediated protection. The type of DC maturation stimulus is another critical factor determining the capacity of DC to instruct protective Th1 cell differentiation.

32.8
Conclusions and Perspectives

Infectious diseases caused by parasites, including leishmaniasis, represent a major world health problem that continues to increase in incidence. Despite considerable efforts, it has not yet been possible to develop any effective vaccine. Moreover, the presently available chemotherapeutic options are unsatisfactory. The rapid progress in the understanding of DC immunobiology and the involvement of DC in the regulation of the *Leishmania*-specific immune response provides a new promise for the development of approaches to manipulate the immune system. DC-based strategies would be of particular relevance to patients in which conventional therapies have failed and to immunocompromised individuals, such as patients with *Leishmania* and HIV co-infections, in which the antimicrobial immune response needs to be enhanced or restored. An ideal scenario would be the antigen loading and activation of DC *in situ*. The approaches currently being explored to target DC in tissues employ DC-specific surface molecules, such as CD205 [101] or

C-type lectins [102], DC-specific promoters [103] and synthetic TLR ligands that specifically interact with DC subpopulations [104]. Additional strategies involve the use of antigen-loaded exosomes derived from DC [105] and antigen delivery by transcutaneous immunization which has been demonstrated to cause the activation of skin DC [106]. Therefore, it is likely that DC-based methods to combat leishmaniasis and other infections caused by intracellular pathogens will become available.

References

1 World Health Organization, *World Health Report* **2002**.

2 L. Rivas, J. Moreno, C. Cañavate, J. Alvar, *Trends Parasitol.* **2004**, 7, 297–301.

3 P. Scott, P. Natovitz, R. L. Coffman, E. Pearce, A. Sher, *J. Exp. Med.* **1988**, 168, 1675–1684.

4 F. P. Heinzel, M. D. Sadick, B. J. Holaday, R. L. Coffman, R. M. Locksley, *J. Exp. Med.* **1989**, 169, 59–72.

5 F. Y. Liew, C. A. O'Donnell, *Adv. Parasitol.* **1993**, 32, 161–259.

6 D. Sacks, N. Noben-Trauth, *Nat. Rev. Immunol.* **2002**, 2, 845–858.

7 H. Moll, M. Röllinghoff, *Eur. J. Immunol.* **1990**, 20, 2067–2074.

8 L. Morris, A. B. Troutt, E. Handman, A. Kelso, *J. Immunol.* **1992**, 149, 2715–2721.

9 P. Scott, A. Eaton, W. C. Gause, X. di Zhou, B. Hondowicz, *Exp. Parasitol.* **1996**, 84, 178–187.

10 F. P. Heinzel, D. S. Schoenhaut, R. M. Rerko, L. E. Rosser, M. K. Gately, *J. Exp. Med.* **1993**, 177, 1505–1509.

11 J. P. Sypek, C. L. Chung, S. E. H. Mayor, J. M. Subramanyam, S. J. Goldman, D. S. Sieburth, S. F. Wolf, R. G. Schaub, *J. Exp. Med.* **1993**, 177, 1797–1802.

12 F. Mattner, J. Magram, J. Ferrante, P. Launois, K. Di Padova, R. Behin, M. K. Gately, J. A. Louis, G. Alber, *Eur. J. Immunol.* **1996**, 26, 1553–1559.

13 A. Y. Park, B. D. Hondowicz, P. Scott, *J. Immunol.* **2000**, 165, 896–902.

14 L. Stobie, S. Gurunathan, C. Prussin, D. L. Sacks, N. Glaichenhaus, C. Y. Wu, R. A. Seder, *Proc. Natl. Acad. Sci. USA* **2000**, 97, 8427–8432.

15 P. Scott, D. Artis, J. Uzonna, C. Zaph, *Immunol. Rev.* **2004**, 201, 318–338.

16 D. J. Matthews, C. L. Emson, G. J. McKenzie, H. E. Jolin, J. M. Blackwell, A. N. McKenzie, *J. Immunol.* **2000**, 164, 1458–1462.

17 Y. Belkaid, C. A. Piccirillo, S. Mendez, E. M. Shevach, D. L. Sacks, *Nature* **2002**, 420, 502–507.

18 D. Sacks, C. Anderson, *Immunol. Rev.* **2004**, 201, 225–238.

19 Y. Belkaid, K. F. Hoffmann, S. Mendez, S. Kamhawi, M. C. Udey, T. A. Wynn, D. L. Sacks, *J. Exp. Med.* **2001**, 194, 1497–1506.

20 J. Banchereau, F. Briere, C. Caux, J. Davoust, S. Lebecque, Y.-J. Liu, B. Pulendran, K. Palucka, *Annu. Rev. Immunol.* **2000**, 18, 767–811.

21 I. Mellman, R. M. Steinman, *Cell* **2001**, 106, 255–258.

22 C. Blank, H. Fuchs, K. Rappersberger, M. Röllinghoff, H. Moll, *J. Infect. Dis.* **1993**, 167, 418–425.

23 M. Colmenares, A. Puig-Kröger, O. Muñiz-Pello, A. L. Corbí, L. Rivas, *J. Biol. Chem.* **2002**, 277, 36766–36769.

24 M. Colmenares, A. L. Corbi, S. J. Turco, L. Rivas, *J. Immunol.* **2004**, 172, 1186–1190.

25 E. Prina, S. Z. Abdi, M. Lebastard, E. Perret, N. Winter, J.-C. Antoine, *J. Cell Sci.* **2004**, 117, 315–325.

26 B. J. Appelmelk, I. van Die, S. J. van Vliet, C. M. J. E. Vandenbroucke-Grauls, T. B. H. Geijtenbeek, Y. van Kooyk, *J. Immunol.* **2003**, 170, 1635–1638.

27 E. von Stebut, Y. Belkaid, T. Jakob, D. L. Sacks, M. C. Udey, *J. Exp. Med.* **1998**, 188, 1547–1552.

28 P. Konecny, A. J. Stagg, H. Jebbari, N. English, R. N. Davidson, S. C. Knight, *Eur. J. Immunol.* **1999**, 29, 1803–1811.

29 S. Henri, J. Curtis, H. Hochrein, D. Vremec, K. Shortman, E. Handman, *Infect. Immun.* **2002**, 70, 3874–3880.

30 D. Chaussabel, R. Tolouei Semnani, M. A. McDowell, D. Sacks, A. Sher, T. B. Nutman, *Blood* **2003**, 102, 672–681.

31 J.-C. Antoine, E. Prina, T. Lang, N. Courret, *Trends Microbiol.* **1998**, 6, 392–401.

32 H. J. Geuze, *Immunol. Today* **1998**, 19, 282–287.

33 S. Flohé, T. Lang, H. Moll, *Infect. Immun.* **1997**, 65, 3444–3450.

34 K. Shortman, Y.-J. Liu, *Nat. Rev. Immunol.* **2002**, 2, 151–161.

35 H. Hochrein, K. Shortman, D. Vremec, B. Scott, P. Hertzog, M. O'Keeffe, *J. Immunol.* **2001**, 166, 5448–5455.

36 H. Moll, H. Fuchs, C. Blank, M. Röllinghoff, *Eur. J. Immunol.* **1993**, 23, 1595–1601.

37 E. Muraille, C. De Trez, B. Pajak, F. Aguilar Torrentera, P. De Baetselier, O. Leo, Y. Carlier, *Infect. Immun.* **2003**, 71, 2704–2715.

38 A. M. Elhassan, A. Gaafar, T. G. Theander, *Clin. Exp. Immunol.* **1994**, 99, 445–453.

39 F. J. Tapia, G. Cáceres-Dittmar, M. A. Sánchez, *Immunol. Today* **1994**, 15, 160–165.

40 N. L. Diaz, O. Zerpa, L. V. Ponce, J. Convit, A. J. Rondon, F. J. Tapia, *Exp. Dermatol.* **2002**, 11, 34–41.

41 T. Baldwin, S. Henri, J. Curtis, M. O'Keeffe, D. Vremec, K. Shortman, E. Handman, *Infect. Immun.* **2004**, 72, 1991–2001.

42 C. Asselin-Paturel, A. Boonstra, M. Dalod, I. Durand, N. Yessaad, C. Dezutter-Dambuyant, A. Vicari, A. O'Garra, C. Biron, G. Trinchieri, *Nat. Immunol.* **2001**, 2, 1144–1150.

43 A. Krug, R. Veeraswamy, A. Pekosz, O. Kanagawa, E. R. Unanue, M. Colonna, M. Cella, *J. Exp. Med.* **2003**, 197, 899–906.

44 H. Moll, *Cell. Microbiol.* **2003**, 5, 493–500.

45 P. Guermonprez, J. Valladeau, L. Zitvogel, C. Théry, S. Amigorena, *Annu. Rev. Immunol.* **2002**, 20, 621–667.

46 N. Romani, G. Ratzinger, K. Pfaller, W. Salvenmoser, H. Stössel, F. Koch, P. Stoitzner, *Int. Rev. Cytol.* **2001**, 207, 237–270.

47 K. Willimann, D. F. Legler, M. Loetscher, R. S. Roos, M. B. Delgado, I. Clark-Lewis, M. Baggiolini, B. Moser, *Eur. J. Immunol.* **1998**, 28, 2025–2034.

48 M. Ato, S. Stäger, C. R. Engwerda, P. M. Kaye, *Nat. Immunol.* **2002**, 3, 1185–1191.

49 M. Steigerwald, H. Moll, *Infect. Immun.* **2005**, 73, 2564–2567.

50 B. Vester, K. Müller, W. Solbach, T. Laskay, *Infect. Immun.* **1999**, 67, 3155–3159.

51 C. Zaph, P. Scott, *Infect. Immun.* **2003**, 71, 1587–1589.

52 T. M. Scharton, P. Scott, *J. Exp. Med.* **1993**, 178, 567–577.

53 N. Sato, S. K. Ahuja, M. Quinones, V. Kostecki, R. L. Reddick, P. C. Melby, W. A. Kuziel, S. S. Ahuja, *J. Exp. Med.* **2000**, 192, 205–218.

54 H. Moll, *Immunol. Today* **1993**, 14, 383–387.

55 H. Moll, *Adv. Exp. Med. Biol.* **2000**, 479, 163–173.

56 A. C. Misslitz, K. Bonhagen, D. Harbecke, C. Lippuner, T. Kamradt, T. Aebischer, *Eur. J. Immunol.* **2004**, 34, 715–725.

57 M. P. Lemos, F. Esquivel, P. Scott, T. M. Laufer, *J. Exp. Med.* **2004**, 199, 725–730.

58 U. Ritter, A. Meißner, C. Scheidig, H. Körner, *Eur. J. Immunol.* **2004**, 34, 1542–1550.

59 T. Aebischer, S. F. Moody, E. Handman, *Infect. Immun.* **1993**, 61, 220–226.

60 C. Bogdan, N. Donhauser, R. Döring, M. Röllinghoff, A. Dieffenbach, M. G. Rittig, *J. Exp. Med.* **2000**, 191, 2121–2130.

61 H. Moll, S. Flohé, M. Röllinghoff, *Eur. J. Immunol.* **1995**, 25, 693–699.

62 S. Stenger, N. Donhauser, H. Thüring, M. Röllinghoff, C. Bogdan, *J. Exp. Med.* **1996**, 183, 1501–1514.

63 S. Akira, K. Takeda, *Nat. Rev. Immunol.* **2004**, 4, 499–511.

64 C. Pasare, R. Medzhitov, *Microbes Infect.* **2004**, 6, 1382–1387.

65 C. De Trez, M. Brait, O. Leo, T. Aebischer, F. Aguilar Torrentera, Y. Carlier, E. Muraille, *Infect. Immun.* **2004**, 72, 824–832.

66 E. Muraille, C. De Trez, M. Brait, P. De Baetselier, O. Leo, Y. Carlier, *J. Immunol.* **2003**, 170, 4237–4241.

67 A. Debus, J. Gläsner, M. Röllinghoff, A. Gessner, *Infect. Immun.* **2003**, 71, 7215–7218.

68 M. J. de Veer, J. M. Curtis, T. M. Baldwin, J. A. DiDonato, A. Sexton, M. J. McConville, E. Handman, L. Schofield, *Eur. J. Immunol.* **2003**, 33, 2822–2831.

69 Y. Belkaid, B. Butcher, D. L. Sacks, *Eur. J. Immunol.* **1998**, 28, 1389–1400.

70 P. M. A. Gorak, C. Engwerda, P. M. Kaye, *Eur. J. Immunol.* **1998**, 28, 687–695.

71 M. A. Marovich, M. A. McDowell, E. K. Thomas, T. B. Nutman, *J. Immunol.* **2000**, 164, 5858–5865.

72 M. A. McDowell, M. Marovich, R. Lira, M. Braun, D. L. Sacks, *Infect. Immun.* **2002**, 70, 3994–4001.

73 H. Qi, V. Popov, L. Soong, *J. Immunol.* **2001**, 167, 4534–4542.

74 E. von Stebut, Y. Belkaid, B. V. Nguyen, M. Cushing, D. L. Sacks, M. C. Udey, *Eur. J. Immunol.* **2000**, 30, 3498–3506.

75 C. Filippi, S. Hugues, J. Cazareth, V. Julia, N. Glaichenhaus, S. Ugolini, *J. Exp. Med.* **2003**, 198, 201–209.

76 M. M. Ellosos, P. Scott, *J. Immunol.* **1999**, 162, 6708–6715.

77 M. L. Mbow, G. K. DeKrey, R. G. Titus, *Eur. J. Immunol.* **2001**, 31, 1400–1409.

78 E. von Stebut, J. M. Ehrchen, Y. Belkaid, S. Lopez Kostka, K. Mölle, J. Knop, C. Sunderkötter, M. C. Udey, *J. Exp. Med.* **2003**, 198, 191–199.

79 A. R. Satoskar, M. Okano, S. Connaughton, A. Raisanen-Sokolwski, J. R. David, M. Labow, *Eur. J. Immunol.* **1998**, 28, 2066–2074.

80 K. Shibuya, D. Robinson, F. Zonin, S. B. Hartley, S. E. Macatonia, C. Somoza, C. A. Hunter, K. M. Murphy, A. O'Garra, *J. Immunol.* **1998**, 160, 1708–1716.

81 U. Eriksson, M. O. Kurrer, I. Sonderegger, G. Iezzi, A. Tafuri, L. Hunziker, S. Suzuki, K. Bachmaier, R. M. Bingisser, J. M. Penninger, M. Kopf, *J. Exp. Med.* **2003**, 197, 323–331.

82 Y. Belkaid, S. Mendez, R. Lira, N. Kadambi, G. Milon, D. Sacks, *J. Immunol.* **2000**, 165, 969–977.

83 C. L. Bennett, A. Misslitz, L. Colledge, T. Aebischer, C. C. Blackburn, *Eur. J. Immunol.* **2001**, 31, 876–883.

84 A. Basu, G. Chakrabarti, A. Saha, S. Bandyopadhyay, *Immunology* **2000**, 99, 305–313.

85 H. Jebbari, A. J. Stagg, R. N. Davidson, S. C. Knight, *Infect. Immun.* **2002**, 70, 1023–1026.

86 A. Ponte-Sucre, D. Heise, H. Moll, *Immunology* **2001**, 104, 462–467.

87 J. L. Matsuda, M. Kronenberg, *Curr. Opin. Immunol.* **2001**, 13, 19–25.

88 J. L. Amprey, J. S. Im, S. J. Turco, H. W. Murray, P. A. Illarionov, G. S. Besra, S. A. Porcelli, G. F. Späth, *J. Exp. Med.* **2004**, 200, 895–904.

89 J. L. Amprey, G. F. Späth, S. A. Porcelli, *Infect. Immun.* **2004**, 72, 589–592.

90 D. J. Costa, C. Favali, J. Clarêncio, L. Afonso, V. Conceição, J. C. Miranda, R. G. Titus, J. Valenzuela, M. Barral-Netto, A. Barral, C. I. Brodskyn, *Infect. Immun.* **2004**, 72, 1298–1305.

91 H. Moll, *Int. J. Med. Microbiol.* **2004**, 294, 337–344.

92 S. B. Flohé, C. Bauer, S. Flohé, H. Moll, *Eur. J. Immunol.* **1998**, 28, 3800–3811.

93 S. S. Ahuja, R. L. Reddick, N. Sato, E. Montalbo, V. Kostecki, W. Zhao, M. J. Dolan, P. C. Melby, S. K. Ahuja, *J. Immunol.* **1999**, 163, 3890–3897.

94 C. Berberich, J. R. Ramírez-Pineda, C. Hambrecht, G. Alber, Y. A. W. Skeiky, H. Moll, *J. Immunol.* **2003**, 170, 3171–3179.

95 P. Tsagozis, E. Karagouni, E. Dotsika, *Int. J. Immunopathol. Pharmacol.* **2004**, 17, 343–352.

96 M. E. Wilson, T. J. Recker, N. E. Rodriguez, B. M. Young, K. K. Burnell, J. A. Streit, J. N. Kline, *Eur. J. Immunol.* **2002**, 32, 3556–3565.

97 Y. Vanloubbeeck, A. E. Ramer, F. Jie, D. E. Jones, *Infect. Immun.* **2004**, 72, 4455–4463.

98 M. Gosh, C. Pal, M. Ray, S. Maitra, L. Mandal, S. Bandyopadhyay, *J. Immunol.* **2003**, 170, 5625–5629.

99 I. B. Kremer, M. P. Gould, K. D. Cooper, F. P. Heinzel, *Infect. Immun.* **2001**, 69, 673–680.

100 J. R Ramírez-Pineda, A. Fröhlich, C. Berberich, H. Moll, *J. Immunol.* **2004**, 172, 6281–6289.

101 L. Bonifaz, D. Bonnyay, K. Mahnke,
M. Rivera, M. C. Nussenzweig,
R. M. Steinman, *J. Exp. Med.* **2002**,
196, 1627–1638.

102 C. G. Figdor, Y. van Kooyk, G. J. Adema,
Nat. Rev. Immunol. **2002**, 2, 77–84.

103 R. Ross, S. Sudowe, J. Beisner,
X. L. Ross, I. Ludwig-Portugall, J. Steitz,
T. Tuting, J. Knop, A. B. Reske-Kunz,
Gene Ther. **2003**, 10, 1035–1040.

104 T. Maurer, A. Heit, H. Hochrein,
F. Ampenberger, M. O'Keeffe, S. Bauer,
G. B. Lipford, R. M. Vabulas, H. Wagner,
Eur. J. Immunol. **2002**, 32, 2356–2364.

105 C. Théry, L. Zitvogel, S. Amigorena,
Nat. Rev. Immunol. **2002**, 2, 569–579.

106 G. M. Glenn, D. N. Taylor, X. Li,
S. Frankel, A. Montemarano, C. R.
Alving, *Nat. Med.* **2000**, 6, 1403–1406.

33

Sentinel and Regulatory Functions of Dendritic Cells in the Immune Response to *Toxoplasma gondii*

Alan Sher, Felix Yarovinsky, Romina Goldszmid, Julio Aliberti and Dragana Jankovic

33.1
Introduction

Toxoplasma gondii is an apicomplexan protozoan that infects felines as its definitive hosts. Cats acquire the parasite through predation of infected rodents and birds. In addition, transmission can occur between these as well as other intermediate hosts in the absence of cats through feeding on infected carrion. Man is an accidental host for *T. gondii* acquiring the infection through contact with cat fecal matter or ingestion of cysts in meat from infected livestock. Most human disease occurs as a consequence of immunodeficiency, typically arising in the unprotected fetus following maternal transplacental infection or in individuals with HIV or drug induced immune suppression [1, 2].

  T. gondii promiscuously infects a wide variety of nucleated host cells. Nevertheless, the organism produces an asymptomatic infection in most susceptible immunocompetent hosts due in large part to the rapid induction of strong cell-mediated immune response that results in the control of the rapidly dividing tachyzoite stage and the establishment of chronic infection mediated by dormant bradyzoites. Host resistance to *T. gondii* is critically dependent on IFN-γ produced by NK, CD4 and CD8 T during the acute and chronic phases of infection. TNF-α (TNF) also participates in host resistance, but appears to be more important in the chronic phase where its effector function appears to be closely linked with nitric oxide production [3].

  IFN-γ synthesis by NK and T lymphocytes during *T. gondii* infection is critically dependent on IL-12 and mice deficient in the latter cytokine succumb to acute toxoplasmosis with the same kinetics as IFN-γ deficient animals. Since mice develop a strong IL-12 response early in infection, it is logical to presume that the induction of this cytokine is a major initiation signal for host resistance to the parasite [3, 4]. IL-12 is also likely to contribute to Th1 effector choice in *T. gondii* infection although no default to a Th2 cytokine production phenotype is seen in IL-12 deficient animals exposed to the parasite [5].

While IL-12 is known to be synthesized by a number of different antigen presenting cells as well as neutrophils [6] during *T. gondii* infection, the most relevant source in terms of immune response initiation is likely to be the dendritic cell (DC) and the evidence that the latter cell population produces high levels of the cytokine upon *in vivo* exposure to the parasite provided the first evidence that *T. gondii* is a potent activator of DC function. In this work it was found that injection of live tachyzoites (the rapidly dividing stage of the parasite) or a soluble tachyzoite extract (STAg) into mice results in the rapid appearance within a few hours in spleen of IL-12 p40 producing cells consisting almost entirely of CD11c$^+$ DC [7]. Staining of the same spleen sections with anti-CD11c mAb revealed a massive redistribution of DC in spleen stimulated by *T. gondii* injection with most cells leaving the red pulp and marginal zone and clustering in the T-cell areas. In this [7] and in follow up studies [8–10], splenic DC were also shown to be highly responsive to STAg *in vitro* producing significant levels of IL-12 at doses as low as 1 pg ml^{-1} (FY, unpublished observations). Together, the above evidence suggested that the production of IL-12 by DC early in infection is the major "ignition signal" for IFN-γ-dependent host resistance to *T. gondii* infection [11].

These early findings, together with *in vitro* studies characterizing the response of human DC to the parasite [12, 13], laid the groundwork for what has become a highly active research area in which a series of central issues dealing with the role of DC in the immune response to *T. gondii* are being addressed. This chapter presents a brief review of the accumulated data on this topic while focusing on the unique features of the *Toxoplasma*/mouse model as a system for studying DC-pathogen interactions.

33.2
Activation of DC by *T. gondii*

33.2.1
Responsive DC Subpopulations

While as noted above, *T. gondii* can provide a potent stimulus for DC function, it is becoming clear that such parasite-driven responses occur under a limited set of conditions. The best stimulation is observed in the model described above in which murine splenic DC are exposed *in vivo* or *in vitro* to live tachyzoites or STAg. In that system it was shown that the responding IL-12 producing DC belong predominantly to the CD8α$^+$ subset [7]. Indeed, bone-marrow (BM)-derived DC which are deficient in this subset respond poorly to *T. gondii* stimulation (JA, FY, unpublished observations). The IL-12 response of splenic CD8α$^+$ DC to STAg *in vivo* was found to be extremely rapid, peaking in 3–6 h and was accompanied by a concomitant increase in cell surface CD40 expression [9]. An immediate fall-off in the response was then observed with a reduction to baseline by 24 h. Why CD8α$^+$ DC selectively respond to this form of stimulation is unclear. The simplest hypothesis is that they uniquely express the pattern recognition receptors which interact with

STAg ligands. It should be pointed out however that under certain conditions CD8α⁻ DC can be triggered to produce IL-12 in response to STAg [8] so this restriction is not absolute.

Although in the initial work in the murine system live tachyzoites were found to be equivalent to STAg in their IL-12 inducing potential, there is some controversy as to whether DC infected with the parasite (rather than exposed to released tachzoite products) are able to mount a cytokine response. This is based on the observation (discussed further below) that infection of immature BM DC with tachzoites inhibits their maturation and cytokine production induced by LPS or CD40L [14].

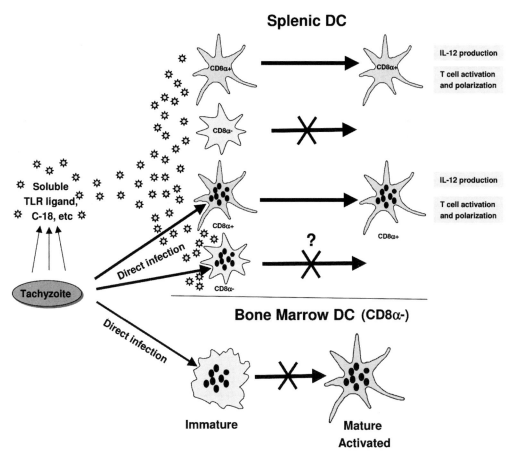

Fig. 33.1 Encounter of DC subsets with *T. gondii* leads to distinct outcomes. CD8α⁺ but not CD8α⁻ DC in spleen and other tissues respond to soluble products released by tachyzoites and present in STAg. Whether intracellular tachyzoites (as opposed to soluble parasite products) can trigger responses in CD8α⁻ DC, particularly after extended incubation, remains to be determined. While direct infection does not appear to influence the activation status of splenic DC, bone-marrow-derived DC not only fail to respond to soluble tachyzoite Ag but when infected with the parasite display a block in maturation and response to exogenous stimulation with LPS or CD40L [14].

While this inhibition is clear cut, it should be noted that BM DC are poorly responsive in terms of IL-12 production even to STAg so the effect of live infection on parasite-induced synthesis of the cytokine is likely to be minor. In contrast, when splenic DC (which represent a more mature population) are exposed to tachyzoites both parasite-infected and non-infected cells respond with vigorous IL-12 production (RM, unpublished). Taken together these observations suggest that while *T. gondii* infection inhibits DC maturation and responsiveness to exogenous non-parasite activation stimuli, already mature DC can nevertheless be stimulated to respond by soluble parasite products regardless of their infection status (Fig. 33.1). Immune response initiation must therefore depend on the latter DC population.

33.2.2
Host Receptors and Parasite Ligands Involved in Triggering of Murine DC

Most studies investigating the ligand–receptor interactions involved in DC activation by *T. gondii* have utilized the STAg/murine DC model because it provides a soluble extract from which to fractionate parasite molecules that stimulate DC function. As mentioned above, the IL-12 response of splenic DC to STAg is unusually potent exceeding that of LPS, CpG oligonucleotides and other stimuli commonly used to trigger pro-inflammatory cytokine production. This implies either the involvement of a novel signaling pathway or the use of additive or synergistic pathways by multiple ligands in the parasite extract. Another distinctive feature of this system is that the IL-12 inducing activity in STAg is protease sensitive [15] suggesting a critical function for protein ligands.

The role of an unconventional signaling pathway in STAg induced DC activation was initially suggested by studies investigating the role of chemokines in the migration of DC to splenic T-cell areas following STAg injection. It was found that spleens of mice lacking the CC chemokine receptor CCR5 not only display impaired DC migration but also exhibited diminished DC IL-12 production [10]. Moreover, DC from naïve CCR5 KO mice showed reduced IL-12 responses when stimulated with STAg in vitro. In agreement with these observations CCR5 KO mice displayed decreased IL-12 and IFN-γ production following live *T. gondii* infection [10].

Since endogenously produced host CCR5 ligands do not possess significant IL-12 inducing activity, it was logical to suspect the involvement of a parasite ligand in this response. This was confirmed when a *T. gondii* molecule with both IL-12 inducing and CCR5 binding activity was identified in tachyzoite supernatants and STAg. This 18-kDa protein was shown to be an isoform of *T. gondii* cyclophilin (C-18) [15]. Cyclophilins are chaperone-like molecules that possess peptidyl-prolyl isomerase activity and bind the drug Cyclosporin A which inhibit this enzymatic function. C-18, unlike many cyclophilins (i.e. human Cyclophilin A) is a secreted molecule. Its release by tachyzoites or infected cells would explain its ability to trigger DC without directly infecting them [15].

Recombinant C-18 was shown to bind to CCR5 with moderate affinity and trigger chemokine receptor signaling as measured by Ca^{++} flux [15]. No CCR5 binding or IL-12 inducing activity is displayed by closely related cyclophilins including the homolog from *P. falciparum* [15] and mutated C-18 molecules which no longer bind to CCR5 fail to induce IL-12 [16]. The latter observations suggest that CCR5 binding by C-18 represents an example of molecular mimicry employed by *T. gondii*. Interestingly, in the same-site directed mutagenesis study the peptidyl-prolyl isomerase activity of C-18 was also shown to be necessary for IL-12 induction suggesting that enzymatic modification of CCR5 itself or another host signaling element may be a component of this mechanism [16].

Although recombinant C-18 possessed significant IL-12 inducing activity it was clearly considerably less potent than the starting tachyzoite material from which it was fractionated. Since the *E. coli* expressed protein retains its peptidyl-prolyl isomerase activity, this was unlikely to be due to improper refolding. Instead, the data pointed to the existence of other important IL-12 inducing ligands missed during the initial fractionation of the parasite material that might function together with C-18 in inducing high level DC activation [15].

Because of their major role in microbial recognition it was logical to test the involvement of receptor pathways involving the TLR/IL-1R superfamily in triggering both DC IL-12 production and host resistance to *T. gondii*. To do so mice lacking MyD88, an adaptor molecule used by most TLR as well as IL-1R and IL-18R were employed. DC from these animals showed a near complete abrogation of the STAg induced IL-12 response and when challenged with *T. gondii* the KO animals displayed a loss in resistance to infection equivalent to that of IL-12 deficient mice [17]. Since DC from IL-18R mice show normal STAg induced IL-12 production (T. Kaisho and A. Sher, unpublished observation) and IL-1R antagonists fail to block the DC IL-12 response (JA, unpublished), the observed MyD88 dependency is likely to reflect TLR involvement. TLR-2 has been shown to contribute to the host resistance of mice to very high challenge doses of *T. gondii* [18] as well as to STAg induced chemokine production by neutrophils [19]. Nevertheless, TLR-2 deficient DC as well as neutrophils produce normal amounts of IL-12 following STAg stimulation and TLR-2$^{-/-}$ mice fully resist more physiological low dose infections [17, 19, 20].

A major clue in identifying the TLR involved in DC activation by *T. gondii* came from the initial observation that the relevant molecules are protease sensitive. Indeed, when the search for parasite ligands stimulating DC IL-12 production was continued, this time using stimulation of CCR5$^{-/-}$ DC as the read-out to avoid the detection of C-18, a second much more potent IL-12 inducing protein was identified [21]. This 17.5-kDa molecule (PFTG) was then tested for its ability to stimulate cell lines transfected with either TLR-5 or TLR-11, the two TLR known to recognize proteins. Only the TLR 11 cells responded to PFTG and a similar response was observed with unfractionated STAg. Subsequent, testing of DC from TLR11$^{-/-}$ mice confirmed the requirement for TLR11 in both STAg and PFTG induced IL-12 production *in vitro* as well as *in vivo* [21]. Based on this new data, it would appear that DC activation by *T. gondii* depends on two signals, a major MyD88-dependent signal provided largely by PFTG-TLR-11 interaction and a second weaker enhancing signal resulting from CCR5 ligation of C-18 (Fig. 33.2). How these distinct ligands, receptors and signaling pathways interact is presently unclear. Receptor clusters

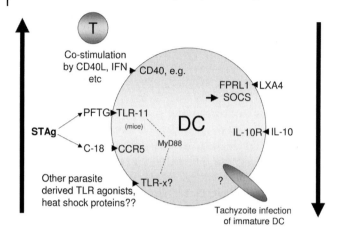

Fig. 33.2 Stimulatory vs downregulatory pathways in the response of DC to *T. gondii*. DC function is stimulated primarily by soluble ligands present in STAg (e.g. PFTG, C-18) although the existence of other agonists in the membrane/insoluble portion of tachyzoites has not been ruled out. These ligands function primarily by triggering MyD88-dependent signaling pathways although an amplifying function for CCR5-dependent signaling stimulated by C-18 has also been demonstrated. TLR-11 has recently been identified as a major TLR in STAg (PFTG)-mediated, MyD88-dependent signaling of murine DC. Other TLR may also participate in mice and those involved in triggering of human DC remain to be identified. DC function following *T. gondii* exposure has been shown to be down regulated by two apparently distinct pathways, the first involving the lipoxin LXA4 and the second IL-10. In addition, parasite infection of immature DC can inhibit the maturation of these cells and their responsiveness to exogenous stimuli.

containing CXCR4 in association with TLR-4 have been observed following LPS stimulation of macrophages [22] and one possibility is that the two *T. gondii* ligands and their corresponding TLR and chemokine receptors stimulate DC by forming a similar membrane complex.

33.2.3
Activation of Human DC

Although much less-well studied, the response of human DC to *T. gondii* appears to have requirements distinct from those described for murine DC. Thus, in two studies involving peripheral blood monocyte derived DC, induction of IL-12 p70 by *T. gondii* was shown to require CD40-CD154 signaling [12, 13], a co-stimulatory mechanism previously shown to enhance, but not to be essential for, murine DC activation [9]. This signal could be provided by co-incubation with CD4+ T cells from seropositive donors. Interestingly, in one of these studies [13], in striking contrast to the murine data, direct infection of the DC was found to be required both for the upregulation of co-stimulatory molecule expression and their CD40L-dependent cytokine expression. Taken together both papers suggest that DC activation in humans requires the presence of pre-existing *T. gondii* specific T cells thereby arguing against the DC as the "ignition signal" for the response of humans against the parasite. An important caveat to this hypothesis is that the DC studied in humans and mice represent distinct subpopulations, the human cells being blood derived and lacking an equivalent of the highly responsive $CD8\alpha^+$ subset studied in mouse spleen. A further complication is that humans do not appear to have a homolog of the TLR11 molecule recently implicated as a major determinant in the responsiveness of murine DC to STAg [21, 23] and may require other parasite ligands and signaling pathways such as those involving heat shock proteins [24]. Such comparisons may be of interest in understanding the evolution of the host–parasite relationship in *T. gondii* since *Toxoplasma* is a natural parasite of rodents and does not depend on human infection for its transmission [2].

33.3
Regulation of DC Activity

The responsiveness of DC is dramatically downmodulated following infection with tachyzoites or exposure to parasite products. This regulation is likely to serve the parasite by preventing clearance of the organism while protecting its host niche from the potentially lethal immunopathologic effects of an uncontrolled inflammatory response. As already introduced above, infection of murine bone-marrow-derived DC appears to both block their maturation and ability to be activated by LPS as well as present the model Ag OVA to $CD4^+$ T cells [14]. A similar suppression of responsiveness has previously been documented in resting macrophages infected with *T. gondii* and linked with an inhibition in nuclear translocation of NF-κB and STAT-1 [25]. It is likely that parallel mechanisms underlie the induction of non-responsiveness in DC.

Suppression of DC function following *T. gondii* stimulation has also been documented *in vivo* in more mature tissue DC populations. As noted above, the IL-12 response of splenic DC to STAg inoculation is short-lived returning to baseline by 24 h. This reduction in IL-12 production occurs concurrently with the development of nonresponsiveness to secondary administration of STAg, a phenomenon which was referred to as "DC paralysis"[8]. A trivial explanation would be that DC paralysis results from the induction of IL-10, a downregulatory cytokine known to be induced by *T. gondii* and which protects infected mice against IL-12 driven tissue inflammation [26]. Nevertheless not only does DC paralysis not require IL-10, but its induction can actually protect IL-10 deficient mice from lethality following subsequent live *T. gondii* challenge [8].

Based on the evidence that DC paralysis is accompanied by down regulation of CCR5, a receptor which as described above participates in STAg induced IL-12 production, a novel mechanism was identified that may explain the loss in DC responsiveness observed following STAg inoculation. This mechanism involves the induction by STAg or live *T. gondii* infection of lipoxin (LX) A_4, an arachidonic acid metabolite generated by a 5-lypoxygenase (LO)-dependent pathway. LXA_4, which is thought to signal through the receptor FPRL1 [27] , had previously been demonstrated to downmodulate CCR5 [28]. In the *T. gondii* studies, an LXA4 analog was found to inhibit STAg induced IL-12 production by DC both *in vivo* and *in vitro* an effect which correlated with reduced CCR5 expression [29]. Moreover, STAg was found to be potent stimulus of LXA_4 synthesis *in vivo*. Since this response was found to be 5-LO-dependent, 5-LO deficient mice were then used to examine the role of LXA_4 during *T. gondii* infection. Consistent with the studies employing STAg stimulation *in vivo*, infection of wild-type mice with live parasites induced 5-LO-dependent LXA_4 production [30]. This response was absent in the 5-LO-deficient mice which exhibited enhanced IL-12/IFN-γ synthesis and reduced parasite (brain cyst) counts but succumbed by 28 days post-infection probably as a consequence of an uncontrolled pro-inflammatory response. The role of LXA_4 in these effects was confirmed in reconstitution experiments using an analog of the mediator with a longer clearance time. *T. gondii* infected 5-LO-deficient mice showed enhanced IL-12 production by CD11c$^+$ as well as CD11c$^-$ cells in brain sections suggesting that *in vivo* LXA4 regulates IL-12 production by other APC in addition to DC [30].

In more recent studies the mechanism by which LXA4 inhibits DC IL-12 responses has been investigated and shown to require suppressor of cytokine signaling (SOCS) but utilizing a pathway independent of that involved in SOCS-3-dependent suppression of DC function by IL-10 (F. Machado and J. Aliberti, unpublished).

33.4
Role of DC in *T. gondii*-induced Immune Polarization

Host control of parasite growth during both the acute and chronic phases of *T. gondii* infection depends on IFN-γ secretion by NK and T cells, rather than on cytotoxicity-based effector functions [3] and both CD4$^+$ and CD8T cells from infected mice and humans display a highly polarized Th1 cytokine production profile. As major sentinels of the innate immune system that interact directly with differentiating T cells, it is logical to hypothesize that DC play an important role in establishing this Th1 polarized state early in infection.

The increased survival of *T. gondii*-infected SCID versus IFN-γ deficient animals [31] clearly demonstrates autonomous and potent activation of the innate immune system by this parasite resulting in temporary control of tachyzoite growth. This response, as discussed above, is likely to be directed by IL-12 producing DC that in turn promote IFN-γ synthesis by NK cells. In T-cell sufficient mice, it has been generally assumed that IL-12 serves as a bridge between innate and adaptive immunity by promoting the development of Th1 effector cells [6, 32, 33] thus ensuring lasting control of infection. Consistent with this concept is the finding that mice deficient in the p40 subunit of IL-12 succumb to acute infection with the same kinetics as infected IFN-γ KO mice [34]. To directly address the role of IL-12 signaling in *T. gondii* induced Th1 polarization in a setting which allows host survival, WT and IL-12p40 deficient mice were repeatedly inoculated with radiation-attenuated tachyzoites or STAg. Although Th1-type cytokine production was diminished in the absence of IL-12, the pathogen-specific CD4$^+$ T cells that emerged nevertheless displayed an IFN-γ-dominated lymphokine profile and failed to default to a Th2 phenotype [5]. The same pattern of cytokine expression in Th lymphocytes was observed in STAg-immunized mice deficient in CCR5 or mice doubly deficient in IL-12 and IL-18 (D Jankovic, unpublished data). Additional studies have failed to reveal a requirement for either STAT-4 [35] or IL-27 [36] in *T. gondii* induced Th1 polarization. Together these findings argue that although clearly promoting the expression of IFN-γ during *T. gondii* infection and critical for host resistance, IL-12 is not essential for Th1 effector choice nor are its sister cytokines IL-23 and IL-27. Indeed, a protective function could be revealed for the Th1 cells that emerge in the absence of IL-12 in mice doubly deficient for both IL-10 and IL-12 [5].

The development of *T. gondii*-specific Th1 cells in the absence of IL-12 production suggests that signals distinct from IL-12 are critical for microbial induced Th1 effector choice. That DC can supply these signals was demonstrated in *in vitro* experiments in which highly purified splenic CD11c$^+$ DC exposed to STAg were shown to efficiently direct naïve DO11.10 transgenic CD4$^+$ T cells towards a Th1 phenotype in the presence of nominal OVA peptide [37]. Importantly, this Th1 biasing occurred efficiently with either wild-type or IL-12 deficient DC consistent again with an IL-12-independent mechanism of Th1 differentiation. Nevertheless, DC from MyD88$^{-/-}$ mice failed to induce Th1 polarization in this system suggesting a major role for TLR-dependent recognition and activation [37]. Also, in common with IL-12 production, Th1 polarization by STAg stimulated DC was restricted to the CD8α$^+$ subset with neither bone marrow derived DC, B cells or macrophages functioning in the assay (S. Steinfelder and D. Jankovic, unpublished). A role for CD8α$^+$ DC in *T. gondii* induced Th1 differentiation *in vivo* was suggested by the observation that T-cell cytokine secretion defaults to a Th2 pattern in *T. gondii*-infected ICSBP-deficient mice [38] that display an impairment in CD8α$^+$ DC generation [39]. Thus, activation of DC for Th1 polarization appears to involve the same DC subset and upstream TLR-dependent signaling pathway as that involved in IL-12 production yet does not require the production of that cytokine.

The encounter of microbes or their products with DC can lead to the upregulation of MHC class II as well as co-stimulatory molecules (e.g. CD40) that together should result in APC–T-cell interaction at the high signal strength level previously shown to be associated with Th1 polarization [40]. Such effects of microbial stimulation on DC are likely to occur independently of IL-12 signaling. For example, the redistribution of dendritic cells in spleen induced by *in vivo* STAg injection is unaltered in IL-12-deficient mice and when stimulated with STAg *in vitro* dendritic cells from these animals show normal upregulation of CD40 as well as chemokine production (D Jankovic, unpublished observations). Thus, there are numerous candidate Th1 polarization signals delivered by DC that would require MyD88 signaling (presumably through PFTG-TLR11 interaction) but occur independently of IL-12 production and the identification of those functionally important is a major goal of current research in this area.

33.5
Mechanisms of Antigen Presentation to T Cells

DC are generally assumed to be highly efficient antigen presenting cells in the response to invading pathogens. Nevertheless, in the case of intracellular parasites such as *T. gondii* that reside in vacuoles and modulate host cellular function, the mechanisms by which DC process and present antigen to T lymphocytes are likely to be complex. A further problem has been the failure to identify a dominant TCR in the response to *T. gondii* that can be used for generating TCR transgenic T cells to assay the activating capacity of DC exposed to the parasite. However, the recent development of recombinant *T. gondii* strains expressing OVA [41, 42] has provided an alternative approach to this problem.

The requirements for antigen presentation to CD4+ T cells have been investigated with bone marrow DC infected with OVA transfected parasites or with splenic DC exposed to STAg or tachyzoites as a consequence of short-term *in vivo* inoculation. In the former studies [41], a comparison of recombinant *T. gondii* expressing OVA in either a cytosolic or secreted form detectable in host parasitophorous vacuoles revealed that activation of OVA-specific TCR transgenic CD4+ T cells occurs only when the DC are infected with the antigen secreting parasite strain. The T-cell responses induced were vigorous and comparable to those obtained with OVA peptide pulsed DC with no evidence of parasite induced suppression of DC function as observed in the work discussed above [14].

As an extension of the work on splenic DC activation in the STAg model, Yarovinsky and colleagues have recently examined the ability of DC populations primed *in vivo* by STAg inoculation to activate STAg specific CD4+ T cells in vitro. They observed that splenic DC acquire the capacity to present Ag with the same kinetics as their expression of IL-12 and that in common with both IL-12 production and Th1 polarization, the DC mediating this function are predominantly CD8α+ and their activity is highly dependent on MyD88 signaling. These findings suggest a role for TLR recognition in promoting efficient processing and presentation of *T. gondii* antigens to CD4+ T cells (F. Yarovinsky, unpublished).

The requirements for DC presentation to CD8+ T cells have been studied primarily with OVA transfected parasites. This work has confirmed that despite the localization of the parasite within parasitophorous vacuoles, *T. gondii* secreted proteins are able to enter the Class I presentation pathway by means of a process dependent on the TAP peptide transporter [42].The latter conclusion was based on studies with OVA transfected parasites as well as experiments using a Cre secreting *T. gondii* transfectant where it was found that infection with this parasite strain results in recombination in the nucleus of infected host cells as detected with a GFP loxP reporter. In additional studies utilizing DC exposed *in vitro* to YFP labeled parasites, it was shown that actively infected (i.e. YFP⁺) cells induce much stronger CD8⁺ T-cell activity than non-infected (YFP⁻) DC arguing against a role of crosspresentation by bystander cells [42].The latter interpretation is supported by recent experiments (RM, unpublished) in which activation of OVA transgenic T cells could not be induced by adding MHC class I incompatible DC infected with irradiated OVA transfected parasites to cultures containing uninfected syngeneic DC. Indeed, in direct contrast to DC presentation to CD4⁺ T cells which can utilize dead organisms, DC presentation of parasite produced OVA to CD8⁺ T lymphocytes was shown in the latter work to require DC infection by live tachyzoites.

33.6
Towards an Understanding of DC Function *in vivo*

The findings summarized in this chapter document a special relationship that has co-evolved between dendritic cells and *T. gondii* that is likely to be of benefit to both host and parasite. Clearly, because of its invasion of multiple host cells and rapid spread into different tissues *T. gondii* is an intrinsically virulent and pathogenic organism. The ability of DC to rapidly detect the parasite and establish cellular responses that limit infection can thus be seen as both protecting the host and promoting successful parasite transmission through extended host survival and dispersion. Nevertheless, as emphasized repeatedly above, careful regulation of DC function is required to prevent complete elimination of the pathogen as well as excessive cytokine production leading to tissue pathology.

While consistent with both the biology of *T. gondii* as a persistent infection and our current knowledge of DC responses to the parasite *in vitro* and *ex vivo* in the mouse model, the above scheme remains to be confirmed *in vivo* particularly in the context of infection by the natural peroral route. In the latter mode of infection, tachyzoites would first encounter DC in the lamina propria or Peyer's patches following invasion of the intestinal mucosa and present antigen to T cells either in those sites or in draining lymph nodes following migration. Whether initial priming of DC in the gut leads to a different outcome than systemic priming is not known. Th2 responses have been argued to protect against *T. gondii* in mucosal sites (as opposed to systemically) and in this light it is of interest that DC isolated from mesenteric lymph nodes pulsed with tachzoite extract conferred a protective Th2 response against peroral challenge [43] in direct contrast to similar experiments employing splenic DC [44]. Since *T. gondii* invades multiple sites in the intestine in an unsynchronized manner, the study of DC activation and function during natural infection will require the development of new, more sensitive tools for both detecting DC in tissues and measuring their responses. The use of mice with flourochrome tagged DC associated genes is one such approach that could used to tackle this problem.

Peroral infection although perhaps initially priming Th2 responses ultimately leads to systemic IL-12-dependent protection. An interesting question concerns the extent to which DC remain an important source of IL-12 once the initial response to the parasite has been initiated. Although requiring stimulation with higher amounts of STAg or IFN-γ priming, neutrophils [45] and macrophages [20, 46] can also produce IL-12 when exposed to *T. gondii* products and it is possible that these cells take over as the major source of the cytokine later in infection. The role of DC as APC during chronic infection also has not yet been carefully addressed.

T. gondii is known to exist in three distinct genetic subtypes which vary in virulence. Essentially all of the work reviewed in this chapter utilized Type I strains such as RH88 for *in vitro* infection studies and as source of tachyzoite extracts while less virulent Type II strains (e.g. ME49) were used for *in vivo* challenge. Although an initial examination of the IL-12 response of splenic DC to Type I and Type II strains failed to reveal a major difference [20], strain distinctions may nevertheless be an important concern in the interpretation of DC functional data. Further studies are needed to examine this issue and to evaluate whether differences in DC reactivity play a role in virulence determination.

Perhaps the ultimate questions concerning the role of DC in the response to *T. gondii* are whether DC are indeed essential for resistance in the murine experimental model and whether a comparable requirement exists in humans. While the former question could be approached in murine studies using newly developed genetic tools for depleting DC *in vivo* [47] our knowledge of the function of DC in the immune response to the parasite in humans is still at a primitive stage. As discussed above, human monocyte derived DC to not appear to be as reactive to *T. gondii* or STAg as murine splenic DC and require co-stimulation in order to display substantial activation responses. Whether this discrepancy reflects a difference in TLR expression between mouse and human (see above) or our inability to study the appropriate DC subset in humans remains to be determined.

Regardless of the inconsistencies with the mouse data, it may be possible to indirectly assess the function of human DC in host resistance through correlative studies in infected subjects. Considerable regional and household clustering of *T. gondii* infection and disease has been observed in countries such as Brazil that cannot be accounted for simply on the basis of parasite exposure [48] (R.T. Gazzinelli, personal communication). As genetic factors are identified which influence these observed differences in the expression of human toxoplasmosis it will be important to include assays of DC function in the group of parameters measured in the population under study. Since disease is relatively rare amongst infected individuals, the discovery of correlations with defective DC response could provide a platform for a more detailed analysis of the influence of DC function on human susceptibility to this widespread pathogen.

Acknowledgements

We thank Caetano Reis e Sousa, Giorgio Trinchieri and Brian Kelsall for their helpful comments and suggestions during the preparation of this chapter.

References

1 Dubey, J. P., Advances in the life cycle of *Toxoplasma gondii*. *Int J Parasitol* **1998**. 28: 1019–1024.

2 Frenkel, J. K., Pathophysiology of toxoplasmosis. *Parasitol Today* **1988**. 4: 273–278.

3 Yap, G. S. and Sher, A., Cell-mediated immunity to *Toxoplasma gondii*: initiation, regulation and effector function. *Immunobiology* **1999**. 201: 240–247.

4 Gazzinelli, R. T., Wysocka, M., Hayashi, S., Denkers, E. Y., Hieny, S., Caspar, P., Trinchieri, G. and Sher, A., Parasite-induced IL-12 stimulates early IFN-gamma synthesis and resistance during acute infection with *Toxoplasma gondii*. *J Immunol* **1994**. 153: 2533–2543.

5 Jankovic, D., Kullberg, M. C., Hieny, S., Caspar, P., Collazo, C. M. and Sher, A., In the absence of IL-12, CD4(+) T cell responses to intracellular pathogens fail to default to a Th2 pattern and are host protective in an IL-10(–/–) setting. *Immunity* **2002**. 16: 429–439.

6 Trinchieri, G., Interleukin-12 and the regulation of innate resistance and adaptive immunity. *Nat Rev Immunol* **2003**. 3: 133–146.

7 Reis e Sousa, C., Hieny, S., Scharton-Kersten, T., Jankovic, D., Charest, H., Germain, R. N. and Sher, A., In vivo microbial stimulation induces rapid CD40 ligand-independent production of interleukin 12 by dendritic cells and their redistribution to T cell areas. *J Exp Med* **1997**. 186: 1819–1829.

8 Reis e Sousa, C., Yap, G., Schulz, O., Rogers, N., Schito, M., Aliberti, J., Hieny, S. and Sher, A., Paralysis of dendritic cell IL-12 production by microbial products prevents infection-induced immunopathology. *Immunity* **1999**. 11: 637–647.

9 Schulz, O., Edwards, A. D., Schito, M., Aliberti, J., Manickasingham, S., Sher, A. and Reis e Sousa, C., CD40 triggering of heterodimeric IL-12 p70 production by dendritic cells in vivo requires a microbial priming signal. *Immunity* **2000**. 13: 453–462.

10 Aliberti, J., Reis e Sousa, C., Schito, M., Hieny, S., Wells, T., Huffnagle, G. B. and Sher, A., CCR5 provides a signal for microbial induced production of IL-12 by CD8 alpha+ dendritic cells. *Nat Immunol* **2000**. 1: 83–87.

11 Sher, A. and Reis e Sousa, C., Ignition of the type 1 response to intracellular infection by dendritic cell-derived interleukin-12. *Eur Cytokine Netw* **1998**. 9: 65–68.

12 Seguin, R. and Kasper, L. H., Sensitized lymphocytes and CD40 ligation augment interleukin-12 production by human dendritic cells in response to Toxoplasma gondii. *J Infect Dis* **1999**. 179: 467–474.

13 Subauste, C. S. and Wessendarp, M., Human dendritic cells discriminate between viable and killed *Toxoplasma gondii* tachyzoites: dendritic cell activation after infection with viable parasites results in CD28 and CD40 ligand signaling that controls IL-12-dependent and -independent T cell production of IFN-gamma. *J Immunol* **2000**. 165: 1498–1505.

14 McKee, A. S., Dzierszinski, F., Boes, M., Roos, D. S. and Pearce, E. J., Functional inactivation of immature dendritic cells by the intracellular parasite *Toxoplasma gondii*. *J Immunol* **2004**. 173: 2632–2640.

15 Aliberti, J., Valenzuela, J. G., Carruthers, V. B., Hieny, S., Andersen, J., Charest, H., Reis e Sousa, C., Fairlamb, A., Ribeiro, J. M. and Sher, A., Molecular mimicry of a CCR5 binding-domain in the microbial activation of dendritic cells. *Nat Immunol* **2003**. 4: 485–490.

16 Yarovinsky, F., Andersen, J. F., King, L. R., Caspar, P., Aliberti, J., Golding, H. and Sher, A., Structural determinants of the anti-HIV activity of a CCR5 antagonist derived from *Toxoplasma gondii*. *J Biol Chem* **2004**. 279: 53 635–53 642.

17 Scanga, C. A., Aliberti, J., Jankovic, D., Tilloy, F., Bennouna, S., Denkers, E. Y., Medzhitov, R. and Sher, A., Cutting edge: MyD88 is required for resistance to *Toxoplasma gondii* infection and regulates parasite-induced IL-12 production by dendritic cells. *J Immunol* **2002**. 168: 5997–6001.

18 Mun, H. S., Aosai, F., Norose, K., Chen, M., Piao, L. X., Takeuchi, O., Akira, S., Ishikura, H. and Yano, A., TLR2 as an essential molecule for protective

immunity against *Toxoplasma gondii* infection. *Int Immunol* **2003**. 15: 1081–1087.

19 Del Rio, L., Butcher, B. A., Bennouna, S., Hieny, S., Sher, A. and Denkers, E. Y., *Toxoplasma gondii* triggers MyD88-dependent IL-12 and CCL2 (MCP-1) responses using distinct parasite molecules and host receptors. *J Immunol* **2004**. 172: 6954–6960.

20 Robben, P. M., Mordue, D. G., Truscott, S. M., Takeda, K., Akira, S. and Sibley, L. D., Production of IL-12 by Macrophages Infected with *Toxoplasma gondii* Depends on the Parasite Genotype. *J Immunol* **2004**. 172: 3686–3694.

21 Yarovinsky, F., Zhang, D., Andersen, J. F., Bannenberg, G. L., Serhan, C. N., Hayden, M. S., Hieny, S., Sutterwala, F. S., Flavell, R. A., Ghosh, S. and Sher, A., TLR11 activation of dendritic cells by a protozoan profilin-like protein. *Science* **2005**. 308: 1626–1629.

22 Triantafilou, K., Triantafilou, M. and Dedrick, R. L., A CD14-independent LPS receptor cluster. *Nat Immunol* **2001**. 2: 338–345.

23 Zhang, D., Zhang, G., Hayden, M. S., Greenblatt, M. B., Bussey, C., Flavell, R. A. and Ghosh, S., A toll-like receptor that prevents infection by uropathogenic bacteria. *Science* **2004**. 303: 1522–1526.

24 Kang, H. K., Lee, H. Y., Lee, Y. N., Jo, E. J., Kim, J. I., Aosai, F., Yano, A., Kwak, J. Y. and Bae, Y. S., *Toxoplasma gondii*-derived heat shock protein 70 stimulates the maturation of human monocyte-derived dendritic cells. *Biochem Biophys Res Commun* **2004**. 322: 899–904.

25 Denkers, E. Y., Kim, L. and Butcher, B. A., In the belly of the beast: subversion of macrophage proinflammatory signalling cascades during *Toxoplasma gondii* infection. *Cell Microbiol* **2003**. 5: 75–83.

26 Gazzinelli, R. T., Wysocka, M., Hieny, S., Scharton-Kersten, T., Cheever, A., Kuhn, R., Muller, W., Trinchieri, G. and Sher, A., In the absence of endogenous IL-10, mice acutely infected with *Toxoplasma gondii* succumb to a lethal immune response dependent on CD4+ T cells and accompanied by overproduction of IL-12, IFN-gamma and TNF-alpha. *J Immunol* **1996**. 157: 798–805.

27 Chiang, N., Fierro, I. M., Gronert, K. and Serhan, C. N., Activation of lipoxin A(4) receptors by aspirin-triggered lipoxins and select peptides evokes ligand-specific responses in inflammation. *J Exp Med* **2000**. 191: 1197–1208.

28 Deng, X., Ueda, H., Su, S. B., Gong, W., Dunlop, N. M., Gao, J. L., Murphy, P. M. and Wang, J. M., A synthetic peptide derived from human immunodeficiency virus type 1 gp120 downregulates the expression and function of chemokine receptors CCR5 and CXCR4 in monocytes by activating the 7-transmembrane G-protein-coupled receptor FPRL1/LXA4R. *Blood* **1999**. 94: 1165–1173.

29 Aliberti, J., Hieny, S., Reis e Sousa, C., Serhan, C. N. and Sher, A., Lipoxin-mediated inhibition of IL-12 production by DCs: a mechanism for regulation of microbial immunity. *Nat Immunol* **2002**. 3: 76–82.

30 Aliberti, J., Serhan, C. and Sher, A., Parasite-induced lipoxin A4 is an endogenous regulator of IL-12 production and immunopathology in *Toxoplasma gondii* infection. *J Exp Med* **2002**. 196: 1253–1262.

31 Gazzinelli, R. T., Hieny, S., Wynn, T. A., Wolf, S. and Sher, A., Interleukin 12 is required for the T-lymphocyte-independent induction of interferon gamma by an intracellular parasite and induces resistance in T-cell-deficient hosts. *Proc Natl Acad Sci U S A* **1993**. 90: 6115–6119.

32 O'Garra, A., Hosken, N., Macatonia, S., Wenner, C. A. and Murphy, K., The role of macrophage- and dendritic cell-derived IL-12 in Th1 phenotype development. *Res Immunol* **1995**. 146: 466–472.

33 Macatonia, S. E., Hosken, N. A., Litton, M., Vieira, P., Hsieh, C. S., Culpepper, J. A., Wysocka, M., Trinchieri, G., Murphy, K. M. and O'Garra, A., Dendritic cells produce IL-12 and direct the development of Th1 cells from naive CD4+ T cells. *J Immunol* **1995**. 154: 5071–5079.

34 Scharton-Kersten, T. M., Yap, G., Magram, J. and Sher, A., Inducible nitric oxide is essential for host control of persistent but not acute infection with the intracellular pathogen *Toxoplasma gondii*. *J Exp Med* **1997**. 185: 1261–1273.

35 Cai, G., Radzanowski, T., Villegas, E. N., Kastelein, R. and Hunter, C. A., Identification of STAT4-dependent and independent mechanisms of resistance to *Toxoplasma gondii*. *J Immunol* **2000**. 165: 2619–2627.

36 Villarino, A., Hibbert, L., Lieberman, L., Wilson, E., Mak, T., Yoshida, H., Kastelein, R. A., Saris, C. and Hunter, C. A., The IL-27R (WSX-1) is required to suppress T cell hyperactivity during infection. *Immunity* **2003**. 19: 645–655.

37 Jankovic, D., Kullberg, M. C., Caspar, P. and Sher, A., Parasite-induced Th2 polarization is associated with down-regulated dendritic cell responsiveness to Th1 stimuli and a transient delay in T lymphocyte cycling. *J Immunol* **2004**. 173: 2419–2427.

38 Scharton-Kersten, T., Contursi, C., Masumi, A., Sher, A. and Ozato, K., Interferon consensus sequence binding protein-deficient mice display impaired resistance to intracellular infection due to a primary defect in interleukin 12 p40 induction. *J Exp Med* **1997**. 186: 1523–1534.

39 Aliberti, J., Schulz, O., Pennington, D. J., Tsujimura, H., Reis e Sousa, C., Ozato, K. and Sher, A., Essential role for ICSBP in the in vivo development of murine CD8alpha+ dendritic cells. *Blood* **2003**. 101: 305–310.

40 Constant, S. L. and Bottomly, K., Induction of Th1 and Th2 CD4+ T cell responses: the alternative approaches. *Annu Rev Immunol* **1997**. 15: 297–322.

41 Pepper, M., Dzierszinski, F., Crawford, A., Hunter, C. A. and Roos, D., Development of a system to study CD4+-T-cell responses to transgenic ovalbumin-expressing *Toxoplasma gondii* during toxoplasmosis. *Infect Immun* **2004**. 72: 7240–7246.

42 Gubbels, M. J., Striepen, B., Shastri, N., Turkoz, M. and Robey, E. A., Class I Major Histocompatibility Complex Presentation of Antigens That Escape from the Parasitophorous Vacuole of *Toxoplasma gondii*. *Infect Immun* **2005**. 73: 703–711.

43 Dimier-Poisson, I., Aline, F., Mevelec, M. N., Beauvillain, C., Buzoni-Gatel, D. and Bout, D., Protective mucosal Th2 immune response against *Toxoplasma gondii* by murine mesenteric lymph node dendritic cells. *Infect Immun* **2003**. 71: 5254–5265.

44 Bourguin, I., Moser, M., Buzoni-Gatel, D., Tielemans, F., Bout, D., Urbain, J. and Leo, O., Murine dendritic cells pulsed in vitro with *Toxoplasma gondii* antigens induce protective immunity in vivo. *Infect Immun* **1998**. 66: 4867–4874.

45 Denkers, E. Y., Del Rio, L. and Bennouna, S., Neutrophil production of IL-12 and other cytokines during microbial infection. *Chem Immunol Allergy* **2003**. 83: 95–114.

46 Li, Z. Y., Manthey, C. L., Perera, P. Y., Sher, A. and Vogel, S. N., *Toxoplasma gondii* soluble antigen induces a subset of lipopolysaccharide-inducible genes and tyrosine phosphoproteins in peritoneal macrophages. *Infect Immun* **1994**. 62: 3434–3440.

47 Jung, S., Unutmaz, D., Wong, P., Sano, G., De los Santos, K., Sparwasser, T., Wu, S., Vuthoori, S., Ko, K., Zavala, F., Pamer, E. G., Littman, D. R. and Lang, R. A., In vivo depletion of CD11c(+) dendritic cells abrogates priming of CD8(+) T cells by exogenous cell-associated antigens. *Immunity* **2002**. 17: 211–220.

48 Portela, R. W., Bethony, J., Costa, M. I., Gazzinelli, A., Vitor, R. W., Hermeto, F. M., Correa-Oliveira, R. and Gazzinelli, R. T., A multihousehold study reveals a positive correlation between age, severity of ocular toxoplasmosis, and levels of glycoinositolphospholipid-specific immunoglobulin A. *J Infect Dis* **2004**. 190: 175–183.

34
Schistosoma

Andrew S. MacDonald and Edward J. Pearce

34.1
Introduction

It has become clear over the past decade that dendritic cells (DC) play a pivotal role in providing the cues that determine the Th1/Th2 effector function bias of CD4 T-cell responses. The early concept of specialized DC subsets, identified in mice on the basis of differential expression of CD4 and CD8α, being restricted in their ability to induce either Th1 or Th2 responses ("DC1" vs. "DC2") has evolved and developed to the point that we now realize that essentially all DC possess the ability to drive both Th1 and Th2 responses. The crucial element that shapes DC plasticity is now recognized to be the nature of the "conditioning" information imparted by the particular pathogen that the DC encounter [1, 2]. The molecular basis of conditioning is most well understood from studies of the maturation of DC exposed to Th1-inducing pathogens, or more usually, defined pathogen-associated molecular patterns (PAMPs) from these organisms. The Th1-biased nature of the data that had been used to construct the current paradigm of DC maturation raised the obvious question of whether DC are activated and function in a similar way in Th2-dominated settings. Several years ago we set out to address this question in the context of an infection, schistosomiasis, in which the immune response is Th2-dominated.

Schistosomes, the causative agents of schistosomiasis, are parasitic helminths (-worms). About 650 million individuals in 76 countries and territories around the world are thought to be are at risk of infection with these complex metazoan pathogens, with at least 190 million people actively infected, according to recent calculations applied to the 1995 world population estimates [3]. In terms of global clinical impact, an estimated 85% of people infected with schistosomes are thought to be located in the African continent [3]. In this area alone, it has been estimated that up to 300 000 people may die each year due to the consequences of severe schistosomiasis.

Four species, *Schistosoma mansoni*, *S. japonicum*, *S. haematobium* and *S. mekongi*, represent the most important agents of human schistosomiasis. The life cycle of these schistosome species is complex. Sexually mature male and female parasites

Handbook of Dendritic Cells. Biology, Diseases, and Therapies.
Edited by M. B. Lutz, N. Romani, and A. Steinkasserer.

live intravascularly in the portal vasculature (*S. mansoni, S. japonicum, S. mekongi*) or in the blood vessels around the bladder (*S. haematobium*), and eggs produced by female worms pass out of the body via the intestine or the bladder (depending on where the adult parasites live), and hatch in fresh water to release a motile stage that finds and invades the snail intermediate host. As a result of differentiation and asexual reproduction in the snail, new life cycle stages (cercariae) are produced that leave the snail and sit in the water in wait for the definitive human host, which they infect by direct skin-penetration. After a few days in the skin, these larval schistosomes (schistosomula) enter the vasculature and over the course of several weeks migrate intravascularly to their final niche. At this site, the parasites mature and mate. During this period, which can take 5–6 weeks, the host is exposed to antigens from successive developmental stages of the parasite. Infections with schistosomes are chronic and although the immune response eventually becomes capable of preventing superinfection (Reviewed in [4]), it is not able to clear established organisms. Pathology during infection is the result of the host's CD4 T cell-mediated granulomatous response to tissue-trapped parasite eggs.

The relative ease with which *S. mansoni* can be maintained in the laboratory, coupled with the susceptibility of the mouse to infection with this parasite species, has provided an accessible system for studying schistosomiasis. Work with this model has revealed that the dominant immune response to these organisms is Th2-like, and that Th2 cells permit host survival but also contribute to the development of the granulomas and fibrosis that envelope trapped eggs [4, 5]. In early experiments monitoring the time-course of Th2 response development during infection it became apparent that the initial stage of infection during which immature parasites migrate from the skin site of entry to the portal vasculature, is associated with a weak Th1 response, and that the Th2 response develops coincidently with the sexual maturation of the parasites and the onset of egg production [4]. Subsequent work has shown that eggs isolated from infected animals induce marked Th2 responses when injected into naïve animals [4]. Importantly, an extract of eggs, SEA (schistosome egg Ag) mimics whole schistosome eggs in that it is also potent at driving Th2 development after injection into naïve mice, even without the need for co-administration of adjuvant. More than this, SEA, like some other helminth products or extracts [6, 7], could itself be described as an adjuvant, being able to promote Th2 development to model antigens [8, 9]. SEA is known to contain the antigens that are secreted by eggs, and which both induce and are the target of the Th2 response under physiological conditions.

34.2
DC Response to Schistosome Ag

The establishment of a link between the ability of pathogens to stimulate IL-12 production by DC and to induce Th1 responses led to the conclusion that DC not only induce T-cell activation by presenting antigenic peptide/MHC complexes in the context of appropriate co-stimulatory signals, but also provide an additional "third" signal that ensures the right type of immune T-cell response develops within a given context [10]. Moreover, especially following the publication of micro-array analyses of gene expression in DC exposed to viral, bacterial, or fungal organisms [11], it is clear that these cells are capable of initiating large scale changes in gene expression in response to PAMPs, and that different pathogens can elicit responses that have both specific elements, and general elements common to the response to other types of pathogens. Thus the "third" signal could comprise the coordinated expression (or repression) of multiple relevant genes. In light of these developments, we were interested in examining the effects of SEA, an inherently Th2-inducing antigen, on DC. We anticipated that SEA would induce some degree of activation, perhaps characterized by the expression of IL-4, a gene known to play a role in Th2 cell development. However, our initial analyses of the expression of cytokines and co-stimulatory molecules by murine bone-marrow-derived DC exposed overnight to SEA or dead *S. mansoni* eggs failed to reveal any sign of activation or maturation (Fig. 34.1) [12]. Specifically, we could find no evidence, either at

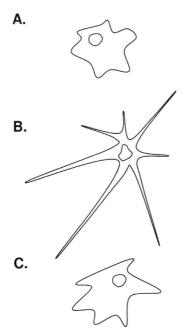

A.

Immature DC:

MHC class II^{lo}
CD40^{lo}, CD54^{lo}, CD80/86^{lo}
No detectable cytokine production
Muted morphology
Excellent phagocytic ability

B.

Conventionally activated DC (e.g. LPS):

MHC class II^{hi}
CD40^{hi}, CD54^{hi}, CD80/86^{hi}
High level cytokine production (e.g. IL-12, TNF, IL-6)
Stellate morphology
Loss of phagocytic ability

C.

Modulated DC (SEA):

MHC class II^{int}
CD40^{lo}, CD54^{lo}, CD80/86^{lo}
No detectable cytokine production (e.g. IL-12 or IL-4)
Muted morphology
No loss of phagocytic ability

Fig. 34.1 The response of DC to schistosomes. (A) Immature DC are specialized at capturing Ag, but display only low-level expression of molecules such as MHCII or CD80/86, and are thus relatively poor at stimulating naïve T cells. (B). On activation or "maturation" with conventional stimuli such as LPS, DC dramatically transform their morphology to the archetypal stellate shape, lose their Ag capturing ability but concurrently up-regulate expression of surface molecules such as MHCII and CD80/86 and cytokines such as IL-12. As such, these conventionally activated DC are extremely potent at activating naïve T cells. (C) DC exposed to SEA have similarity to immature, rather than conventionally mature, DC. In comparison to immature cells SEA-treated DC display only minor surface MHCII upregulation, retain the ability to phagocytose (unpublished data), and fail to produce detectable cytokine.

the protein or mRNA level, for significant up-regulation of activation markers such as CD80, CD86, CD54 or CD40 or production of cytokines such as IL-12 (p40 or p70), IL-10, IL-6, or TNF. Indeed, the only evidence that we could find for a response of the cells to SEA was a minor up-regulation of surface MHC Class II expression. This contrasted markedly with the effect of *E. coli* LPS or heat killed *Propionibacterium acnes* (a gram+ bacterium) on DC, which we found to be as expected, inducing strongly increased expression of most of the markers examined. Notably, we have never observed IL-4 production by DC, no matter which stimulus we have examined (data not shown). Comprehensive micro-array analyses of gene expression in DC indicate that the expression of very few genes is changed in response to SEA, and analyses of those few that are affected have revealed no currently appreciated links to Th2 response induction [13].

Based on the available data at the time (our study and another which had shown a similar lack of response of murine bone marrow DC to an Ag derived from the parasitic nematode *Acanthocheilonema viteae* [14]) we proposed that a lack of conventional DC activation might be a feature of DC responses to helminths, and perhaps of Th2 development in general. Follow-up studies in other helminth systems have prompted a reassessment of this original idea, indicating that gastrointestinal nematodes such as *Toxocara canis* [15] and *Trichuris muris* (Richard Grencis, personal communication) may "selectively" mature DC, inducing up-regulation of some, but not all, of the hallmarks of conventional activation. However, it needs to be stressed that a complicating issue of these types of studies is that the preparation of Ag from helminths first requires the isolation of the helminth life stage of interest from their hosts or, in some cases, from the environment, and thus any work with these pathogens must include careful controls to rule out the possibility that observed effects on DC are due to contaminating PAMPS such as LPS. In the case of the SEA that we have prepared, due to what we believe are important precautions taken during the preparation of the Ag, which include the inclusion of polymyxin B in the buffers used to isolate the eggs, and the final passage of the isolated eggs through a Percoll density gradient, we have routinely failed to identify any contamination with LPS. Recent work has indicated that SEA not only fails to conventionally activate DC, but additionally is able to induce the downregulation of the steady-state expression levels of some genes [16], and suppress the activation of DC induced by TLR ligands such as *E. coli* LPS, *P. acnes*, CpG, poly I:C or soluble *T. gondii* [13, 17, 18]. This is in part due to the fact that SEA promotes TLR ligand-induced production of IL-10 [13]. However, SEA inhibits TLR-initiated DC activation even in situations where IL-10 is absent. Although initial analyses focused on the ability of SEA to inhibit IL-12 production, or co-stimulatory molecule upregulation, detailed analyses have revealed that SEA inhibits TLR-initiated MAPK and NF-κB signaling, and micro-array based analyses show significant IL-10-independent effects of SEA on the expression of over 100 LPS-regulated genes [13]. These findings indicate that SEA exerts potent anti-inflammatory effects by directly regulating the ability of DC to respond to pro-inflammatory TLR ligands.

How SEA is able to suppress TLR-initiated signaling is the focus of ongoing research. One possibility is that SEA ligates receptors that initiate anti-inflammatory pathways. A candidate for such a receptor is the C-type lectin DC-SIGN or its mouse homolog. DC-SIGN was recently shown to recognize the major Lewis^x[Galβ1,4(Fucα1,3)GlcNAc] glycan epitope of SEA [19]. This is of considerable interest, since binding of the *Mycobacterium tuberculosis* component manLAM to DC-SIGN has been shown to inhibit the production of IL-12, and promote the production of IL-10 [20].

One caveat to the accumulating data indicating muted activation of DC by schistosomes has been presented by recent studies focusing on the effect of live schistosome eggs on murine dendritic cells. In these studies, live *S. mansoni* eggs derived from infected hamsters were found to provoke the transcription of a range of proinflammatory genes, including type I IFN (IFNβ) TNFα and IL-12p40, that one might at first glance associate with Th1, rather than Th2 response induction [21]. Recent work has identified double strand RNA, released from eggs and binding to TLR3 on DC, as the likely initiator of this response [22]. It remains to be seen whether dsRNA is also present within SEA, and/or whether the anti-inflammatory mediator in SEA is actually released from living eggs.

The influence of SEA on human DC appears subtly different to that seen with murine cells. Available data suggest the human monocyte-derived DC (mDC) exposed to SEA *in vitro* display a degree of activation at the level of phenotype, although less than that induced by conventional maturation stimuli such as LPS [23, 24]. Similar to what is seen with murine DC, however, human mDC produce little detectable IL-12p70, TNF or IL-10 in response to SEA [23–25].

More refined dissection of which components of SEA are responsible for affecting DC function has revealed that lipid and carbohydrate components may play important roles. Fractionation of the lipid components of *S. mansoni* eggs or adult worms by TEAE-cellulose column chromatography has identified that lysophosphatidylserine (lyso-PS) isolated from either life cycle stage confers on human mDC the ability to skew allogeneic T cells towards Th2 development *in vitro*. Lyso-PS from either stage also appears to induce IL-10-producing T cells in such conditions. Strikingly, the ability of DC exposed to lyso-PS to modulate DC to induce IL-10 producing "regulatory" T cells requires TLR2, whereas induction of IL-4 producing Th2 cells does not [25]. In contrast to this, somewhat confusingly, TLR4 has been implicated in the ability of lacto-*N*-fucopentaose III (which contains the Lewis^x epitope alluded to above)-pulsed DC to promote Th2 cell development [26, 27]. While these reports and that identifying schistosome egg-derived dsRNA as a ligand for TLR3 [22], implicate TLRs in the induction of Th2 responses, it is important to emphasize that the SEA-specific Th2 response can develop in the absence of MyD88 [66], and that the ability of murine splenic DC to be conditioned by SEA to drive Th2 responses *in vitro* is MyD88-independent [16]. Thus if, as several reports now indicate [23, 28], TLRs are involved in the recognition of SEA by DC, they should be envisaged as activating MyD88-independent rather than MyD88-dependent signaling pathways leading to the conditioning of the DC to promote Th2 cell development. This is consistent with the generally accepted notion that MyD88-dependent signaling leads to the development of DC capable of inducing Th1 rather than Th2 responses [29].

Few studies detail the response of DC to other (i.e. non-egg) schistosome life cycle stages. A recent report indicates that secretions of the infective larvae of *S. mansoni* appear to activate or mature murine DC in a more conventional manner than is true for SEA [30], while still conferring on DC the ability to direct DO11.10 Tg T cells towards a Th2 development. It should be born in mind, however, that the prepatent phase of schistosome infection in mice has not been clearly defined in terms of what type of T-cell response dominates, with a strong Th2 response only really becoming evident after the onset of egg laying (see above) [31, 32].

34.3
Th2 Induction by DC in Response to Schistosome Ag

An emerging idea is that DC that are not fully activated (immature or "semi-mature"), may be tolerogenic rather than immunogenic [33, 34]. It has been proposed that this might provide a mechanism for tolerance induction when DC encounter "innocuous" Ag in the periphery in the form of commensal bacteria, or apoptotic cells. However, the universal applicability of this theory has been called into question by studies of the immunogenic properties of DC exposed to schistosome Ag. As introduced above, DC exposed to SEA effectively retain an immature phenotype and are inhibited from responding maximally to co-pulsed inflammatory TLR-ligands. The obvious question that arises given that SEA pulsed DC could arguably be described as immature or "semi-mature" [34] by standard criteria, is: Are these DC immunogenic or tolerogenic?

If fact SEA-primed DC do not tolerize T cells but instead drive a potent Th2 response, following their transfer into naïve recipient mice [12, 35], and *in vitro* during culture with naïve ova-specific T-cell receptor-transgenic T cells [16]. The Th2-driving ability of SEA-pulsed DC is lost when transferred DC are MHC class II-deficient [12], implying that T-cell responses in this system are induced by the injected DC, and not simply by Ag transfer to recipient cells. These kinds of experiment suggest that DC can play a, perhaps surprisingly, autonomous role in provoking SEA-specific Th2 responses, with no requirement for additional innate cell recognition of any of the components of SEA in this process. This is particularly interesting given that a range of other innate cells including mast cells, basophils and eosinophils, have been shown to produce IL-4 rapidly after exposure to schistosome eggs or egg Ag [36–38]. However, eggs can induce equivalent Th2 responses in the presence or absence of mast cells or eosinophils [36, 39], suggesting that IL-4 derived from these innate sources is contributory, rather than central, for the Th2 process in this setting. Together, these points raise the obvious question of mechanism: how do SEA-exposed DC, activate naïve Th cells and provoke Th2 development?

As alluded to above, our initial studies failed to find any evidence, either at the protein or the message level, for IL-4 production by bone marrow DC in response to SEA [12], even though DC IL-4 has been proposed as a mechanism by which fetal mouse skin DC induce Th2 polarization [40]. Subsequently, we and others have shown that IL-4-deficient DC induce an equivalent SEA-specific Th2 response when compared to wild-type DC, both *in vivo* [41] and *in vitro* [16], formally ruling out a requirement for IL-4 production during Th2 induction in this context. Similar experiments transferring wild-type or gene-deficient DC into wild-type recipients have revealed that neither IL-12 nor IL-10 production by DC influences their ability to drive an SEA-specific Th2 response ([41] and unpublished data).

Much attention has been afforded to the potential regulatory role of endogenous IL-10 produced during DC activation, and how this might influence their subsequent ability to direct T-cell responses [42, 43]. Although DC-derived IL-10 may affect Th-cell polarization in certain settings, we have found that SEA-pulsed IL-10-deficient murine bone marrow derived DC induce equivalent Th2 response development to WT DC after transfer into recipient mice (Perona-Wright and MacDonald, unpublished observations). Moreover, we have also found that when SEA-pulsed DC (whether WT or IL-10-deficient) are transferred into IL-10-deficent recipient mice, Th2 development is significantly reduced in comparison to WT recipient animals, and accompanied by the emergence of SEA-specific IFNγ production (Perona-Wright and MacDonald, unpublished observations). The implication of this is that IL-10 derived from a source other than the initiating DC acts in the SEA-driven Th2 setting to limit IFNγ production. Thus IL-10, rather than directly promoting Th2 development, appears to be acting to constrain Th1 development to SEA.

An alternative possibility is that SEA-primed DC may in fact play no active role in Th2 development, with Th2 response simply emerging in the absence of DC production of Th1-skewing mediators such as IL-12 ⁻ the so-called "default hypothesis" [44, 45]. Although some evidence to support this passive model exists, such data has generally been derived from experimental systems using Ag that are not pathogen-derived, so the relevance of this concept must be re-evaluated in the light of current understanding of the important contribution that the type of Ag encountered makes to DC function. The recent description of the powerful ability of SEA to actively interfere with TLR-mediated signaling in DC (see above) discounts a passive default model of Th2 induction to SEA, but does suggest that suppression of TLR-initiated activation events may favor Th2 development. On this basis, perhaps "default" Th2 development should be qualified as "active" (e.g. SEA) or "passive" (e.g. ova), based on whether or not the Ag in question displays any measurable effect (positive or negative) on TLR-mediated DC activation. Additionally, the possibility of a novel DC mechanism that actively promotes Th2 cell development cannot yet be excluded. It is possible that such a mechanism may only be revealed after "secondary" stimulation of SEA-primed DC via CD40. Indeed, we recently reported the surprising finding that CD40-deficient murine DC lose the ability to stimulate Th2 development to SEA [46], even though SEA does not itself actively induce expression of CD40 on exposed DC [12]. CD40-mediated "licensing" of DC function is not a new concept, but has focused on provision of help for CD8 T-cell responses [47, 48], and Th1 amplification through increased IL-12 production [49, 50]. However, in addition to provoking elevated IL-12 production, CD40 signaling can exert a range of activation effects on DC, including enhancing expression of a wide range of co-stimulatory molecules such as CD80, CD86, CD70, 41BBL and OX40L [49, 51–53]. This prompts the question of what molecules downstream of DC activation via CD40 might be involved in the Th2 induction process to SEA. Although not upregulated on the surface of murine DC after exposure to SEA, OX40L has been implicated in the ability of human mDC to direct Th2 polarization of naïve T cells *in vitro* [24], making this a prime candidate for a molecule that could provide a positive signal from DC for Th2 induction. Whether this is also the case *in vivo* remains to be determined, as does identifying whether OX40L represents the sole positive signal for SEA driven Th2 induction, or one of a team of collaborators. Of considerable interest in this regard is the recent report that the expression of the Notch ligand Jagged by DC is important for their ability to induce Th2 cell development [54], and current work aims to examine Jagged expression on DC exposed to SEA plus or minus CD40 agonist.

As mentioned previously, there are only limited reports of how other (non SEA) Ag influence the ability of DC to direct T-cell responses, and this area requires further work. It appears that while the secretions of mechanically transformed cercariae confer on murine DC the ability to polarize DO11.10 T cells *in vitro* and *in vivo* [30], soluble schistosomulum Ag does not act to do the same when pulsed DC are transferred into naïve recipient mice [35].

It is clear that several factors, including human DC versus murine DC differences, source/subset of DC used, method of DC generation, etc., collude to confound

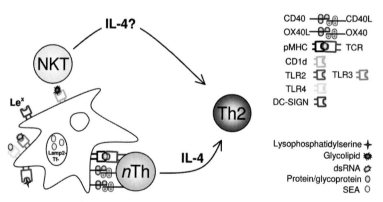

Fig. 34.2 Th2 induction by DC in response to SEA. It is likely that multiple components of SEA bind to a range of pattern recognition receptors on DC, including C-type lectins [19] and TLRs [22, 25, 27]. Although several TLRs have been implicated in this process, MyD88-deficient mice are still capable of making Th2 responses to SEA [66], suggesting either a degree of redundancy in the system or, more likely, that MyD88-independent TLR-initiated signaling is important. Once internalized SEA is distributed to Lamp2⁻, Transferrin⁻ vesicular compartments within the DC [18].

CD40 expression is critical for DC induction of Th2 response to SEA [46], suggesting downstream involvement of CD40-mediated activation events in this process, with OX-40L being one potential candidate that could fulfill this function. IL-4-deficient DC show no deficiency in Th2 induction to SEA [41], although IL-4 from a source other than the initiating DC is paramount to sustain the developing CD4 T-cell response. NKT cells may provide one source of such IL-4, as CD1d-deficient DC display impaired Th2-induction abilities to SEA [35].

attempts to build a coherent understanding of how DC function during Th2 development to SEA. Perhaps the single most important factor to consider may be whether studies have been carried out *in vitro* or *in vivo*. While *in vitro* DC:T cell coculture experiments undeniably provide a finely controlled system, they exclude the potentially central input and influence of other cell types during the T-cell activation and polarization process [55]. For example, once activated, CD4 T cells migrate to B cell follicles [56], wherein the B cells can provide "help" to support T-cell expansion, effector function and memory formation, via mechanisms such as costimulation [57]. It could be argued that the interaction of T cells with other cell types may not be important during T-cell priming, where DC:T cell dialog has been thought to be key. However, it appears that even in DC-driven Ag-delivery systems (i.e. in the absence of free Ag) innate cells such as NK cells can be intricately involved in T-cell polarization *in vivo* [58]. In relation to this, several recent reports have promoted a role for CD1d-restricted NKT cells in SEA-driven Th2 response development *in vivo* [17, 35]. It is likely, therefore, that during both induction and amplification of the developing T-cell response, a "network" of additional cell types contributes towards the emergent T-cell polarization phenotype.

A summary of the issues discussed in this section is presented in Fig. 34.2.

34.4
DC During Schistosome Infection

Understanding of how DC function during active schistosome infection is still in its infancy. Dissecting the role of DC in the complex setting of an ongoing infection with an organism that has several developmental stages and locations within the mammalian host presents a somewhat daunting challenge. Added to these factors, chronic schistosome infection is in no way as controlled a setting as direct, one-shot Ag administration, with ongoing production and release of eggs providing a constant and accumulating source of Ag. How can we begin to evaluate DC activation and function in this context? The limited data on this topic that does exist has been generated in rodent models of schistosome infection.

A key role has been proposed for cercarial-derived prostaglandins (PGD_2) in preventing migration of Langerhans cells (LC) from the site of larval penetration to the skin-draining lymph nodes during infection of mice with *S. mansoni* [59]. Whether other stages of the parasite exert similar modulatory effects on LC (or other DC subsets) remains to be determined. These results expand upon earlier experiments assessing cellular recruitment in skin-draining lymph nodes of guinea pigs exposed to *S. mansoni* cercariae. In this case, enhanced LC recruitment was seen following percutaneous exposure of guinea pigs to UV-attenuated compared to normal *S. mansoni* cercariae [60], raising the interesting possibility that attenuated cercariae may themselves exhibit impaired PGE_2 production or secretion as well as being less effective at provoking PGE_2 from host skin cells [61].

As yet, studies assessing DC function during infection beyond the early stages of cercarial penetration are limited. We have assessed the activation phenotype of DC isolated from the spleen or MLN over the course of murine infection with *S. man-*

soni, and observed only minor up-regulation of expression of conventional activation markers by both CD8α$^+$ and CD8α$^-$ subsets, but most marked in the CD8α$^+$ cells, that are not apparent in SEA-pulsed bone marrow DC [62]. Another intriguing difference between SEA-pulsed bone marrow DC and DC isolated from the spleens of infected mice is that, while SEA-pulsed bone marrow DC do not make IL-12 following CD40 ligation [12], total CD11c$^+$ splenic DC isolated from infected mice are primed to make this cytokine and indeed produce more of it in response to CD40 ligation than do CD11c$^+$ splenic DC from uninfected mice [62]. These bone marrow/splenic DC differences could be explained by the heterogenous nature of CD11c$^+$ DC isolated *ex vivo*. Splenic CD11c$^+$ DC comprise several subsets, not all of which are similar to the CD8α$^-$ bone marrow DC most extensively studied in schistosome research [63]. Compared to CD8α$^-$ DC, CD8α$^+$ DC are more potent IL-12 producers and while both subsets from the spleens of infected mice can be activated to produce IL-12 in response to CD40 ligation, it is likely that it is the CD8α$^+$ subset that is capable of making the most IL-12.

The low-level activation phenotype of DC isolated from schistosome infected mice appears to depend upon CD40:CD154 interaction, as it is not apparent in DC from infected CD154$^{-/-}$ mice [62]. Further, maintenance of low-level DC activation status during murine schistosome infections may be controlled, at least in terms of MHC Class II and co-stimulatory molecule expression, by IL-10, since DC isolated from schistosome infected IL-10$^{-/-}$ mice display a hyperactivated phenotype [64]. Together, these data support the idea of a coordinated sequence of events centering on (low level) DC activation mediated by CD40 ligation, and under the ultimate control of IL-10.

Recent developments in the area of intravital imaging, using techniques such as 2-photon microscopy [65], should prove invaluable in furthering our currently rudimentary understanding of DC activation and function during active infection. An important step towards realizing such experiments will be the generation of transgenic mice to be able to track schistosome-specific lymphocytes during infection. Additionally, characterization of human DC isolated from schistosome-infected individuals will be the vital "next step" in extending the conclusions gained from murine studies to a broader stage, and ultimately towards the rational design and development of vaccines and therapies.

34.5
Discussion

How does all of the above information fit together in terms of what is actually likely to be occurring *in vivo* during active schistosome infection? It is becoming increasingly clear that in order to be able to assemble this increasingly complex puzzle, we will first need to gather more pieces. Determining how DC respond, both *in vitro* and *in vivo*, to the many and varied immune stimuli provided by the different schistosome life cycle stages, is the vital first step towards dissecting this process. Although technically challenging, translating that information into understanding its actual influence over effector T-cell fate *in vivo*, a process that is currently very much the realm of potentially misleading speculation, is the next crucial step that will require thinking outside the 96-well plate. Only once we have a firm grasp of these basic events will we be able to attempt to integrate available information into a unifying model of DC function during schistosomiasis.

Acknowledgements

Dendritic cell research in EJP's laboratory has been supported by grants from the NIH. EJP is a Burroughs Wellcome Fund Scholar in Molecular Parasitology. ASM's laboratory is supported by funding provided by The Medical Research Council (UK) and The Wellcome Trust.

References

1 Manickasingham, S. P., A. D. Edwards, O. Schulz, C. Reis e Sousa **2003**. The ability of murine dendritic cell subsets to direct T helper cell differentiation is dependent on microbial signals. *Eur. J. Immunol. 33:* 101.

2 Kapsenberg, M. L. **2003**. Dendritic-cell control of pathogen-driven T-cell polarization. *Nat. Rev. Immunol. 3:* 984.

3 Engels, D., L. Chitsulo, A. Montresor, L. Savioli. **2002**. The global epidemiological situation of schistosomiasis and new approaches to control and research. *Acta Trop. 82:* 139.

4 Pearce, E. J., A. S. MacDonald. **2002**. The immunobiology of schistosomiasis. *Nat. Rev. Immunol. 2:* 499.

5 Wynn, T. A. **2004**. Fibrotic disease and the T(H)1/T(H)2 paradigm. *Nat Rev Immunol 4:* 583.

6 Holland, M. J., Y. M. Harcus, P. L. Riches, R. M. Maizels. **2000**. Proteins secreted by the parasitic nematode Nippostrongylus brasiliensis act as adjuvants for Th2 responses. *Eur. J. Immunol. 30:* 1977.

7 Macedo, M. S., E. Faquim-Mauro, A. P. Ferreira, I. A. Abrahamsohn. **1998**. Immunomodulation induced by Ascaris suum extract in mice: effect of anti-interleukin-4 and anti-interleukin-10 antibodies. *Scand. J. Immunol. 47:* 10.

8 Okano, M., A. R. Satoskar, K. Nishizaki, M. Abe, D. A. Harn, Jr. **1999**. Induction of Th2 responses and IgE is largely due to carbohydrates functioning as adjuvants on Schistosoma mansoni egg antigens. *J. Immunol. 163:* 6712.

9 Okano, M., A. R. Satoskar, K. Nishizaki, D. A. Harn, Jr. **2001**. Lacto-N-fucopentaose III found on Schistosoma mansoni egg antigens functions as adjuvant for proteins by inducing Th2-type response. *J. Immunol. 167:* 442.

10 Kalinski, P., C. M. Hilkens, E. A. Wierenga, M. L. Kapsenberg. **1999**. T-cell priming by type-1 and type-2 polarized dendritic cells: the concept of a third signal. *Immunol. Today 20:* 561.

11 Huang, Q., D. Liu, P. Majewski, L. C. Schulte, J. M. Korn, R. A. Young, E. S. Lander, N. Hacohen. **2001**. The plasticity of dendritic cell responses to pathogens and their components. *Science 294:* 870.

12 MacDonald, A. S., A. D. Straw, B. Bauman, E. J. Pearce. **2001**. CD8-dendritic cell activation status plays an integral role in influencing Th2 response development. *J. Immunol. 167:* 1982.

13 Kane, C. M., L. Cervi, J. Sun, A. S. McKee, K. S. Masek, S. Shapira, C. A. Hunter, E. J. Pearce. **2004**. Helminth antigens modulate TLR-initiated dendritic cell activation. *J. Immunol. 173:* 7454.

14 Whelan, M., M. M. Harnett, K. M. Houston, V. Patel, W. Harnett, K. P. Rigley. **2000**. A filarial nematode-secreted product signals dendritic cells to acquire a phenotype that drives development of Th2 cells. *J. Immunol. 164:* 6453.

15 Balic, A., Y. Harcus, M. J. Holland, R. M. Maizels. **2004**. Selective maturation of dendritic cells by Nippostrongylus brasiliensis-secreted proteins drives Th2 immune responses. *Eur. J. Immunol. 34:* 3047.

16 Jankovic, D., M. C. Kullberg, P. Caspar, A. Sher. **2004**. Parasite-induced Th2 polarization is associated with down-regulated dendritic cell responsiveness to Th1 stimuli and a transient delay in T lymphocyte cycling. *J. Immunol. 173:* 2419.

17 Zaccone, P., Z. Fehervari, F. M. Jones, S. Sidobre, M. Kronenberg, D. W. Dunne, A. Cooke. **2003**. Schistosoma mansoni antigens modulate the activity of the innate immune response and prevent onset of type 1 diabetes. *Eur. J. Immunol. 33:* 1439.

18 Cervi, L., A. S. MacDonald, C. Kane, F. Dzierszinski, E. J. Pearce. **2004**. Cutting edge: dendritic cells copulsed with microbial and helminth antigens undergo modified maturation, segregate the antigens to distinct intracellular compartments, concurrently induce microbe-specific Th1 and helminth-specific Th2 responses. *J. Immunol. 172:* 2016.

19 van Die, I., S. J. van Vliet, A. K. Nyame, R. D. Cummings, C. M. Bank, B. Appelmelk, T. B. Geijtenbeek, Y. van Kooyk. **2003**. The dendritic cell-specific C-type lectin DC-SIGN is a receptor for Schistosoma mansoni egg antigens and recognizes the glycan antigen Lewis x. *Glycobiology 13:* 471.

20 van Kooyk, Y., T. B. Geijtenbeek. **2003**. DC-SIGN: escape mechanism for pathogens. *Nat. Rev. Immunol. 3:* 697.

21 Trottein, F., N. Pavelka, C. Vizzardelli, V. Angeli, C. S. Zouain, M. Pelizzola, M. Capozzoli, M. Urbano, M. Capron, F. Belardelli, F. Granucci, P. Ricciardi-Castagnoli. **2004**. A type I IFN-dependent pathway induced by Schistosoma mansoni eggs in mouse myeloid dendritic cells generates an inflammatory signature. *J. Immunol. 172:* 3011.

22 Aksoy, E., C. S. Zouain, F. Vanhoutte, J. Fontaine, N. Pavelka, N. Thieblemont, F. Willems, P. Ricciardi-Castagnoli, M. Goldman, M. Capron, B. Ryffel, F. Trottein. **2005**. Double-stranded RNAs from the Helminth Parasite Schistosoma Activate TLR3 in Dendritic Cells. *J. Biol. Chem. 280:* 277.

23 Agrawal, S., A. Agrawal, B. Doughty, A. Gerwitz, J. Blenis, T. Van Dyke, B. Pulendran. **2003**. Cutting Edge: different toll-like receptor agonists instruct dendritic cells to induce distinct Th responses via differential modulation of extracellular signal-regulated kinase-mitogen-activated protein kinase and c-Fos. *J. Immunol. 171:* 4984.

24 de Jong, E. C., P. L. Vieira, P. Kalinski, J. H. Schuitemaker, Y. Tanaka, E. A. Wierenga, M. Yazdanbakhsh, M. L. Kapsenberg. **2002**. Microbial compounds selectively induce Th1 cell-promoting or Th2 cell-promoting dendritic cells in vitro with diverse th cell-polarizing signals. *J. Immunol. 168:* 1704.

25 van der Kleij, D., E. Latz, J. F. Brouwers, Y. C. Kruize, M. Schmitz, E. A. Kurt-Jones, T. Espevik, E. C. de Jong, M. L. Kapsenberg, D. T. Golenbock, A. G. Tielens, M. Yazdanbakhsh. **2002**. A novel host-parasite lipid cross-talk. Schistosomal lyso-phosphatidylserine activates toll-like receptor 2 and affects immune polarization. *J. Biol. Chem. 277:* 48122.

26 Hokke, C. H., A. M. Deelder. **2001**. Schistosome glycoconjugates in host-parasite interplay. *Glycoconj. J. 18:* 573.

27 Thomas, P. G., M. R. Carter, O. Atochina, A. A. Da'Dara, D. Piskorska, E. McGuire, D. A. Harn. **2003**. Maturation of dendritic cell 2 phenotype by a helminth glycan uses a toll-like receptor 4-dependent mechanism. *J. Immunol. 171:* 5837.

28 Thomas, P. G., D. A. Harn, Jr. **2004**. Immune biasing by helminth glycans. *Cell Microbiol. 6:* 13.

29 Iwasaki, A., R. Medzhitov. **2004**. Toll-like receptor control of the adaptive immune responses. *Nat. Immunol. 5:* 987.

30 Jenkins, S. J., A. P. Mountford. **2005**. Dendritic cells activated with products released by schistosome larvae drive Th2-type immune responses, which can be inhibited by manipulation of CD40 costimulation. *Infect. Immun. 73:* 395.

31 Pearce, E. J., P. Casper, J.-M. Grzych, F. A. Lewis, A. Sher. **1991**. Downregulation of Th1 cytokine production accompanies

induction of Th2 responses by a parasitic helminth, *Schistosoma mansoni. J. Exp. Med. 173:* 159.

32 Grzych, J. M., E. Pearce, A. Cheever, Z. A. Caulada, P. Caspar, S. Heiny, F. Lewis, A. Sher. **1991**. Egg deposition is the major stimulus for the production of Th2 cytokines in murine schistosomiasis mansoni. *J. Immunol. 146:* 1322.

33 Steinman, R. M., D. Hawiger, K. Liu, L. Bonifaz, D. Bonnyay, K. Mahnke, T. Iyoda, J. Ravetch, M. Dhodapkar, K. Inaba, M. Nussenzweig. **2003**. Dendritic cell function in vivo during the steady state: a role in peripheral tolerance. *Ann. NY Acad. Sci. 987:* 15.

34 Lutz, M. B., G. Schuler. **2002**. Immature, semi-mature and fully mature dendritic cells: which signals induce tolerance or immunity? *Trends Immunol. 23:* 445.

35 Faveeuw, C., V. Angeli, J. Fontaine, C. Maliszewski, A. Capron, L. Van Kaer, M. Moser, M. Capron, F. Trottein. **2002**. Antigen presentation by CD1d contributes to the amplification of Th2 responses to Schistosoma mansoni glycoconjugates in mice. *J. Immunol. 169:* 906.

36 Sabin, E. A., M. A. Kopf, E. J. Pearce. **1996**. *Schistosoma mansoni* egg-induced early IL-4 production is dependent upon IL-5 and eosinophils. *J. Exp. Med. 184:* 1871.

37 Haisch, K., G. Schramm, F. H. Falcone, C. Alexander, M. Schlaak, H. Haas. **2001**. A glycoprotein from Schistosoma mansoni eggs binds non-antigen-specific immunoglobulin E and releases interleukin-4 from human basophils. *Parasite Immunol. 23:* 427.

38 Schramm, G., F. H. Falcone, A. Gronow, K. Haisch, U. Mamat, M. J. Doenhoff, G. Oliveira, J. Galle, C. A. Dahinden, H. Haas. **2003**. Molecular characterization of an interleukin-4-inducing factor from Schistosoma mansoni eggs. *J. Biol. Chem. 278:* 18384.

39 Brunet, L. R., E. A. Sabin, A. W. Cheever, M. A. Kopf, E. J. Pearce. **1999**. Interleukin 5 (IL-5) is not required for expression of a Th2 response or host resistance mechanisms during murine schistosomiasis mansoni but does play a role in development of IL-4-producing non-T, non-B cells. *Infect. Immun. 67:* 3014.

40 d'Ostiani, C. F., G. Del Sero, A. Bacci, C. Montagnoli, A. Spreca, A. Mencacci, P. Ricciardi-Castagnoli, L. Romani. **2000**. Dendritic cells discriminate between yeasts and hyphae of the fungus Candida albicans. Implications for initiation of T helper cell immunity in vitro and in vivo. *J. Exp. Med. 191:* 1661.

41 MacDonald, A. S., E. J. Pearce. **2002**. Cutting edge: Polarized Th cell response induction by transferred antigen-pulsed dendritic cells is dependent on IL-4 or IL-12 production by recipient cells. *J. Immunol. 168:* 3127.

42 Lavelle, E. C., E. McNeela, M. E. Armstrong, O. Leavy, S. C. Higgins, K. H. Mills. **2003**. Cholera toxin promotes the induction of regulatory T cells specific for bystander antigens by modulating dendritic cell activation. *J. Immunol. 171:* 2384.

43 Dillon, S., A. Agrawal, T. Van Dyke, G. Landreth, L. McCauley, A. Koh, C. Maliszewski, S. Akira, B. Pulendran. **2004**. A Toll-like receptor 2 ligand stimulates Th2 responses in vivo, via induction of extracellular signal-regulated kinase mitogen-activated protein kinase and c-Fos in dendritic cells. *J. Immunol. 172:* 4733.

44 Moser, M., K. M. Murphy. **2000**. Dendritic cell regulation of TH1-TH2 development. *Nat. Immunol. 1:* 199.

45 Jankovic, D., A. Sher, G. Yap. **2001**. Th1/Th2 effector choice in parasitic infection: decision making by committee. *Curr. Opin. Immunol. 13:* 403.

46 MacDonald, A. S., A. D. Straw, N. M. Dalton, E. J. Pearce. **2002**. Cutting edge: Th2 response induction by dendritic cells: a role for CD40. *J. Immunol. 168:* 537.

47 Ridge, J. P., F. Di Rosa, P. Matzinger. **1998**. A conditioned dendritic cell can be a temporal bridge between a CD4+ T-helper and a T-killer cell. *Nature 393:* 474.

48 Bennett, S. R., F. R. Carbone, F. Karamalis, R. A. Flavell, J. F. Miller, W. R. Heath. **1998**. Help for cytotoxic-T-cell responses is mediated by CD40 signalling. *Nature 393:* 478.

49 Cella, M., D. Scheidegger, K. Palmer-Lehmann, P. Lane, A. Lanzavecchia, G. Alber. **1996**. Ligation of CD40 on dendritic cells triggers production of high levels of interleukin-12 and enhances T cell stimulatory capacity: T-T help via APC activation. *J. Exp. Med. 184:* 747.

50 Reis e Sousa, C., S. Hieny, T. Scharton-Kersten, D. Jankovic, H. Charest, R. N. Germain, A. Sher. **1997**. In vivo microbial stimulation induces rapid CD40 ligand-independent production of interleukin 12 by dendritic cells and their redistribution to T cell areas. *J. Exp. Med. 186:* 1819.

51 Chen, A. I., A. J. McAdam, J. E. Buhlmann, S. Scott, M. L. Lupher, Jr., E. A. Greenfield, P. R. Baum, W. C. Fanslow, D. M. Calderhead, G. J. Freeman, A. H. Sharpe. **1999**. Ox40-ligand has a critical costimulatory role in dendritic cell: T cell interactions. *Immunity 11:* 689.

52 Diehl, L., G. J. van Mierlo, A. T. den Boer, E. van der Voort, M. Fransen, L. van Bostelen, P. Krimpenfort, C. J. Melief, R. Mittler, R. E. Toes, R. Offringa. **2002**. In vivo triggering through 4-1BB enables Th-independent priming of CTL in the presence of an intact CD28 costimulatory pathway. *J. Immunol. 168:* 3755.

53 Tesselaar, K., Y. Xiao, R. Arens, G. M. van Schijndel, D. H. Schuurhuis, R. E. Mebius, J. Borst, R. A. van Lier. **2003**. Expression of the murine CD27 ligand CD70 in vitro and in vivo. *J. Immunol. 170:* 33.

54 Amsen, D., J. M. Blander, G. R. Lee, K. Tanigaki, T. Honjo, R. A. Flavell. **2004**. Instruction of distinct CD4 T helper cell fates by different notch ligands on antigen-presenting cells. *Cell 117:* 515.

55 Jenkins, M. K., A. Khoruts, E. Ingulli, D. L. Mueller, S. J. McSorley, R. L. Reinhardt, A. Itano, K. A. Pape. **2001**. In vivo activation of antigen-specific CD4 T cells. *Annu. Rev. Immunol. 19:* 23.

56 Garside, P., E. Ingulli, R. R. Merica, J. G. Johnson, R. J. Noelle, M. K. Jenkins. **1998**. Visualization of specific B and T lymphocyte interactions in the lymph node. *Science 281:* 96.

57 Linton, P. J., B. Bautista, E. Biederman, E. S. Bradley, J. Harbertson, R. M. Kondrack, R. C. Padrick, L. M. Bradley. **2003**. Costimulation via OX40L expressed by B cells is sufficient to determine the extent of primary CD4 cell expansion and Th2

cytokine secretion in vivo. *J. Exp. Med. 197:* 875.

58 Martin-Fontecha, A., L. L. Thomsen, S. Brett, C. Gerard, M. Lipp, A. Lanzavecchia, F. Sallusto. **2004**. Induced recruitment of NK cells to lymph nodes provides IFN-gamma for T(H)1 priming. *Nat. Immunol. 5:* 1260.

59 Angeli, V., C. Faveeuw, O. Roye, J. Fontaine, E. Teissier, A. Capron, I. Wolowczuk, M. Capron, F. Trottein. **2001**. Role of the parasite-derived prostaglandin D2 in the inhibition of epidermal Langerhans cell migration during schistosomiasis infection. *J. Exp. Med. 193:* 1135.

60 Sato, H., H. Kamiya. **1998**. Accelerated influx of dendritic cells into the lymph nodes draining skin sites exposed to attenuated cercariae of Schistosoma mansoni in guinea-pigs. *Parasite Immunol. 20:* 337.

61 Ramaswamy, K., P. Kumar, Y. X. He. **2000**. A role for parasite-induced PGE2 in IL-10-mediated host immunoregulation by skin stage schistosomula of Schistosoma mansoni. *J. Immunol. 165:* 4567.

62 Straw, A. D., A. S. MacDonald, E. Y. Denkers, E. J. Pearce. **2003**. CD154 plays a central role in regulating dendritic cell activation during infections that induce Th1 or Th2 responses. *J. Immunol. 170:* 727.

63 Shortman, K., Y. J. Liu. **2002**. Mouse and human dendritic cell subtypes. *Nat. Rev. Immunol. 2:* 151.

64 McKee, A. S., E. J. Pearce. **2004**. CD25+CD4+ cells contribute to Th2 polarization during helminth infection by suppressing Th1 response development. *J. Immunol. 173:* 1224.

65 Germain, R. N., M. K. Jenkins. **2004**. In vivo antigen presentation. *Curr. Opin. Immunol. 16:* 120.

66 Jankovic, D., M. C. Kullberg, S. Hieny, P. Caspar, C. M. Collazo and A. Sher. **2002**. In the absence of IL-12, CD4(+) T cell responses to intracellular pathogens fail to default to a Th2 pattern and are host protective in an IL-10(−/−) setting. *Immunity 16:* 429–439.

XII
Bacteria

35
Dendritic Cells and Immunity to Salmonella

Mary Jo Wick

35.1
Introduction

An early and necessary event required for pathogen elimination is phagocytosis and destruction of the bacteria. This must be coupled to processing and presentation of bacterial antigens to ensure generation of adaptive immunity. Initiating an adaptive response in a primary infection requires that the bacteria-derived peptides are presented under conditions that lead to activation of naïve T cells, eliciting the appropriate effector function in the responding cells, and developing a memory T-cell pool. This requires sufficient co-stimulation, production of appropriate response-skewing cytokines such as IL-12, and localization in the appropriate place, that is, in secondary lymphoid organs. To achieve this, the cells that initially phagocytize the bacteria must migrate from the peripheral infected tissues to a secondary lymphoid organ.

There is only one cell type capable of performing all of these functions, dendritic cells (DC). DC fulfill a unique niche in immunity to bacterial pathogens and are the cornerstone in the transition from innate to adaptive immunity. This chapter summarizes work underlying the current understanding of the role of DC in the immune response to *Salmonella*. Studies performed *in vitro* that provide insight into bacterial uptake, processing and presentation of *Salmonella* by DC, and DC maturation in response to *Salmonella* are summarized first. Findings on the role of DC during *Salmonella* infection assessed in murine infection models will subsequently be discussed.

Handbook of Dendritic Cells. Biology, Diseases, and Therapies.
Edited by M. B. Lutz, N. Romani, and A. Steinkasserer.
Copyright © 2006 WILEY-VCH Verlag GmbH & Co. KGaA, Weinheim
ISBN: 3-527-31109-2

35.2
Dendritic Cell Subsets, Short and Sweet

Although presented in great detail and scope elsewhere in this book, a very brief description of murine DC subsets may be useful here to facilitate discussions of *Salmonella*-DC interactions in this chapter. Conventional DC in mice express high levels of the p150/90 integrin CD11c as well as high basal levels of MHC-II. They can thus conveniently be identified as CD11c^{hi}MHC-II$^+$ cells. These DC from secondary lymphoid organs can be further divided into subsets based on expression of CD11b, CD8α, and CD4, among other molecules [1–3]. For example, conventional DC in the spleen can be divided into CD8α^+CD4$^-$CD11b$^-$, CD8α^-CD4$^+$CD11b$^+$ and CD8α^-CD4$^-$CD11b$^+$ subsets that comprise approximately 25%, 50% and 25% of total CD11c^{hi}MHC-II$^+$ cells in this organ, respectively. In contrast to the spleen, CD4$^+$ DC are rare in mesenteric lymph nodes (MLN), Peyer's patches (PP) and the liver of mice. CD11c^{hi}MHC-II$^+$ cells present in these organs can be described as CD8α^+CD4$^-$CD11b$^-$, CD8α^-CD4$^-$CD11b$^+$ and CD8α^-CD4$^-$CD11b$^-$ populations [4–7]. These subsets are present in somewhat different relative proportions depending on the organ [8].

35.3
Dendritic Cells and Salmonella: Lessons from in vitro Studies

35.3.1
Bacterial Uptake and the Fate of Internalized Bacteria

Salmonella enterica Serovar Typhimurium (*S. typhimurium*) is a facultative, Gram negative intracellular pathogen that has developed strategies allowing it to survive in phagocytic cells, both in macrophages and DC, despite landing in what can be a rather harsh vacuolar environment after phagocytosis. Studies using murine DC, either derived from bone marrow cultured in the presence of GM-CSF or CD11c^{hi} DC enriched from the spleen, liver or MLN of normal mice, have demonstrated that these cells can indeed internalize *Salmonella* [7, 9–15]. Internalization requires only a fairly short pulse of DC with bacteria and occurs even at low bacteria to DC infection ratios [7, 9, 10, 12–15]. In addition, active opsonization is not required, although opsonization increases the number of bacteria internalized [11]. With respect to a differential capacity of DC subsets to internalize *Salmonella*, no major differences have thus far been reported. For example, a similar percent of freshly isolated CD8α^+, CD8α^-CD4$^-$ and CD8α^-CD4$^+$ splenic DC were positive for green fluorescent protein (GFP) after co-culture with GFP-expressing *Salmonella* [10]. This suggests no major differences among the splenic DC subsets to internalize bacteria at a given infection ratio. Similar analysis of DC isolated from the liver, however, revealed that CD8α^- DC were perhaps slightly more effective than CD8α^+ DC in internalizing *Salmonella*, while cells separated on the bases of MHC-II expression rather than CD8α showed that MHC-II^{hi} cells were more efficient at internalizing the bacteria than MHC-II^{low} cells [7]. Functional consequences of this are presently not known.

Salmonella enter DC in a process that involves actin cytoskeletal rearrangements [7, 10, 16] and reside in vacuolar compartments termed *Salmonella*-containing vacuoles (SCVs) [17]. Unlike other intracellular bacteria such as *Listeria* and *Shigella*, *Salmonella* remain confined in vacuoles and do not produce a pore-forming protein allowing their active escape into the cytosol. Instead they encode proteins, particularly SifA, that are required to maintain the integrity of the SCVs and keep *Salmonella* physically separated from the cytosol [15, 17, 18]. Thus, only *Salmonella* lacking SifA readily access the cytosol of DC whereas wild type bacteria remain confined in vacuoles [15].

The fate of *Salmonella* after phagocytosis by macrophages has been studied for decades and has resulted in a wealth of knowledge on this topic. For example, *Salmonella* residence in macrophage SCVs diverts the normal maturation of phagosomes, prevents vesicle fusion with lysosomes, and interferes with delivery of antimicrobial effector proteins, such as NADPH oxidase and inducible nitric oxide synthase (iNOS) that otherwise contribute to the demise of the bacteria, to bacteria-containing compartments [17, 18].

In contrast to the extensive literature on *Salmonella* trafficking in macrophages, however, relatively little is known about the fate of this bacterium after phagocytosis by DC. The few studies on the intracellular fate of *Salmonella* in DC available thus far suggest similar manipulation of phagosomes in DC as reported for macrophages. For instance, SCVs in DC are capable of inducing the *Salmonella* genes necessary for altered SCV trafficking and function [14]. In addition, the vacuoles containing *Salmonella* in DC have features consistent with SCVs defined in macrophages, such as the presence of LAMP-1 and MHC-II [14]. However, in contrast to the increase in bacterial number that occurs after *Salmonella* phagocytosis by macrophages, studies from several groups have shown that *Salmonella* do not increase in number after internalization by DC [12, 14, 15, 19]. Although details such as the initial multiplicity of infection used, time frame analyzed and number of bacteria recovered differs in these studies, the consensus appears to be that the number of *Salmonella* in infected DC remains relatively constant over the first 48 h of infection. A recent report further showed that the maintenance of constant bacterial numbers during the first 24 h in SCVs of infected DC is due to the static, non-dividing nature of intracellular *Salmonella* rather than the alternate explanation that bacterial multiplication occurs at a rate equivalent to bacterial killing [14]. Although static in nature, recent gene expression profiling studies suggest that the bacteria in SCVs are not in the stationary growth phase classically defined by *in vitro* culturing of bacteria in liquid medium [20]. These profiling studies were performed on infected macrophages, however, so whether the same is true in infected DC remains to be directly addressed.

The mechanisms underlying the capacity of *Salmonella* to remain in a viable, static state in SCVs of infected DC are not known. What is known from studies in macrophages, however, is that production of reactive oxygen and nitrogen intermediates by inducible nitric oxide synthase (iNOS) and NADPH oxidase, respectively, have a role in controlling *Salmonella* replication [21, 22]. Indeed, *Salmonella* has tackled this problem by preventing the co-localization of these enzymes to SCVs [22–24]. Precisely how products of these pathways work to control *Salmonella* replication in infected macrophages is complex [25–27]. There is a sequential contribution of an NADPH oxidase–dependent oxidative bactericidal phase followed by an iNOS–dependent bacteriostatic phase. Moreover, the mediators generated by these pathways can interact in several ways and multiple antimicrobial effects are possible [25–27].

The contribution of reactive oxygen and nitrogen intermediates to controlling *Salmonella* growth in DC is not well understood. Murine DC appear capable of producing nitric oxide in response to *Salmonella*, despite some inconsistency in reports where two groups found production [11, 28] and one did not [19]. The precise reason for this discrepancy is not clear. Possibilities include that the infection conditions resulted in different quantities of bacteria internalized per DC and a threshold number of bacteria is required to trigger iNOS, or that the use of opsonized versus non-opsonized bacteria results in altered intracellular targeting of the bacteria which could be linked to a difference in the capacity to divert iNOS localization [24]. Alternatively, different times of bacteria-DC co-culture used or the presence of residual extracellular bacteria in cultures that provided a stimulus that influenced iNOS production may contribute to the observed difference. Nevertheless, murine DC are capable of producing nitric oxide in response to *Salmonella*. Furthermore, nitric oxide produced by DC co-cultured with opsonized bacteria reduced intracellular bacterial yields and had a bactericidal effect [11]. This was in contrast to the effect of nitric oxide produced by *Salmonella*-infected macrophages, which has a bacteriostatic effect [11, 22, 26]. This difference may indicate different environmental conditions in SCVs of macrophages versus DC [25–27].

In addition to their capacity to produce reactive nitrogen intermediates, both murine and human DC have NADPH oxidase components and can generate reactive oxygen products in response to bacterial ligands such as LPS and flagella [29, 30]. Although not yet addressed for *Salmonella*, Vulcano et al showed NADPH-dependent killing of *E. coli* after co-culture with LPS-treated DC [30]. Interestingly, DC appear to have a NADPH oxidase inhibitory factor located in the membrane whose inhibition of oxidase activation is relieved after exposure to microbial components such as LPS [29]. This suggests that DC may exert additional regulatory control on NADPH oxidase activity. Despite these data, however, the role of NADPH oxidase-dependent mechanisms in controlling replication in *Salmonella*-infected DC is not yet established. Likewise, if SCVs of DC have a capacity similar to that seen in macrophages to modulate the localization of NADPH oxidase or iNOS awaits investigation. More detailed studies of the environment of *Salmonella*-containing vacuoles in infected DC and the contribution of NADPH oxidase- and iNOS-dependent control of *Salmonella* replication by DC warrant further study. Moreover, the mechanisms that maintain the viable but nondividing nature of *Salmonella* in DC SCVs [14], and the host and bacterial factors that regulate this, promise to be interesting areas of investigation.

35.3.2
Presentation of Salmonella Antigens by Dendritic Cells

35.3.2.1 **Processing of Salmonella for Direct Presentation on MHC-II by Infected Dendritic Cells**

CD4 T-cell responses, particularly induction of IFNγ-producing cells, are critical to controlling and eradicating *Salmonella* in infected hosts [31–33]. To initiate CD4[+] T-cell responses, antigens must be processed and presented on major histocompatibility complex (MHC) class II molecules (MHC-II). The type of antigens typically processed for MHC-II presentation are exogenous in nature, that is, are extracellular antigens that are taken into a cell by pinocytosis, endocytosis, or phagocytosis, depending on the nature of the antigen. Despite that *Salmonella* reside intracellularly in infected hosts [34–38], they spend part of their time extracellularly, such as before the bacteria are phagocytosed by, or invade, host cells. They also have access to the extracellular environment after death and lysis of an infected cell [39, 40].

It thus follows that antigens from *Salmonella* are processed and presented on MHC-II for recognition by CD4[+] T cells. Indeed, immature DC phagocytose and present *Salmonella*-derived antigens on MHC-II to CD4[+] T cells, as has been shown using bone-marrow-derived DC [12, 16, 19]. It has also been demonstrated in assays using DC freshly isolated from the spleen, MLN and liver of naïve mice [7, 10]. Presentation of *Salmonella* antigens on MHC-II by *Salmonella*-infected DC requires actin-driven cytoskeletal rearrangements to internalize the bacteria and bacterial passage through acidic compartments [10, 16]. Furthermore, CD8α[+] and CD8α[−] splenic DC appear to have a similar capacity to internalize and process *Salmonella* for peptide presentation on MHC-II [10]. Thus, DC in the spleen, liver and MLN can internalize and process *Salmonella* for peptide presentation on MHC-II for recognition by CD4[+] T cells.

35.3.2.2 **Processing of Salmonella for Direct Presentation on MHC-I by Infected Dendritic Cells**

As discussed above, wild type (SifA-expressing) *Salmonella* internalized by DC remain confined in vacuolar compartments [15, 17, 18]. However, DC can process internalized *Salmonella* and present peptides derived from bacteria-encoded proteins on MHC-I [7, 10, 12, 16, 19, 41, 42]. At first glance this appears incongruous with respect to the classical MHC-I antigen presentation pathway where endogenous, cytosolic proteins are processed in a proteasome-dependent fashion, transported into the endoplasmic reticulum by the transporter associated with antigen processing (TAP) and loaded on newly synthesized MHC-I for ultimate display on the cell surface [43]. However, data reported three decades ago showing that exogenous antigens can be presented on MHC-I [44] were the basis for solving this paradox and the starting point for what is now extensive literature documenting and characterizing presentation of exogenous antigens on MHC-I [45–47].

Presentation of antigens by DC that have phagocytosed *Salmonella* has been used by several investigators both *in vitro* and *in vivo* as a model to understand presentation of vacuole-confined bacteria. The relevance of studying MHC-I presentation of *Salmonella* antigens is underscored by the development of CD8+ T cells, in addition to CD4+ T cells, to bacteria-encoded antigens in infected mice and humans [36, 48–53]. Indeed, recombinant attenuated *Salmonella* is a potential vaccine delivery system to elicit CD4 and CD8 T-cell responses to cloned and expressed antigens.

Similar to the results for MHC-II presentation, DC isolated from the spleen, MLN and liver of naïve mice as well as bone-marrow-derived DC are capable of presenting *Salmonella* antigens on MHC-I [7, 10, 12, 16, 19, 41, 42]. Likewise, both CD8α+ and CD8α− splenic DC subsets internalize *Salmonella* and process the bacteria for peptide presentation on MHC-I [10]. Moreover, despite the vacuolar localization of *Salmonella* after phagocytosis by DC, the data available thus far suggest that components of the cytosolic MHC-I antigen presentation pathway are used. For example, DC have a strict requirement for the TAP peptide transporter to present *Salmonella* antigens on MHC-I [54]. In addition, although not formally demonstrated for *Salmonella*, studies using *E. coli* demonstrate that newly synthesized MHC-I molecules and the proteasome are also required for DC presentation of bacteria-encoded antigens [41]. It thus appears that the cytosolic antigen presentation machinery is used to for MHC-I presentation of bacterial antigens after phagocytic uptake of *Salmonella* by DC. This mechanism appears distinct from that used by macrophages to present *Salmonella* antigens. In the case of macrophages, the TAP transporter and proteasomes are not required [46, 55, 56]. It thus appears that macrophages and DC have different requirements to present *Salmonella* antigens on MHC-I. However, relatively little is known about the pathway(s) used for MHC-I presentation of *Salmonella* antigens by infected DC, and additional experiments that characterize this would be informative. This is particularly true in light of recent findings describing the intersection of the endoplasmic reticulum with phagosomes that results in phagosomes capable of TAP- and proteasome-dependent presentation of exogenous antigens on MHC-I [57, 58].

35.3.2.3 Modulating of Antigen Presentation by *Salmonella*

Salmonella is a facultative intracellular bacterium that has evolved strategies to survive in phagosomal environments, which are otherwise designed to kill phagocytosed microbes. It thus follows that the intracellular survival strategies used by *Salmonella* may influence the capacity of an infected DC to process and present bacterial antigens. Indeed, the *phoP/phoQ* regulatory system, which controls the expression of over 40 genes and is involved in bacterial survival in phagosomal compartments [59, 60], can influence the ability of infected DC to present *Salmonella*-encoded antigens. Antigens from *Salmonella* constitutively expressing *phoP*, so that the set of *phoP*-activated genes are switched on and the *phoP*-repressed genes are off, are more efficiently presented on MHC-II after bacterial phagocytosis by DC [19]. The effect of *phoP/phoQ* was observed when antigen presentation was quantitated after a short (2-h) but not a longer (24-h) exposure to bacteria [12, 19]. The effect of the *phoP* locus on antigen presentation by infected DC was abrogated when the bacteria were heat-killed, demonstrating that bacterial gene expression was required for the effect [19]. Together these data suggest that the *phoP/phoQ* regulatory locus can influence the capacity of DC to present *Salmonella* antigens on MHC-II during a short time frame after bacterial infection. Despite the effect of *phoP/phoQ* on presentation of *Salmonella* antigens on MHC-II by infected DC, an effect of this locus on presentation of *Salmonella* antigens on MHC-I was not detected [19].

Other factors, such as the ease of bacterial uptake and the amount of antigen present in the bacteria that are internalized, can influence the efficiency of antigen presentation. Relatively little data is available that directly addresses these issues for the presentation of *Salmonella* antigens by DC. However, a recent report showed that directing *Salmonella* to Fcγ receptors on DC by opsonization with *Salmonella*-specific IgG enhanced the presentation of a *Salmonella*-encoded antigen on MHC-I and MHC-II [61]. In contrast to previous reports [7, 10, 12, 16, 19, 42], Tobar et al. could only detect presentation of antigens encoded in wild type *Salmonella* when the bacteria were opsonized. Based on this they concluded that virulent *Salmonella* interferes with the capacity of DC to process the bacteria for antigen presentation, and this could be overcome by targeting the bacteria to Fcγ receptors. However, antigen abundance in the two strains compared in the study was not assessed, and direct evidence for active inhibition of antigen presentation by *Salmonella* was lacking [61]. Thus, the alternate explanation that increasing antigen load in the DC by opsonization, perhaps combined with altered intracellular trafficking when internalized as immune complexes, was responsible for the observed presentation cannot be excluded. Indeed, opsonization increases the number of bacteria per DC [11, 62], and targeting antigens to Fcγ receptors on DC has been shown to increase antigen presentation in several other settings [63–66].

35.3.2.4 Waste not, Want not: Dendritic Cells as Bystander Antigen-presenting Cells

The presentation of *Salmonella* antigens by DC discussed above focused on direct antigen presentation by infected cells. In this pathway, DC internalize *Salmonella*, process the bacteria and display MHC molecules containing bacterial antigens on their cell surface for recognition by T cells. In other words, the *Salmonella*-infected DC directly process the bacteria and present bacterial antigens to T cells. However, DC can also present bacterial antigens when they themselves are not infected by the bacteria in a process called indirect presentation or, in the case of MHC-I, crosspresentation. In indirect presentation, the DC that present *Salmonella* antigens are not infected by the bacteria *per se*. Instead, the DC are non-infected bystander cells that acquired *Salmonella* antigens by internalizing neighboring cells that have undergone death due to *Salmonella* infection. *Salmonella* expressing the type III secretion system is cytotoxic to infected cells [39], and dead cells can not productively interact with T cells [42]. However, the indirect antigen presentation pathway provides a safety valve where DC "mop up" cell debris containing *Salmo*-

nella antigens and use this material to stimulate T cells [42]. In the case of *Salmonella* and other intracellular bacteria, this property could be very useful to the immune system and allow detection of microbes that could otherwise be elusive [67]. Indeed, bystander presentation of antigenic material from cells induced to undergo apoptosis due to infection with *Mycobacterium tuberculosis* has also been shown [68]. In the case of this vacuole-confined bacterium, apoptotic macrophages shuttle vesicles containing mycobacterial lipids and proteins to DC which in turn present the material to CD1b- and MHC-I-restricted T cells, respectively [68].

Salmonella-induced cell death has been best studied in infected macrophages and epithelial cells [39]. However, *Salmonella* can also kill infected DC by a mechanism dependent on the type III secretion system [40, 54]. Whether DC that have undergone *Salmonella*-mediated death are also a reservoir of cell debris containing *Salmonella* antigens that can be presented by neighboring, bystander DC is presently not known. Interestingly, the capacity to act as a bystander antigen presenting cell appears to be a unique feature of DC, as bystander macrophages ingest *Salmonella*-induced apoptotic cells but do not present peptides from *Salmonella* antigens [42]. Preliminary data suggest that macrophages compete for apoptotic material and limit the antigen available for presentation by bystander DC [42].

The precise nature of the material in the cell debris responsible for the observed bystander presentation of *Salmonella* antigens from apoptotic macrophages is not known. However, neither peptides released into the environment that bind preformed surface MHC molecules on bystander DC nor bacteria released into the surroundings that are subsequently phagocytosed and processed by the bystander cells account for the observed presentation [42]. Additional experiments are needed to characterize the antigenic material derived from the dying, *Salmonella*-infected cells.

Thus, DC can either be direct or indirect presenters of *Salmonella* antigens. They directly present bacterial antigens to T cells upon phagocytic processing of *Salmonella* that does not induce their death. They can also present bacterial antigens to T cells as bystander antigen presenting cells that engulf antigenic material from neighboring cells that have undergone *Salmonella*-induced apoptotic death.

35.4

Time to go to Work: Salmonella-induced Dendritic Cell Maturation

A steady-state population of DC is found in essentially all tissues of the body. DC situated in non-inflamed, non-infected tissues are in a so-called immature state, poised to respond to infection and capture antigen. Immature DC in peripheral tissues have a high capacity to internalize and process antigens, including bacteria, but a relatively poor capacity to stimulate naïve T cells [69, 70]. Immature DC express significant levels of surface MHC-II and MHC-I and low to intermediate levels of CD86, CD80, CD40 and CD54 (ICAM-I). They also are quiescent with respect to cytokine production. Upon receiving signals indicating inflammation and/or infection, immature DC undergo a series of phenotypic and functional changes and become mature, or activated, DC. This transition changes the DC from cells programmed for capturing and processing antigens into ones specialized in presenting antigens and activating T cells, particularly naïve T cells. To accomplish this, DC maturation includes transient stimulation of endocytic capacity followed by downregulation of antigen capture [71], optimizing MHC synthesis, trafficking and stability, and increasing co-stimulatory molecule expression [69, 70, 72–74]. The maturation process also involves enhancing cytokine secretion and altering chemokine responsiveness, the latter directing DC migration from infected tissues to draining lymphoid organs [69, 70, 75]. The net result of DC maturation is that antigen-laden DC with optimized capacity to activate antigen-specific T cells are located in secondary lymphoid tissues. This underscores the importance of DC in initiating adaptive immunity.

As microbial compounds that signal through Toll-like receptors, such as LPS, and the cytokines IL-1β and TNF-α are well established inducers of DC maturation [69, 70], it is no surprise that *Salmonella* contact with DC initiates maturation. For example, murine DC pulsed briefly with *Salmonella* have a decreased capacity to present antigens upon subsequent encounter with bacteria [19]. This is similar to the progressive reduction in endocytic capacity of DC after exposure to TLR ligands such as LPS [71, 76, 77]. Although the mechanism of down regulation of endocytosis in response to LPS has been eloquently studied [71, 76, 77], how *Salmonella* reduces presentation from a second exposure to bacteria remains to be investigated.

Co-culture of *Salmonella* with murine DC [11, 16, 28] or with human monocyte-derived DC [78] also up regulates CD86, CD80, MHC-I, MHC-II, CD40 and CD54. The up regulation of molecules occurs rapidly and on the entire population of CD11c^{hi} cells. Up regulation of surface molecules does not require live bacteria or active bacterial internalization by the DC [16, 28]. This suggests that the presence of the bacteria in the culture was sufficient to trigger the population to increase expression of the molecules. With the exception of CD80, increased expression of the molecules examined was apparent within 8 hours after infection and was maximal within 24 hours [11, 16, 28]. Upregulation of CD80 had a slower kinetics and more stringent requirements. For instance, while upregulation of MHC molecules, CD86, CD40 and CD54 occurred in a similar fashion after a 2-h pulse with live or heat-killed bacteria, CD80 responded poorly [28] or not at all to heat-killed bacteria even when examined 48 h post infection [16].

Consistent with the induction of cytokines when DC are exposed to purified microbial ligands such as LPS, DC also activate NF-κB and produce several pro-inflammatory cytokines, as well as chemokines, in response to *Salmonella*. For instance, murine DC from the spleen, liver and cultured from bone marrow precursors produce TNF-α, IL-1β and IL-12p40 after a brief co-culture with *Salmonella* [7, 10, 13, 16, 28, 79]. However, despite robust IL-12p40 production by murine DC upon exposure to *Salmonella*, little IL-12p70 is detected [19]. This is in contrast to human monocyte-derived DC which produce IL-12p70 as well as TNF-α upon co-culture with *Salmonella* [78]. The inability to detect significant amounts of the IL-12p70 heterodimer in the murine system may reflect the absence of required augmentory signals such as CD40 engagement or IFNγ, or could reflect the presence of inhibitory cytokines such as IL-10 [80–82].

The ability of DC subsets to produce cytokines in response to *Salmonella,* as well as the relationship between bacterial uptake and cytokine production, have been studied by infecting DC with GFP-expressing *Salmonella* and assessing cytokine production by flow cytometry. These data showed that both CD8α^+ and CD8α^- DC from the spleen and liver of mice produce IL-12p40 in response to *Salmonella,* and that the CD8α^+ subset contained a somewhat higher fraction of IL-12p40$^+$ cells relative to CD8α^- DC [7, 10]. A higher fraction of splenic DC in the CD8α^- subset produce TNF-α compared to their CD8α^+ counterparts while this trend was not apparent in liver DC, where the CD8α^+ subset contained a slightly higher fraction of TNF-α-producing cells following co-culture with *Salmonella* [7, 10]. Production of IL-12p40 and TNF-α by DC did not require bacterial internalization but did require physical contact with the bacteria. Moreover, bacterial internalization did not necessarily result in production of either TNF-α or IL-12p40, as not all GFP$^+$ cells produced cytokine [7, 10].

These studies also showed that a significant fraction of cytokine positive cells were not associated with GFP-expressing bacteria [7, 10]. Indeed, the highest fraction of cytokine positive cells was among GFP$^-$ cells. Analysis of bacterial factors that induce cytokine production by DC showed that a diffusible product in the bacteria-DC culture could not induce cytokine production. This demonstrates that DC-bacteria contact is required, as mentioned above. Although TLR4-mediated signaling by LPS on *Salmonella* was not required for increased expression of MHC and co-stimulatory molecules by *Salmonella*-infected DC, LPS was involved in triggering cytokine production by DC [10, 13].Thus, LPS is an important bacterial ligand for cytokine production by DC, but other TLR ligands appear to be sufficient to induce up regulation of surface molecules in the absence of LPS.

Additional studies have addressed whether *Salmonella* with mutations resulting in LPS alterations influence DC maturation. For example, the effects of *Salmonella* with mutations in *phoP/phoQ* or *lpxM,* which have alterations in the lipid A portion of LPS relative to wild type bacteria [28, 83, 84] on down regulation of antigen presentation capacity, up regulation of co-stimulatory molecule expression and T-cell stimulatory capacity as well as cytokine production have been analyzed using murine DC [19, 28]. These studies showed that mutant bacteria expressing LPS with modified lipid A, or purified LPS containing a lipid A modification, had little if any alteration in their capacity to influence these aspects of DC maturation. However, given the observation that human but not murine TLR4 can discriminate between LPS containing lipid A modifications [85], it would be interesting to address the influence of lipid A modifications on maturation of human DC.

Although culture systems represent an environment not entirely similar to the milieu in an organ, particularly during infection where numerous cell types are activated and the cytokine environment is complex, they are valuable sources of information due to the capacity to manipulate parameters in a controlled fashion. However, *in vivo* infection models are an irreplaceable compliment to *in vitro* studies and have provided much insight into the role of DC and DC subsets during the course of *Salmonella* infection. Findings from murine infection models are summarized below.

35.5

Murine Infection Models to Study Dendritic Cell Interaction with *Salmonella in vivo*

35.5.1

***Salmonella* Infection and Penetration of the Intestinal Epithelium**

Salmonella spp. cause a variety of diseases, from localized gastroenteritis to systemic illnesses. While *S. typhimurium* infection of humans remains localized and results in gastroenteritis, mice infected with this species get a systemic infection with pathogenesis resembling that of typhoid fever in humans, which is caused by *S. typhi*. Thus, a murine infection model is a useful tool to study systemic *Salmonella* infection.

    Salmonella is transmitted by the oral route, and these bacteria must penetrate the intestinal epithelium to cause systemic illness. Data from murine infection models suggest that *Salmonella* may use more than one mechanism to cross the intestinal barrier. For example, penetration through M cells as well as via enterocytes in the epithelium overlying Peyer's patches can contribute to *Salmonella* penetration of the intestinal epithelium [86]. In addition, DC that breach the epithelial layer and sample the gut luminal bacteria also appear to have a role in transporting *Salmonella* across the epithelium [87–89]. Extending the original findings of Rescigno et al. [87], a recent report showed that lamina propria DC, particularly in the villi of the terminal ileum, extend dendrites between epithelial cells and are involved in the uptake of bacteria *in vivo* including orally acquired *Salmonella* [89]. When mice with lamina propria DC capable of forming transepithelial dendrites (CX_3CR1^+ mice) were infected with *Salmonella*, bacteria were recovered only from DC. In contrast, *Salmonella* are found in DC as well as in other phagocytes after oral infection of mice defective in CX_3CR1, which is required for DC to extend dendrites across the intestinal epithelium into the gut lumen [89]. The invasive nature of the bacteria may also influence the cell population(s) that harbor them after ingestion [88–90].

    Evidence available thus far suggests that DC and/or other $CD18^+$ cells are responsible for initial bacterial internalization and dissemination of *Salmonella* in the infected host [87–92]. Additional studies showing the presence of $CD8\alpha^-CD11b^-$ DC in the follicle-associated epithelium overlying PP [5], which would be situated to sample intestinal bacteria, and that $CD11c^+$ DC in the subepithelial dome of PP contain microparticles that were orally administered to mice [93] further support a role of DC in mediating bacterial transit from the intestinal lumen. However, the presence of *Salmonella* in phagocytes other than DC ($CD11c^-$ cells) in the MLN and PP of orally challenged mice suggests that DC may not be the only cell involved in transporting invasive bacteria acquired orally [88, 89]. A caveat to this, however, is that the bacteria could have been internalized by other phagocytes in the MLN after transport to this organ in DC from the lamina propria or PP. Finally, although neither neutrophils nor macrophages appear to be resident populations in the subepithelial dome of PP [94, 95], they infiltrate PP and MLN in response to oral *Salmonella* infection (A. Rydström and M. J. Wick, unpublished data). These cells could participate in microbial uptake in PP and MLN of infected mice, a possibility supported by the studies finding *Salmonella* in CD11c⁻ cells of MLN and PP after oral infection of mice with invasive bacteria [88, 89]. But whether the non-DC function mainly to kill bacteria or have additional functions such as shuttling bacteria in the host, presenting antigens to effector T cells, or providing bacterial antigens for bystander presentation by neighboring DC remain to be determined.

35.5.2
Dendritic Cell Take-up *Salmonella in vivo*

Murine models have shown that DC in the spleen, MLN, liver and PP harbor *Salmonella* during infection [7, 10, 36, 88, 89, 91]. A similar fraction of the three major splenic DC subsets (CD8α⁺, CD8α⁻CD4⁺ and CD8α⁻CD4⁻) contain *Salmonella* after administration of a given bacterial dose [10]. Likewise, roughly equal percentages of CD8α⁺ and CD8α⁻ DC in the liver of infected mice contain *Salmonella* after infection [7]. However, similar to the studies performed *in vitro* assessing bacterial uptake by freshly isolated splenic or liver DC discussed above, *in vivo* studies also showed that CD8α⁺ DC in the spleen had a higher capacity than the other two subsets to associate with bacteria at lower bacterial doses while at the highest dose tested CD8α⁻CD4⁻ DC had the highest fraction of cells containing *Salmonella* [10]. Thus, the DC subsets may have differential access to bacteria depending on the bacterial load in the organ with CD8α⁺ DC possibly being better at scavenging a limiting number of bacteria.

Although DC internalize *Salmonella in vivo*, it is important to note that the number of DC containing *Salmonella* is very low, particularly in orally-infected mice. Whereas intravenous injection of *Salmonella* results in 1–10% of splenic CD11cʰⁱ cells containing bacteria shortly after administration [10, 36], depending on the dose administered, splenic or liver DC associated with GFP-expressing *Salmonella* given orally are in such low numbers that their quantitation by flow cytometry is not reliable (our unpublished data; see also [7, 91]). In PP and MLN, however, quantitating cell-associated GFP⁺ *Salmonella* in the early stages of oral infection is possible [88, 91, 115]. In these organs, the number of DC containing *Salmonella* is ~5 bacteria/1000 DC in PP and ~1–2 bacteria/1000 DC in MLN [88, 115]. Cells other than conventional DC can also harbor bacteria after oral administration of invasive *Salmonella* as mentioned above [88, 115].

35.5.3
Getting the Game Started: Dendritic Cells Initiate Adaptive Immunity to *Salmonella*

35.5.3.1 *Salmonella*-induced Dendritic Cell Maturation During Infection
Total splenic CD11c⁺ cells respond to *Salmonella* infection by increasing surface expression of CD86, CD80 and CD40 beginning approximately one week after infection, depending on the bacterial dose and strain used [36, 115]. The appearance of DC exhibiting higher surface levels of co-stimulatory molecules corresponds to

the time post infection when *Salmonella*-specific T cells are beginning to appear [96]. Recent data also suggest that expression of these molecules occurs in a subset-specific fashion on splenic, MLN and PP DC in response to oral *Salmonella* infection. For example, CD80 is preferentially up regulated on CD8α^- DC in these organs, while the greatest increase in surface CD40 is on CD8α^+ DC and both CD8α^+ and CD8α^- up regulate CD86 [115]. Synchronous up regulation of co-stimulatory molecules occurs on the entire population of CD11c^{hi} cells in a given organ despite the low number of DC that contain bacteria. Moreover, up regulation of co-stimulatory molecules occurs simultaneously in PP, MLN and spleen despite the sequential seeding of these organs by orally administered bacteria [115]. These data suggest that indirect signaling by soluble mediator(s) rather than direct signaling by bacterial association/uptake are responsible for *Salmonella*-induced DC maturation *in vivo*.

In addition to the subset-specific differences in up regulation of co-stimulatory molecules, splenic DC subsets are differentially modulated with respect to number, distribution, and cytokine production in the early stages of oral infection [97]. For instance, splenic CD8α^+ and CD8α^-CD4$^-$ DC double in number five days after infection, while CD8α^-CD4$^+$ DC numbers are unchanged. The increase in CD8α^+ and CD8α^-CD4$^-$ DC was reflected by an influx of these subsets in the splenic red pulp of *Salmonella*-infected mice [97]. Quantitating changes in DC subsets in the MLN of infected mice revealed little change in the number of CD8α^+ and CD8α^- during the first five days of infection, while both subsets increased in this organ 2–3 weeks after infection in long term kinetic studies of mice infected with a strain of somewhat reduced virulence [115]. Thus, DC respond with quantitative changes in these organs in response to oral *Salmonella* infection.

An increase in CD8α^+ DC producing TNF-α, a cytokine critical for host survival to *Salmonella*, was also detected among splenic DC of mice orally infected with *Salmonella* [97]. However, the number of TNF-α^+ DC in the spleen during the first few days of infection is quite low relative to the number of TNF-α-producing neutrophils and macrophages in this organ [97, 98]. Given the relatively minor contribution DC appear to make to the overall TNF-α response during *Salmonella* infection, the function of DC-derived TNF-α comes into question. The low number of TNF-α-producing DC suggests that the major contribution of this population is not controlling bacterial replication through their bactericidal activity *per se*, a task likely attributed to the macrophages and neutrophils that respond to the infection [98]. However, TNF-α produced by DC during *Salmonella* infection may work locally in an autocrine or paracrine fashion to orchestrate DC maturation and migration. In this way, DC-derived TNF-α could facilitate linking innate and adaptive immunity.

In addition to TNF-α, IL-12 and IFNγ are also important for host survival to *Salmonella* [33, 99]. However, in contrast to the capacity of some microbial stimuli to elicit IL-12 by DC [5, 80, 81, 100–104], no significant increase in DC producing IL-12 was apparent in the first few days following oral infection [97]. Moreover, finding significant IL-12p70 in organ lysates or serum during the early stages of infection has been elusive despite finding increased IL-12p40 in infected mice (our unpublished data; [105]). These data would be consistent with a situation where orally acquired *Salmonella* induces a slow IL-12p70 response by relatively few cells. This could make detection of significant numbers of DC producing IL-12 difficult, and IL-12p70 may not readily accumulate to levels reliably detected in orally infected mice. The inability to detect significant IL-12$^+$ DC in *Salmonella*-infected mice may also reflect the limited numbers of bacteria in the spleen early during infection or the absence of additional signals necessary to enhance IL-12 production by DC in response to bacteria, such as those mediated through CD40 or by IFNγ [80, 81, 106]. It is also worth noting that even in systems where rather robust production of IL-12 by DC occurs in response to intravenous administration of microbial stimuli, the response is rapid and transient [80, 101, 102, 104, 107].

Thus, the "window" for finding IL-12p70, particularly if made by a nonabundant cell type such as DC, during oral *Salmonella* infection appears to be quite narrow. The lack of DC producing IL-12 does not seem to be due to production of IL-10 by these cells, however, [97], but it remains possible that IL-10 is produced by non-DC and reduces the capacity of DC to make IL-12 during infection [81]. Alternatively, the possibility that cells other than DC, such as macrophages and neutrophils, produce IL-12 during *Salmonella* infection requires further investigation [108, 109]. Thus, despite the role of IL-12 in host survival to *Salmonella* [110–112], the cell(s) responsible for its production, their relative contribution to the overall IL-12 response and the magnitude and kinetics of IL-12p70 production require further investigation.

35.5.3.2 Presentation of *Salmonella* Antigens by Dendritic Cells *in vivo*
As discussed above, DC in infected tissues (spleen, MLN, liver and PP) contain *Salmonella* during infection and respond by increasing co-stimulatory molecules and producing TNF-α. They are thus capable of interacting with T cells and participating in the adaptive immune response to this bacterium. Support for this comes from data showing that primary, antigen-specific CD4$^+$ and CD8$^+$ T cells are stimulated to proliferate upon co-culture with splenic DC isolated from *Salmonella*-infected mice, and that DC loaded with *S. typhimurium* can elicit bacteria-specific CD4$^+$ and CD8$^+$ T cells after transfer into naïve animals [10, 36].

Although these data indeed support a role for DC in stimulating *Salmonella*-specific T cells during infection, no additional information on, for example, the capacity of DC subsets in different organs to stimulate bacteria-specific CD4$^+$ or CD8$^+$ T cells during infection is presently available. Nor is there direct evidence that DC in infected animals are required to prime naïve bacteria-specific T cells, as has been shown for priming CD8$^+$ T cells during *Listeria* infection [113]. We are also lacking *in vivo* data addressing the relative role of direct presentation by *Salmonella*-infected DC to that of indirect presentation by non-infected bystander cells. The only thing that is known thus far is that both direct presentation of *Salmonella* antigens by infected DC and indirect presentation of *Salmonella* antigens by noninfected bystander DC can occur, and this data comes from *in vitro* systems (see section 35.3.2). No study has yet directly addressed the relative contribution of direct versus indirect presentation during *Salmonella* infection, as has been done for *Listeria*

[114]. In *Listeria*-infected mice, neutrophils were shown to be an important source of bacterial antigens that were crosspresented to CD8⁺ T cells [114]. Similar studies assessing the capacity of cells that contain *Salmonella in vivo* but can not directly present antigens to T cells, such as epithelial cells and neutrophils, to be sources of bacterial antigens that can be crosspresented to CD4⁺ and CD8⁺ T cells by DC during infection certainly warrants investigation.

35.6
Concluding Remarks

Our knowledge concerning the role of DC and DC subsets in the immune response to *Salmonella* is advancing. However, there are many aspects of DC function during infection that remain to be elucidated. For example, little is known about the host and bacterial factors involved in DC activation and recruitment during infection. In addition, although recent studies have shed great insight into the cells used and mechanisms involved in transporting orally acquired *Salmonella* from the intestinal lumen to downstream lymphoid organs, this intriguing process is not yet fully understood. Elucidating the relative contribution of direct versus indirect presentation of bacterial antigens in activating CD4⁺ and CD8⁺ T cells during *Salmonella* infection, as well as the cellular reservoirs of bacterial antigens involved in indirect presentation *in vivo*, are also exciting aspects of anti-*Salmonella* immunity that need further study. Finally, determining the consequences of the differential response of DC subsets during *Salmonella* infection on adaptive immunity to this bacterium remains a challenge to tackle. Fortunately, studies using *Salmonella* infection models provide the exciting opportunity to address these issues and decipher the role of DC in the immune response to this pathogen in the complex cellular and cytokine milieu of infected lymphoid organs.

Acknowledgements

Work from our laboratory is supported by funding from the Swedish Research Council (project 621-2004-1738), the Swedish Foundation for Strategic Research (project A3 01:93/01/01) and the Swedish Cancer Foundation (project 4884-B03-01XAB). Malin Sundquist, Miguel Tam, Anna Rydström, Stina Lindgren, and former laboratory members are gratefully acknowledged for their enthusiastic and valuable contributions to understanding the immune response during bacterial infection.

References

1 Vremec, D., Shortman, K., Dendritic cell sybtypes in mouse lymphoid organs. Cross-correlation of surface markers, changes with incubation, and differences among thymus, spleen and lymph nodes. *J. Immunol.* **1997**. 159: 565–573.

2 Vremec, D., Pooley, J., Hochrein, H., Wu, L., Shortman, K., CD4 and CD8 expression by dendritic cell sybtypes in mouse thymus and spleen. *J. Immunol.* **2000**. 164: 2978–2986.

3 Shortman, K., Liu, Y.-J., Mouse and human dendritic cell subsets. *Nat. Rev. Immunol.* **2002**. 2: 151–161.

4 Iwasaki, A., Kelsall, B. L., Localization of distinct Peyer's patch dendritic cell subsets and their recruitment by chemokines macrophage inflammatory protein (MIP)-3α, MIP-3β, and secondary lymphoid organ chemokine. *J. Exp. Med.* **2000**. 191: 1381–1393.

5 Iwasaki, A., Kelsall, B. L., Unique functions of CD11b⁻ CD8α⁺ and double-negative Peyer's patch dendritic cells. *J. Immunol.* **2001**. 166: 4884–4890.

6 Henri, S., Vremec, D., Kamath, A., Waithman, J., Williams, S., Benoist, C., Burnham, K., Saeland, S., Handman, E., Shortman, K., The dendritic cell populations of mouse lymph nodes. *J. Immunol.* **2001**. 167: 741–748.

7 Johannson, C., Wick, M. J., Liver dendritic cells present bacterial antigens and produce cytokines upon *Salmonella* encounter. *J. Immunol.* **2004**. 172: 2496–2503.

8 Sundquist, M., Johansson, C., Wick, M. J., Dendritic cells as inducers of antimicrobial immunity in vivo. *APMIS* **2003**. 111: 715–724.

9 Mariott, I., Hammond, T. G., Thomas, E. K., Rost, K. L., *Salmonella* efficiently enter and survive within cultured CD11c⁺ dendritic cells initiating cytokine expression. *Eur. J. Immunol.* **1999**. 29: 1107–1115.

10 Yrlid, U., Wick, M. J., Antigen presentation capacity and cytokine production by murine splenic dendritic cell subsets upon *Salmonella* encounter. *J. Immunol.* **2002**. 169: 108–116.

11 Eriksson, S., Chambers, B. J., Rhen, M., Nitric oxide produced by murine dendritic cells is cytotoxic for intracellular *Salmonella enterica* sv. Typhimurium. *Scand. J. Immunol.* **2003**. 58: 493–502.

12 Niedergang, F., Sirard, J.-C., Tallichet Blanc, C., Kraehenbuhl, J.-P., Entry and survival of *Salmonella typhimurium* in dendritic cells and presentation of recombinant antigens do not require macrophage-specific virulence factors. *Proc. Natl. Acad. Sci. USA* **2000**. 97: 14650–14655.

13 Rescigno, M., Urbano, M., Rimoldi, M., Valzassina, B., Rotta, G., Granucci, F., Ricciardi-Castagnoli, P., Toll-like receptor 4 is not required for the full maturation of dendritic cells or for the degradation of Gram-negative bacteria. *Eur. J. Immunol.* **2002**. 32: 2800–2806.

14 Jantsch, J., Cheminay, C., Chakravortty, D., Lindig, T., Hein, J., Hensel , M., Intracellular activities of *Salmonella enterica* in murine dendritic cells. *Cellular Microbiol.* **2003**. 5: 933–945.

15 Petrovska, L., Aspinall, R. J., Barber, L., Clare, S., Simmons, C. P., Stratford, R., Khan, S. A., Lemoine, N. R., Frankel, G., Holden, D. W., Dougan, G., *Salmonella enterica* serovar Typhimurium interaction with dendritic cells: impact of the *sifA* gene. *Cellular Microbiol.* **2004**. 6: 1071–1084.

16 Svensson, M., Stockinger, B., Wick, M. J., Bone marrow-derived dendritic cells can process bacteria for MHC-I and MHC-II presentation to T cells. *J. Immunol.* **1997**. 158: 4229–4236.

17 Brumell, J. H., Grinstein, S., *Salmonella* redirects phagosomal maturation. *Curr. Opin. Microbiol.* **2004**. 7: 78–84.

18 Waterman, S. R., Holden, D. W., Functions and effectors of the *Salmonella* pathogenicity island 2 type III secretion system. *Cellular Microbiol.* **2003**. 5: 501–511.

19 Svensson, M., Johansson, C., Wick, M. J., *Salmonella enterica* serovar Typhimurium-induced maturation of bone marrow-derived dendritic cells.

Infect. Immun. **2000**. 68: 6311–6320.

20 Eriksson, S., Lucchini, S., Thompson, A., Rhen, M., Hinton, J. C. D., Unravelling the biology of macrophage infection by gene expression profiling of intracellular *Salmonella enterica. Mol. Microbiol.* **2003**. 47: 103–118.

21 Shiloh, M. U., MacMicking, J. D., Nicholson, S., Brause, J. E., Potter, S., Marino, M., Fang, F. C., Dinauer, M., Nathan, C., Phenotype of mice and macrophages deficient in both phagocyte oxidase and inducible nitric oxide synthase. *Immunity* **1999**. 10: 29–38.

22 Vazquez-Torrez, A., Jones-Carson, J., Mastroeni, P., Ischiropoulos, H., Fang, F. C., Antimicrobial actions of the NADPH phagocyte oxidase and inducible nitric oxide syntase in experimental Salmonellosis. I. Effects on microbial killing by activated peritoneal macrophages *in vivo. J. Exp. Med.* **2000**. 192: 227–236.

23 Gallois, A., Klein, J. R., Allen, L.-H. H., Jones, B. D., Nauseef, W. M., *Salmonella* pathogenicity island 2-encoded type III secretion system mediates exclusion of NADPH oxidase assembly from the phagosomal membrane. *J. Immunol.* **2001**. 166: 5741–5748.

24 Chakravortty, D., Hansen-Wester, I., Hensel , M., *Salmonella* pathogenicity island 2 mediates protection of intra-cellular *Salmonella* from reactive nitrogen intermediates. *J. Exp. Med.* **2002**. 195: 1155–1166.

25 De Groote, M. A., Granger, D., Xu, Y., Campbell, G., Prince, R., Fang, F. C., Genetic and redox determinants of nitric oxide cytotoxicity in a *Salmonella typhimurium* model. *Proc. Natl. Acad. Sci. USA* **1995**. 92: 6399–6403.

26 Vazquez-Torres, A., Fang, F. C., Oxygen-dependent anti-*Salmonella* activity of macrophages. *Trends Microbiol.* **2001**. 9: 29–33.

27 Fang, F. C., Antimicrobial reactive oxygen and nitrogen species: concepts and controversies. *Nature Rev. Microbiol.* **2004**. 2: 820–832.

28 Kalupahana, R., Emilianus, A. R., Maskell, D., Blacklaws, B., *Salmonella enterica* Serovar Typhimurium epressing mutant lipid A with decreased endo-toxicity causes maturation of murine dendritic cells. *Infect. Immun.* **2003**. 71: 6132–6140.

29 Elsen, S., Doussiere, J., Villiers, C. L., Faure, M., Berthier, R., Papaioannou, A., Grandvaux, N., Marche, P. N., Vignais, P. V., Cryptic O_2-generating NADPH oxidase in dendritic cells. *J. Cell Sci.* **2004**. 117: 2215–2226.

30 Vulcano, M., Dusi, S., Lissandrini, D., Badolato, R., Mazzi, P., Riboldi, E., Borroni, E., Calleri, A., Donini, M., Plebani, A., Notarangelo, L., Musso, T., Sozzani, S., Toll receptor-mediated regulation of NADPH oxidase in human dendritic cells. *J. Immunol.* **2004**. 173: 5749–5756.

31 Hess, J., Ladel, C., Miko, D., Kaufmann, S. H. E., *Salmonella typhimurium aroA*- infection in gene-targeted immuno-deficient mice. Major role of CD4+ TCR-αβ and IFN-γ in bacterial clearance independent of intracellular location. *J. Immunol.* **1996**. 156: 3321–3326.

32 Nauciel, C., Role of CD4+ T cells and T-independent mechanisms in acquired resistance to *Salmonella typhimurium* infection. *J. Immunol.* **1990**. 145: 1265–1269.

33 Mittrücker, H.-W., Kaufmann, S. H. E., Immune response to infection with *Salmonella typhimurium* in mice. *J. Leukoc. Biol.* **2000**. 67: 457–462.

34 Richter-Dahlfors, A., Buchan, A. M. J., Finlay, B. B., Murine salmonellosis studied by confocal microscopy: *Salmonella typhimurium* resides intra-cellularly inside macrophages and exerts a cytotoxic effect on phagocytes in vivo. *J. Exp. Med.* **1997**. 186: 569–580.

35 Salcedo, S. P., Noursadeghi, M., Cohen, J., Holden, D. W., Intracellular replica-tion of *Salmonella typhimurium* strains in specific subsets of splenic macrophages in vivo. *Cell. Microbiol.* **2001**. 3: 587–597.

36 Yrlid, U., Svensson, M., Chambers, B., Ljunggren, H.-G., Wick, M. J., *In vivo* activation of dendritic cells and T cells during *Salmonella enterica* serovar Typhimurium infection. *Infect. Immun.* **2001**. 69: 5726–5735.

37 Sheppard, M., Webb, C., Heath, F., Mallos, V., Emilianus, R., Maskell, D., Mastroeni, P., Dynamics of bacterial

growth and distribution within the liver during *Salmonella* infection. *Cellular Microbiol.* **2003**. 5: 593–600.

38 Monack, D. M., Bouley, D. M., Falkow, S., *Salmonella typhimurium* persists within macrophages in the mesesnteric lymph nodes of chronically infected *Nramp1+/+* mice and can be reactivated by IFNγ neutralization. *J. Exp. Med.* **2004**. 199: 231–241.

39 Monack, D. M., Navarre, W. W., Falkow, S., *Salmonella*-induced macrophage death: the role of caspase-1 in death and inflammation. *Microb. Infect.* **2001**. 3: 1201–1212.

40 van der Velden, A. W. M., Velassquez, M., Starnbach, M. N., *Salmonella* rapidlly kill dendritic cells via a caspase-1-dependent mechanism. *J. Immunol.* **2003**. 171: 6742–6749.

41 Svensson, M., Wick, M. J., Classical MHC-I presentation of a bacterial fusion protein by bone marrow-derived dendritic cells. *Eur. J. Immunol.* **1999**. 29: 180–188.

42 Yrlid, U., Wick, M. J., *Salmonella*-induced apoptosis of infected macrophages results in presentation of a bacteria-encoded antigen after uptake by bystander dendritic cells. *J. Exp. Med.* **2000**. 191: 613–623.

43 Shastri, N., Schwab, S., Serwold, T., Producing nature's gene-chips: the generation of peptides for display by MHC class I molecules. *Annu. Rev. Immunol.* **2002**. 20: 463–493.

44 Bevan, M. J., Cross-priming for a secondary cytotoxic response to minor H antigens with H-2 congenic cells which do not cross-react in the cytotoxic assay. *J. Exp. Med.* **1976**. 143: 1283–1288.

45 Yewdell, J. W., Norbury, C. C., Bennink, J. R., Mechanisms of exogenous antigen presentation by MHC class I molecules *in vitro* and *in vivo*: Implications for generating CD8+ T cell responses to infectious agents, tumors, transplants and vaccines. *Adv. Immunol.* **1999**. 73: 1–77.

46 Wick, M. J., Ljunggren, H.-G., Processing of bacterial antigens for peptide presentation on MHC class I molecules. *Immunol. Rev.* **1999**. 172: 153–162.

47 Ackerman, A. L., Cresswell, P., Cellular mechanisms governing cross-presentation of exogenous antigens. *Nat. Immunol.* **2004**. 5: 678–684.

48 Sztein, M. B., Tanner, M. K., Polotsky, Y., Orenstein, J. M., Levine, M. M., Cytotoxic T lymphocytes after oral immunization with attenuated vaccine strains of *Salmonella typhi* in humans. *J. Immunol.* **1995**. 155: 3987–3993.

49 Salerno-Goncalves, R., Pasetti, M. R., Sztein, M. B., Characterization of CD8+ effector T cell responses in volunteers immunized with *Salmonella enterica* serovar Typhi strain Ty21a typhoid vaccine. *J. Immunol.* **2002**. 169: 2196–2203.

50 Salerno-Goncalves, R., Wyant, T. L., Pasetti, M. R., Fernandez-Vina, M., Tacket, C. O., Levine, M. M., Sztein, M. B., Concomitant induction of CD4+ and CD8+ T cell responses in volunteers immunized with *Salmonella enteric* serovar Typhi strain CVD 908-htrA. *J. Immunol.* **2003**. 170: 2734–2741.

51 Lo, W.-F., Ong, H., Metcalf, E. S., Soloski, M. J., T cell responses to Gram-negative intracellular bacterial pathogens: a role for CD8+ T cells in immunity to *Salmonella* infection and the involvement of MHC class Ib molecules. *J. Immunol.* **1999**. 162: 5398–5406.

52 Wijburg, O. L., van Rooijen, N., Strugnell, R. A., Induction of CD8+ T lymphocytes by *Salmonella typhimurium* is independent of Salmonella pathogenicity island 1-mediated host cell death. *J. Immunol.* **2002**. 169: 3275–3283.

53 Kirby, A. C., Sundquist, M., Wick, M. J., In vivo compartmentalization of functionally distinct, rapidly responsive antigen-specific T-cell populations in DNA-immunized or *Salmonella enterica* serovar Typhimurium-infected mice. *Infect. Immun.* **2004**. 72: 6390–6400.

54 Yrlid, U., Svensson, M., Kirby, A. C., Wick, M. J., Antigen-presenting cells and anti-*Salmonella* immunity. *Microb. Infect.* **2001**. 3: 1239–1248.

55 Wick, M. J., Pfeifer, J. D., MHC-I presentation of OVA(257–264) from exogenous sources: protein context influences the degree of TAP-independent presentation. *Eur. J. Immunol.* **1996**. 26: 2790–2799.

56 Song, R., Harding, C. V., Roles of proteasomes, transporter for antigen presentation (TAP), and β2-micro-globulin in the processing of bacterial or particulate antigens via an alternate class I MHC processing pathway. *J. Immunol.* **1996**. 156: 4182–4190.

57 Guermonprez, P., Saveanu, L., Kleijmeer, M., Davoust, J., van Endert, P., Amigorena, S., ER-phagosome fusion defines an MHC class I cross-presentation compartment in dendritic cells. *Nature* **2003**. 425: 397–402.

58 Houde, M., Bertholet, S., Gagnon, E., Brunet, S., Goyette, G., Laplante, A., Princiotta, M. F., Thibault, P., Sacks, D., Desjardins, M., Phagosomes are competent organelles for antigen cross-presentation. *Nature* **2003**. 425: 402–406.

59 Ohl, M. E., Miller, S. I., *Salmonella*: A model for bacterial pathogenesis. *Annu. Rev. Med.* **2001**. 52: 259–274.

60 Groisman, E. A., The pleiotropic two-component regulatory system PhoP-PhoQ. *J. Bacterol.* **2001**. 183: 1835–1842.

61 Tobar, J. A., González, P. A., Kalergis, A. M., *Salmonella* escape from antigen presentation can be overcome by targeting bacteria to Fcγ receptors on dendritic cells. *J. Immunol.* **2004**. 173: 4058–4065.

62 Hu, P. Q., Tuma-Warrino, R. J., Bryan, M. A., Mitchell, K. G., Higgins, D. E., Watkins, S. C., Salter, R. D., *Escherichia coli* expressing recombinant antigen and listeriolysin O stimulate class I-restricted CD8⁺ T cells following uptake by human APC. *J. Immunol.* **2004**. 172: 1595–1601.

63 Regnault, A., Lankar, D., Lacabanne, V., Rodriguez, A., Théry, C., Rescigno, M., Saito, T., Verbeek, S., Bonnerot, C., Ricciardi-Castagnoli, P., Amigorena, S., Fcγ receptor-mediated induction of dendritic cell maturation and major histocompatibility complex class I-restricted antigen presentation after immune complex internalization. *J. Exp. Med.* **1999**. 189: 371–380.

64 Machy, P., Serre, K., Leserman, L., Class I-restricted presentation of exogenous antigen acquired by Fcγ receptor-mediated endocytosis is regulated in

dendritic cells. *Eur. J. Immunol.* **2000**. 30: 848–857.

65 Guyre, C. A., Barreda, M. E., Swink, S. L., Fanger, M. W., Colocalization of FcγRI-targeted antigen with class I MHC: Implications for antigen processing. *J. Immunol.* **2001**. 166: 2469–2478.

66 Kalergis, A. M., Ravetch, J. V., Inducing tumor immunity through the selective engagement of activating Fcγ receptors on dendritic cells. *J. Exp. Med.* **2002**. 195: 1653–1659.

67 Winau, F., Kaufmann, S. H. E., Schaible, U. E., Apoptosis paves the detour path for CD8 T cell activation against intracellular bacteria. *Cellular Microbiol.* **2004**. 6: 599–607.

68 Schaible, U. E., Winau, F., Sieling, P. A., Fischer, K., Collins, H. L., Hagens, K., Modlin, R. L., Brinkmann, V., Kaufmann, S. H. E., Apoptosis facilitates antigen presentation to T lymphocytes through MHC-I and CD1 in tuberculosis. *Nature Med.* **2003**. 9: 1039–1046.

69 Banchereau, J., Steinman, R. M., Dendritic cells and the control of immunity. *Nature* **1998**. 392: 245–252.

70 Banchereau, J., Briere, F., Caux, C., Davoust, J., Lebecque, S., Liu, Y.-J., Pulendran, B., Palucka, K., Immuno-biology of dendritic cells. *Annu. Rev. Immunol.* **2000**. 18: 767–811.

71 West, M. A., Wallin, R. P. A., Matthews, S. P., Svensson, H. G., Zaru, R., Ljunggren, H.-G., Prescott, A. R., Watts, C., Enhanced dendritic cell antigen capture via toll-like receptor-induced actin remodeling. *Science* **2004**. 305: 1153–1157.

72 Trombetta, E. S., Ebersold, M., Garrett, W., Paypaert, M., Mellman, I., Activation of lysosomal function during dendritic cell maturation. *Science* **2003**. 299: 1400–1403.

73 Lelouard, H., Gatti, E., Cappello, F., Gresser, O., Camosseto, V., Pierre, P., Transient aggregation of ubiquitinated proteins during dendritic cell maturation. *Nature* **2002**. 417: 177–182.

74 Mellman, I., Steinman, R. M., Dendritic cells: specialized and regulated antigen processing machines. *Cell* **2001**. 106: 255–258.

75 Sozzani, S., Allavena, P., D'Amico, G., Luini, W., Bianchi, G., Kataura, M., Imai, T., Yoshie, O., Bonecchi, R., Mantovani, A., Differential regulation of chemokine receptors during dendritic cell maturation: A model for their trafficking properties. *J. Immunol.* **1998**. 161: 1083–1086.

76 Garrett, W. S., Chen, L.-M., Kroschewski, R., Ebersold, M., Turley, S., Trombetta, S., Galán, J. E., Mellman, I., Developmental control of endocytosis in dendritic cells by Cdc42. *Cell* **2000**. 102: 325–224.

77 West, M. A., Prescott, A. R., Eskelinen, E.-L., Ridley, A. J., Watts, C., Rac is required for constitutive macropinocytosis by dendritic cells but does not control its downregulation. *Curr. Biol.* **2000**. 10: 839–848.

78 Dreher, D., Kok, M., Cochand, L., Kiama, S. G., Gehr, P., Perchère, J.-C., Nicod, L. P., Genetic background of attenuated *Salmonella typhimurium* has profound influence on infection and cytokine patterns in human dendritic cells. *J. Leukocyte Biol.* **2001**. 69: 583–589.

79 Hofer, S., Rescigno, M., Granucci, F., Citterio, S., Francolini, M., Ricciardi-Castagnoli, P., Differential activation of NF-κB subunits in dendritic cells in response to Gram-negative bacteria and to lipopolysaccharide. *Microb. Infect.* **2001**. 3: 259–265.

80 Schulz, O., Edwards, A. D., Schito, M., Aliberti, J., Manickasingham, S., Sher, A., Reis e Sousa, C., CD40 triggering of heterodimeric IL-12p70 production by dendritic cells in vivo requires a microbial priming signal. *Immunity* **2000**. 13: 453–462.

81 Maldonado-López, R., Maliszewski, C., Urbain, J., Moser, M., Cytokines regulate the capacity of CD8α$^+$ and CD8α$^-$ dendritic cells to prime Th1/Th2 cells in vivo. *J. Immunol.* **2001**. 167: 4345–4350.

82 Edwards, A. D., Manickasingham, S. P., Spörri, R., Diebold, S. S., Schultze, O., Sher, A., Kaiso, T., Akira, S., Reis e Sousa, C., Microbial recognition via toll-like receptor-dependent and -independent pathways determines the cytokine response of murine dendritic cell subsets to CD40 triggering. *J. Immunol.* **2002**. 169: 3652–3660.

83 Guo, L., Lim, K. B., Gunn, J. S., Bainbridge, B., Darveau, R. P., Hackett, M., Miller, S. I., Regulation of lipid A modifications by *Salmonella typhimurium* virulence genes *phoP-phoQ*. *Science* **1997**. 276: 250–253.

84 Kawasaki, K., Ernst, R. K., Miller, S. I., 3-O-deacylation of lipid A by PagL, a PhoP/PhoQ-regulated deacylase of *Salmonella typhimurium*, modulates signaling through toll-like receptor 4. *J. Biol. Chem* **2004**. 279: 20044–20048.

85 Hajjar, A. M., Ernst, R. K., Tsai, J. H., Wilson, C. B., Miller, S. I., Human toll-like receptor 4 recognizes host-specific LPS modifications. *Nat. Immunol.* **2002**. 3: 354–359.

86 Jepson, M. A., Clark, M. A., The role of M cells in *Salmonella* infection. *Microb. Infect.* **2001**. 3: 1183–1190.

87 Rescigno, M., Urbano, M., Valzasina, B., Francolini, M., Rotta, G., Bonasio, R., Granucci, F., Kraehenbuhl, J.-P., Ricciardi-Castagnoli, P., Dendritic cells express tight junction proteins and penetrate gut epithelia monolayers to sample bacteria. *Nature Immunol.* **2001**. 2: 361–367.

88 Macpherson, A. J., Uhr, T., Induction of protective IgA by intestinal dendritic cells carrying commensal bacteria. *Science* **2004**. 303: 1662–1665.

89 Niess, J. H., Brand, S., Gu, X., Landsman, L., Jung, S., McCormick, B. A., Vyas, J. M., Boes, M., Ploegh, H. L., Fox, J. G., Littman, D. R., Reinecker, H.-C., CX$_3$CR1-mediated dendritic cell access to the intestinal lumen and bacterial clearance. *Science* **2005**. 307: 254–258.

90 Vasquez-Torres, A., Jones-Carson, J., Bäumler, A. J., Falkow, S., Valdivia, R., Brown, W., Le, M., Berggren, R., Parks, W. T., Fang, F. C., Extraintestinal dissemination of *Salmonella* by CD18-expressing phagocytes. *Nature* **1999**. 401: 804–808.

91 Hopkins, S. A., Niedergang, F., Corthesy-Theulaz, I. E., Kraehenbuhl, J.-P., A recombinant *Salmonella typhimurium* vaccine strain is taken up and survives within murine Peyer's patch dendritic cells. *Cell. Microbiol.* **2000**. 2: 59–68.

92 Nagler-Anderson, C., Man the barrier! Strategic defences in the intestinal mucosa. *Nat. Rev. Immunol.* **2001**. 1: 59–67.

93 Shreedhar, V. K., Kelsall, B. L., Neutra, M. R., Cholera toxin induces migration of dendritic cells from the subeithelial dome region to T-and B-cell areas of Peyer's Patches. *Infect. Immun.* **2003**. 71: 504–509.

94 Kelsall, B. L., Strober, W., Distinct populations of dendritic cells are present in the subepithelial dome and T cell regions of the murine Peyer's patch. *J. Exp. Med.* **1996**. 183: 237–247.

95 Pulendran, B., Lingappa, J., Kennedy, M. K., Smith, J., Teepe, M., Rudensky, A., Maliszewski, C. R., Maraskovsky, E., Developmental pathways of dendritic cells in vivo. Distinct function, phenotype, and localization of dendritic cell subsets in FLT3 ligand-treated mice. *J. Immunol.* **1997**. 159: 2222–2231.

96 McSorley, S. J., Cookson, B. T., Jenkins, M. K., Characterization of CD4+ T cell responses during natural infection with *Salmonella typhimurium*. *J. Immunol.* **2000**. 164: 986–993.

97 Kirby, A. C., Yrlid, U., Svensson, M., Wick, M. J., Differential involvement of dendritic cell subsets during acute *Salmonella* infection. *J. Immunol.* **2001**. 166: 6802–6811.

98 Kirby, A. C., Yrlid, U., Wick, M. J., The innate immune response differs in primary and secondary *Salmonella* infection. *J. Immunol.* **2002**. 169: 4450–4459.

99 Eckmann, L., Kagnoff, M. F., Cytokines in host defense against *Salmonella*. *Microb. Infect.* **2001**. 3: 1191–1200.

100 Aliberti, J., Reis e Sousa, C., Schito, M., Hieny, S., Wells, T., Huffnagle, G. B., Sher, A., CCR5 provides a signal for microbial induced production of IL-12 by CD8α+ dendritic cells. *Nature Immunol.* **2000**. 1: 83–87.

101 Reis e Sousa, C., Hieny, S., Scharton-Kersten, T., Jankovic, D., Charest, H., Germain, R. N., Sher, A., In vivo microbial stimulation induces rapid CD40 ligand-independent production of interleukin 12 by dendritic cells and their redistribution to T cell areas. *J. Exp. Med.* **1997**. 186: 1819–1829.

102 Reis e Sousa, C., Yap, G., Schulz, O., Rogers, N., Schito, M., Aliberti, J., Hieny, S., Sher, A., Paralysis of dendritic cell IL-12 production by microbial products prevents infection-induced immuno-pathology. *Immunity* **1999**. 11: 637–647.

103 Hochrein, H., Shortman, K., Vremec, D., Scott, B., Hertzog, P., O'Keeffe, M., Differential production of IL-12, IFN-α, and IFN-γ by mouse dendritic cell subsets. *J. Immunol.* **2001**. 166: 5448–5455.

104 Huang, L.-Y., Reis e Sousa, C., Itoh, Y., Inman, J., Scott, D. E., IL-12 induction by a Th1-inducing adjuvant in vivo: dendritic cell subsets and regulation by IL-10. *J. Immunol.* **2001**. 167: 1423–1430.

105 Brigl, M., Bry, L., Kent, S. C., Gumperz, J. E., Brenner, M. B., Mechanism of CD1d-restricted natural killer T cell activation during microbial infection. *Nat. Immunol.* **2003**. 4: 1230–1237.

106 Vieira, P. L., de Jong, E. C., Wierenga, E. A., Kaspenberg, M. L., Kaliński, P., Development of a Th1-inducing capacity in myeloid dendritic cells requires environmental instruction. *J. Immunol.* **2000**. 164: 4507–4512.

107 Jiao, X., Lo-Man, R., Guermonprez, P., Fiette, L., Dériaud, D., Burgaud, S., Gicquel, B., Winter, N., Leclerc, C., Dendritic cells are host cells for myco-bacteria in vivo that trigger innate and acquired immunity. *J. Immunol.* **2002**. 168: 1294–1301.

108 Skeen, M. J., Miller, M. A., Shinnick, T. M., Ziegler, H. K., Regulation of murine macrophage IL-12 production. Activation of macrophages in vivo, restimulation in vitro and modulation by other cytokines. *J. Immunol.* **1996**. 156: 1196–1206.

109 Bliss, S. K., Butcher, B. A., Denkers, E. Y., Rapid recruitment of neutrophils containing prestored IL-12 during microbial infection. *J. Immunol.* **2000**. 164: 4515–4521.

110 Kincy-Cain, T., Clemens, J. D., Bost, K. L., Endogenous and exogenous inter-leukin-12 augment the protective immune response in mice orally challenged with *Salmonella dublin*. *Infect. Immun.* **1996**. 64: 1437–1440.

111 Mastroeni, P., Harrison, J. A., Robinson, J. H., Clare, S., Khan, S., Maskell, D. J., Dougan, G., Hormaeche, C. E., Interleukin-12 is required for control of the growth of attenuated aromataic-compound-dependent salmonellae in Balb/c mice: role of gamma interferon and macrophage activation. *Infect. Immun.* **1998**. 66: 4767–4776.

112 Lehmann, J., Bellmann, S., Werner, C., Schröder, R., Schütze, N., Alber, G., IL-12p40-dependent agonistic effects on the development of protective innate and adaptive immunitiy against *Salmonella enteritidis. J. Immunol.* **2001**. 167: 5304–5315.

113 Jung, S., Unutmaz, D., Wong, P., Sano, G.-I., De los Santos, K., Sparwasser, T., Wu, S., Vuthoori, S., Ko, K., Zavala, F., Pamer, E. G., Littman, D. R., Lang, R. A., In vivo depletion of CD11c⁺ dendritic cells abrogates priming of CD8⁺ T cells by exogenous cell-associated antigens. *Immunity* **2002**. 17: 211–220.

114 Tvinnereim, A. R., Hamilton, S. E., Harty, J. T., Neutrophil involvement in cross-priming CD8⁺ T cell responses to bacterial antigens. *J. Immunol.* **2004**. 173: 1994–2002.

115 Sundquist, M., Wick, M. J., TNF-α-dependent and -independent maturation of dendritic cells and recruited CD11c^int CD11b⁺ cells during oral *Salmonella* infection. *J. Immunol.* **2005**. 175: 3287–3298.

36
Dendritic Cells in Tuberculosis

Ulrich E. Schaible and Florian Winau

36.1
Introduction

Tuberculosis (TB) is world-wide the most important bacterial infection in humans. The "white plague", as it was called in the 19th century, is part of the triad of most prevalent infectious diseases that also includes AIDS and malaria. The causative agent, *Mycobacterium (M.) tuberculosis*, was identified in 1882 by Robert Koch (1843–1910; Nobel prize 1905). However, the biology of the host–pathogen interaction is far from being conclusively elucidated, and current vaccination and therapeutic measures require significant improvement.

Dendritic cells (DC) are professional antigen-presenting cells (APC) with the potency to activate T cells through MHC class I, class II and CD1 molecules, which render DC indispensable for T-cell priming in TB. DC take up mycobacterial antigens for subsequent processing and T-cell activation [1]. Moreover, because they are able to phagocytose, DC engulf live mycobacteria and can thus serve as host cells for these intracellular pathogens. Here we discuss the multifaceted role of DC as APC in immunity against mycobacterial infection [2].

36.2
Tuberculosis

Tuberculosis is caused by *M. tuberculosis* and, in some African regions, *M. africanum*. *M. bovis*, primarily a pathogen of ruminants causing tuberculosis in cattle, can also infect humans, and the current live attenuated vaccine strain, *M. bovis* Bacille Calmette–Guerin (BCG) was derived from this species. *M. tuberculosis* is spread through aerosols from an infected individual by coughing. The mycobacteria primarily settle in lung tissue where they are engulfed by resident phagocytes such as alveolar macrophages and immature DC. Subsequently, infected cells migrate to the draining lung-associated lymph nodes (LAL) to mediate T-cell priming. Through DC, mycobacteria also reach the LAL but do not disseminate (Fig. 36.1) [1].

Handbook of Dendritic Cells. Biology, Diseases, and Therapies.
Edited by M. B. Lutz, N. Romani, and A. Steinkasserer.
Copyright © 2006 WILEY-VCH Verlag GmbH & Co. KGaA, Weinheim
ISBN: 3-527-31109-2

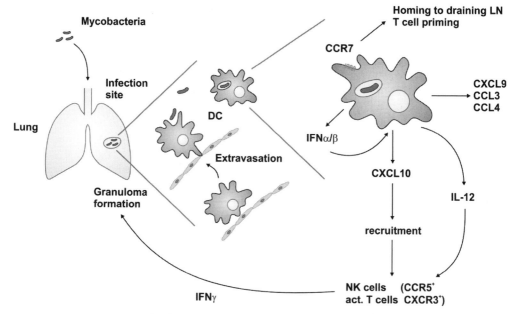

Fig. 36.1 Role of DC in tuberculosis. Inhalation of mycobacteria by aerosol leads to the infection of macrophages and DC in the lung. Infection activates DC to express CCR7 and to subsequently migrate to draining lymph nodes for T-cell priming. Concurrent release of type I IFN-α/β starts an autocrine activation loop initiating chemokine secretion (CXCL10, 9, CCL3, 4) to recruit activated T cells and NK cells to the site of infection. Onset of a protective T-cell response with IFNγ and TNFα as essential cytokines leads to macrophage/DC activation and granuloma formation to restrict spread of mycobacteria.

Cells of the innate immune system are activated by mycobacterial pathogen-associated molecular patterns (PAMP) through their pattern recognition receptors (PRR) such as the Toll-like receptors (TLR). Mycobacterial PAMP include lipoproteins, peptidoglycans, glycolipids, and low-methylated DNA (CpG). Subsequent chemo- and cytokine release by activated DC and macrophages finally recruits primed T cells and other lymphocytes to the site of infection [1]. In the immunocompetent host, the onset of the antimycobacterial T-cell response marks the limitation of infection within a histomorphological structure, the granuloma (tubercle). The granuloma is histologically characterized by concentric layers of immune cells. T and B cells surround an area consisting of macrophage-derived multinucleated Langhans giant and epitheloid cells harboring mycobacteria bordering a necrotic center [3, 4]. DC as defined by the expression of CD11c also promote granuloma formation but harbor less mycobacteria than macrophages [5, 6, 7]. DC from infected lungs, granuloma and LAL together are termed the Ghon complex. This restricted form represents the latent, clinically non-apparent stage of infection, which is estimated by the World Health Organization (WHO) to be present in one third of the world's human population. Latency at this stage of infection is due to a well-regulated protective immune response controlling mycobacterial growth and spread. Less than 10% of infected individuals develop the disease [2]. The inner part of the granuloma liquefies and the containment of infection is abrogated. Subsequently, the mycobacteria grow unrestricted and spread hemo- and lymphogenically to other organs. Co-infection with HIV increases the risk of developing the disease 800-fold [2]. Apart from severe immunocompromising conditions, other reasons for the transition from the latent to the clinically apparent stage are not clear yet. Thus, in more than 90% of infected humans, immunity controls the infection. However, *M. tuberculosis* still kills 2 million patients annually world-wide [1]. The current vaccine in use, BCG, is only protective against systemic tuberculosis in children but not against lung infection in adults. The antibiotic therapy requires 6 to 12 months to be effective and depends on high compliance of patients [2]. Therefore, more efficient vaccination strategies and therapeutic protocols are needed to conquer tuberculosis in the future. Understanding the biology of the infectious agent and its survival strategies, as well as the pathways which lead to a strong protective immune response, i.e. antigen-processing/presentation and subsequent T-cell activation, are prerequisites to improve treatment. Thus, the biology of the DC as a central cell in host response to infection is an essential issue to unravel the pathogenesis of tuberculosis.

36.3
Mycobacteria are Intracellular Pathogens

M. tuberculosis and its relatives are facultative intracellular pathogens. Apart from those mycobacteria causing tuberculosis, other members of this genus have adopted a similar life style: *M. leprae,* the agent of leprosy; *M. marinum*, a fish pathogen sometimes causing skin ulcers in humans; *M. avium-intracellulare*-complex and *M. kansasii,* both opportunists mainly infecting immunocompromised patients. Their preferential host cells, macrophages, are primarily infected *in vivo* [8]. Moreover, mycobacteria also bind to receptors on DC, which facilitate entry into these cells. A C-type lectin molecule, DC-specific ICAM-3-grabbing nonintegrin (DC-SIGN), is a surface protein on DC. DC-SIGN is induced by IL-13 and IL-4 but negatively regulated by LPS and TNFα [9]. DC-SIGN binds mannose-residues and allows mycobacterial uptake by DC [10, 11]. However, other receptors have also been involved in this process such as the mannose receptor (MR), DEC205, scavenger receptors, and the complement receptor (C3R) [12, 13].

Upon uptake by macrophages, mycobacteria inhibit the maturation of phagosomes [14]. Thus, mycobacteria create their own intracellular niche with characteristics of an early endosomal compartment. The mycobacterial phagosome acidifies only mildly (pH 6.4) due to paucity of the vesicular H$^+$ATPase facilitating endosomal acidification. The early endosomal stage enables mycobacterial survival through access to iron, an essential growth factor. Iron is imported into early endosomes bound to transferrin (Tf) by the transferrin receptor (TfR). Tf/TfR intersects with the mycobacterial phagosome and delivers iron to the bacteria. Other marker-molecules which characterize the mycobacterial phagosome as an early endosomal compartment comprise the small GTPase Rab5, and the vesicle fusion initiator molecule syntaxin 13 [14]. Mycobacterial phagosomes have only minute amounts of the lysosome-associated-membrane-protein-1 (LAMP-1) and lack late endosomal/lysosomal features, i.e. acidic pH, Rab7, active cathepsin D and the mannose-6-phosphate receptor. Mycobacterial trafficking inside DC is a controversial issue. One study suggests that mycobacteria grow as well in murine DC as in macrophages, whereas another report claims survival in DC but no growth [15, 16]. Experiments in the human system revealed that macrophages but not DC are permissive for *M. tuberculosis* [17]. Upon activation by IFNγ, which enables macrophages to kill mycobacteria efficiently, DC restricted mycobacterial growth but were unable to kill them. In DC, the compartment harboring mycobacteria appears to be different from the early endosome-type compartment in macrophages. Inhibition of phagosome maturation seems to be incomplete in DC since less than 50% of the phagosomes contain the early endosomal markers Tf, TfR and the antigen-presenting molecule CD1a [18]. An independent study demonstrated only limited presence of the early endosomal markers Tf and Rab11 (a recycling endosome specific GTPase) but also the lack of lysosomal features such as a low pH [17].

M. tuberculosis biases IFNα-induced differentiation of human monocytes towards CD14$^+$ macrophages instead of generation of DC [19]. This strategy probably provides more host cells favoring mycobacterial multiplication. Taken together, trafficking of mycobacteria within DC is guided by unique mechanisms when compared to macrophages.

Mycobacteria are covered by a rigid and hydrophobic cell wall consisting of an array of genus-specific glycolipids, glycophospholipids and waxes including 80-carbon-long fatty acids, the mycolates. This causes the acid-fastness of mycobacteria facilitating diagnostic staining (Ziehl–Neelsen stain) and makes dryness, organic solvents and disinfectants less harmful to mycobacteria then to other microbes. Through ester bonds, mycolates form arabinogalactan mycolates (AGM), trehalose dimycolate (TDM; Cord factor) and glucose monomycolate (GMM). Other compounds are the phosphatidylinositol-anchored lipoarabinomannans (LAM), phosphatidylinositol mannosides (PIM) forming an outer capsule-like layer as well as mycocerosates which are putatively involved in virulence [20]. Mycobacterial lipids are T-cell antigens presented by CD1 molecules [21]. Mycobacterial proteins including ESAT-6, CFP-10, Ag85 and the p19 lipoprotein have been characterized so far as T-cell antigens. The most potent T-cell antigens are either secreted or released proteins [2]. Some of these have recently been further tested successfully as novel vaccine candidates. Although mycobacteria reside in non-mature phagosomes, they release lipid and protein antigens, which enter the endosomal system of infected cells. Thus, antigens reach hydrolytic compartments where processing and loading onto MHC-II, CD1b, CD1c and CD1d molecules occurs [18, 22, 23].

36.4
Dendritic Cells Present Antigens in Tuberculosis

The immune response elicited by mycobacteria comprises a multitude of T-cell subpopulations: MHC class II-restricted CD4+ T helper cells, MHC class I-restricted CD8+ cytotoxic T cells, CD1-restricted T cells expressing $\alpha\beta$ T-cell receptors (TCR), and T cells expressing a $\gamma\delta$ TCR. T cells primarily participate in protection by IFNγ release to activate the antimicrobial effector mechanisms of macrophages. Moreover, CD8 T cells can kill infected cells and deliberate mycobacteria [1, 2]. Apart from human $\gamma\delta$ T cells not requiring classical antigen presentation for activation, DC represent the prime APC in TB. DC express sufficient amounts of MHC-I and MHC-II as well as CD1 molecules on their surface. Additionally, potent T-cell priming requires co-stimulatory molecules such as CD80 (B7.1) and CD86 (B7.2) as well as CD40 present on DC (Fig. 36.2).

BCG-pulsed DC present antigens and activate T cells from vaccinated donors [24]. Using CD11c as a distinctive marker for selection, DC have been isolated from infected lungs. Lung-derived DC were able to stimulate T cells despite expression of low levels of MHC-II and co-stimulatory molecules [5]. Due to the potent antigen-presenting function, DC have been used as vaccine carriers and natural adjuvants to protect against *M. tuberculosis* infection. Murine DC pulsed with either whole mycobacteria (either *M. tuberculosis* or BCG) or CD8 and CD4 T-cell epitopes from Ag85 protected mice against *M. tuberculosis* challenge by inducing mycobacteria-specific IFNγ-secreting CD4 and CD8 T cells [25, 26, 27]. Co-stimulation through CD40 prior to vaccination further enhanced the capacity of these DC to stimulate protective immunity [28].

Dendritic cells are the only cells in humans expressing all 5 CD1 molecules. Group I CD1 molecules comprise CD1a, CD1b, CD1c and CD1e, and group II consists of CD1d. CD1 molecules are homologues to MHC-I molecules and also associate with β2-microglobulin, but exhibit only minor polymorphism [21]. CD1a is primarily expressed on Langerhans cells, a DC subtype within epithelia, and to some extent on immature and mature DC. CD1a is a residential protein of the early/recycling endosome pathway involving the GTPase ARF6. Therefore, CD1a does not require an acidified compartment for antigen acquisition. CD1b and CD1c are present on immature and mature DC. However, the half-life of surface presence of CD1b and CD1c is largely increased on mature DC, similarly to what is seen for MHC molecules. CD1b and CD1c traffic to late endosomes and lysosomes due to a tyrosine-containing endosomal-sorting sequence (YXXΘ). This sequence allows CD1b to interact with adaptor protein (AP) 3 for delivery to lysosomes [21, 29].

CD1 molecules present lipid antigens to T cells. The lipids identified as CD1 ligands are mainly derived from the mycobacterial cell wall [21, 29]: CD1a: sulfatides, lipopeptides [30]; CD1b: LAM, PIM, GMM, mycolic acid, sulfoglycolipid (such as Ac$_2$SGL) [31]; CD1c: isoprenoids; CD1d: PIM [32]. Most peripheral, CD8 T-cell specific for mycobacteria (BCG) in humans are restricted by group I CD1 molecules [33]. Freshly converted skin-test-positive donors indicating a recent contact with mycobacteria respond more frequently to CD1b– and CD1c-presented

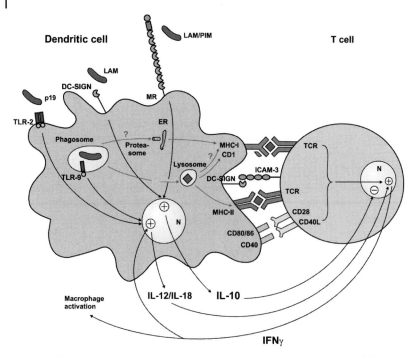

Fig. 36.2 The DC as immune-modulating APC. Mycobacteria and their pathogen associate molecular patterns (PAMP) such as lipoarabinomannan (LAM), the 19-kDa lipoprotein (p19) and low-methylated bacterial DNA (CpG) bind various pattern recognition receptors (PRR) on DC. Engagement of DC-SIGN and the mannose receptor (MR) induce release of the anti-inflammatory/immuno-suppressive cytokine IL-10. In contrast, ligands for Toll-like receptors (p19 – TLR-2, CpG – TLR-9) activate DC to secrete the pro-inflammatory cytokines IL-12/IL-18. Thereby an IFNγ-dominated T-cell response (T helper type I) is initiated leading to macrophage activation and mycobactericidal effector mechanisms. DC are potent antigen-presenting cells employing MHC-I, -II, CD1 and co-stimulatory molecules (CD80, CD86, CD40) to prime mycobacterium-specific T cells. Mycobacterial antigens are delivered to the lysosomes for processing and binding to MHC-II and CD1 molecules. Loading of lipids onto CD1 molecules involves the lipid transfer proteins saposins. The pathway leading to MHC-I presentation within infected cells – whether through processing and loading within late endosomes/lysosomes, processing by proteasomes and loading in the ER, or by both – is yet unclear.

mycobacterial lipid antigens compared to skin-test-negative donors [31, 34, 35]. The question of enzymatic processing of lipid antigens within lysosomes remains unsolved. However, presentation of mycobacterial lipids by CD1b and CD1c requires transport through an acidified compartment. Loading of LAM, GMM as well as mycolic acid is facilitated through saposin C, a lysosomal lipid transfer protein (LTP) involved in lipid metabolism of mammalian cells [36]. Saposin C facilitates extraction of lipid antigens from intra-endosomal membranes and interacts with CD1b to allow lipid loading. Thus, the hydrophilic gap between membranes and the hydrophobic antigen-binding groove of CD1b is bridged by saposins.

CD1e is exclusively found intracellularly with a distinctive stage-dependent localization within the Golgi apparatus of immature DC and in late endosomal/lysosomal compartments of mature DC [37, 38]. The presence of CD1e only in DC suggests a specific function in antigen-presentation and lipid transfer. Indeed, CD1e binds glycolipids, and its soluble form is required for processing of high-mannosylated PIM_6 by lysosomal α-mannosidase to generate PIM_2 for recognition by CD1b-restricted T cells [82].

CD1d restricts natural killer (NK) T cells expressing NK1.1. CD1d has a broader cellular distribution including macrophages, certain B cells and epithelial cells, but is also expressed on DC [21]. CD1d also requires trafficking through acidified compartments for loading of exogenous ligands. However, CD1d also presents endogenous lipid antigens such as iGb3 (the lysosomal β-hexosaminidase degradation product of iGb4) [39]. Presentation of endogenous lipids and subsequent NKT cell activation requires induction of DC maturation through TLR engagement by PAMP [40, 41]. Purified mycobacterial PIM directly activates NKT cells in a CD1d-dependent manner in the absence of APC [32]. However, due to multiple PAMP present in mycobacteria, they could also activate APC to present self-lipids to NKT cells. Loading of CD1d with exogenous (as shown for the nominal NKT cell antigen α-galactosyl-ceramide derived from a marine sponge) and probably also endogenous lipids like iGb3 depends to some extent on LTP, mainly saposin A, C and the GM2 activator protein [42, 43].

DC express an array of surface PRR which serve as antigen-capturing receptors and/or signal transmitters. The TLR bind PAMP which in turn activate immature DC to migrate to draining lymphnodes and to express co-stimulatory molecules and cytokines. Langerin is a C-type lectin expressed by Langerhans cells. Langerin is localized in Birbeck granules, specialized early endosomal compartments carrying CD1a [44]. Birbeck granules probably function as antigen-loading compartments for CD1a since langerin facilitates lipid transfer to CD1a for T-cell activation. TLR present on DC and triggered by PAMP include (i) TLR-2 recognizing peptidoglycan, LAM and lipoproteins such as p19 lipoprotein [45, 46], and (ii) the endosomally located TLR-9 binding bacterial CpG. Interaction of mycobacterial ligands with TLR leads to the maturation of DC as shown by increased expression of CD83, CD80, CD86, CD54, CD58 and MHC-II [45]. This function of mycobacterial PAMP has been used to generate potent adjuvants for immunization as exemplified by the complete Freund's adjuvant and more sophisticated ones using liposomes containing purified PIM [47]. Further PRR on DC comprise DC-SIGN, DEC205 and the MR, which bind mannose-capped LAM and other mannose-containing mycobacterial lipids and proteins (see Section 36.3). Apart from binding and uptake of these ligands, these PRR appear also to mediate signals to the DC. Thus, interaction of mycobacteria with PRR activates DC and ultimately leads to the release of secreted mediators, i.e. cytokines and chemokines.

36.5
Dendritic Cells are Regulatory Cells in Tuberculosis

Upon interaction with mycobacteria, DC secrete cytokines and chemokines, which are important in determining the subsequent T-cell responses. Moreover, DC-derived chemokines play a pivotal role in tuberculosis by recruiting inflammatory cells to the site of infection and promoting tissue remodeling for granuloma formation. Mycobacteria induce DC to secrete cytokines such as IL-12, IL-18 and IFNα. Induction of IFNα is preceded by activation of an autocrine loop of IFNβ stimulation involving the transcription factors NF-kB and IRF-3 [48]. Type I IFNα and β subsequently activate DC to secrete the chemokine CXCL10 in order to attract NK cells and activated T cells expressing the respective receptors, CCR5 and CXCR3, to the site of infection [49]. CCR5-deficient mice control *M. tuberculosis* equally as wild-type mice but exhibit exacerbated pulmonary pathology [50]. This observation also points out that there is a certain redundancy with respect to the function of individual chemokines in tuberculosis.

Other chemokines induced in DC within the first 8 h of mycobacterial infection include CCL3, CCL4, CCL5 and CXCL9, which also participate in the recruitment and homing of activated effector T cells and NK cells [48, 49]. Mycobacteria also trigger DC to express CCR7, the chemokine receptor required for the homing of mature DC to draining lymphnodes in order to prime T cells [51]. Moreover, the chemokine receptor CCR2 is instrumental in recruiting DC but also macrophages and T cells to the mycobacteria-infected lung [52]. As a consequence, mice lacking CCR2 succumb much earlier to *M. tuberculosis* infection and show 100 times higher bacterial burden than wild-type mice. This suggests an additional defect in T-cell priming in the absence of CCR2. Moreover, migration of CCR2-KO DC to the infected lung is strongly delayed [53]. The importance of an early recruitment of DC to the infection site is supported by mathematical modeling of DC and T-cell turnover within the Ghon complex upon *M. tuberculosis* infection. This study points to the time range between infection and arrival of DC in the LAL and T-cell recruitment to the lung as a critical factor to tip the scales between latency and disease [54, 55].

Importantly, DC and NK cells establish a mutual relationship in tuberculosis. Upon mycobacterial infection, DC-derived IL-12 and IFNα stimulates NK cells as indicated by expression of the activation marker CD69, and enhance their cytolytic activity. In turn, NK cells trigger DC to mature and function as proper APC through signals involving direct cell-cell contact as well as TNFα and IFNγ secretion [56]. A similar reciprocal stimulation has been described for DC and γδ T cells involving DC-secreted IL-12 and T cell-derived IFNγ. This *pas de deux* however, does not require cell–cell contact [57].

Ligand-binding to DC-SIGN and MR induces the inhibitory cytokine IL-10 counteracting inflammatory responses induced by IL-12 [10, 11, 58, 59]. DC also produce IL-1-R antagonist, CCL22 and CCL17 but none of the pro-inflammatory cyto- and chemokines thus biasing the T helper 2 circuit [59]. Adoptive transfer of DC from IL-10 KO mice revealed that autocrine IL-10-production impairs both, trafficking of DC to the draining lymph nodes and local IL-12 production [60]. However, mycobacterial stimulation of anti-inflammatory signals in DC must not only be seen as beneficial for the survival of the pathogen but is also of benefit to the host due to limiting inflammation-mediated pathology in tuberculosis lesions [61].

Engagement of TLR, mainly TLR-2, leads to the release of the pro-inflammatory cytokines IFNα, IL-1β, IL-12, IL-18 which biases towards IFNγ secretion by T helper 1, NKT- and γδ T cells as well as NK cells. IFNγ is essential for protective immunity in tuberculosis because it activates macrophages to optimally express their mycobactericidal effector mechanisms [1]. IFNγ induces the expression of inducible nitric oxide synthase (iNOS or NOS2) to generate toxic reactive nitrogen intermediates (RNI) and decreases TfR-expression to limit iron access to mycobacteria. In contrast to DC, mycobacteria-infected macrophages enter a different cytokine-expression program predominated by the pro-inflammatory cytokines IL-1, IL-6 and TNFα instead of IL-12. Macrophages also produce IL-10 suppressing IL-12-triggered T-helper 1 type responses [62]. A controversial report however showed that pro-inflammatory and immunosuppressive cytokines can be induced at the same time. *M. tuberculosis* induced IL-6 as well as IL-10 secretion in DC, which was mainly dependent on TLR-2 [63]. IL-10 converts DC into macrophage-like cells with increased capability of killing mycobacteria [64]. In contrast, TNFα-matured DC are even less able to control mycobacterial growth than immature DC [65]. Thus, DC and macrophages play distinct roles in the host response against mycobacteria: macrophages act first as pro-inflammatory cells and subsequently as mycobactericidal effector cells. DC however are primarily involved in priming and maintaining of antimycobacterial T-cell responses. Both cells are keeping an equilibrium between pro-inflammatory and anti-inflammatory responses to limit bacterial growth without overwhelming pathology [66].

36.6
Dendritic Cells and Cross-Priming

Live mycobacteria reside within early endosomal phagosomes. However, the phagosomal membrane secludes mycobacteria and their antigens from the cytosolic MHC-I pathway. Thus, activation of MHC-I-restricted CD8 T cells in tuberculosis remains elusive. Moreover, macrophages do not express group I CD1 molecules precluding CD1-restricted T-cell activation by primary infected cells. In addition, mycobacteria-infected macrophages as well as DC loose their antigen-presenting capacity quickly after infection [19, 67, 68].

Recently, the detour pathway in tuberculosis has been described involving apoptotic blebs from infected macrophages, which carry mycobacterial antigens to DC for subsequent CD8 T-cell activation (Fig. 36.3) [67, 69]. *M. tuberculosis* triggers apoptosis of infected macrophages, a process which has also been observed in tuberculosis lesions. Apoptosis of infected host cells leads to disintegration into and release of apoptotic blebs. Upon engulfment of these blebs, non-infected DC present mycobacterial antigens to CD8 and CD4 T cells through MHC-I, MHC-II

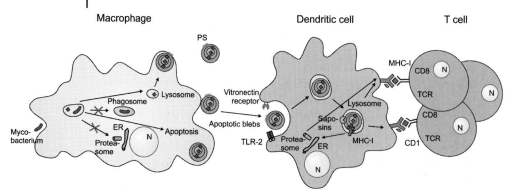

Fig. 36.3 The detour pathway of MHC-I and CD1 presentation. In macrophages, myco-bacteria are segregated within phagosomes from the classical MHC class I pathway. Moreover, macrophages do not express group I CD1 molecules (CD1a, b, c). Finally, mycobacteria-infected cells loose their ability to present antigens to T cells. These hinderances to induce proper T-cell immunity are overcome by the detour pathway in tuber-culosis. Infection-induced apoptosis leads to the release of phosphatidyl-serine (PS)-positive apoptotic blebs from infected cells.

Thereby, mycobacterial antigens are carried to non-infected DC for presentation. Apoptotic blebs are engulfed by the vitronectin receptor (VR) and probably the PS-receptor and reach the endosomal system of the DC. DC matura-tion is initiated upon engagement of TLR-2 by mycobacterial PAMP (such as p19). DC subsequently prime T cells through MHC-I and CD1, but also MHC-II molecules. Processing of mycobacterial antigens is predominantly dependent on the lysosomal pathway. Saposins are involved in this process.

and CD1 molecules. Blebs from infected cells carry a triad consisting of antigens (proteins and lipids such as Ag85 and LAM), adjuvants (p19 lipoprotein and PIM) and antigen-presenting molecules (MHC-I) [70]. The adjuvant activates DC through TLR-2 [71]. Notably, the main adaptor molecule for the TLR signaling cas-cade, myeloid differentiation factor 88 (MyD88), is a prerequisite for cross-priming of CD8 T cells due to its function in DC maturation [72]. MHC-I molecules can be transferred by apoptotic blebs from macrophages to DC [70]. Thus, apoptotic blebs represent an autonomous immunological entity and mediate potent protection against tuberculosis upon vaccination in the mouse model [70]. In conclusion, host cell apoptosis represents an essential prerequisite for cross-priming of T cells in tu-berculosis [69].

In DC, apoptotic blebs are targeted to lysosomes. Processing of the antigenic content for MHC-I presentation primarily depends on functionally intact lyso-somes and less on proteasomal activity [68]. Saposins are instrumental in the pro-cessing of apoptotic blebs probably by disrupting the bleb membranes to release antigens [70]. The question as to how antigens subsequently intersect with MHC-I molecules in DC is not solved yet. However, a transport pathway for protein anti-gens from lysosomes to cytosol was described, which appears to be restricted to DC and absent from macrophages [73].

36.7
Mycobacteria Interfere with Antigen Presenting Cell Function

In tuberculosis, DC are essential APC and cross-priming appears to be a prerequisite for T-cell activation due to the interference of *M. tuberculosis* with APC functions [67]. Upon infection with mycobacteria, DC loose surface expression of MHC as well as CD1 molecules and become unable to activate T cells. The mycobacterial lipoproteins p19 and LprG (24 kDa) are responsible for interference with antigen presentation [74, 75, 76]. These lipoproteins are released from intracellular mycobacteria and shuttle to various compartments of the host cell including the ER. Engagement of TLR-2 by these lipoproteins causes down-regulation of MHC-II surface expression and activation of CD4 T cells in a MyD88-dependent manner. Moreover, mycobacterial lipoproteins inhibit MHC-I cross-presentation of phagosomal particle-associated antigens [77]. However, these studies have been conducted so far only in macrophage populations. Thus, the question is still to be asked whether mycobacterial lipoproteins have similar effects on DC.

Full maturation of DC is a prerequisite for potent induction of T-cell immunity. *M. tuberculosis*, even in the presence of strong triggers for DC maturation such as TNFα, IL-1β and prostaglandin E2 (PGE2), hampers maturation of human DC and consequently T-cell stimulation [78]. Moreover, upon infection of mice with ovalbumin-expressing BCG or listeria followed by the transfer of ovalbumin-specific T cells, antigen-specific activation of T cells was delayed and weaker in BCG-infected mice when compared to those infected with listeria [79]. *M. tuberculosis*-infected monocytes differentiate into CCR7- and CD83-positive DC upon GM-CSF/IL-4 stimulation, but express only low levels of CD1, MHC-II and CD80 molecules and fail to secrete IL-12. Consequently, these cells are unable to activate T cells *in vivo*. This has been correlated with the observation that in mycobacteria-infected mice DC were apoptotic for a prolonged period of time. Similarly, BCG hampers DC to develop their full APC potential, i.e. proper CD1, MHC-II, CD80 and CD40 expression [80]. These DC were unable to prime naïve T cells to undergo a T helper type I polarization, i.e IFNγ-secretion suggesting a putative reason for the ineffectiveness of BCG as a vaccine against *M. tuberculosis*. However, one study should be noted claiming that *M. tuberculosis*–derived secreted antigens initiate monocyte differentiation into DC in the absence of GM-CSF [81].

In conclusion, viable mycobacteria interfere with antigen-presentation through MHC-I, MHC-II and CD1 molecules there by hampening activation of CD4 and CD8 T cells. Inhibition of antigen presentation by mycobacteria renders the detour pathway of cross presentation through apoptotic blebs essential for cell priming. Activated mycobacteria-specific T cells can subsequently detect freshly infected cells before mycobacterial inhibition of antigen presentation occures.

36.8
Conclusion

DC are the main APC in tuberculosis to activate CD4 and CD8 T cells through antigen presentation by MHC-I, MHC-II and CD1 molecules. APC function however is targeted by mycobacterial virulence factors. This interference with T-cell activation is circumvented by cross priming through apoptotic blebs. Moreover, by releasing chemokines and cytokines, DC modulate subsequent T-cell responses and recruit immune cells to the site of infection. Thus, DC are central in the host response to mycobacteria, and unraveling their functions in protective immunity will allow better measures to control tuberculosis.

Acknowledgement

The authors thank the Deutsche Forschungsgemeinschaft DFG for financial support (SFB421, SPP1130) and Luise Fehlig for the graphics.

References

1 Kaufmann S H. *Nat Rev Immunol.* **2001**. 1(1), 20–30.

2 Kaufmann S H, McMichael AJ. *Nat Med.* **2005**. 11(4 Suppl), S33–44

3 Ulrichs T, Kosmiadi G A, Trusov V, Jorg S, Pradl L, Titukhina M, Mishenko V, Gushina N, Kaufmann S H. *J Pathol.* **2004**. 204(2), 217–228.

4 Ulrichs T, Lefmann M, Reich M, Morawietz L, Roth A, Brinkmann V, Kosmiadi G A, Seiler P, Aichele P, Hahn H, Krenn V, Gobel U B, Kaufmann SH. *J Pathol.* **2005**. Apr; 205(5):633–640.

5 Gonzalez-Juarrero M, Orme I M. *Infect Immun.* **2001**. 69(2), 1127–1133

6 Garcia-Romo G S, Pedroza-Gonzalez A, Aguilar-Leon D, Orozco-Estevez H, Lambrecht B N, Estrada-Garcia I, Flores-Romo L, Hernandez-Pando R. *Immunology* **2004**. 112(4), 661–668.

7 Pedroza-Gonzalez A, Garcia-Romo G S, Aguilar-Leon D, Calderon-Amador J, Hurtado-Ortiz R, Orozco-Estevez H, Lambrecht B N, Estrada-Garcia I, Hernandez-Pando R, Flores-Romo L. *Int J Exp Pathol.* **2004**. 85(3), 135–145.

8 Alaniz R C, Sandall S, Thomas E K, Wilson C B. *J Immunol.* **2004**. 172(6), 3725–3735.

9 Puig-Kroger A, Serrano-Gomez D, Caparros E, Dominguez-Soto A, Relloso M, Colmenares M, Martinez-Munoz L, Longo N, Sanchez-Sanchez N, Rincon M, Rivas L, Sanchez-Mateos P, Fernandez-Ruiz E, Corbi A L. *J Biol Chem.* **2004**. 279(24), 25 680–25 688.

10 Geijtenbeek T B, Van Vliet S J, Koppel E A, Sanchez-Hernandez M, Vandenbroucke-Grauls C M, Appelmelk B, Van Kooyk Y. *J Exp Med.* **2003**. 197(1), 7–17.

11 Tailleux L, Schwartz O, Herrmann J L, Pivert E, Jackson M, Amara A, Legres L, Dreher D, Nicod L P, Gluckman JC, Lagrange P H, Gicquel B, Neyrolles O. *J Exp Med.* **2003**. 197(1), 121–127.

12 Villeneuve C, Gilleron M, Maridonneau-Parini I, Daffe M, Astarie-Dequeker C, Etienne G. *J Lipid Res.* **2005**. 46(3),475–83.

13 Ernst J D. *Infect Immun.* **1998** 66(4), 1277–1281.

14 Vergne I, Chua J, Singh S B, Deretic V. *Annu Rev Cell Dev Biol.* **2004**. 20, 367–94.

15 Bodnar KA, Serbina N V, Flynn J L. *Infect Immun.* **2001**. 69(2), 800–809.

16 Jiao X, Lo-Man R, Guermonprez P, Fiette L, Deriaud E, Burgaud S, Gicquel B, Winter N, Leclerc C. *J Immunol.* **2002**. 168(3), 1294–301.

17 Tailleux L, Neyrolles O, Honore-Bouakline S, Perret E, Sanchez F, Abastado J P, Lagrange P H, Gluckman JC, Rosenzwajg M, Herrmann J L. *J Immunol*. **2003**. 170(4), 1939–1948

18 Schaible U E, Hagens K, Fischer K, Collins H L, Kaufmann S H. *J Immunol*. **2000**. 164(9), 4843–4852

19 Mariotti S, Teloni R, Iona E, Fattorini L, Romagnoli G, Gagliardi M C, Orefici G, Nisini *Infekt Immun*. **2004**. 72(8), 4385–92

20 Brennan PJ. *Tuberculosis* (Edinb). **2003**. 83(1–3), 91–97.

21 Brigl M, Brenner M B. *Annu Rev Immunol*. **2004**. 22, 817–890.

22 Beatty W L, Rhoades E R, Ullrich H J, Chatterjee D, Heuser J E, Russell D G. *Traffic*. **2000**. 1(3), 235–247.

23 Beatty W L, Russell D G. *Infect Immun*. **2000**. 68(12), 6997–7002.

24 Inaba K, Inaba M, Naito M, Steinman R M. *J Exp Med*. **1993** 178(2), 479–488.

25 Demangel C, Bean A G, Martin E, Feng C G, Kamath A T, Britton W J. *Eur J Immunol*. **1999** 29(6), 1972–1979

26 Tascon R E, Soares C S, Ragno S, Stavropoulos E, Hirst E M, Colston M J. *Immunology*. **2000**. 99(3), 473–480.

27 McShane H, Behboudi S, Goonetilleke N, Brookes R, Hill A V. *Infect Immun*. **2002**. 70(3), 1623–1626.

28 Demangel C, Palendira U, Feng C G, Heath A W, Bean A G, Britton W J. *Infect Immun*. **2001**. 69(4), 2456–2461.

29 Kaufmann S H, Schaible U E. *Curr Opin Immunol*. **2005**. 17(1), 79–87.

30 Moody D B, Young D C, Cheng T Y, Rosat J P, Roura-Mir C, O'Connor P B, Zajonc D M, Walz A, Miller M J, Levery S B, Wilson I A, Costello C E, Brenner M B. *Science* **2004**. 303(5657), 527–531.

31 Gilleron M, Stenger S, Mazorra Z, Wittke F, Mariotti S, Bohmer G, Prandi J, Mori L, Puzo G, De Libero G. *J Exp Med*. **2004**. 199(5), 649–659.

32 Fischer K, Scotet E, Niemeyer M, Koebernick H, Zerrahn J, Maillet S, Hurwitz R, Kursar M, Bonneville M, Kaufmann S H, Schaible U E. *Proc Natl Acad Sci USA* **2004**. 101(29), 10685–10690.

33 Kawashima T, Norose Y, Watanabe Y, Enomoto Y, Narazaki H, Watari E, Tanaka S, Takahashi H, Yano I, Brenner M B, Sugita M. *J Immunol*. **2003**. 170(11), 5345–5348.

34 Ulrichs T, Moody D B, Grant E, Kaufmann S H, Porcelli SA. *Infect Immun*. **2003**. 71(6), 3076–87.

35 Moody D B, Ulrichs T, Muhlecker W, Young D C, Gurcha S S, Grant E, Rosat J P, Brenner M B, Costello CE, Besra G S, Porcelli SA. *Nature*. **2000**. 404(6780), 884–888.

36 Winau F, Schwierzeck V, Hurwitz R, Remmel N, Sieling P A, Modlin RL, Porcelli S A, Brinkmann V, Sugita M, Sandhoff K, Kaufmann S H, Schaible U E. *Nat Immunol*. **2004**. 5(2), 169–74.

37 Angenieux C, Salamero J, Fricker D, Cazenave J P, Goud B, Hanau D, de La Salle H. *J Biol Chem*. **2000**. 275(48), 37757–37764.

38 Angenieux C, Fraisier V, Maitre B, Racine V, van der Wel N, Fricker D, Proamer F, Sachse M, Cazenave J P, Peters P, Goud B, Hanau D, Sibarita JB, Salamero J, de la Salle H. *Traffic* **2005**. 6(4), 286–302.

39 Zhou D, Mattner J, Cantu C 3rd, Schrantz N, Yin N, Gao Y, Sagiv Y, Hudspeth K, Wu YP, Yamashita T, Teneberg S, Wang D, Proia R L, Levery S B, Savage P B, Teyton L, Bendelac A. *Science* **2004**. 306(57), 525–529.

40 Brigl M, Bry L, Kent S C, Gumperz J E, Brenner M B. *Nat Immunol*. **2003**. 4(12), 1230–1237.

41 Mattner J, Debord K L, Ismail N, Goff RD, Cantu C 3rd, Zhou D, Saint-Mezard P, Wang V, Gao Y, Yin N, Hoebe K, Schneewind O, Walker D, Beutler B, Teyton L, Savage P B, Bendelac A. *Nature* **2005**. 434(7032), 525–529.

42 Zhou D, Cantu C 3rd, Sagiv Y, Schrantz N, Kulkarni A B, Qi X, Mahuran D J, Morales C R, Grabowski G A, Benlagha K, Savage P, Bendelac A, Teyton L. *Science* **2004**. 303(5657), 523–527.

43 Kang S J, Cresswell P. *Nat Immunol*. **2004**. 5(2), 175–181.

44 Hunger R E, Sieling P A, Ochoa M T, Sugaya M, Burdick A E, Rea T H, Brennan P J, Belisle J T, Blauvelt A, Porcelli S A, Modlin R L. *J Clin Invest*. **2004**. 113(5), 701–708.

45 Hertz C J, Kiertscher S M, Godowski P J, Bouis D A, Norgard M V, Roth M D, Modlin R. *J Immunol*. **2001**. 166(4), 2444–50.

46 Uehori J, Matsumoto M, Tsuji S, Akazawa T, Takeuchi O, Akira S, Kawata

T, Azuma I, Toyoshima K, Seya T. *Infect Immun.* **2003**. 71(8), 4238–4249.

47 Sprott G D, Dicaire C J, Gurnani K, Sad S, Krishnan L. *Infect Immun.* **2004**. 72(9), 5235–5246.

48 Remoli M E, Giacomini E, Lutfalla G, Dondi E, Orefici G, Battistini A, Uze G, Pellegrini S, Coccia E M. *J Immunol.* **2002**. 169(1), 366–374.

49 Lande R, Giacomini E, Grassi T, Remoli M E, Iona E, Miettinen M, Julkunen I, Coccia E. *J Immunol.* **2003**. 170(3), 1174–1182.

50 Algood H M, Flynn J L. *J Immunol.* **2004**. 173(5), 3287–3296.

51 Bhatt K, Hickman S P, Salgame P. *J Immunol.* **2004**. 172(5), 2748–2751.

52 Peters W, Scott H M, Chambers H F, Flynn J L, Charo I F, Ernst J D. *Proc Natl Acad Sci U S A.* **2001**. 98(14), 7958–7963.

53 Peters W, Cyster J G, Mack M, Schlondorff D, Wolf A J, Ernst J D, Charo I F. *J Immunol.* **2004**. 172(12), 7647–7653.

54 Marino S, Pawar S, Fuller C L, Reinhart T A, Flynn J L, Kirschner D E. *J Immunol.* **2004**. 173(1), 494–506.

55 Marino S, Kirschner D E. *J Theor Biol.* **2004**. 227(4), 463–486.

56 Gerosa F, Baldani-Guerra B, Nisii C, Marchesini V, Carra G, Trinchieri G. *J Exp Med.* **2002**. 195(3):327–333.

57 Dieli F, Caccamo N, Meraviglia S, Ivanyi J, Sireci G, Bonanno C T, Ferlazzo V, La Mendola C, Salerno A. *Eur J Immunol.* **2004**. 34(11), 3227–3235.

58 Nigou J, Zelle-Rieser C, Gilleron M, Thurnher M, Puzo G. *J Immunol.* **2001**. 166(12), 7477–7485.

59 Chieppa M, Bianchi G, Doni A, Del Prete A, Sironi M, Laskarin G, Monti P, Piemonti L, Biondi A, Mantovani A, Introna M, Allavena P. *J Immunol.* **2003**. 171(9), 4552–4560.

60 Demangel C, Bertolino P, Britton W J. *Eur J Immunol.* **2002**. 32(4), 994–1002.

61 Kaufmann S H, Schaible U E. *J Exp Med.* **2003**. 197(1), 1–5.

62 Giacomini E, Iona E, Ferroni L, Miettinen M, Fattorini L, Orefici G, Julkunen I, Coccia E M. *J Immunol.* **2001**. 166(12), 7033–7041.

63 Jang S, Uematsu S, Akira S, Salgame P. *J Immunol.* **2004**. 173(5), 3392–3397.

64 Förtsch D, Rollinghoff M, Stenger S. *J Immunol.* **2000**. 165(2), 978–987.

65 Buettner M, Meinken C, Bastian M, Bhat R, Stossel E, Faller G, Cianciolo G, Ficker J, Wagner M, Rollinghoff M, Stenger S. *J Immunol.* **2005**. 174(7), 4203–4209.

66 Murray P J. *Trends Microbiol.* **1999** 7(9), 366–372.

67 Schaible U E, Winau F, Sieling P A, Fischer K, Collins H L, Hagens K, Modlin R L, Brinkmann V, Kaufmann SH. *Nat Med.* **2003**. 9(8), 1039–1046.

68 Stenger S, Niazi K R, Modlin RL. *J Immunol.* **1998** 161(7), 3582–3588.

69 Winau F, Kaufmann S H, Schaible U E. *Cell Microbiol.* **2004**. 6(7), 599–607.

70 Winau F, Weber S, Sad S, de Diego J, Locatelli Hoops S, Breiden B, Sandhoff Brinkmann V, Kaufmann SH, Schaible U E. **2005**. Submitted.

71 Heldwein K A, Fenton M J. *Microbes Infect.* **2002**. 4(9), 937–944.

72 Palliser D, Ploegh H, Boes M. *J Immunol.* **2004**. 172(6), 3415–3421.

73 Rodriguez A, Regnault A, Kleijmeer M, Ricciardi-Castagnoli P, Amigorena S. *Nat Cell Biol.* **1999**. 1(6), 362–368.

74 Gehring A J, Dobos K M, Belisle J T, Harding C V, Boom W H. *J Immunol.* **2004**. 173(4), 2660–2668.

75 Gehring A J, Rojas R E, Canaday D H, Lakey D L, Harding C V, Boom W H. *Infect Immun.* **2003**. 71(8), 4487–4497.

76 Fulton S A, Reba S M, Pai R K, Pennini M, Torres M, Harding CV, Boom W H. *Infect Immun.* **2004**. 72(4), 2101–2110

77 Tobian A A, Potter N S, Ramachandra L, Pai R K, Convery M, Boom W H, Harding C V. *J Immunol.* **2003**. 171(3), 1413–1422.

78 Hanekom W A, Mendillo M, Manca C, Haslett P A, Siddiqui M R, Barry C 3rd, Kaplan G J. *Infect Dis.* **2003**. 188(2), 257–266.

79 van Faassen H, Dudani R, Krishnan L, Sad S. *J Immunol.* **2004**. 172(6), 3491–3500.

80 Gagliardi M C, Teloni R, Mariotti S, Iona E, Pardini M, Fattorini L, Orefici G, Nisini R. *Vaccine* **2004**. 22(29–30), 3848–3857

81 Latchumanan V K, Singh B, Sharma P, Natarajan K. *J Immunol.* **2002**. 169(12), 6856–6864.

82 de la Salle H, Mariotti S, Angenieux C, Gilleron M, Garcia-Alles L-F, Malm D, Berg T, Paoletti S, Maître B, Mourey L, Salamero J, Cazenave J P, Hanau D, Mori L, Puzo G, De Libero G. *Science* **2005** in press.

37

Dendritic Cell–Epithelial Cell Interactions in Response to Intestinal Bacteria

Maria Rescigno

37.1
The Intestinal Epithelium and the Gut-associated Lymphoid Tissue (GALT)

The intestinal epithelium is the first line of defense towards dangerous microorganisms [1, 2]. It opposes a physical, electric and chemical barrier against luminal bacteria. The permeability of the barrier is regulated by the presence of both tight junctions (TJ) between epithelial cells (ECs) and a negatively charged mucous glycocalix. TJ seal adjacent ECs to one another and regulate solute and ion flux between cells [3]. The glycocalix sets the size of macromolecules that can reach the apical membrane of ECs [4] and opposes an electric barrier to bacteria. Finally, ECs and Paneth cells, specialized cells located at the base of the crypt of intestinal villi, release antimicrobial peptides including defensins and cathelicidins that target broad classes of microorganisms [5]. The intestinal epithelial barrier is further complicated by the presence of two important cell types that are interspersed between ECs and play a crucial role in sampling the luminal content: (microfold) M cells [6] and DCs [7–9]. M cells are found primarily in the follicle-associated epithelium (FAE) of Peyer's patches (PP) but they have also recently been described as being scattered among the absorptive epithelium, where they could potentially transport antigens to the lamina propria (LP) [10]. M cells, differently from ECs, do not have an organized brush border and are more permissive to antigen uptake [4]. DCs are phagocytic cells that are scattered throughout the intestinal epithelium [11]. We have recently reported that DCs are able to send dendrites out like periscopes into the lumen for bacterial uptake [12, 13]. The integrity of the epithelial barrier is preserved because DCs express TJ proteins and can establish new TJ-like structure with adjacent ECs [12]. These 'creeping' DCs are characterized by the expression of the myeloid marker CD11b and the lack of CD8α [13, 14]. Their presence in the terminal ileum, where the gradient of bacteria gradually increases, suggests they may be recruited by the presence of luminal bacteria. Interestingly, DCs in CX3CL1 (fractalkine) receptor-deficient mice are unable to spread their dendrites across the epithelial barrier, indicating the involvement of CX3CL1 in driv-

Handbook of Dendritic Cells. Biology, Diseases, and Therapies.
Edited by M. B. Lutz, N. Romani, and A. Steinkasserer.
Copyright © 2006 WILEY-VCH Verlag GmbH & Co. KGaA, Weinheim
ISBN: 3-527-31109-2

ing the extension of the dendrites [14]. It is not known whether bacteria can directly drive fractalkine production by epithelial cells nor whether fractalkine modulates TJ protein expression in DCs. Interestingly, bacteria lacking LPS are unable to recruit DCs in *in-vitro* generated epithelial cell monolayers suggesting that bacteria play an active role in the induction of DC migration across the epithelial barrier [15].

The GALT can be divided into inductive sites where the immune response is initiated and effector sites where immune cells carry out their function [2, 16]. PP, mesenteric lymph nodes (MLN) and isolated lymphoid follicles are important inductive sites for mucosal immune responses whereas the epithelium and the lamina propria of the mucosa are considered effector sites for antibody production and T-cell responses.

37.2
Antigen Uptake in the Gut and DC Populations

Antigen uptake in the gut depends on the nature of the antigen. In fact, soluble antigens like digested food can penetrate through the meshes of the glycocalix and can be internalized by ECs throughout the intestinal wall. However, because absorptive epithelial cells rapidly degrade ingested proteins, it is likely that additional mechanisms of antigen uptake like the DC-mediated mechanism [12] are important in the mucosa. After internalization of soluble antigen either through ECs or through LP-DCs, the latter are probably involved in the induction of oral tolerance. In fact expansion of DCs *in vivo* enhanced tolerance induction after antigen feeding [17]. It is possible that antigen-loaded DCs migrate to MLN which is the preferential site for naïve T-cell activation and expansion after oral feeding of soluble antigen [18]. Conversely, particulate antigen is most likely taken up in PP as mice lacking PP are perfectly competent to induce antibody response towards soluble but not towards particulate (microsphere) antigen [19].

The mechanisms of bacterial entrance depend on their pathogenicity (Fig. 37.1). Most of the pathogens have developed strategies to penetrate ECs or to facilitate M-cell invasion (for a review see [1]), whereas noninvasive bacteria can enter mucosal surfaces either through M cells or DCs. M cells can release their 'cargo' to underlying phagocytic cells, including DCs, that can migrate to the interfollicular region of PP for T- and B-cell interactions, whereas DCs that take up bacteria directly across mucosal surfaces are likely to migrate to MLN. Interestingly, MLN set the border for mucosal compartment avoiding systemic spread of commensal-loaded DCs [20]. Neither mechanism discriminates between invasive pathogenic and non-invasive commensal bacteria. An alternative mechanism for antigen entry across a mucosal surface that also targets DCs and could be used for bacterial internalization, has recently been described [21]. It is mediated by neonatal Fc receptors (FcRn) expressed by adult human (but not mouse) intestinal epithelial cells that transport IgG across the intestinal epithelial barrier, and after binding with cognate antigen in the intestinal lumen, recycles the immune complexes back to the

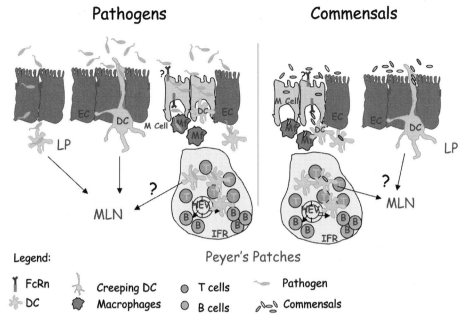

Fig. 37.1 Mechanisms of bacterial uptake. The mechanisms of bacterial entrance depend on their pathogenicity. Most of the pathogens have developed strategies to penetrate ECs or to facilitate M-cell invasion, alternatively they are captured by creeping DCs (left). Commensal bacteria can enter mucosal surfaces either through M cells or DCs (right). M cells can release their 'cargo' to underlying phagocytic cells, including DCs, that can migrate to the interfollicular region (IFR) of Peyer's Patches for T and B-cell interactions, whereas DCs that take up bacteria directly across mucosal surfaces are likely to migrate to MLN. Alternatively, PP-DCs could migrate to MLN. An alternative mechanism for antigen entry across a mucosal surface that also targets DCs and could be used for bacterial internalization is mediated by neonatal Fc receptors (FcRn) expressed by adult human (but not mouse) intestinal epithelial cells. FcRn transport directs and delivers the antigens in the form of immune complexes directly to underlying DCs. (HEV: high endothelial venules).

LP [21]. Antigens bound by IgG are less susceptible to degradation within the epithelial cells because endosomes formed after uptake by FcRn do not readily fuse with lysosomes. FcRn transport directs and delivers the antigens in the form of immune complexes directly to DCs lying in the LP. As DCs can be activated by immune complexes, it would be interesting to know whether DCs internalize the immune complexes via the FcγRs or via FcRn (both of which are expressed by DCs) and whether these receptors differentially affect DC function. Finally, DCs can process antigens from apoptotic intestinal epithelial cells, both in the steady state [22] and following reovirus infection [23], which constitutes another mechanism of DC antigen uptake that directly involves interactions with the epithelium.

The uptake route together with the nature of the ingested antigens dictates the type of immune response that is generated, whether this is related to the subtype of DCs that is targeted by each route or to their location remains to be established. In fact at least four DC populations in the mouse intestine have been described. They are all characterized by the expression of CD11c but differ for the expression of the surface markers CD11b, CD8α and B220 (for a review see [8, 24]) as well as for the expression of chemokine receptors CCR6 and CCR7 [25]. Interestingly, the different DC populations have particular locations in PP [26]. In fact, it is important to say that in PP two important functions are carried out by DCs: uptake of antigen after its transcytosis across the FAE and T and B-cell activation. Therefore, differently from other peripheral tissues, it is possible to find in the PP both immature DCs that are mainly localized in the sub epithelial dome, below the FAE and mature DCs that are found in interfollicular T-cell areas. Two additional DC subsets have been described in MLN that are characterized by the differential expression of CD4 and DEC-205 [8, 24]. The characterization of human intestinal DCs is still very poor, but at least two DC cell types have been described in the colon: a CD11c+HLA-DR+ population and a CD11c– population [27] that we have identified as CD83+CD123+, possibly plasmacytoid DCs (our unpublished observations). Hence, scattered throughout mucosal tissues it is possible to find the same DC subsets present in other nonmucosal tissues.

37.3
Cross-talk between Bacteria and Epithelial Cells

The major interaction between mucosal tissues and luminal bacteria occurs at the level of ECs that are the most representative cell type of the epithelium. Both pathogens and commensal bacteria have been described to undertake an active cross-talk with ECs [1]. Whereas the first are primarily involved in the activation of an inflammatory cascade of events, the latter seem to downregulate the ability of ECs to initiate inflammatory responses. The mechanisms through which pathogens can activate ECs are similar to those used by monocytes and DCs to sense the presence of bacteria. In fact ECs express a series of pathogen recognition receptors (PRRs) including Toll-like receptors (TLRs) and NOD proteins that are also expressed by phagocytes [1]. The major difference stands in the location of these receptors. In fact ECs seem to express these receptors either intracellularly (like TLR-4) or in a polarized fashion leaving the apical surface nearly free of PRR expression. Therefore only invasive bacteria or those equipped with type three or four secretion systems [28] that act as syringes to pump DNA or effector proteins directly into the cytoplasm of host cells, are sensed by PRRs for activation of the inflammatory cascade. Moreover, some of the receptors (like NOD2) are constitutively expressed only in Paneth cells [29] that reside at the base of the crypts and are induced in ECs only after bacterial encounter [30, 31]. A typical indicator of epithelial infection by invasive bacteria is the expression of the chemokine CXCL-8 (IL-8) which is a strong chemoattractant for neutrophils [32–35]. A more debated issue relates to the expression of TLR-5, the receptor for flagellin [36]. It has recently been described that flagellin-dependent stimulation of intestinal ECs results in triggering of

CCL20 via a TLR-5 dependent mechanism [37]. CCL20 is responsible for the recruitment of CCR6– expressing immature DCs [38]. However, some authors suggest that TLR-5 is expressed only basolaterally of ECs [39, 40], whereas others have described it also apically [37, 41]. We favor the second hypothesis because we have evidence that invasive-deficient mutant of *Salmonella* and the flagellated noninvasive soil bacterium *Bacillus subtilis* induce the expression of CCL-20 by polarized ECs [15, 42]. Our experiments in the mouse also confirm that noninvasive flagellated bacteria can induce the expression of CCL-20 suggesting the possibility that different responses might depend on the EC cell line used for *in vitro* experiments [42].

How commensals can downregulate the inflammatory response induced by pathogen associated molecular patterns (PAMPs) has only recently started to be unraveled. It is becoming clear that recognition of commensal flora via TLRs is required for intestinal homeostasis [43] and that commensal bacteria can interfere at different levels of TLR signaling. Expression and activation of IRAK-M [44] or of a truncated version of the TLR adaptor protein MyD88 [45] that both interfere with TLR signaling have been described. Along the same line, the interaction of ECs with the commensal *Bacteroides thetaiotamicron* or with nonvirulent mutants of *Salmonella typhimurium* interfere with the activation of NF-κB that is downstream of TLR signaling either by triggering binding of peroxisome-proliferator-activated receptor γ (PPAR-γ) with the NF-κB subunit Rel-A in the nucleus [46] or by blocking the degradation of IκBα, an intracellular inhibitor of NF-κB [47]. Therefore, the induction of an inflammatory response in ECs depends on the ability of invasive pathogens to activate PRR signaling pathways and on that of commensals to perturb the same signaling pathways.

37.4
Unique Functions of Mucosal DCs

DCs isolated from a variety of mucosal sites (PP, LP, mesenteric lymph nodes (MLN), lung) have the natural propensity to induce T_H2 responses in *in vitro* T-cell priming assays, and to express cytokines such as IL-10, and possibly TGF-β [2, 48–51]. Interestingly, the same CD11c$^+$CD11b$^+$CD8α$^-$ DC subset isolated from PP but not from spleen preferentially polarizes antigen-specific T cells to produce T_H2 cytokines and IL-10 *in vitro* [52], suggesting that the observed differences are not attributable to subset-intrinsic properties but most likely to the local mucosal microenvironment. Further, the same PP but not spleen DC subset is able to promote IgA production by naïve B cells, which is mediated by a higher release of IL-6 [53] and T-cell help. These data suggest that mucosal DCs may be specialized in inducing a noninflammatory environment and in providing help to B cells via the activation of T_H2 T cells. This is consistent with the fact that many "tolerogenic" responses to mucosal antigens, for example to commensal organisms, are associated with the generation of antibody responses [20, 49], rather than with a broad immunological unresponsiveness. In addition, CD8$^+$ CD11c^{lo} plasmacytoid DCs may also be important for maintaining tolerance to innocuous antigens since this population can induce the differentiation of IL-10 producing regulatory T cells (Treg) *in vitro* [8].

Another important feature of DCs isolated from mucosal tissues is that they have the unique ability to selectively imprint gut-homing T cells [54–56]. Moreover, naïve CD8[+] T cells primed by PP-DCs acquire gut tropism [55], despite showing similar patterns of activation markers and effector activity as those primed by DCs isolated from other nonmucosal lymphoid organs. PP-DCs induced high expression of the intestinal homing integrin $\alpha_4\beta_7$ and the chemokine receptor CCR9 in primed CD8[+] T cells. Interestingly, reactivation of skin-committed memory T cells with DCs isolated from gut changed T-cell tissue tropism, suggesting that memory T cells are relocated according to the tissue where they are needed [57]. Finally, mucosal DCs have been shown to continually migrate to draining lymph nodes in the "steady", or unperturbed state with a rapid turnover rate (2–4 days in the intestinal wall). In the rat, two types of migrating DCs could be identified, both of which are positive for the αE integrin CD103, but only the fraction that expresses low levels of CD172 (SIRPα), has features of immature cells and carries apoptotic enterocytes to MLNs [58]. Because these DCs process apoptotic epithelial cells in the steady state [22], this CD103[+]CD172[lo] DC population may be involved in tolerance to self-proteins, although this hypothesis remains to be tested. DC emigration from the gut can be greatly enhanced by systemic LPS injection which does not change the proportion of SIRPα^{hi}/SIRPα^{lo} populations as well as their activation state [59]. Interestingly, whereas SIRPα^{lo} DCs migrate to T-cell areas of MLNs under steady-state conditions, SIRPα^{hi} DCs do so only after intravenous LPS injection suggesting that LPS injection facilitates antigen presentation by this DC subset.

37.5
Intestinal Immune Homeostasis is Regulated by the Cross-talk between ECs and DCs

DCs play an active role in bacterial uptake across mucosal surfaces and have unique functions that allow the generation of mucosal immune responses. Moreover, DCs can intercalate between ECs and can interact directly with the luminal bacteria and with all the TLR ligands that are carried by commensal or pathogenic bacteria. Therefore, three important questions arise: what is the role played by the local microenvironment in driving mucosal DC differentiation? How can DCs avoid the induction of exaggerated inflammatory responses towards commensal bacteria? Is there any relationship between the unique phenotype of mucosal DCs and the regulation of gut immune homeostasis? It is becoming clear that the relationship between DCs and the microenviroment profounds affects the functional properties of tissue DCs. This has been demonstrated in the spleen [60, 61], but there are strong evidences that a similar situation is occurring also in the gut. In fact, the ability of intestinal DCs to induce gut-tropism during T-cell priming [54–56] and reactivation [57] and to promote T$_H$2 T-cell responses [2, 48–52], as well as IgA antibody production [53] strongly favors this hypothesis. As intestinal ECs are in close contact with DCs, they could play an active role in driving mucosal DC differentiation. We found that this is indeed the case and that ECs release constitutively TSLP, a molecule involved in driving T_H2 differentiation by DCs [62, 63]. Interestingly, DCs exposed to EC-conditioning are unable to release IL-12 and to drive T_H1 type of T-cell responses even after activation with T_H1-inducing pathogens (Fig. 37.2) [64]. Moreover, TSLP acts in a very narrow window of concentrations: at lower or higher TSLP concentrations, DCs reacquire the ability to release IL-12 and to drive T_H1 T-cell responses. Therefore, we believe that resident DCs

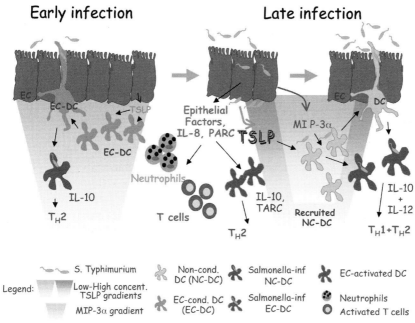

Fig. 37.2 Early *Salmonella typhimurium* infection: resident DCs are conditioned by EC-released TSLP (EC-DC). EC-DC release IL-10 after bacterial exposure and drive default T_H2 responses to *S. typhimurium*.

Late infection: since *S. typhimurium* is an invasive bacterium, it induces ECs to release pro-inflammatory chemokines like IL-8 (CXCL-8) and PARC (CCL-18), which attract neutrophils, granulocytes and activated T cells that generate an inflamed site. The binding of *Salmonella* to the basolateral membrane of ECs induces the upregulation of TSLP. TSLP at this concentration drives T_H1 rather than T_H2 promoting DCs in response to bacteria. Unidentified EC-derived factors can also activate 'bystander' DCs that have not been in contact directly with the bacteria. DCs activated in this way release IL-10 and TARC (CCL-17) but not IL-12, thus driving and recruiting T_H2 T cells. *Salmonella* also induces the release of MIP-3α (CCL-20) that recruits CCR6-expressing immature DCs.

Most likely, recruited DCs are not subjected to EC-conditioning, rather they could find increased TSLP concentrations in the infected site. Newly recruited DCs (NC-DC) can either creep between ECs to take up bacteria or they can phagocytose bacteria that have breached across the epithelial barrier and release both IL-10 and IL-12, thus promoting T_H1 and T_H2 responses. This allows the establishment of protective anti-*Salmonella* responses.

even though they have the chance to contact directly the bacteria, they are unable to activate inflammatory cells and this can help maintaining the homeostasis of the gut. In fact, nearly 70% of individuals affected by a T_H1-mediated chronic inflammatory disease like in Crohn's disease [65] have undetectable levels of TSLP and this correlates with the inability of intestinal ECs to regulate DC function [64]. Therefore, resident DCs that are actively involved in taking up bacteria at steady state do not drive inflammatory responses and this can explain why the intestinal immune homeostasis is preserved even though DCs are continuously exposed to TLR ligands.

37.6
Cross-talk between ECs and DCs in Bacterial Handling

Despite this propensity for the induction of T_H2 and Tregs by mucosal DCs, T_H1 and CTL responses are effectively generated to mucosal pathogens and are required to fight intracellular microorganisms [66–70]. Whether this involves the same or different DC subsets as those responsible for mucosal responses and tolerance induction, remains to be established. However, it is conceivable that resident mucosal DCs are 'educated' by ECs to initiate noninflammatory responses, whereas DCs recruited after bacterial invasion might retain their ability to respond in an inflammatory mode. In fact, infection by flagellated bacteria like *Salmonella* spp. induces the recruitment of DCs in the intestinal epithelium [12, 14] via the release of CCL-20 by ECs [37]. These nonconditioned newly recruited DCs might be responsible for the induction of T_H1 responses to invasive bacteria (Fig. 37.2). This hypothesis is supported by *in vitro* three-part studies in which DCs were seeded from the basolateral membrane of EC monolayers shortly before apical bacterial application [64]. Interestingly, due to their ability to creep between ECs and to contact bacteria directly, DCs were 'qualitatively' similarly activated regardless of the invasiveness or pathogenicity of the apical bacteria. Bacteria-activated DCs produced both IL-12 and IL-10 and skewed towards a T_H1 phenotype [64]. This suggests that nonconditioned DCs can drive the induction of inflammatory responses provided that they are not subject to EC conditioning before their encounter with bacteria. Moreover, bacteria invading ECs induce the upregulation of TSLP thus switching to DCs that have the propensity to induce T_H1- rather than T_H2-T cells in response to bacteria. Interestingly, bystander DCs that do not contact directly the bacteria are activated by EC-derived factors to noninflammatory DCs producing IL-10 and TARC (CCL-17) and inducing or recruiting T_H2 T cells, probably as a feedback mechanism to turn off the inflammatory response [42].

Another possibility is that epithelial cell derived factors, such as TNF or type 1 IFNs, produced during pathogen invasion may directly affect DC activation. This hypothesis is supported by studies of murine intestinal infection with type-1 reovirus [23]. Reovirus productively infects epithelial cells overlying PPs, yet viral antigen associated with apoptotic epithelial cells is avidly taken up by CD11c$^+$ CD8α^- CD11b$^-$ DCs in the subepithelial dome [23]. The observation that reovirus neither productively infects DCs *in vivo* or *in vitro*, nor activates DCs to mature or produce cytokines *in vitro*, suggests a role for environmental factors, possibly derived from infected epithelial cells, in driving DCs to induce T_H1 responses to the virus. Interestingly, IFNαβR-deficient mice, but not MyD88-deficient or TLR3-deficient mice have an increased susceptibility to reovirus infection. In addition, MyD88-deficient mice mount normal IgG1, IgG2a/c and IgG2b responses, suggesting that type 1 IFN, possibly derived in the early stages of infection from infected epithelial cells, but not signaling via at least a single TLR pathway is important for inducing protection from reovirus infection.

Taken together, these studies highlight an important emerging relationship between DCs and epithelial cells in the maintenance of mucosal homeostasis and the induction of innate and adaptive immunity to mucosal infection with pathogens such as *Salmonella* and reovirus.

37.7
Conclusions

In conclusion, mucosal DCs have specialized functions that allow establish mucosal immune responses, including the induction of T_H2 T-cell responses and IgA antibody production. DCs play a crucial role both in the uptake of intestinal bacteria and in the induction of tolerance and immunity towards them. However, it is not yet fully clarified whether different DC subsets have clearly distinct functions *in vivo* or whether the local microenvironment is responsible to control DC function. Important issues that also need to be addressed are where DCs interact with T cells for the induction of regulatory or effector immune responses and if there are specialized induction sites that allow the generation of tolerance versus immunity, or systemic versus mucosal immune responses.

References

1 Sansonetti, P. J., War and peace at mucosal surfaces. *Nat Rev Immunol* **2004**. 4: 953–964.

2 Mowat, A. M., Anatomical basis of tolerance and immunity to intestinal antigens. *Nat Rev Immunol* **2003**. 3: 331–341.

3 Schneeberger, E. E., Lynch, R. D., The tight junction: a multifunctional complex. *Am J Physiol Cell Physiol* **2004**. 286: C1213–1228.

4 Frey, A., Giannasca, K. T., Weltzin, R., Giannasca, P. J., Reggio, H., Lencer, W. I., Neutra, M. R., Role of the glycocalyx in regulating access of microparticles to apical plasma membranes of intestinal epithelial cells: implications for microbial attachment and oral vaccine targeting. *J Exp Med* **1996**. 184: 1045–1059.

5 Ganz, T., Defensins: antimicrobial peptides of innate immunity. *Nat Rev Immunol* **2003**. 3: 710–720.

6 Kraehenbuhl, J. P., Neutra, M. R., Epithelial M cells: differentiation and function. *Annu Rev Cell Dev Biol* **2000**. 16: 301–332.

7 Banchereau, J., Briere, F., Caux, C., Davoust, J., Lebecque, S., Liu, Y. J., Pulendran, B., Palucka, K., Immunobiology of dendritic cells. *Annu Rev Immunol* **2000**. 18: 767–811.

8 Bilsborough, J., Viney, J. L., Gastrointestinal dendritic cells play a role in

immunity, tolerance, and disease. *Gastroenterology* **2004**. 127: 300–309.

9 Kelsall, B. L., Rescigno, M., Mucosal dendritic cells in immunity and inflammation. *Nat Immunol* **2004**. 5: 1091–1095.

10 Jang, M. H., Kweon, M. N., Iwatani, K., Yamamoto, M., Terahara, K., Sasakawa, C., Suzuki, T., Nochi, T., Yokota, Y., Rennert, P. D., Hiroi, T., Tamagawa, H., Iijima, H., Kunisawa, J., Yuki, Y., Kiyono, H., Intestinal villous M cells: an antigen entry site in the mucosal epithelium. *Proc Natl Acad Sci USA* **2004**. 101: 6110–6115.

11 Maric, I., Holt, P. G., Perdue, M. H., Bienenstock, J., Class II MHC antigen (Ia)-bearing dendritic cells in the epithelium of the rat intestine. *J Immunol* **1996**. 156: 1408–1414.

12 Rescigno, M., Urbano, M., Valzasina, B., Francolini, M., Rotta, G., Bonasio, R., Granucci, F., Kraehenbuhl, J. P., Ricciardi-Castagnoli, P., Dendritic cells express tight junction proteins and penetrate gut epithelial monolayers to sample bacteria. *Nat Immunol* **2001**. 2: 361–367.

13 Rescigno, M., Rotta, G., Valzasina, B., Ricciardi-Castagnoli, P., Dendritic cells shuttle microbes across gut epithelial monolayers. *Immunobiology* **2001**. 204: 572–581.

14 Niess, J. H., Brand, S., Gu, X., Landsman, L., Jung, S., McCormick, B. A., Vyas, J. M., Boes, M., Ploegh, H. L., Fox, J. G., Littman, D. R., Reinecker, H. C., CX3CR1-Mediated Dendritic Cell Access to the Intestinal Lumen and Bacterial Clearance. *Science* **2005**. 307: 254–258.

15 Rimoldi, M., Chieppa, M., Vulcano, M., Allavena, P., Rescigno, M., Intestinal epithelial cells control DC function. *Ann NY Acad Sci* **2004**. 1029: 1–9.

16 Nagler-Anderson, C., Man the barrier! Strategic defences in the intestinal mucosa. *Nat Rev Immunol* **2001**. 1: 59–67.

17 Viney, J. L., Mowat, A. M., O'Malley, J. M., Williamson, E., Fanger, N. A., Expanding dendritic cells in vivo enhances the induction of oral tolerance. *J Immunol* **1998**. 160: 5815–5825.

18 Kunkel, D., Kirchhoff, D., Nishikawa, S., Radbruch, A., Scheffold, A., Visualization of peptide presentation following oral application of antigen in normal and Peyer's patches-deficient mice. *Eur J Immunol* **2003**. 33: 1292–1301.

19 Kunisawa, J., Takahashi, I., Okudaira, A., Hiroi, T., Katayama, K., Ariyama, T., Tsutsumi, Y., Nakagawa, S., Kiyono, H., Mayumi, T., Lack of antigen-specific immune responses in anti-IL-7 receptor alpha chain antibody-treated Peyer's patch-null mice following intestinal immunization with microencapsulated antigen. *Eur J Immunol* **2002**. 32: 2347–2355.

20 Macpherson, A. J., Uhr, T., Induction of protective IgA by intestinal dendritic cells carrying commensal bacteria. *Science* **2004**. 303: 1662–1665.

21 Yoshida, M., Claypool, S. M., Wagner, J. S., Mizoguchi, E., Mizoguchi, A., Roopenian, D. C., Lencer, W. I., Blumberg, R. S., Human neonatal Fc receptor mediates transport of IgG into luminal secretions for delivery of antigens to mucosal dendritic cells. *Immunity* **2004**. 20: 769–783.

22 Huang, F. P., Platt, N., Wykes, M., Major, J. R., Powell, T. J., Jenkins, C. D., MacPherson, G. G., A discrete sub-population of dendritic cells transports apoptotic intestinal epithelial cells to T cell areas of mesenteric lymph nodes [see comments]. *J Exp Med* **2000**. 191: 435–444.

23 Fleeton, M. N., Contractor, N., Leon, F., Wetzel, J. D., Dermody, T. S., Kelsall, B. L., Peyer's patch dendritic cells process viral antigen from apoptotic epithelial cells in the intestine of reovirus-infected mice. *J Exp Med* **2004**. 200: 235–245.

24 Shortman, K., Liu, Y. J., Mouse and human dendritic cell subtypes. *Nature Rev Immunol* **2002**. 2: 151–161.

25 Iwasaki, A., Kelsall, B. L., Localization of distinct Peyer's patch dendritic cell sub-sets and their recruitment by chemokines macrophage inflammatory protein (MIP)-3alpha, MIP-3beta, and secondary lymphoid organ chemokine. *J Exp Med* **2000**. 191: 1381–1394.

26 Niedergang, F., Didierlaurent, A., Kraehenbuhl, J. P., Sirard, J. C., Dendritic cells: the host Achille's heel for mucosal pathogens? *Trends Microbiol* **2004**. 12: 79–88.

27 Bell, S. J., Rigby, R., English, N., Mann, S. D., Knight, S. C., Kamm, M. A., Stagg, A. J., Migration and maturation of human colonic dendritic cells. *J Immunol* 2001. 166: 4958–4967.

28 Viala, J., Chaput, C., Boneca, I. G., Cardona, A., Girardin, S. E., Moran, A. P., Athman, R., Memet, S., Huerre, M. R., Coyle, A. J., DiStefano, P. S., Sansonetti, P. J., Labigne, A., Bertin, J., Philpott, D. J., Ferrero, R. L., Nod1 responds to peptidoglycan delivered by the Helicobacter pylori cag pathogenicity island. *Nat Immunol* 2004. 5: 1166–1174.

29 Lala, S., Ogura, Y., Osborne, C., Hor, S. Y., Bromfield, A., Davies, S., Ogunbiyi, O., Nunez, G., Keshav, S., Crohn's disease and the NOD2 gene: a role for paneth cells. *Gastroenterology* 2003. 125: 47–57.

30 Rosenstiel, P., Fantini, M., Brautigam, K., Kuhbacher, T., Waetzig, G. H., Seegert, D., Schreiber, S., TNF-alpha and IFN-gamma regulate the expression of the NOD2 (CARD15) gene in human intestinal epithelial cells. *Gastroenterology* 2003. 124: 1001–1009.

31 Gutierrez, O., Pipaon, C., Inohara, N., Fontalba, A., Ogura, Y., Prosper, F., Nunez, G., Fernandez-Luna, J. L., Induction of Nod2 in myelomonocytic and intestinal epithelial cells via nuclear factor-kappa B activation. *J Biol Chem* 2002. 277: 41701–41705.

32 Eckmann, L., Kagnoff, M. F., Fierer, J., Epithelial cells secrete the chemokine interleukin-8 in response to bacterial entry. *Infect Immun* 1993. 61: 4569–4574.

33 Jung, H. C., Eckmann, L., Yang, S. K., Panja, A., Fierer, J., Morzycka-Wroblewska, E., Kagnoff, M. F., A distinct array of proinflammatory cytokines is expressed in human colon epithelial cells in response to bacterial invasion. *J Clin Invest* 1995. 95: 55–65.

34 McCormick, B. A., Colgan, S. P., Delp-Archer, C., Miller, S. I., Madara, J. L., Salmonella typhimurium attachment to human intestinal epithelial monolayers: transcellular signalling to subepithelial neutrophils. *J Cell Biol* 1993. 123: 895–907.

35 McCormick, B. A., Hofman, P. M., Kim, J., Carnes, D. K., Miller, S. I., Madara, J. L., Surface attachment of Salmonella typhimurium to intestinal epithelia imprints the subepithelial matrix with gradients chemotactic for neutrophils. *J Cell Biol* 1995. 131: 1599–1608.

36 Hayashi, F., Smith, K. D., Ozinsky, A., Hawn, T. R., Yi, E. C., Goodlett, D. R., Eng, J. K., Akira, S., Underhill, D. M., Aderem, A., The innate immune response to bacterial flagellin is mediated by Toll- like receptor 5. *Nature* 2001. 410: 1099–1103.

37 Sierro, F., Dubois, B., Coste, A., Kaiserlian, D., Kraehenbuhl, J. P., Sirard, J. C., Flagellin stimulation of intestinal epithelial cells triggers CCL20-mediated migration of dendritic cells. *Proc Natl Acad Sci USA* 2001. 98: 13722–13727.

38 Sozzani, S., Allavena, P., D'Amico, G., Luini, W., Bianchi, G., Kataura, M., Imai, T., Yoshie, O., Bonecchi, R., Mantovani, A., Differential regulation of chemokine receptors during dendritic cell maturation: a model for their trafficking properties. *J. Immunol.* 1998. 161: 1083–1086.

39 Gewirtz, A. T., Simon, P. O., Jr., Schmitt, C. K., Taylor, L. J., Hagedorn, C. H., O'Brien, A. D., Neish, A. S., Madara, J. L., Salmonella typhimurium translocates flagellin across intestinal epithelia, inducing a proinflammatory response. *J Clin Invest* 2001. 107: 99–109.

40 Lyons, S., Wang, L., Casanova, J. E., Sitaraman, S. V., Merlin, D., Gewirtz, A. T., Salmonella typhimurium transcytoses flagellin via an SPI2-mediated vesicular transport pathway. *J Cell Sci* 2004. 117: 5771–5780.

41 Ramos, H. C., Rumbo, M., Sirard, J. C., Bacterial flagellins: mediators of pathogenicity and host immune responses in mucosa. *Trends Microbiol* 2004. 12: 509–517.

42 Rimoldi, M., Chieppa, M., Vulcano, M., Allavena, P., Rescigno, M., Dendritic cells activated by bacteria or by bacteria-stimulated epithelial cells are functionally different. *Blood* 2005. 106: 2818–2826.

43 Rakoff-Nahoum, S., Paglino, J., Eslami-Varzaneh, F., Edberg, S., Medzhitov, R., Recognition of commensal microflora by toll-like receptors is required for intestinal homeostasis. *Cell* 2004. 118: 229–241.

44 Kobayashi, K., Hernandez, L. D., Galan, J. E., Janeway, C. A., Jr., Medzhitov, R., Flavell, R. A., IRAK-M is a negative regulator of Toll-like receptor signaling. *Cell* **2002**. 110: 191–202.

45 Janssens, S., Burns, K., Tschopp, J., Beyaert, R., Regulation of interleukin-1- and lipopolysaccharide-induced NF-kappaB activation by alternative splicing of MyD88. *Curr Biol* **2002**. 12: 467–471.

46 Kelly, D., Campbell, J. I., King, T. P., Grant, G., Jansson, E. A., Coutts, A. G., Pettersson, S., Conway, S., Commensal anaerobic gut bacteria attenuate inflammation by regulating nuclear-cytoplasmic shuttling of PPAR-gamma and RelA. *Nat Immunol* **2004**. 5: 104–112.

47 Neish, A. S., Gewirtz, A. T., Zeng, H., Young, A. N., Hobert, M. E., Karmali, V., Rao, A. S., Madara, J. L., Prokaryotic regulation of epithelial responses by inhibition of IkappaB-alpha ubiquitination. *Science* **2000**. 289: 1560–1563.

48 Akbari, O., DeKruyff, R. H., Umetsu, D. T., Pulmonary dendritic cells producing IL-10 mediate tolerance induced by respiratory exposure to antigen. *Nat Immunol* **2001**. 2: 725–731.

49 Alpan, O., Rudomen, G., Matzinger, P., The role of dendritic cells, B cells, and M cells in gut-oriented immune responses. *J Immunol* **2001**. 166: 4843–4852.

50 Iwasaki, A., Kelsall, B. L., Freshly isolated Peyer's patch, but not spleen, dendritic cells produce interleukin 10 and induce the differentiation of T helper type 2 cells. *J Exp Med* **1999**. 190: 229–239.

51 Williamson, E., Bilsborough, J. M., Viney, J. L., Regulation of mucosal dendritic cell function by receptor activator of NF-kappa B (RANK)/RANK ligand interactions: impact on tolerance induction. *J Immunol* **2002**. 169: 3606–3612.

52 Iwasaki, A., Kelsall, B. L., Unique functions of cd11b(+), cd8alpha(+), and double-negative Peyer's patch dendritic cells. *J Immunol* **2001**. 166: 4884–4890.

53 Sato, A., Hashiguchi, M., Toda, E., Iwasaki, A., Hachimura, S., Kaminogawa, S., CD11b+ Peyer's patch dendritic cells secrete IL-6 and induce IgA secretion from naive B cells. *J Immunol* **2003**. 171: 3684–3690.

54 Stagg, A. J., Kamm, M. A., Knight, S. C., Intestinal dendritic cells increase T cell expression of alpha4beta7 integrin. *Eur J Immunol* **2002**. 32: 1445–1454.

55 Mora, J. R., Bono, M. R., Manjunath, N., Weninger, W., Cavanagh, L. L., Rosemblatt, M., Von Andrian, U. H., Selective imprinting of gut-homing T cells by Peyer's patch dendritic cells. *Nature* **2003**. 424: 88–93.

56 Johansson-Lindbom, B., Svensson, M., Wurbel, M. A., Malissen, B., Marquez, G., Agace, W., Selective generation of gut tropic T cells in gut-associated lymphoid tissue (GALT): requirement for GALT dendritic cells and adjuvant. *J Exp Med* **2003**. 198: 963–969.

57 Mora, J. R., Cheng, G., Picarella, D., Briskin, M., Buchanan, N., von Andrian, U. H., Reciprocal and dynamic control of CD8 T cell homing by dendritic cells from skin- and gut-associated lymphoid tissues. *J Exp Med* **2005**. 201: 303–316.

58 Yrlid, U., Macpherson, G., Phenotype and function of rat dendritic cell subsets. *Apmis* **2003**. 111: 756–765.

59 Turnbull, E. L., Yrlid, U., Jenkins, C. D., Macpherson, G. G., Intestinal Dendritic Cell Subsets: Differential Effects of Systemic TLR4 Stimulation on Migratory Fate and Activation In Vivo. *J Immunol* **2005**. 174: 1374–1384.

60 Svensson, M., Maroof, A., Ato, M., Kaye, P. M., Stromal cells direct local differentiation of regulatory dendritic cells. *Immunity* **2004**. 21: 805–816.

61 Zhang, M., Tang, H., Guo, Z., An, H., Zhu, X., Song, W., Guo, J., Huang, X., Chen, T., Wang, J., Cao, X., Splenic stroma drives mature dendritic cells to differentiate into regulatory dendritic cells. *Nat Immunol* **2004**. 5: 1124–1133.

62 Soumelis, V., Liu, Y. J., Human thymic stromal lymphopoietin: a novel epithelial cell-derived cytokine and a potential key player in the induction of allergic inflammation. *Springer Semin Immunopathol* **2004**. 25: 325–333.

63 Soumelis, V., Reche, P. A., Kanzler, H., Yuan, W., Edward, G., Homey, B., Gilliet, M., Ho, S., Antonenko, S., Lauerma, A., Smith, K., Gorman, D., Zurawski, S., Abrams, J., Menon, S., McClanahan, T.,

de Waal-Malefyt Rd, R., Bazan, F., Kastelein, R. A., Liu, Y. J., Human epithelial cells trigger dendritic cell mediated allergic inflammation by producing TSLP. *Nat Immunol* **2002**. 3: 673–680.

64 Rimoldi, M., Salucci, V., Chieppa, M., Avogadri, F., Sonzogni, A., Sampietro, G. M., Nespoli, A., Viale, G., Allavena, P., Rescigno, M., Intestinal immune homeostasis is regulated by the cross-talk between epithelial cells and dendritic cells. *Nat Immunol* **2005**. 6: 507–514.

65 Kosiewicz, M. M., Nast, C. C., Krishnan, A., Rivera-Nieves, J., Moskaluk, C. A., Matsumoto, S., Kozaiwa, K., Cominelli, F., Th1-type responses mediate spontaneous ileitis in a novel murine model of Crohn's disease. *J Clin Invest* **2001**. 107: 695–702.

66 Mastroeni, P., Menager, N., Development of acquired immunity to Salmonella. *J Med Microbiol* **2003**. 52: 453–459.

67 Hess, J., Ladel, C., Miko, D., Kaufmann, S. H., Salmonella typhimurium aroA-infection in gene-targeted immuno-deficient mice: major role of CD4+ TCR-alpha beta cells and IFN-gamma in bacterial clearance independent of intracellular location. *J Immunol* **1996**. 156: 3321–3326.

68 George, A., Generation of gamma interferon responses in murine Peyer's patches following oral immunization. *Infect Immun* **1996**. 64: 4606–4611.

69 Liesenfeld, O., Kosek, J. C., Suzuki, Y., Gamma interferon induces Fas-dependent apoptosis of Peyer's patch T cells in mice following peroral infection with Toxoplasma gondii. *Infect Immun* **1997**. 65: 4682–4689.

70 Vossenkamper, A., Struck, D., Alvarado-Esquivel, C., Went, T., Takeda, K., Akira, S., Pfeffer, K., Alber, G., Lochner, M., Forster, I., Liesenfeld, O., Both IL-12 and IL-18 contribute to small intestinal Th1-type immunopathology following oral infection with Toxoplasma gondii, but IL-12 is dominant over IL-18 in parasite control. *Eur J Immunol* **2004**. 34: 3197–3207.